BATTERY L
1st UNITED STATES ARTILLERY
1860–1865

BATTERY L
1st UNITED STATES ARTILLERY
1860–1865

Robert C. Simmonds

Peter E. Randall Publisher

Portsmouth, New Hampshire

2022

Copyright © 2017, 2022 by Robert Simmonds.

Second Edition

Softcover ISBN: 978-1-937721-84-8

All rights reserved. No part of this book may be reproduced or transmitted in any form or by any means, electronic or mechanical, including photocopying, recording, or by any information storage and retrieval system, without permission in writing from the copyright owner.

Published by:

Peter E. Randall Publisher
Portsmouth, NH 03801
www.perpublisher.com

Cover illustration: Gordon Carlisle

Book design: Grace Peirce

First Edition
Hardcover ISBN: 978-1-937721-36-7
Library of Congress Control Number: 2017944671

Contents

Preface	viii
Acknowledgments	ix
Introduction	xi
Chapter 1	1

Fort Brown; "Record" 2/60; "Record" 4/60; "Record" 10/60; Fort Duncan; "Record" 12/60; A Pivotal Year

Chapter 2	12

The Texas Surrender; "Record" 2/61; "Record" 4/61

Chapter 3	24

Garrisoning the Florida Forts; Fort Taylor; Fort Jefferson; "Record" 6/61; Fort Pickens; The Sumter and Pickens "Truce"; Secret Expeditions; The Anaconda Plan; Secret Board of Officers; Pickens Armament Complete; Battery L to Pickens; "Record" 8/61; Events Elsewhere; The "Truce" Ends; "Record" 10/61

Chapter 4	50

Raid on Santa Rosa; "Record" 12/61; Bombardment of November 22nd & 23rd; Confederate Concerns; "Record" 2/62; Bombardment of January 1st, 1862; Confederate Withdrawals; "Record" 4/62; Reconnaissance on Santa Rosa

Chapter 5	72

The Secret Navy Plan; Troops to Ship Island; A Promising Outlook; Louisiana; Farragut and Butler; New Orleans; Vicksburg I; Failure at Corinth; "Record" 6/62; Occupation of Pensacola; "Record" 8/62; Reassignments; "Record" 10/62

Chapter 6	90

Vicksburg II; Wavering in Washington; Halleck General-in-Chief; Baton Rouge; The Department of the Gulf; Banks; "Record" 12/62; The Close of 1862

Chapter 7	108

"Record" 2/63; The Nineteenth Army Corps; Engagement at Bayou Teche; Grant; The Mississippi Marine Brigade; Return of the 42nd; Indianola Lost; Banks' Plan; Farragut Passes Port Hudson

Chapter 8 126
 "Record" 4/63; The Teche Campaign; Chasing the Fox; Opelousas; Cooperation with Grant; Grierson's Raid; Alexandria; Farragut Departs; Turn to Port Hudson

Chapter 9 162
 "Record" 6/63; Port Hudson; Demonstrations; Plains Store; Closing the Ring; May 23rd; May 24th; May 25th; May 26th; The Assault; Weitzel Advances; Sherman Advances; Augur Advances; Cease Fire

Chapter 10 198
 The Siege Begins; June 10th; June 14th; Paine's Assault; Weitzel's Assault; Augur's Feint; Dwight's Assault; June 15th; Jackson; Taylor Again; The Last Days; Finally; "Record" 8/63; The Next Step

Chapter 11 240
 The Sabine Pass Expedition; "Record" 10/63; A Land Route to Texas; The Rio Grande Expedition; "Record" 12/63; The Battle of Grand Coteau; "Record" 2/64; "Record" 4/64; 1864 Roster; Winter Quarters; The Red River Campaign

Chapter 12 268
 Wilson's Farm; Mansfield (Sabine Crossroads & Pleasant Grove); Pleasant Hill

Chapter 13 298
 April 10th; Grand Ecore; Cane River; Porter's Passage to Alexandria; The Dam at Alexandria; "Record" 6/64; Marksville & Mansura; Last Days in the Gulf

Chapter 14 314
 "Record" 8/64; Camp Barry; Washington Threatened; The 19th Corps Arrives; After Early; Sheridan Appointed; The Army of the Shenandoah; Weapons; First Moves; Early Reinforced

Chapter 15 337
 Cedarville/Winchester/Summit Point; The Public Mind; Berryville; Smithfield Crossing; "Record" 10/64; Anderson; Winchester

Chapter 16 366
    Fisher's Hill; The March Up The Valley; Terminated;
    Tom's Brook/Strasburg/Woodstock Races

Chapter 17 385
    Decisions, Decisions; Sheridan's Ride; Cedar Creek; Mosby;
    "Record" 12/64

Chapter 18 416
    Rosser; "Record" 2/65; Sheridan Leaves the Valley; Richmond;
    Surrender; "Record" 4/65; The Faithful Few; The Three Cooks

Appendix... 432
Sources 439
Bibliography & Key to Footnotes.............................. 439
Index.. 449
About the Author... 473

Preface

In 1876, Capt. William L. Haskin published *The History of the First Regiment of Artillery*. His preface begins: "It was my intention to omit the preface." Haskin explains that he did not wish to dictate to the reader how he "should read the pages which follow." I am in sympathy with Haskin, yet compelled to articulate some points drawn from *Battery L*.

The war was protracted. The Confederacy was able to raise large armies despite a population less than one-third that of the Union. Had they enlisted their slaves, the number would still have been less than one-half. An explanation may lie in a line in Grant's *Memoirs*, which labeled the South a "military despotism."

It took time for the Union to construct the hundreds of ships and rivercraft required to carry out its primary strategy—the blockade. Unfortunately, the Union suffered from failed generals and misguided political priorities. For almost three years, Lincoln was not favored with either good advice or luck.

I am pleased to note the close cooperation between our army and navy. The entire war on the western rivers and along the coasts was a joint army/navy affair. The opening of the Mississippi and *perhaps* the repulse of Lee at Gettysburg (he retreated into Virginia unmolested) finally brought the war to a turning point in July of 1863.

Low pay, dreadful medical care, and a poor diet initially characterized the soldier's lot, but as the war wore on, not the government, but the Sanitary Commission—the predecessor of the Red Cross—made a difference. Even so, by 1864, huge bonuses were felt necessary to reenlist almost all of those tired-yet-proud three-year volunteers.

Weapons technology improved over the course of the war, but little of it was seen by the infantry soldier, and subsequent battle descriptions largely ignore it. Did the Spencer repeating carbine (issued to the cavalry) make a difference in the Shenandoah Valley in 1864?

Acknowledgements

Paul Callsen—A descendent of Battery L's Patrick Donnelly, Paul has been involved with the preparation of this and the earlier book from the very beginning, including introducing me to the fact that the battery records were available at the National Archives.

T. Jeff Driscoll—Read the manuscript and provided the glowing review on the back cover.

Janet Simmonds—My late wife supported me through more than a decade of gathering the material for this and the earlier book.

Nancy Coleman—A dear friend, read the manuscript and provided valuable input.

Introduction

This is the Civil War odyssey of a small regular army unit. It is based on their bi-monthly muster roll "Record of Events" section – a synopsis of each appears chronologically throughout.

Four artillery regiments, each with nine batteries, were established by the Act of Congress of March, 1821. In 1838, the number of batteries authorized for each became ten, and during the Mexican War, it became twelve, all designated by the letters A-M, omitting the J. Hence, Battery L was mustered and equipped in 1847.

Finally, a fifth regiment was authorized in May of 1861.

The artillery regiment was unique. Unlike an infantry or cavalry regiment which took the field as a fighting entity, the artillery regiment did not. It had no control over its battery assignments which were made directly by the Office of the Adjutant-General in Washington. Hence, it was left with little more than to manage its recruitment and the monthly and muster roll reports of its batteries – assigned individually and widely scattered.

The size and equipage of a battery depended upon its assignment, and differed in times of peace and war. The battery would have to transform itself as required to man a fort, or support a mobile unit such as infantry or cavalry, thus becoming a "foot" or "field" battery.

In peacetime, a battery had four guns and 76 men. "Preparation for Service" called for 100 men and "Full War Organization," six guns and 150 men. A battery assigned to infantry was known as "mounted" and as "light" if to cavalry. Regarding horses, a battery in any of the given situations might have had none or more than 125. Battery L played all three roles.

In 1860, there were 16,387 officers and men in the U.S Army – often referred to as the "old army," or the Regulars. When war threatened, it was clear that the under-funded and ill-equipped old army was not up to the task. Hence, Lincoln's first call for 75,000 troops in April of 1861 was to the state militias. They were, in theory at least, organized. Subsequent calls were all to the states and the regular army was all but ignored. In the second year of the war there were 775,336 Volunteers and 26,255 Regulars.

Chapter 1

Fort Brown; "Record" 2/60; "Record" 4/60;
"Record" 10/60; Fort Duncan; "Record" 12/60; A Pivotal Year

Fort Brown

In the winter of 1860, Battery L found itself as part of the garrison of Fort Brown, at Brownsville, Texas, figure 1.[1] Their mission here was to aid in the chase after the so-called marauder Cortina, who had threatened the Brownsville area and briefly held Brownsville. Cortina's threats were derived from land disputes, he and his fellow Mexicans claiming to have been cheated out of their ancestral holdings by crooked American officials.

FIGURE 1

"Record" 2/60
31 DECEMBER 1859–29 FEBRUARY 1860, FORT BROWN, TEXAS

"The Co. returned from a scout of 15 days after the marauder Cortina, January 4th."
Samuel K. Dawson   Capt. Commanding.
William Silvey   1st Lt. Reg. Adjutant O. No. 7 Head Qrs. 1st Artillery, Fort Dallas, Fla, Aug. 13, '57. Left Co. April 22, 1854, S.O. No. 62, Head Qrs. New York, April 18, 1854.

---

1. *Harper's Weekly*, March 23rd, 1861.

James W. Robinson 1st Lt. On leave of absence, left Co. Sept. 5, 1859.
Loomis L. Langdon 2nd Lt. On detached service at Brazos Santiago, in charge of Government stores from abandoned forts. Left Co. Feb. 25, '59.

Strength: 39. Sick: 1.
Detached: 5 enlisted at Ringgold Barracks, Texas.

Note that those present for duty were only 1 officer and 30 enlisted men.

Captain Dawson, being the senior officer present, commanded the fort. Company M, 1st Lt. Bennett H. Hill commanding, was also a part of the garrison.

The regimental return for January lists Battery L as requiring 50 recruits. As far as resources are concerned, the company had no guns (cannons) or horses, though the men of L carried the Maynard patent model 1855 rifled musket. In fact, only two of the 12 companies in the 1st Regiment had any guns or horses at all. Company I at Fort Leavenworth, Kansas, John B. Magruder commanding, reported 4 guns and 63 horses, 1 unserviceable; and Company K, at Fort Clark, Texas, William H. French, commanding, had 4 guns and 24 horses, with 20 unserviceable. Only Company I was in a condition approaching the "preparation for service" category. Ten out of the 12 batteries of the 1st Regiment fell into the definition of a "foot" battery, supposed to garrison a fort. Battery A and Battery D were stationed at Fort Monroe, Virginia; B was at Key West Barracks, nearby to Fort Taylor (though the fort itself was unoccupied); C was at Ringgold Barracks, (Rio Grande City) Texas; E and H were at Fort Sumter, South Carolina; and G was at Barrancas Barracks, across Pensacola Bay, Florida, from Fort Pickens, which also was unoccupied.

Thus, we see that none of the batteries of the regiment were actually in a genuine brick-and-mortar fort. The typical "fort" was simply an open cantonment or camp, lucky to have some permanent structures. By these standards, the majority of the army artillery batteries, including Battery L, at this time were de facto infantry.

"Record" 4/60
29 FEBRUARY–30 APRIL 1860, FORT BROWN, TEXAS

[No entry in Record of Events]

Samuel K. Dawson	Capt. In Command
William Silvey	1st Lt. Reg'l Adjt. O. No. 7 Hdqrs. 1st Arty. Ft. Dallas, Fla. Aug. 13, 1857, Left Co. Apr. 22,'54 S.O. No. 62 Hdqrs. N.Y. Apr. 18,'54
James W. Robinson	1st Lt. Present
Loomis Langdon	2nd Lt. Present
Lewis Keller	1st Sgt. 1 Dec.'59, Newport, KY. Transferred to Comp'y L pr. Ord. No. 3, Hdqrs. Gen'l Rect'g Serv. War Dept., Mar. 5, 1860. Appt'd. Apr. 7th 1860, to date from 1st Apr. Appt'd. 1st Sgt. Apr.7th, 1860.

Thomas Conroy	Sgt.	1 Jan.'59, San Antonio, TX, 1st re-enlistment
James Flynn	Sgt.	12 Sept.'59, Fort Clark, TX, 1st re-enlistment
Thomas Newton	Sgt.	13 Dec. '58, Ft. Brown, TX, 2nd re-enlistment
Alexander Livingston	Cpl.	22 Sept.'58, Syracuse, NY, sick
Lewis Lighna	Musician	13 Oct.'58, Ft. Brown, TX
Francis Hagan	"	21 Jan.'59,

Strength: 80, Sick: 1.
Died:
John Murtaugh, Pvt. 10 Oct.'59, NewYork. Joined from General Recruiting Depot, Ft. Columbus, New York, April 15, 1860. Drowned in the Rio Grande, Apr. 27, 1860.

On April 15th, 46 recruits of the 50 called for arrived at Fort Brown, table 1. Their first appearance on the muster roll came at the end of April, and some of them, those listed as "in confinement," had already gotten into unspecified trouble, with unspecified disciplinary action. Most of the recruits in this first group to arrive were foreign born, as obtained from the 1860 census of "Garrison Fort Brown," which was taken on June 25th.

Analyzing the locations from where all the enlisted men in the battery had signed up, in addition to the recruits (except where *reenlistments* were made, which was usually at the duty station), produces a list of mostly Northern addresses: Buffalo, Rochester, Syracuse, New York, Boston, Philadelphia, Baltimore, Detroit, St. Louis, and Newport, Kentucky, which was across the Ohio River from Cincinnati. Note that the overwhelming majority signed up at either New York or Boston.

NAME	ENLIST. DATE	ENLIST. PLACE	ORIGIN	REMARKS
Anglin, Edmund	19 October '59	New York	Maine	In confinement
Anderson, James	9 November '59	New York	Ireland	
Becker, Julius	12 October '59	New York	Prussia	
Brunskill, William C.	19 October '59	New York	England	
Bissel, John	3 November '59	New York	Ireland	
Brown, William F.	1 November '59	Boston	England	
Baby, Alexander, J.	10 February '60	Boston	Canada	In confinement
Beeler, Andrew J.	9 February '60	Boston	Ireland	
Brook, Thomas	9 February '60	Boston	Massachusetts	
Benjamin, George	13 February '60	New York		In confinement
Casey, John	15 October '59	New York	Ireland	
Cain, Isaac T.	4 October '59	Boston	New Mexico	
Carr, Edward	1 November '59	Boston	Ireland	
Donnelly, Patrick	1 March '60	Boston	Ireland	In confinement
Demarest, William	15 October '59	New York	Maryland	
Dodge, Charles E.	3 November '59	Boston	Massachusetts	
Duffy, Thomas	12 November '59	Boston	Ireland	In confinement
Flynn, Patrick	3 November '59	New York		
Foley, Christopher	3 November '59	Boston	Ireland	
Friedman, George	7 February '60	New York	Bavaria	
Farrell, Bernard	11 November '59	New York	Ireland	
Golden, James	2 February '60	New York	Ireland	
Hadley, George	1 March '60	Boston	Massachusetts	
Harrison, John	25 October '60	New York	Pennsylvania	

Jaecke, Daniel	11 February '60	New York		
Kenny, Michael	8 February '59	New York	Ireland	
Kinney, Joseph	19 October '59	New York	Newfoundland	
Kutschor, Joseph	13 February '60	New York	Austria	
Morris, John	22 October '59	New York	France	
Myers, John	1 March '60	New York	Baden, Ger.	
Murphy, Michael	8 February '60	New York	Ireland	
Murphy, John	21 February '60	Boston	Rhode Island	
Murtaugh, John	10 October '59	New York		
Nitschke, John G.	6 February '60	New York	Prussia	
Riley, Charles	7 October '59	Boston	New York	
Ryan, James	14 October '59	New York		
Rupprecht, Ludwig	7 February '60	New York	Prussia	
Smith, Joseph	11 October '60	New York	Bavaria	
Schoenfeld, Charles	6 February '60	New York	Prussia	
Straub, Amelius	8 February '60	New York	Baden, Ger.	
Scott, William E.	9 February '60	Boston	England	
Stone, Richard	16 February '60	Boston	Vermont	In confinement
Thayer, Henry B.	9 February '60	Boston	Massachusetts	
White, Michael	7 October '59	Boston	Ireland	
Wicks, David J.	25 October '59	New York	Sweden	
Wilkinson, Joseph	4 February '60	New York	Ireland	

TABLE 1

It is striking that there was only one Southerner (though hardly south, St. Louis was in a slave state) in the ranks and none from the Deep South. Do these statistics tell us that for years before the Civil War the country had become polarized to the point that a Southerner would disdain to enlist in the "Yankee" army? There might have been some of that sentiment, but the basic pattern of few Southern enlistments had been true for years. Even when Jefferson Davis of Mississippi was secretary of war[2] (1853–1857), there appears no evidence that the recruitment of Southerners was greater, i.e., more popular. In the report of the secretary of war for 1854, the general recruiting for the army resulted in the enlistment of 2,365 men. Included was a breakdown, as compiled by the adjutant general, which is summarized below:[3]

Eastport, Maine	18		Ohio	25
Boston, Massachusetts	221		St. Louis	134
New York, NY	951		Chicago	87
New York State (remainder)	265		Detroit	1
Pennsylvania	419		Warrington, Florida	1
Maryland	73		Fort Ripley, Minnesota	1
Kentucky	169			

---

2. *Encyclopedia Britannica*, Vol. 7, p. 867.
3. A. O. P. Nicholson, 1854, p. 68. Whether some of the numbers include reenlistments is not specified.

The number from Kentucky included 166 from Newport, across the river from Cincinnati, suggesting that many came from Ohio. Note that most of the enlistments were from two cities – which were major points of entry for immigration, Boston and New York. Clearly, the immigrant was the staple of the "old" army.

A second group of recruits, table 2, was requested after the battery experienced a large number of desertions over the course of the summer. On December 5th, 25 arrived at Fort Duncan, Texas, to where the battery had moved that September.

	Enlisted	Place		Enlisted	Place
Ahern, James	18 Oct.'60	Boston	McDonough, Miles	17 Sept.'60	New York
Beglan, James	25 Oct.'60	New York	McGaley, Terence	18 Sept.'60	New York
Burke, John	7 Oct.'60	New York	Olvany, Michael	25 Oct.'60	New York
Craffy, Patrick	27 Sept.'60	Boston	O'Sullivan, Michael	26 Sept.'60	Boston
Creed, William	27 Sept.'60	Boston	Parketton, William	4 Oct.'60	New York
Cummings, Patrick	25 Oct.'60	Boston	Roper, John	22 Oct.'60	New York
Ferrari, Prosper	22 Oct.'60	New York	Schmidt, Heinrick	26 Oct.'60	New York
Flint Charles, A.	22 Sept.'60	Boston	Shaw, Warren P.	26 Oct.'60	Boston
Harkins, James	10 Oct.'60	New York	Stoll, Andrew	24 Oct.'60	New York
Howard, George	25 Oct.'60	New York	Thompson, Wm. V.	13 Sept.'60	Roch. NY
Jackel, Charles	26 Oct.'60	New York	Townsend, Reuben	27 Sept.'60	Boston
McCarthy, James	4 Oct.'60	Boston	William, Henry	28 Sept.'60	Roch. NY
McCoy, Daniel	24 Oct.'60	New York			

TABLE 2

The desertions that had taken place are listed in table 3, below:

Name and Rank	Enlistment Date	Enlistment Place	Deserted From & Date
Alexander Livingston, Cpl.	22 Sept.'58	Syracuse, NY	Ft. Brown 6 Aug.'60
James Anderson, Pvt.	9 Nov.'59	New York, NY	" 3 Aug.'60
Edmund Anglin, Pvt.	19 Oct.'59	"	" 30 May'60
George Benjamin, Pvt	13 Feb.'60	"	" 1 May'60
John Bissel, Pvt.	3 Nov.'59	"	" 30 May'60
Edward Carr, Pvt.	1 Nov.'59	Boston, MA	" 2 Aug.'60
Peter Cunningham, Pvt.	1 Aug.'58	Ft. Clark, TX	" 11 Mar.'60
Charles E. Dodge, Pvt.	3 Nov.'59	Boston, MA	" 2 Aug.'60
Thomas Duffey, Pvt.	12 Nov.'59	"	" 2 Aug.'60
Patrick Flynn, Pvt.	3 Nov.'59	New York, NY	" 16 Mar.'60
Francis Hagan, Pvt.	21 Jan.'59	"	" 25 Sept.'60
John Harrison, Pvt.	25 Oct.'59	"	" 2 Aug.'60
James Haynes, Pvt.	25 Nov.'57	St. Louis, MO	" 26 June '60
Daniel Jaecke, Pvt.	11 Feb.'60	New York, NY	" 24 Apr.'60
Joseph Kenny, Pvt.	19 Oct.'59	"	" 16 May'60
James McKenzie, Pvt.	23 Sept.'58	Buffalo, NY	" 2 Aug.'60
Francis Merle, Pvt.	30 Sept.'58	New York, NY	" 17 Mar.'60
John Morris, Pvt.	22 Oct.'59	"	" 6 Aug.'60
John Murphy, Pvt.	21 Feb.'60	Boston, MA	" 30 May'60
Michael Murphy, Pvt.	8 Feb.'60	"	Duncan 20 Feb.'60
James Ryan, Pvt.	14 Oct.'59	New York, NY	Fort Brown 16 May'60
Charles Schonfeld, Pvt.	6 Feb.'60	"	" 4 Aug.'60

Robert E. Schlatter, Pvt.	30 Sept. '58	"	"	11 Mar. '60
Richard Stone, Pvt.	16 Feb. '60	Boston, MA	"	3 Aug. '60
Henry B. Thayer, Pvt.	9 Feb. '60	"	"	18 July '60
Richard Walsh, Pvt.	5 Oct. '58	"	[Record obscured.]	
James Williams, Pvt.	22 Oct. '58	New York, NY	Ft. Brown 11 Mar. '60	

TABLE 3

By the end of the year, there had been 26 desertions—19 from the group of new recruits that had arrived at Fort Brown on April 15th.

An important observation regarding desertions is the fact that in August the company was likely to have been under orders to move to Fort Duncan, more than 400 long dusty miles north.[4] Any potential deserters would thus have known that the company was soon to travel into a remote and hostile area and that surviving after desertion would be very difficult. The month of August would be now or never.

The fate of only four of the deserters is known, and that only partially. Edmund Anglin, John Bissel, and John Murphy were apprehended and rejoined on June 4th, 1860, after only five days and were court-martialed. They were apparently unlucky or foolish enough to have remained in Texas. Had they crossed the Rio Grande they would have been free of any U.S. authority, and with money to spend, they would have been welcomed. As well, the many ships reaching Matamoras would have been an option to go elsewhere.

Edmund Anglin, though in and out of trouble in the ensuing years, remained to serve out the full term of his five-year enlistment and was discharged on October 12th, 1864. He apparently had had enough at the end of his five years because he did not take advantage of the substantial reenlistment bonus offered in July of that war year.

John Murphy also remained on duty but was drowned at Carrollton, Louisiana, on August 1st, 1863.

Francis Hagan turned himself in on January 2nd, 1861, after an absence of three months! He must have presented a convincing argument as he was restored to duty without trial on January 23rd. Bissel and Hagan again deserted on March 10th, 1861, just as the command was leaving Fort Brown to embark at the mouth of the Rio Grande. They were never heard from again.

It is interesting to note that the army thought it could afford to put soldiers in confinement and lose the benefit of their services as useful people. This would change soon. During the war, most of the deserters that were apprehended, and only a few were, were only fined. At that time, their services as soldiers were badly needed.

The muster rolls for this period indicate that most of the deserters vanished owing money to the United States, the sutler, and the laundresses. The money owed to the United States was for extra clothing, "camp and garrison equipment," and

---

4. They left on September 29th. Record of events, 31 August–31 October muster roll.

"ordnance." In 1860, there was a $45.97 allowance for a private's clothing in his first year of service, with lesser amounts in the subsequent four years of his enlistment. Assessments for clothing were for required extra-issue items needing replacement. Assessments for camp and garrison equipment and ordnance were supposed to be only for that which was lost or damaged through neglect. Most of the deserters owed from $2 to $3, out of their $11 monthly pay. This would be $306–$460, if based on a comparison with a private's pay for 2019.[5] Typical debt for clothing was $8 and for camp and garrison equipment, $3. Patrick Flynn owed the most of any individual: $7.70 for clothing, $4.24 for camp and garrison equipment, and $16.91 for ordnance, not to mention 50¢ to one of the laundresses. No doubt, he felt *put upon*, owing rather more than earning, as this was almost three months' pay. No mystery that he deserted on May 16th,[6] and never was heard from again.

Fort Duncan

"Record" 10/60
31 AUGUST–31 OCTOBER 1860, FORT DUNCAN, TEXAS

"Company left Fort Brown, Texas, Sept. 29, 1860 (by Steamboat) and arrived at Ringgold Barracks, Tex. Oct. 2nd 1860. Left Ringgold Barracks, Octr. 3rd and marched to Laredo (Ft. McIntosh) Texas, Octr. 8, 1860. Left Fort McIntosh Octr. 9th and arrived at Fort Duncan, Texas Octr. 14th 1860. Distance by steamboat 240 miles and from Ringgold Barracks, to Fort Duncan, Texas about 240 miles more."

Samuel K. Dawson	Capt. Commanding
William Silvey	1st Lt. Reg. Adjt. O. No. 7 Hdqrs. 1st Arty Ft. Dallas, Fla. Aug. 14,'57. Left Co. Apr. 22,'54. S.O. No. 62 Hdqrs. N.Y. April 18,'54
James W. Robinson	1st Lt.

Transferred: Loomis Langdon, by promotion to 1st Lt. Company A, 1st Artillery, Fort Monroe, Virginia
Strength: 61. Sick: none.

When the battery left Fort Brown, they probably left any possibility of discovering the whereabouts of their 23 deserters. They were transported, as noted, by shallow draft riverboat to Ringgold Barracks (Rio Grande City, the steamboat terminus) where they disembarked on October 3rd. They reached Fort Duncan on October 14th. This 11-day march amounted to an average of 22 miles per day, excessive compared to the 16 miles per ten-hour day recommended for artillery.[7]

---

5. In 2019, monthly pay for a private with less than two years of service was $1,680.
6. From the muster roll ending in June 1860. The money owed for ordnance was likely for the 1855 rifled musket, which featured the Maynard tape roll of caps. This feature was proven unsuitable for field conditions and was soon discontinued.
7. Journal of the Military Service Institute of the U.S., November 1889, p. 614; *Board of Artillery Officers*, p. 48.

Fort Duncan, Maj. William H. French commanding, was already occupied by Companies F and K of the 1st Artillery. Ordered abandoned by Secretary of War Floyd in 1859, Fort Duncan was ordered to be regarrisoned by Lt. Col. Robert E. Lee, commander, Department of Texas, in March of 1860. F had arrived from Fort Clark, at Bracketville, 60 miles north, on March 18th, after Fort Duncan's lease from the landowner John Twohig had been signed,[8] and K arrived on August 1st.

Founded in 1849, Fort Duncan was one of the largest posts in the west. It was important in the early fifties because of its location on the California Road (for those headed to the gold fields) and as a base for scouting against the Lipan Apache and Comanche Indians.

A survey map of "Cantonment Duncan" was prepared by the Corps of Engineers in 1853, figure 2. Shown are five officer's quarters, some with kitchens nearby, and three soldier's quarters; also, a bakery, a quartermaster and commissary store, guardhouse, sutler's store, and a hospital. It also had stables, a magazine, a quartermaster's office, and a commandant's office.[9] The fort had gone from abandoned to a sudden occupancy of some 144 soldiers, not including wives, children, laundresses, cooks, and servants. The arrival of Battery L and its final 25 recruits would have made up a total of about 234.

Adjacent was the town of Eagle Pass, population 520,[10] originally established south of the fort on the Rio Grande near the mouth of the Mexican Rio Escondido and opposite the Mexican town of Piedras Negras, a "double row of huts along the river . . . where the men smoked *cigarritos*, and the women made tortillas, and the dogs howled all night long."[11]

---

8. Notes from displays, and the docent's presentation at the Fort Duncan Commandant's office in 2009.
9. National Archives. *Cantonment Duncan,* sheet 34; *U.S. Census*, Ft. Duncan, Eagle Pass Post Office, 4th August 1860.
10. U.S. Census, July 1860.
11. Haskin, p. 352.

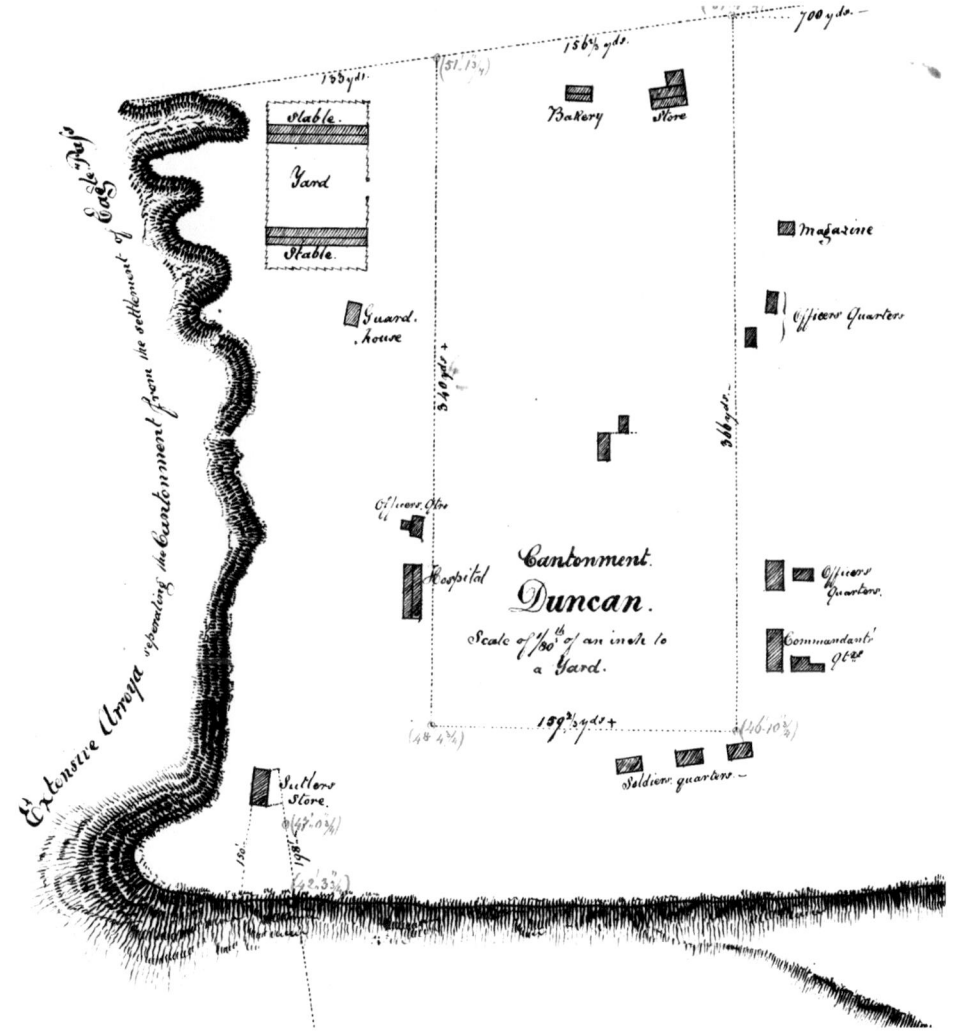

FIGURE 2

The commandant's office is shown in figure 3, as it was in 2009. It has been expanded considerably since 1860. Back then it would have had a thatched roof.

The new inhabitants to the area were undoubtedly hoping to make something of a home here; likely, they would be here for five years. Could anyone have guessed what was to happen in a mere four months? Their hope now, with the winter season coming on, was that it would be pleasant, compared to the 100°F summer heat, which had to be endured mostly by rationalizing that "it was dry and there was always a pleasant southwesterly breeze."

FIGURE 3

Perhaps the fort would once more be the scene of a social life to while away the time as there had been back in 1854 when Phil Sheridan and the cavalry were stationed here. Sheridan's memoirs[12] mention that the area was filled with antelope, deer, and wild turkeys and that hunting was a happy diversion. He also mentions horseback riding, races, and invitations to dances at the home of the Mexican commandant at Piedras Negras.

Sheridan also comments on the diet:

> . . . the food was the soldier's ration . . . flour, pickled pork, nasty bacon – cured in the dust of ground charcoal – and fresh beef . . . supplemented with game . . . The sugar, coffee, and small parts of the ration were good, but we had no vegetables and the few jars of preserves kept by the sutler were too expensive to be indulged in. So all during the period I lived at Fort Duncan and its sub-camps, nearly sixteen months, fresh vegetables were practically unobtainable. To prevent scurvy we used the juice of the maguey plant (*agave*), called pulque . . .

He goes on to describe how his company was detailed to ride some forty miles out to cut and gather the stalks, return, and press the mass to obtain the juice. It was then bottled and allowed to ferment, creating the traditional sour Mexican drink. Everyone was required to drink a cup at roll call.

Credit Sheridan and associates with an enlightened view of dietary requirements, recognizing the deficiencies of the army ration and taking action accordingly; though cactus beer was a quaint way to provide it.

Sheridan also cites the general lawlessness of the area.[13]

---

12. Sheridan, Vol. 1, pp. 21, 27, 28. Figure 3, photo by the author.
13. Sheridan, Vol. 1, pp. 33–34.

"Record" 12/60
31 OCTOBER–31 DECEMBER 1860, FORT DUNCAN, TEXAS

[No entry in the "Record of Events"]

Samuel K. Dawson	Capt. On leave, since May 2nd.
William Silvey	1st Lt. Reg. Adj. O. No. 7 hdqrs. 1st Arty, Ft. Dallas, Florida, Aug. 13,'57. Left Co. Apr. 22, 1854. S.O. No. 62 Hdqrs. N.Y. Apr. 18,'54.
James W. Robinson	1st Lt. In temporary Command.

Strength: 87. Sick: 6.

A Pivotal Year

The 1860 presidential campaign brought the slavery issue glaringly to the forefront. The relatively new Lincoln Republicans were antislavery. The Democrats were divided between proslavery and "popular sovereignty," the idea of leaving the choice of slavery to the inhabitants of those individual states or territories. The Democratic Convention fell apart, with the contending factions retiring to nominate separate candidates. A Southern faction nominated John C. Breckinridge, then vice president under Buchanan, and a Northern faction nominated Stephen A. Douglas. Split, the Democrats lost the election.

Inflamed by the election of that "Black Republican"[14] Lincoln, the Southern members of congress, and the South in general, now saw that it must find a remedy before Lincoln could take control, i.e., before March 4th, 1861 (at that time, the date of the presidential inauguration).

Long before this, a weak President Buchanan had done little to control the situation. Earlier that month, in his message to Congress, he had stated that under the Constitution no authority was given for any state to secede, though he had no power to prevent it.

Encouraged by this diffidence, Southern boldness only grew, and his administration began to fall apart. Howell Cobb, the treasury secretary, resigned and returned to Georgia to promote secession. Jacob Thompson, the interior secretary, returned to Mississippi to do likewise. John B. Floyd, the war secretary, was finally forced to resign after financial irregularities and his treasonous shipment of weapons from Northern arsenals to Southern ones were discovered.

Secretary of State Lewis Cass of Michigan resigned for yet another reason—he had failed in his efforts to convince President Buchanan to reinforce the forts in Charleston Harbor.

---

14. O.R. Vol. 1, pp. 115–118, 273; The term "Black Republican" is found to have been widely used in Southern correspondence. Nicolay & Hay, Vol. II, 1909, pp. 323, 325, 392, 397; *Encyclopedia Britannica*: *Cobb*, Vol. VI, p. 606, *Cass*, Vol. V, p. 455, *Black*, Vol. IV, p. 18, *Floyd*, Vol. X, p. 573.

Chapter 2

The Texas Surrender; "Record" 2/61; "Record" 4/61

The Texas Surrender

Hardly having settled into the routine of army life at their new post, events elsewhere were heralding the beginning of the Civil War. Within weeks, War Department orders to the three artillery companies at Fort Duncan would be received with instructions to turn the fort over to the infantry and march out of Texas. On December 6th, 1860, only one day after the last of the new recruits had arrived at Fort Duncan, delegates were elected to a convention called by the governor of South Carolina to consider an ordinance of secession. The "fire eaters" were in control, and on December 20th, the ordinance was adopted unanimously.[15] The people never had a direct say; no popular vote was ever held to ratify the action of the convention.

The first act of force occurred on December 27th, with the occupation of Fort Moultrie in Charleston Harbor[16] by an armed group of militia directed there on orders from the governor.

Though there was mail service to Eagle Pass from San Antonio, it is likely that news of these events did not reach Battery L until late January. However, the tenor of the times prompted the new[17] commander of the Department of Texas to write the following letter to the general-in-chief:[18]

> San Antonio, *December 13, 1860*
> Lieut. Gen. W. Scott, *Commanding U.S. Army, New York:*
>
> General: I think there can be no doubt that many of the Southern States will secede from the Union. The State of Texas will be among the number, and,

---

15. *Encyclopedia Britannica*, Vol. 27, p. 705, and Vol. 25, p. 501. The convention delegates were "elected" by the legislature. At this time, South Carolina was operating under its Constitution of 1790, which did not require a popular vote to ratify the action of the convention.
16. Fort Moultrie and Castle Pinckney, unoccupied, in Charleston Harbor were occupied by South Carolina troops on the 27th and the Palmetto flag was flown over the Federal Customs House and the Post Office. O.R. Vol. 1, pp. 6, 118.
17. Twiggs assumed command on December 8th, 1860, Nicolay & Hay, Vol. IV, p. 179. The authors speculate that this assignment was part of Floyd's War Department conspiracy to place officers of Southern birth in positions of trust.
18. O.R. Vol. 1, p. 579.

from all appearances at present, it will be at an early day, certainly before the 4th of March next. What is to be done with the public property in charge of the Army? . . .

I am, general, with sentiments of respect and regard yours, &c.,

D. E. TWIGGS

Scott, born in 1786 and commissioned as a captain of artillery in the U.S. Army in 1808, had personally experienced nearly every military action that had involved the army since the Revolution. By a half century of experience, he had foreseen the events leading up to the current situation. He remembered the resolute action of President Jackson in 1832, in quelling the nullification question, and was appalled at the diffidence of President Buchanan on this new challenge from South Carolina. On October 29th, 1860, he violated the chain of command by writing directly to the President. In part, his letter:[19]

> From a knowledge of our Southern population it is my solemn conviction that there is some danger of an early act of rashness proclaiming secession, viz., the seizure of some or all of the following posts: Forts Jackson and St Philip, on the Mississippi, below New Orleans, both without garrisons; Fort Morgan below Mobile, without a garrison; Forts Pickens and McRee, Pensacola Harbor; with an insufficient garrison of one, Fort Pulaski, below Savannah, without a garrison; Forts Moultrie and Sumter, Charleston Harbor, the former with an insufficient garrison and the latter without any; and Fort Monroe, Hampton Roads, without a sufficient garrison. In my opinion, all these works should be immediately so garrisoned as to make any attempt to take any one of them, by surprise or *coup de main* ridiculous.

He then requested of the secretary of war, Floyd, permission to warn those few installations that had garrisons to be alert to attack. Floyd denied permission, and the President did not act.

Scott finally obtained a personal interview with the President on December 15th, when he repeated his views, with specific emphasis on reinforcing forts Sumter and Moultrie, but the President said, in substance, that "the time had not arrived for doing so." Scott perhaps had not dared to mention President Jackson's actions on the nullification question; but on reflection, that same night, after the meeting, Scott wrote a note to the President in which he recounted the decisive action President Jackson had taken to reinforce the forts in Charleston Harbor. (At that time, the reinforcements had been under Scott's personal command.)

Thus, the answer to Twiggs' letter came with no surprises. It came from George W. Lay, Scott's chief of staff, dated December 28th, stating in part that "The President has listened to him [Scott] with due friendliness and respect, but

---

19. Scott, W., pp. 610, 613, 614.

the War Department has been little communicative."[20]

Scott's hands were tied as long Floyd's policies ruled. Finally, after a stormy cabinet meeting on December 13th, which had centered on the question of the relief of the Charleston forts,[21] Floyd recommending against, the path to his resignation on December 29th was opened. The postmaster general, Joseph Holt, was appointed as interim secretary on December 31st.[22] The general-in-chief, though ill, now finally had the opportunity to begin the action that he had recommended in October, though it was still to be limited by the caution of Buchanan.

Events at Fort Sumter were a priority. Major Anderson, the commander, later explained that he felt Moultrie was not defensible with its small garrison and that a combined one in Sumter could present a better defense. Thus, on December 26th, 1860, he abandoned Moultrie and occupied Sumter.

Anderson's move offered the new governor of South Carolina a technicality over which to stir the pot. He sent an aide to Washington to complain that a previous understanding with the President had been violated in that "no reenforcements [sic] were to be sent to any of the forts, and particularly this one . . ."[23] As a consequence, the South Carolina militia occupied Fort Moultrie the day after it was abandoned, and on December 30th seized the Charleston Arsenal.

The same day,[24] not knowing of Floyd's resignation, but having long recognized his damaging policies, Scott directly requested of the President that he be allowed to prepare "without reference to the War Department and otherwise, as secretly as possible, to send two hundred and fifty recruits from New York Harbor to re-enforce Fort Sumter." The next day Scott ordered that a force of 200 troops be made ready to board the sloop-of-war *Brooklyn* to be sent to Fort Sumter.[25] The President's fear of armed conflict blunted such boldness. Quoting Scott:[26] "afterward, Secretary Holt and myself endeavored, in vain, to obtain a ship of war for the purpose, and were finally obliged to employ the passenger steamer Star of the West."

By January 5th, 1861, the plan[27] was initiated. The only advantage that the *Star of the West* offered was a cover for its purpose. Because it regularly ran between New York and New Orleans, its departure might not be regarded as a special expedition by ubiquitous Southern sympathizers. Southern spies, however, took due note of

---

20. O.R. Vol. 1, pp. 579, 580; see also, pp. 113, 114.
21. Nicolay & Hay, Vol. II, p. 394.
22. O.R. Ser. III, Vol. 1, p. 21.
23. O.R. Vol. 1, p. 3. Report of Robert Anderson.
24. O.R. Vol. 1, p. 114.
25. O.R. Vol. 1, p. 119.
26. Scott, W., p. 621. Holt replaced Floyd in January.
27. O.R. Vol. 1, p. 9.

the 200 soldiers that were brought on board and hidden.[28] Forewarned, South Carolina fired upon the unarmed steamer as it approached Charleston Harbor, and it returned to New York.

On January 3rd, and 4th, Scott had the opportunity to take his long since recommended action to occupy other coastal forts. Orders[29] were issued to Lt. A. J. Slemmer, and his Company G, of the 1st Artillery at Barrancas Barracks, to occupy Fort Pickens; to Capt. J. M. Brannan and his Company B of the 1st Artillery at Key West, to occupy Fort Taylor; to Major L. G. Arnold and his Company C of the 2nd Artillery at Fort Independence, Boston, to occupy Fort Jefferson, Tortugas, Florida; and to Major Z. B. Tower to take command of all the facilities "in and about" Pensacola, Florida. These actions were perhaps a few moments past "not a moment too late" but, at least, they were significant directives from whence there had been none. Some parts of a deteriorating situation might now be looked to be salvaged, since southern states were beginning to seize the Federal property within their borders.

The clouds of secession gathered rapidly:

Jan. 9, 1861	Mississippi
Jan. 10, 1861	Florida
Jan. 11, 1861	Alabama
Jan. 19, 1861	Georgia
Jan. 26, 1861	Louisiana
Feb. 1, 1861	Texas, ratified on Feb. 23

The next wave of secessions followed a month later:

March 16, 1861	Arizona Territory
April 17, 1861	Virginia [divided, as forty-eight Western counties split off to form West Virginia]
May 6, 1861	Arkansas
May 20, 1861	North Carolina
June 8, 1861	Tennessee
October 31, 1861	Missouri [divided, with an unelected pro-Union government]
November 20, 1861	Kentucky [both pro-Union and pro-Confederate factions]

The events[30] bearing on the future of Battery L and its contemporaries in Texas were a study in slow communications and interference by the Texas authorities. There was not, however, any lack of initiative on the part of either the new

---

28. O.R. Vol. 1, p. 253.
29. O.R. Vol. 1, pp. 334, 345, 350.
30. O.R. Vol. 1, pp. 581, 584. Communications took anywhere from 12 to 19 days, from a perusal of Union correspondence. There was a telegraph between New York and New Orleans but apparently not to San Antonio.

secretary of war, Holt, or General Scott. When Major General Twiggs revealed in a letter to the War Department, dated January 15th, that he would follow the South, he being a "Georgia man," he was promptly relieved by special orders no. 22 dated January 28th, 1861, and replaced by Colonel Carlos A. Waite of the 1st Infantry, stationed at Camp Verde, Texas, some 60 miles from San Antonio.

On January 31st, orders[31] directed the commanding officer of the Department of Texas, "to take immediate measures for replacing the five companies of artillery on the Rio Grande . . ." Twiggs, unaware of his having been replaced, duly issued special orders no. 25[32] on February 14th, the same day that the dispatch was received:

> Companies F, K, and L, First Artillery at Fort Duncan, Company M, First, and Company M, Second Artillery, at Fort Brown, will march, immediately upon receipt of this order, for Brazos Santiago, at which place a steamer has been directed to be in readiness to receive them for transportation out of Texas. The light companies will take their guns, ammunition, and equipments with them, but will leave their horses on embarkation.

Twiggs received the news of his being replaced the next day.[33] Colonel Waite did not arrive at San Antonio to assume command until the 19th, only to discover that events in Texas had run faster than anyone in Washington could have contemplated. On the morning of February 16th, the Alamo and other of the facilities of the headquarters of the Department of Texas, at the plaza in San Antonio, figure 1, were surrounded by some 1,000 volunteers under the command of Maj. Benjamin McCulloch of the First Regiment of Texas Mounted Riflemen, and two days later, Twiggs agreed to give up all Federal property and to withdraw all of the 2,684[34] Federal troops stationed in Texas.

Three plans for the withdrawal developed: one put forward by Colonel Waite, one by the commissioners, and one by Washington. The plan from Washington was immediately shattered and unworkable, and the one from Colonel Waite soon became subordinate to the whims of the Texans. At first, the withdrawal was rather orderly and relations between the Texas volunteers and the U.S. Army were correct and courteous. However, as time dragged on, the impatience of the Texans grew. News from elsewhere began to harden their attitude, and some began to regard the Federal troops not so much as departing colleagues but as an escaping group of traitors. On February 25th, Major McCulloch urged that, "This force ought to be disorganized before it leaves this State."[35] He had the audacity to suggest that

---

31. O.R. Ser. I, Vol. 1, p. 585.
32. O.R. Vol. 1, p. 589.
33. O.R. Vol. 1, p. 590.
34. O.R. Ser. II, Vol. 1, p.8.
35. O.R. Vol. 1, p. 609.

"the proper bounty" would recruit many to remain and join Texas' service.

FIGURE 1

Only one day after Battery L and its colleagues received their orders to leave Fort Duncan, the following was sent from Washington (and received on March 1st):[36]

> Headquarters, Department of the Army,
> *Washington, February 15, 1861*
>
> Col. C. A. Waite,
> *First Infantry, Commanding Dep't of Texas, San Antonio:*
> Sir: In the event of the secession of the State of Texas, the General-in-Chief directs that you will without unnecessary delay, put in march for Fort Leavenworth the entire military force of your department.
> [The letter goes on to describe how to accomplish this, in considerable detail, here omitted.]
>
> L. THOMAS
> *Assistant Adjutant General*

The remainder of the letter was moot because the plan was not what was to be dictated by the Texas commissioners. By February 23rd, the Ordinance of Secession was ratified, and the Texas commissioners took full control – even over Governor Sam Houston.

---

36. O.R. Vol. 1, p. 589.

On March 19th, General Scott at Washington had had another idea. Orders were sent[37] to Col. Waite, to the effect that, if there were still sufficient numbers of troops that had not left – not less than 500 and preferably 1,200 – they were not to embark but to form

> . . . a strongly-entrenched camp at some suitable point convenient to and covering the post of Indianola [to keep a foothold in Texas until the question of secession may be settled between competing factions and] to give such aid and support to General Houston or other head of authority in defense of the federal government as may be within your power. . . .
>
> PS. – If on receipt of this duplicate it should be well known that neither Governor Houston nor any other executive authority of Texas has any considerable number of men up in arms in defense of the Federal Government . . . you will consider the foregoing instructions withdrawn.

A special messenger was sent to Sam Houston to discuss the plan. He declined any such assistance,[38] and the perplexed governor then urged: "by all means take no action towards hostile movements . . ."

Scott's order was obsolete the day it was sent. Before Colonel Waite received it, he had sent a note to headquarters in Washington that he had no choice but to embark the troops at Indianola as instructed, "under [the orders of] the date of the 12th, ultimo"[39] and to proceed to New York. Scott's letter did not arrive at Waite's headquarters until April 1st, and by that time more than 500 U.S. troops had already gathered at Green Lake, near Indianola, prepared for departure.

The idea of a strong foothold, however, did not die. The retention of, and subsequently the taking of, strong footholds at coastal points along the shores of the Confederacy were to become a central strategy in the Union prosecution of the war. It would become known as the "Anaconda Plan."

"Record" 2/61
31 DECEMBER 1860–29 FEBRUARY 1861, FORT DUNCAN, TEXAS

The Company left Fort Duncan, Texas, February 20 inst. In pursuance of S.O. No. 25, Headquarters Dept. of Texas, San Antonio, Feb. 14 inst. and arrived at Camp Alburquielas, Texas[40] at 1 o'clock today. [29 Feb.]

Samuel K. Dawson	Capt. Leave of absence for 2 months S.O. no. 4 Dept. of Texas, April 30, 1860 extended For 6 mo. S.O. no. 124, A.G.O. June 20, 1860. Left Co. May 2nd 1860.

---

37. O.R. Vol. 1, p. 598.
38. O.R. Vol. 1, p. 551. Houston later refused to swear allegiance to the Confederacy, and on March 16th, the Texas Legislature declared the office of governor vacant.
39. O.R. Vol. 1, p. 598.
40. Unknown location, estimated at fifty miles northwest of Rio Grande City. Questionable spelling from handwritten copy.

William Silvey	1st Lt. Reg. Adjutant O. No. 7 Head Qrs. 1st Artillery, Fort Dallas, Fla, Aug. 13, '57. Left Co. April 22, 1854, S.O. No. 62, Head Qrs. New York, April 18, 1854.			
James W. Robinson	1st Lt. Det. svc. on a train to the coast, left co. Feb. 14, 1861.			
Richard H. Jackson	2nd Lt. In command of Co. since Feb. 14, 1861.			

Detached:
   Wallace Wright Pvt. Escort duty to San Antonio, Texas. Absent from Co. since Febr'y 14, 1861.

Deserted:

Michael Murphy	Pvt.	Deserted near Fort Duncan, Texas,	Febr'y 20, 1861

Discharged:

Henry Hehn	Pvt.	By reason of expiration of service,	Jan. 10, 1861
Henry Lundenberg	Pvt.	" " "	Jan. 25, 1861
Edward O'Donnell	Pvt.	" " "	Jan. 5, 1861
William Robinson	Pvt.	" " "	Jan. 26, 1861

Strength: 83 Sick: 1
Present Sick: none
Absent Sick: Charles Riley Sgt. Sick at Fort McIntosh, Texas. Absent since February 25th 1861.

The muster roll mentions the detached service of 1st Lt. James Robinson who was "on a train to the coast." His mission was the result of a personal note received earlier by the commander of the three batteries at Fort Duncan, Maj. William French. The note came from a friend on the general staff in San Antonio, and it had warned, in part, that: "The matter is arranged. All will be surrendered to the State, and that in its sovereign capacity."[41]

This provided French with confirmation of the plans of the secessionists. The surrender would soon be announced, and there was no time to be wasted in planning to be ready to move the artillery out. His first move was to get the women and children away.

In peacetime, a soldier could make arrangements for his family to accompany him, largely on the basis of what he could afford. Thus, officers often had their family and servants living on the post. Enlisted families came along if somehow their income could be supplemented. The families of record were Lt. J. W. Robinson and his wife, two sons and a servant; Sgt. Thomas Newton's wife, Ann, and daughter, Mary; Sgt. Thomas Conroy, with wife, Elizabeth; Pvt. Amelius Straub, with wife, Margaret; Pvt. Louis Lighna, with wife, Julia, and son, Louis; and Pvt. James Flynn with wife, Mary, his son, Thomas, and daughter, Emily.

A "train" was put together, consisting of the few available wagons, Mexican carts, and ambulances that could be found, considering that the means of transport had been gradually taken away by the Texas authorities on one pretext or another; clearly their plan being to immobilize them and subject them to threat of capture.

---

41.   Haskin, pp. 136–138, 353, 354.

The women and children on board, their train set out on February 14th for San Antonio and Indianola, with Lieutenant Robinson in command and Pvt. Wallace Wright as escort.

Twiggs' order for the five artillery companies to move to Brazos Santiago had been issued on the same day. It was rushed to Fort Duncan by "private hand" but did not reach French until February 20th. W. A. Nichols, on Waite's staff at San Antonio, had added a note to it, dated February 16th, 1861: "This order was to be cut off[42] yesterday, move rapidly; the object of the authorities of Texas is to demand the surrender of the guns of the light batteries."

This was all the warning that was needed. The command left Fort Duncan that day. The scarcity of wagons meant that nearly all their personal property was left behind. They moved rapidly, expecting to be intercepted at any of a number of points advantageous for ambush. At Willow Pond, Sgt. Charles Riley was thrown from his horse and his right leg broken in two places. Luckily, the route they had taken led them past Fort McIntosh (Laredo). There was a surgeon at the fort, and he could set the leg. Since Riley could not be moved, he would have to be left behind; and since it was likely that the infantry stationed at McIntosh would soon be forced to leave, a collection was taken up to pay for Riley's keep with a Laredo family, who were allegedly sympathetic to the Union. Bidding farewell on February 25th, the march continued, with Riley left behind in the Fort McIntosh hospital.

"Record" 4/61
29 FEBRUARY–30 APRIL 1861, FORT JEFFERSON, FLORIDA

The Company left Camp Alburquielas on the 1st and arrived at Ringgold Barracks, Texas on the 3rd of March. The Company left Ringgold Barracks, Texas on the 4th and arrived at Fort Brown, Texas on the 9th of March. The Company left Fort Brown on the 10th and arrived at the mouth of the Rio Grande (Brazos Santiago) on the 11th March. The Company left the mouth of the Rio Grande and embarked on board the Steamship "Gen'l Rusk" on the afternoon of the 19th and arrived at Fort Jefferson, Tortugas Island, Florida, on the 24th of March, 1861.

Samuel K. Dawson   Capt. Joined Co. from absence with leave Mar. 24, 1861.
William Silvey   1st Lt. Reg. Adjutant O. No. 7 Head Qrs. 1st Artillery, Fort Dallas, Fla, Aug. 13,'57. Left Co. April 22, 1854, S.O. No. 62, Head Qrs. New York, April 18, 1854
James W. Robinson   1st Lt. Joined Company from det. svc. Mar. 9,'61.
Richard H. Jackson   2nd Lt.

Detached:
James Ahern	Pvt.	On det. svc. at Fort Pickens, left Co.	April 14, 1861.
Thomas Brook	Pvt.	" "	"
Wm, F, Brown	Pvt.	" "	"
Patrick Cummings	Pvt.	" "	"

---

42. Meaning intercepted and kept from circulation; Riley Pension File no. 476558, National Archives.

Christopher Foley	Pvt.	"	"		"
James McCarthy	Pvt.	"	"		"
James McCoy	Pvt.	"	"		"
John Meyer	Pvt.	"	"		"
Michael O'Sullivan	Pvt.	"	"		"
William Parketton	Pvt.	"	"		"
Thomas Poole	Pvt.	"	"		"

Joined:

Henry A. Ward	Pvt.	From General Recruiting Depot, Fort Columbus, New York,	Mar. 21, 1861
Henry Wilkson	Pvt.	" " "	"
Owen A. Wren	Pvt.	" " "	"

Deserted:
  John Bissell Pvt. At Fort Brown, Texas, March 10, 1861
  Francis Hagan Pvt. At Fort Brown, Texas, March 10, 1861

Extra Duty with Engineer Dept.: Patrick Craffy, Patrick Donnelly, John Holland, Wm. Scott, Joseph Smith

Strength: 84, Sick: 7

Sick Present: William Brunskill, Bernard Farrell, Joseph Kutschor, Edward McLaughlin, Henry Wilkson, (one not listed).

Sick Absent: Charles Riley Sick at Fort McIntosh, Texas. Absent from Company since Febr'y 25, 1861.

Though the "Record" never mentions such details, a salute was fired in honor of the inauguration of President Lincoln as they departed Ringgold Barracks at dawn on the 5th.[43] By 11:00 a.m., they had marched 22 miles. During the noon halt, an express message came to Major French from Brownsville, with the news that the assistant adjutant general of the army, Maj. Fitz John Porter, had been sent there with instructions for the command and that Porter would await the arrival of the column. The march was then resumed for another seven miles, reaching 29 miles in total – a frantic pace, enough to exhaust both men and animals. On the 6th, they marched 26 miles; on the 7th, 27; and on the 8th, a march of 23 miles brought them into Brownsville. Here they joined companies M, 1st Artillery and M, 2nd Artillery, including several infantry companies.

Though the garrison now consisted of two cavalry, five artillery and the infantry companies, they were far outnumbered by the Texas troops that were gathering and organizing in the vicinity. The Texans had previously approached Capt. Bennett H. Hill, commanding, to attempt to enlist him into their cause. He had completely rebuffed them and had replied that he considered it his duty to send traitors to Washington in irons.

---

43. Haskin, p. 138.

FIGURE 2. *DANIEL WEBSTER*, OFF PORT ISABEL, TEXAS

On March 3rd, Porter had chartered the steamer *Daniel Webster*, figure 2, and it was waiting off the bar of the Rio Grande. The command arrived at Brazos on March 11th,[44] whereupon Porter deemed that it was not a secure place to use for embarkation. He was concerned, not that a shooting incident would occur, but that the undisciplined Texas volunteers and "minute men" would loot the provisions placed there for the use of those awaiting embarkation.[45] In any event, the steamer did not cross the bar and tie up at the wharves. It remained anchored off the mouth of the Rio Grande, enduring high northeast winds and the delay caused by the difficulty of lighters to ferry supplies out to it. Porter's report later hinted that the Texans were using any excuse to make things difficult. Loading did not begin until the 16th.

Major Porter had been instructed by Washington to load additional troops, infantry if available, regardless of prior orders from headquarters at San Antonio. He thus took the initiative to charter an additional steamer, the 750-ton sidewheeler *General Rusk*. Porter's sense of urgency was justified, as later events proved. It was not long after that the *General Rusk*, based at Galveston, was seized by the state of Texas and converted into a gunboat.

Both the *Rusk* and the *Daniel Webster*[46] got off late on the 19th, the *Rusk* to deliver batteries F and K to Fort Taylor at Key West, and batteries L and M to Fort Jefferson. The *Daniel Webster*, with M of the 2nd Artillery and C and E of the 3rd Infantry, continued on to Fort Hamilton, New York, arriving there on March 30th.

---

44. O.R. Vol. 52, p. 128; Vol. 1, pp. 587, 588. Additional narrative of departure from Haskin, pp. 136, 137.
45. O.R. Ser. II, Vol. 1, p.13.
46. Figure 2, *Harper's Weekly*, April 13, 1861. *Harper's* intimation that the troops were embarked at Port Isabel does not agree with the muster roll record.

Having gotten out of Texas by the "skin of their teeth," a now-distant memory was the fate of Battery L's 22 deserters,[47] those left behind prior to September 29th, when the company left Brownsville for Fort Duncan. Then there was Michael Murphy, who had disappeared on February 20th, the very day that they marched out of Duncan. After deserting at Fort Brown and being restored to duty, one could have hardly imagined that John Bissel and Francis Hagan would again walk away on March 10th, just as the command left Brownsville for the mouth of the Rio Grande.

Most of the army succeeded in leaving Texas, but a few from headquarters at San Antonio, and six companies of the 8th infantry, who were marching from New Mexico did not. They were apprehended on May 9th and held prisoner for 20 months until exchanged in December of 1862. The events in April—Lincoln's declaration of a blockade of the South on April 9th, and the fall of Fort Sumter on the 12th had signalled that a war was on.

---

47. The 1860 muster rolls. The April 13th issue of *Harper's* mentions the Indian danger: "The Indians followed the march of the troops, and committed great havoc among the people, killing some and running off their stock." Murphy had taken a great risk in deserting.

Chapter 3

Garrisoning the Florida Forts; Fort Taylor;
Fort Jefferson; "Record" 6/61; Fort Pickens;
The Sumter and Pickens "Truce"; Secret Expeditions;
The Anaconda Plan; Secret Board of Officers; Pickens Armament Complete;
Battery L to Pickens; "Record" 8/61; Events Elsewhere; The "Truce" Ends;
"Record" 10/61

Garrisoning the Florida Forts

The events of December – the secession of South Carolina and a cabinet shakeup, elevating Holt and Black and adding Stanton, all decisive men and against secession, finally helped President Buchanan to decide that it was his constitutional duty to protect Federal property. His confusion over his power to deal with secession remained, but at least that might be dealt with as a separate issue.

A decision was made to occupy three strategic Florida coastal forts long abandoned, or nearly so:[48] Taylor at Key West, Jefferson at Dry Tortugas, and Pickens at Pensacola. The orders were issued on January 3rd and 4th. The occupation of Jefferson went as planned. The occupation of Taylor did not go as planned, but went smoothly nevertheless, due to the initiative of the officer involved. The occupation of Pickens became a long and complicated drama that would hatch treasonous intrigue on the part of Southern members of Congress, result in a falsely authorized secret naval expedition, and cause the stonewalling of the facts before Congress by the Lincoln administration.

Fort Taylor

The occupation of Fort Taylor was accomplished first and its existence in the Union fold allowed it to be used as a key base of supply and support for subsequent activity at Jefferson and Pickens. Fort Taylor was in the best condition of the three.

Capt. John M. Brannan, commanding Company B of the 1st U.S. Artillery at Key West, had heard of "the recent seizure by unauthorized persons of several

---

48. Fort McCree, on Perdido Key, opposite Pickens, was occupied by a lone ordnance sergeant. Haskin, p. 487.

forts and arsenals in the Southern States"[49] and in the absence of any answer to a letter he wrote to the adjutant general on December 11th, 1860, asking what action he should take, he acted on his own initiative. In a report dated January 14th, he notes that he has moved his "scanty" force of 44 men from the Key West barracks, located on the northeast side of the island, to the fort, a distance of about two miles. Thus, Key West Harbor was secured from "privateers."

On January 31st, he blandly acknowledges the receipt, on the 29th, of General Scott's orders of January 4th, which he "had anticipated some time ago."[50]

Fort Jefferson[51]

FIGURE 1

To occupy the then-unmanned Fort Jefferson became the object of Maj. Lewis G. Arnold, commander of Company C of the 2nd Artillery, stationed at Fort Independence, Boston Harbor.[52] Acknowledging the receipt of his telegraphed orders, he sailed on January 10th on the steamer *Joseph Whitney* and arrived at Fort Jefferson on the 18th.[53]

Jefferson is located 68 miles west of Key West, on Garden Key in the Dry Tortugas. As of January of 1861, it was less than half completed.[54] In fact, it would later be abandoned as obsolete before it was ever completed. Capt. Montgomery C. Meigs of the Army Corps of Engineers had been supervising its construction since November 8th, 1860. Eventually rising to a height of 50 feet, the walls at this time averaged 35 feet and only the 35 embrasures (gun openings) of the lower tier were

---

49. O.R. Vol. 1, p. 342.
50. O.R. Vol. 1, p. 344; Haskin, p. 135.
51. Figure 1, photo by the author's wife Janet..
52. O.R. Ser. III, Vol. 1, p. 23.
53. O.R. Vol. 1, p. 346.
54. O.R. Ser. III, Vol. 1, p. 47.

ready.[55] There were no guns or barracks, and the water in the cisterns was brackish.[56]

Unlike Brannan at Fort Taylor, Arnold would be totally isolated at Jefferson. There was no land approach to make the fort vulnerable from that direction. The original assumption in the design and construction of these coastal forts was that the threat was from the water. Now, at Taylor – and as we shall see, Pickens – the adversaries were on the land. Only Jefferson was functioning as intended; and with the U.S. Navy virtually ruling the sea, there was little danger of a large enough force arriving that could be successful in any attack.

"Record" 6/61
30 APRIL–30 JUNE 1861, FORT JEFFERSON, FLORIDA

[No entry in the "Record of Events"]

Samuel K. Dawson	Capt. Sick
William Silvey	1st Lt. Reg. Adj. O. No. 7, Hdqtrs. 1st Arty. Ft. Dallas, Florida Aug. 13,'57 Left Co. Apr. 22,'54. S.O. No. 62 Hdqrs. N.Y. Apr. 18 '54.
Richard H. Jackson	2nd Lt. On detached service Fort Pickens, Florida S.O. No. 24 Hdqrs. Dept of Florida May 29,'61. Left Co. June 11,'61.

Strength: 83, Sick: 5.

Sick Present: Samuel K. Dawson, James Golden, Andrew Stoll, Wallace Wright

Sick Absent: Charles Riley Sick at Ft. McIntosh, Tex. Absent from Company since Feb. 25th 1861

Temp. Attached: M.M. Blunt 1st Lt. June 28th S.O. No. 32 Hdqtrs Fort Jef. Tortugas

Resigned: James W. Robinson 1st Lt. May 15, 1861, as per notice from A.G.O. Washington, June 13, 1861.

On May 29th, 1st Lt. Richard H. Jackson, in command of Battery L at Fort Jefferson, was ordered to be transferred to Fort Pickens. This left the company with no one in command from June 11th, when Jackson left. Capt. Samuel Dawson was sick, and William Silvey, long since assigned as the regimental adjutant of the 1st Artillery, had not been with the company since 1854, and 1st Lt. James W. Robinson,[57] 13th in his class at West Point in 1852 and the next in line for command, had earlier written to the adjutant general's office resigning his commission. His original intention was to join the Confederate service in his native state of Virginia.

---

55. O.R. Vol. 52/I, pp. 4, 5.
56. O.R. Ser. III, Vol. 1, p. 691. Dated November 30th, 1861, this report indicates that much of the work of "temporary construction of wooden buildings for storage and shops, and casemates and wooden buildings fitted for barracks and quarters" were constructed after the arrival of the troops. Appropriations for permanent barracks, etc. were requested for the next year. The report also refers to the fact that 79 of the cisterns were "tightened" because of the leakage of seawater into them.
57. Cullum, Vol. I, no. 1548.

Though he had changed his mind, probably after the protestations of his Massachusetts-born wife, his resignation was accepted. Hoping to convince the secretary of war that he was loyal and that he wished to continue in the regular army, he had remained at Fort Jefferson. He never was reinstated. He briefly stayed on, however, as a sutler. Finally, on June 28th, Lt. Mathew M. Blunt, an infantry officer and member of Arnold's staff, was attached to command the company.

Fort Pickens

This huge brick structure was built on Santa Rosa Island, opposite Pensacola. The ravages of time have left the structure in a much poorer condition than Jefferson, and several historic views, keyed to the plan view in figure 2[58], offer the reader a feel for the original structure. In 1861, a person standing at "A" would have seen the view in figure 3, the sally port, which faces north. To the right, the northeast bastion, with the flag on it ("E" in the diagram) is no longer standing, destroyed in an explosion of ammunition stored there in 1899. The sloped embankment to the left is a portion of the glacis, the wide shaded area in figure 2, which faces east, toward the only land accessible to attackers.

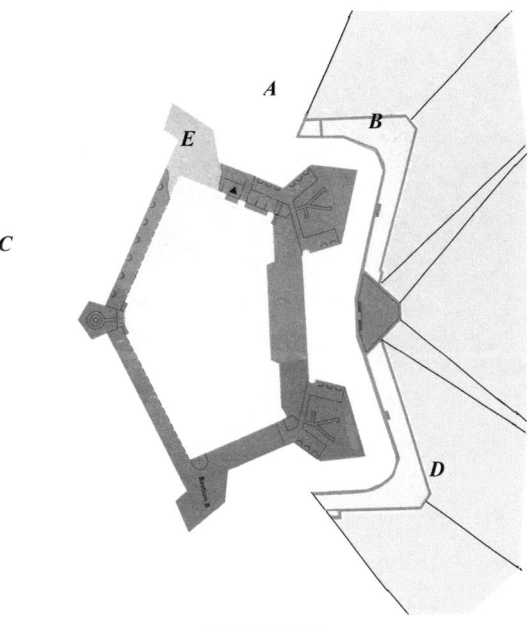

FIGURE 2

FIGURE 3

---

58. Figure 2, *Atlas*, plate 5, sketch 6, altered by the author; Figure 3, *Harper's Weekly*, March 9th, 1861, p. 156.

A person standing at "B" would see the view in figure 4, which is the author's photo taken in 2010. From left to right, the top of the glacis, the covered way, the counterscarp, and ditch are seen, all of which are still intact today.

FIGURE 4

The west walls are shown in figure 5, which would be as viewed from "C." In 1861, the gun ports were seven feet above the ground, but now, fill from later construction has been dumped to a depth of some four feet or more. In 1916, extensive modifications were made to the southwest curtain walls at the right. The parapet was removed to clear the field of fire for heavy coastal guns installed on the parade ground in anticipation of World War I.

FIGURE 5

Pensacola Bay was the site of three unoccupied forts: Pickens, McCree, and Barrancas. Pickens, begun in 1829 and completed in 1834, was the largest and most important. Located on the outer harbor, it could command the entrance to the harbor and most of the inner harbor, regardless of the lesser two. Pensacola was also the site of a large naval base, at Warrington, on the mainland of the inner harbor, east of Barrancas, figure 6.[59]

In January of 1861, this entire complex was manned by only one army artillery company, 1st Lt. Adam J. Slemmer's Company G, 1st U.S. Artillery (in Barrancas Barracks and understrength at 48 officers and men) and 4 officers, 70 ordinary seamen, and 48 marines in the navy yard at Warrington.[60]

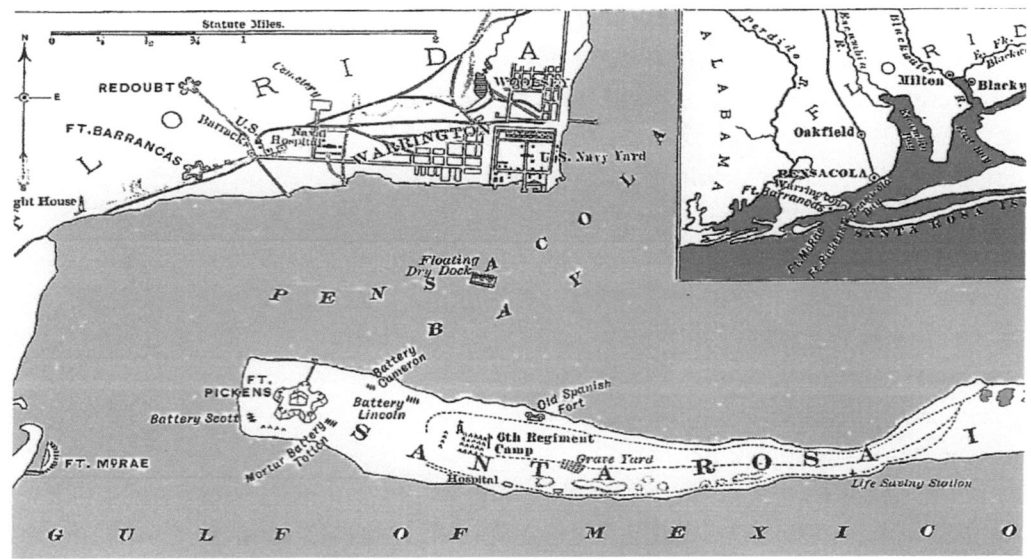

FIGURE 6

Scott's orders from Washington warning of the danger of seizure went out to Pensacola in two sets on January 3rd.[61] One was sent by special messenger and the other by telegraph, in cipher. The telegraphic message was "stopped" at Montgomery, Alabama, by the Confederates.[62] As a consequence, the hand-carried orders were not received by Slemmer until January 9th.

Fortunately, Slemmer, like Brannan at Key West, took the initiative. He had sensed the danger after receiving word of the seizure of Fort Morgan at nearby Mobile on January 5th. On January 7th, he contacted the commander of the navy yard, Cdr. James Armstrong. Like Slemmer, Armstrong had not received orders of any kind. He declined to cooperate with Slemmer, who planned to move stores from the mainland, those at forts McRee and Barrancas, across the harbor to Fort Pickens. Slemmer correctly viewed the mainland installations to be vulnerable and

---

59. Morris, p. 70, revised. The sand batteries and encampments shown were constructed later.
60. Scharf, p. 601; O.R. Vol. 1, p. 341; monthly return, 1st Regiment of Artillery, February 1861.
61. O.R. Vol. 1, p. 334.
62. O.R. Vol. 52/II, p. 27.

concluded that the best defense of the area, with his tiny command, could only be made from Fort Pickens.

On the morning of the 9th, when the orders from army headquarters arrived, Armstrong haltingly cooperated, and Slemmer completed moving stores from Fort McCree and Barrancas Barracks to Fort Pickens on January 11th, the very day that a Confederate banner was seen flying from the navy yard. Armstrong had been betrayed by his staff, who favored the Confederacy, and having made no defensive preparations, it was surrendered to Alabama and Florida militia.

Repeated demands were made for the surrender of Pickens, but Slemmer rebuffed them all, when finally on the 18th he made it clear that he would have to be driven out.[63] A standoff had begun, which looked grim for Slemmer and his company – soon they would be starved out.

Then, on January 26th, the most unexpected thing happened. Colonel Chase, the commander of the Confederate forces, announced that he was going to allow the mail that had accumulated at the Warrington Post Office to be delivered, and that it would not be interrupted again. He also offered to supply fresh provisions and to allow the company G laundresses, who had been taken prisoner, to be transferred to the fort.[64]

What had happened?

The Confederate Government had not yet been formed, and would not be, until after February 4th, when the Weed Convention met at Montgomery, Alabama, for that purpose. It would be February 12th before the Confederate congress took full official control of war policies, including the seizure of U.S. government property,[65] and Jefferson Davis would not be inaugurated President until the 18th.

The diverse seizure efforts that the individual seceding states had earlier taken were so successful that it appeared that anything could be accomplished by brashness with little fear of bloodshed. However, the passivity of President Buchanan was briefly shown to have reached a limit when, influenced by his new secretary of war, Holt, he had allowed plans to go forward to send troops to Fort Sumter on the *Star of the West*, though little does history know that he soon became unnerved and had attempted to recall it.[66].

The Sumter and Pickens "Truce"

After the *Star* had been fired upon as it entered Charleston Harbor on January 9th, it had become clear to the Confederacy how ill-prepared it was, without a central government, to fight a war.

Thus, a cynical strategy was developed by several Southern senators still in

---

63. O.R. Vol. 1, p. 388.
64. O.R. Vol. 1, p. 340, Haskin, p. 500.
65. O.R. Vol. 1, p. 254.
66. ORN Ser. I, Vol. 4, p. 220; O.R. Vol. 1, pp. 352, 355, 357, 360.

Washington that would avoid any act on the part of the South that could result in immediate open conflict.[67] On the 21st of January, senators Mallory of Florida, Fitzpatrick of Alabama and Slidell of Louisiana, had a personal interview with President Buchanan and War Secretary Toucy, in which it was agreed that no attack would be made on Sumter or Pickens so long as no further attempts were made to reinforce them.

This was briefly imperiled when word that an artillery company had embarked at Fort Monroe. This was Capt. Israel Vodges' Company A of the 1st U.S. Artillery, which had been detailed by the commander of Fort Monroe in compliance with orders from General Scott on January 4th. The delay from then to the 21st was due to finding sufficient men to bring the company up to strength, and in waiting for the warship *Brooklyn* to become available – it was the ship that had carried Buchanan's failed message of recall to the *Star of the West*.[68]

The problem was solved by orders sent out to Slemmer and the navy that Vodges' company were not to land.

Thus the "Truce" began, to be honored both at Sumter and Pickens. It would allow mail, food and essentials to be delivered, while all telegraphic dispatches were monitored, and the Confederacy's own buildup was accelerated. Their attitude had not changed, as exemplified by a resolution made by the Confederate Congress on the 15th of February which said that: "immediate steps should be taken to obtain possession of forts Sumter and Pickens . . . either by negotiations or force . . ."[69]

On March 4th, President Lincoln was sworn into office. When briefed on the situation at Fort Pickens, he was taken aback by the "quasi armistice," as he later referred to it.[70] However, he was not deterred, and, on March 5th, reversing Buchanan, he verbally directed General Scott to keep "all possible vigilance for the maintenance of all the places."

Hence, Scott issued orders to Vodges to land his company "At the first favorable moment."[71]

The orders, by a series of incredible foulups and misunderstandings, did not arrive at Pensacola until April 13th when Vodges and 110 marines[72] of the squadron were finally landed at 2:00 a.m. They had been aboard ship for 64 days, not including their two weeks passage from Fort Monroe.

On March 1st, the Confederate government assumed control of the operations

---

67. O.R. Vol. 1, pp. 442, 443, 445-446.
68. Scott W., pp. 622, 623; O.R. Vol. 1, pp. 345, 350, 354.
69. O.R. Vol. 1, p. 258.
70. O.R. Vol. 1, p. 440.
71. O.R. Vol. 1, p. 360; Welles, Vol. I, p. 29; Nicolay & Hay, Vol. III, p. 393. They incorrectly identify the *Mohawk* as the carrier of the order.
72. O.R. Vol. 1, pp. 355, 360, 363, 395, ORN Ser. I, Vol. 4, pp. 90-95, 100, 103, 110.

in the Charleston area by naming P. G. T. Beauregard to the command, and on March 7th did likewise for the Pensacola area, by assigning Braxton Bragg.[73] Bragg assumed command on the 11th and was informed on the 14th that a requisition for 5,000 troops from Alabama, Mississippi, Louisiana, Georgia, and Florida had been made by the adjutant general's office. They could be expected within ten days. On March 16th, orders were given by the Confederate war department to the Baton Rouge arsenal to supply General Bragg with an impressive list of thousands and thousands of shot, shell, and related supplies.[74]

Bragg soon discovered the effects of the "truce" in that citizens of the area were selling supplies of fuel, food, and water to the Fort Pickens garrison; and on March 14th, he issued orders to put a stop to it.[75] By the end of March, Bragg had an impressive number of troops, as is indicated in his return, figure 7.[76]

Troops.	Present for duty.				Total present.	Aggregate present.
	Infantry.		Cavalry.			
	Officers.	Men.	Officers.	Men.		
General staff						17
First Regiment Alabama, twelve months' volunteers	46	762			839	885
Georgia, twelve months' volunteers	4	106			109	113
Louisiana Zouaves	5	95			97	101
Total	55	963			1,045	1,116

FIGURE 7

On March 30th, Slemmer made a routine report to the assistant adjutant general wherein he states that he estimates that nearly 1,000 men are in Bragg's forces (remarkably accurate) with 5,000 expected. He goes on to say that the Confederates have fully armed Fort Barrancas and that guns have been mounted in the navy yard. He mentions that he has protested the continuation of this work, stopped by Chase and reinitiated by Bragg, and that it is important to "be informed of passing events" since he has had no important communication from headquarters since February 23rd.[77]

---

73. O.R. Vol. 1, pp. 259, 449.
74. O.R. Vol. 1, p. 450.
75. O.R. Vol. 1, p. 451.
76. O.R. Vol. 1, p. 455.
77. O.R. Vol. 1, p. 365.

Secret Expeditions

It is noted that April 12th was the day that Confederate firing on Fort Sumter began. It fell on the 14th, rendering moot any plans to reinforce it.

It soon was evident in Florida that there had been clandestine activity in Washington. Col. Harvey Brown and a large force would appear off the outer beach at Fort Pickens only five days after Vodges had landed. We must step back in time to examine a most bizarre story.

The attempts for the relief of Fort Pickens – begun by the Buchanan administration, as we have seen – were in a state of suspension at the commencement of the Lincoln administration. Taken aback by the revelation of the "quasi-armistice" at Pensacola – and given the advice that it was now too late to reinforce Fort Sumter – Lincoln was "much distressed."[78] The exhortations of Postmaster General Blair, however, saved the idea of an expedition to Sumter.

Thus, a split developed between Secretary Seward and the position of Lincoln and the rest of the cabinet. Seward favored more effort at saving Pickens and the complete abandonment of Sumter. His logic was that the reinforcements to Sumter would have to fight their way in, whereas, those to Pickens likely would not. Thus, if the Lincoln administration continued with the Buchanan policy of no reinforcement of Sumter, it would soon be starved out without a fight; and the two parties, Union and Confederate, could be reconciled. As the head of the department of state, Seward would be in a position to carry out these conciliatory negotiations, though he should have stopped to realize that this policy was de facto recognition of the legitimacy of the rebels – anathema to Lincoln.

We recall that Lincoln had decided, on March 5th, that regardless of the "armistice," Vodges should land, though Scott's order did not even begin its journey to Florida until the 16th. Having the notion that the Pickens issue was resolved, Lincoln returned to the discussion of Sumter. After listening to Blair's declarations that the abandonment of Sumter would be regarded as treason to the country, Lincoln decided to take a firm stand at Sumter as well. Inquiries were made, among them to Gustavus Fox, Blair's brother-in-law.[79] Having served 19 years as a naval officer, though now in private life, Fox had earlier proposed a plan for the relief of Sumter to the Buchanan administration.

Fox was now summoned to plan the task.[80] On March 19th, he was authorized by General Scott to visit Sumter. He did so, and returned with the assurance of its commander, Major Anderson, that Sumter could hold out until April 15th, and plans went ahead. At the cabinet meeting of March 29th, Blair, Welles, and Treasury

---

78. Welles, Vol. I, pp. 8–14.
79. Nicolay & Hay, Vol. VI, p. 383.
80. Welles, Vol. I, pp. 14–25; O.R. Vol. 1, pp. 208, 209, 235.

Secretary Chase were in favor of the Fox expedition, with Attorney General Bates on the fence and Seward and Interior Secretary Smith against.[81]

On April 4th, transports to deliver subsistence were chartered. Fox was given his instructions from War Secretary Cameron on the same day, which said, in part:

> If you are opposed in this you are directed to report the fact to the senior naval officer of the harbor, who will use his entire force to open a passage, when you will, if possible, effect an entrance and place both troops and supplies in Fort Sumter.

The captains of the *Powhatan*, *Pawnee*, *Pocahontas*, and *Harriet Lane* were given their sealed orders on the 5th.

Fox arrived off the Charleston bar, as planned, on April 12th.[82] However, his most important ship – the fighting portion of the expedition, the *Powhatan* – had not arrived. They waited all the day of the 13th and finally learned "that the Powhatan was withdrawn from duty" *and had been since the 7th*! Nevertheless, Fox decided to "go in" on the night of the 13th. However, Sumter's commander, Major Anderson, had made arrangements to surrender the fort, which was done on April 15th.

As a part of Secretary Seward's position against the Fox expedition, he had countered with his own plan. He would

> call in Captain Meigs [of the engineers, familiar with Pickens] forthwith. Aided by his counsel, I would prepare for war at Pensacola and Texas, to be taken, however, only as a consequence of maintaining the possessions and authority of the United States.

A meeting with Lincoln ensued, and as a nervous Seward stood by, Lincoln questioned Meigs as to whether Fort Pickens could be held. "Certainly" was the answer.

Finally, Lincoln offered to consider the idea and in a few days he would let them know.

On Sunday morning, March 31st, General Scott's military secretary, Erasmus Keyes, called on Meigs and took him to Seward, who requested them to "go forthwith" to Scott, to put a detailed plan on paper, and to have it in the President's hands before four o'clock that afternoon.

Their first stop was at the engineer office to study the charts of Pensacola and the drawings of the fort. This consumed so much time that they were compelled to go directly to the President for the four o'clock meeting without having conferred with General Scott. Having listened to their presentation, Lincoln said: "Tell him that I wish this thing done . . ." General Scott then detailed the plan the same

---

81. Nicolay & Hay, Vol. III, pp. 429–433.
82. O.R. Vol. 1, pp. 11, 12, 235, 236.

evening. It consisted of the assignment of a senior commander with a force some four, or five, times the size of Vodges' company, which would land on the outer beach at Santa Rosa Island. It would be supported by a powerful naval force, which would steam into the inner harbor and take up a position to prevent the crossing of the Confederate forces to Fort Pickens. In Meigs' handwriting,[83] the orders were approved by Lincoln without reading them. It would entirely bypass the chain of command of the navy, calling for Lt. D. D. Porter to take command of the *Powhatan*, earlier designated as the warship to support Fox in the run into Sumter.

The left hand not aware of what the right hand had done thus ruined the Sumter expedition.

On April 2nd, Bvt. Col. Harvey Brown, commanding at Fort McHenry, Maryland, received a letter[84] from General Scott, outlining the first part of the plan and endorsed "APPROVED" by Abraham Lincoln. It began:

> You have been designated to take command of an expedition to re-enforce and hold Fort Pickens, in the harbor of Pensacola. You will proceed with the least possible delay to that place, and you will assume command of all land forces of the United States within the limits of the State of Florida . . . Should a shot be fired at you, you will defend yourself and your expedition at whatever hazard . . .

Brown, West Point class of 1818,[85] was a veteran of 43 years of service dating from the Blackhawk War, the Florida War, and meritorious service in the Mexican War. He would bring experience, organizing skills, and judgment to the "farce of peace"[86] at Pickens; but he would fail to bring initiative.

The first ship of Brown's expedition, the *Atlantic*, carried two artillery companies, A and M of the 2nd Regiment (M fresh out of Texas); two infantry companies, C and E of the 3rd (also fresh out of Texas); Duane's Engineer Company; and 20 hired carpenters, a total of about 450 men.[87] After the men, equipment, and 73 horses were loaded, the *Atlantic* left the wharf at Fort Hamilton, New York, on April 6th. It remained in the harbor loading more supplies overnight and got underway in the morning. The dock was left covered with equipment to be loaded on the steamer *Illinois*, carrying companies H and K of the 2nd Artillery[88] and due to follow on the evening of the 8th.

The *Atlantic* arrived off Key West on April 12th, and Brown landed on the 13th to consult with Major French and selected local citizens. He then detached 33

---

83. Welles, Vol. I, p. 17.
84. O.R. Vol. 1, p. 365.
85. Cullum, Vol. I, pp. 189–197.
86. O.R. Vol. 1, p. 398; M. C. Meigs.
87. O.R. Vol. 1, p. 399.
88. O.R. Vol. 1, pp. 366, 393, 394.

men from the Fort Taylor garrison to make up for deficiencies caused by desertion or absence at New York. Taking some mortars and a field battery, and evidently satisfied with the conditions at Key West, Brown departed for Fort Jefferson.

Brown arrived at Jefferson at 1:00 p.m. on the 14th. He complimented Arnold on the "good order" of the fort. He detached 31 volunteers[89] from the garrison to complete the roster of his forces for Pickens and also "took twenty negro laborers" for the engineer department.

Twelve of those detached were from Battery L, as is seen on the muster roll summary for the period: James Ahern, Thomas Brook, William F. Brown, Patrick Cummings, Christopher Foley, James McCarthy, James McCoy, John Meyer, David Meyers, Michael O'Sullivan, William Parketton, and Thomas Poole.

After a stormy passage, Brown was off Pensacola on the 17th, only five days after the first relief expedition (Vodges) had landed. Unloading across the outer beach, through the surf, began. The signal that an attack was imminent was received from the fort.[90] Though the landing party congratulated themselves on the fact that their arrival prevented the attack, the Confederate torches and hurrahs around Fort Barrancas, at the arrow in figure 8, were probably a celebration of the news of the secession of Virginia, promptly telegraphed to Bragg. We remember that Virginia seceded on the 17th.[91]

FIGURE 8

Brown found Fort Pickens in "miserable condition" and the guns and carriages that were in place of "indifferent" condition.[92] He overestimated the numbers of the Confederates at 7,000, though their plans only called for 5,000[93]; and at this time, there were only about 2,700 actually in the area. He appreciated that the

---

89. Haskin, p. 400. Nine were from M, 1st Artillery.
90. O.R. Vol. 1, p. 396.
91. O.R. Vol. 1, pp. 386, 397.
92. O.R Vol. 1, p. 378.
93. O.R. Vol. 1, pp. 454, 379; Figure 8, photo by the author.

distances across the harbor, one-and-a-quarter to one-and-one-half miles, were beyond the useful range of most of his and the Confederate guns alike, and that the new 42-pounder rifled gun would be a requirement for reducing the rebel forts and the navy yard. He had requested six on April 12th.[94]

He was apprehensive of an engagement with the Confederates. He viewed his forces at this time as too weak and any provocation as inconsistent with his orders.[95] Ever true to his orders to conduct operations only on the defensive, Brown thought it proper to inform "the secession general" of his arrival and of the defensive posture. The message was delivered by Captain Vodges, who did not obtain a written reply. Bragg merely muttered that the truce had been broken.[96] On the 18th, Brown announced Fort Pickens as his headquarters.

By the end of April, there were over 1,000[97] troops under Brown's command, not counting the crews of the various ships of the fleet: *St. Louis, Crusader, Powhatan, Sabine, Minnesota, Brooklyn,* and *Wyandotte.*

Fort Sumter having fallen, some Confederate troops, including workmen and artillery, were freed by P. G. T. Beauregard, its new commander, to come to Pensacola. Bragg had recommended a reserve force of 3,000 men be held "on the railroad"[98] between Pensacola and Mobile. On April 20th, Bragg declared martial law in the Pensacola area and declared all intercourse among Santa Rosa Island, Fort Pickens, and the United States fleet prohibited. On May 6th, Bragg[99] wrote to the Confederate Secretary of War L. P. Walker, noting Brown's efforts at the extending of facilities outside of the fort and that: "Every hour will add seriously to the difficulties to be overcome." He added that his "works" for offense or defense were nearly complete but that he still lacked many essentials. "Our best defense against the fleet – shells – cannot be used for want of fuses. Not one has yet reached me." At this rate, it was going to be a long war.

One news story that may have been eagerly awaited from Fort Pickens was the report of the visit of British journalist William Russell and the artist who accompanied him, Theodore Davis. As neutrals writing for the *London Times*, Russell and Davis were allowed to tour *both* Fort Pickens and the Confederate installations on the opposite shore. On May 14th, Russell's tour of the Confederate batteries shook his confidence in American journalism. He had read elsewhere that there was a formidable array of hundreds of guns, "better described as tens," figure 9.[100] He toured ten of the thirteen of Bragg's batteries and estimated that there were but five heavy

---

94. O.R. Vol. 1, p. 373.
95. O.R. Vol. 1, p. 366.
96. O.R. Vol. 1, p. 379.
97. O.R. Vol. 1, p. 386.
98. O.R. Vol. 1, p. 464.
99. O.R. Vol. 1, p. 465.
100. Miller, ed., Vol. 5, p. 56. Note that the guns are offset, or "barbet" mounted.

siege guns capable of doing damage to Pickens. "I was quite satisfied Gen. Bragg was perfectly correct in refusing to open his fire on Pickens and the fleet, which ought certainly to have knocked his works about his ears."[101]

Russell recognized what the Confederate Chief of Ordnance Maj. Josiah Gorgas, apparently did not – that few, if any, of the 169 guns, howitzers and heavy mortars sent to Bragg had sufficient range to reach Fort Pickens.

FIGURE 9

Near the end of May, Bragg received a letter from Jefferson Davis. The information it contained showed that the Confederate leaders were looking east. The capital was being moved[102] from Alabama to Virginia, and Bragg was asked how many troops he could *spare*. This must have come as quite a shock since almost every other communication he had received up to that date had asked the opposite. He responded[103] on the 28th, "I can spare twenty-five hundred men for Virginia and can start them immediately, well armed." If Bragg had not the heart to attack Pickens earlier, he would not do so now. He was still making requests for more troops and ammunition at the end of June.[104]

Though the blockade had been declared on April 19th, no specific action to enforce it had been taken by the navy squadron standing offshore. On May 8th, Brown[105] requested that the navy prevent vessels "loaded with forage or provisions . . . or contraband of war" from entering the harbor.

The blockade, though declared openly, was a part of the broader plan

---

101. Russell, p. 213.
102. O.R. Ser. IV, Vol. 1, p. 255.
103. O.R. Vol. 1, p. 468.
104. O.R. Vol. 1, p. 469.
105. O.R. Vol. 1, p. 409.

outlined originally by Gen. Winfield Scott[106] to secure "posts" at critical points on the coast to deny them to the Confederacy and provide places to support the ships maintaining the blockade. In addition, these were to be places of strength to support possible moves inland. Recall that Scott had attempted to hold such a post in Texas. This concept of coastal occupation and envelopment, supported by the navy, was Scott's broad strategy for the prosecution of the war. It was eventually dubbed "The Anaconda Plan" or "Scott's Great Snake" by the press, mocking the idea that the Confederacy could be defeated by strangling it as a constrictor snake kills its victim, figure 10.

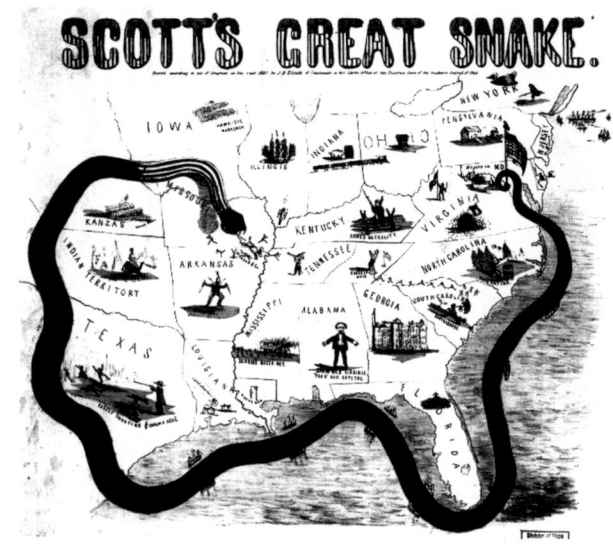

FIGURE 10

On May 13th, the "old company," that is, Slemmer's Company G of the 1st Artillery that had been there since the beginning, was ordered to Fort Hamilton, New York. Almost all were suffering from scurvy; and upon the post surgeon's recommendation, Brown gave the order, noting that a "northern climate" was necessary to restore their health.

The Anaconda Plan

The form the plan took on the western rivers would be unique. The navy assigned Cdr. John Rodgers to cooperate with the army in the blockade of the Ohio and Mississippi Rivers in May of 1861.[107] He soon purchased three small steamboats for conversion into gunboats.[108] After some controversy as to whether he had exceeded his authority, his purchases were approved and paid for by the army. Rodgers' steamers were modified by adding guns and heavy timbers for armor. An example is the USS *Tyler*, figure 11, which he later commanded.[109]

Indicative of the growing importance of the mission, Rodgers was replaced by a more senior officer, Capt. A. H. Foote., on August 30th.[110]

Any new construction was specifically for inland waters and it took on a

---

106. Nicolay & Hay, Vol. IV, pp. 302–307; Figure 10: Elliot, J. B.
107. ORN Ser. I, Vol. 22, p. 280
108. ORN Ser. I, Vol. 22, pp. 283, 286.
109. Figure 11, ORN Ser. II, Vol. 1, p. 226b; Figure 12, Miller, Vol. 1, p. 223.
110. ORN Ser. I, Vol. 22, p. 307.

radically different form, an example being the ironclad Cairo, figure 12. The flotilla was unique in yet another way. Though funded by the army, through Fremont's Department of the West, it was under the overall command of a navy officer.

Later, a second flotilla consisting of commercial steamers was converted not into gunboats, but into unarmed rams, under the direction of Charles Ellet, who conceived of the idea. These were initially under army control but were eventually transferred to the navy, in a political fight described later.

It is remarkable that Scott's Anaconda Plan remained as the set policy of the war, and by-and-large, it was consistently followed. Control of the Mississippi would be a problem, and it would not be achieved until the fall of Port Hudson, after a sweaty and bloody siege all too familiar to Battery L, on July 9th, 1863. Vicksburg, that objective better known to history, had surrendered five days earlier, more than a year after it was first bombarded by Union gunboats.

FIGURE 11

FIGURE 12

Secret Board of Officers

To implement the Anaconda Plan, a secret board of officers was named and convened by Navy Secretary Gideon Welles on June 25th, 1861. The board consisted of Capts. S. F. DuPont and C. H. Davis of the navy; Prof. A. D. Bache, chief of the coast survey; and Maj. J. G. Barnard of the Army Corps of Engineers.[111] They identified Port Royal, South Carolina, and New Orleans; but an attack on the Confederate batteries at Hatteras Inlet, North Carolina, could be launched quickly from Fort Monroe, Virginia, the planned base of operations. Further, it would only require an army contingent of an estimated 860[112] men. Therefore, Hatteras Inlet was chosen as a first trial. On August 13th, General Scott directed the commander of Fort Monroe,[113] Maj. Gen. John E. Wool, to supply the necessary army contingent. Due to the byzantine command arrangements there, this duty would fall to the

---

111. Welles, *Galaxy Magazine*, Vol. 12, issue 5, November 1871, p. 672.
112. O.R. Vol. 4, p. 580.
113. O.R. Vol. 4, p. 579.

prominent Democrat and major general of volunteers, Benjamin F. Butler.[114]

Butler, Waterville College (Colby) class of 1838, began a successful law practice in Massachusetts, was elected to the Massachusetts legislature, and was later appointed to the Massachusetts militia. From his position as a brigadier general, he was called upon by Massachusetts Governor Andrew to lead a Massachusetts regiment to Maryland in April of 1861. His actions, along with those of another Massachusetts militiaman, Nathaniel Banks, are largely credited with preventing the secession of that state.

On May 18th he was given command of the Department of Virginia and the volunteer forces at Fort Monroe. His first action, the first of any in Virginia (the Battle of Bull Run or First Manassas took place in July) he was repulsed near Big Bethel Church on the Yorktown Peninsula on June 10th, 1861. However, this did not slow Butler's enthusiasm for mounting other operations, though the peninsula area north of Fort Monroe, including the Norfolk Navy Yard, would remain in Confederate hands until May 10th, 1862.[115]

He was soon "cheerfully granted" permission to recruit 5,000 volunteers from Massachusetts and the "Eastern States" by Assistant Secretary of War Thomas Scott.[116]

On August 26th, Butler was ready and a joint amphibious expedition led by Flag Officer Silas Stringham carrying General Butler and his troops embarked for Hatteras Inlet. On the 28th, supported by a naval bombardment on forts Clark and Hatteras, Butler's forces landed behind the Confederate batteries. The Confederate garrisons surrendered the next day.[117] It was a resounding success and was received with enthusiasm in the north. The *Harper's Weekly* of September 14th labeled it "brilliant," and put Stringham's and Butler's portraits on the cover.

Pickens Armament Complete

It is noted that on May 30th, Lt. Henry W. Closson and Company F, 1st Artillery, arrived at Pickens from Key West.[118] Closson would become commander of Battery L on November 21st, 1861, after his promotion to captain on May 14th, and remain its commander throughout the rest of the war. Graduated from West Point, class of 1854, he remained in the army after the war, eventually achieving the rank of brigadier general. He retired in 1896 and died in 1917. He is buried in Arlington National Cemetery.[119]

---

114. O.R. Vol. 4, p. 602; *Encyclopedia Britannica*, Vol. IV, p. 881.
115. O.R. Vol. 11/I, p. 2.
116. O.R. Ser. III, Vol. 1, pp. 423, 498.
117. O.R. Vol. 4, pp. 581–587; Butler, B. F., pp. 282–286; CWSAC no. NC001.
118. O.R. Vol. 1, p. 424.
119. Cullum, Vol. II, no. 1638, pp. 580, 581; Powell, p. 134; Powell, *Army List*, p. 249; www.arlingtoncemetery.net/hwclosson.htm.

On June 12th, the steamers *Star of the South*, the *South Carolina*, and the *Massachusetts* arrived at Pickens with more guns and stores. Brown notes that he had not received orders of any kind since arriving, and given the bombardment and surrender of Fort Sumter, "I do not feel under any obligations to confine myself to defensive measures . . ."[120] On the 14th, he reports: "The weather is now intensely hot, and will probably soon be, to men of Northern constitutions exposed to the miasma of the swamps, very unhealthy. I would not, therefore, throw any large body of troops here until after the September gales are over." He puts forward a scheme for raiding the mainland near Fort McCree with six or eight shallow draft gunboats and another plan for negotiating the channel at the east end of Santa Rosa Island. For this, however, he would need gunboats and some 500 more troops.

On June 22nd, Brown notes that he has received seven of the 42-pounder rifled guns which would be necessary to effectively fire across the bay.[121] He fails to even note the arrival of the 6th New York Volunteers who were on the steamer *Vanderbilt*.[122]

On June 26th, he complained in a letter to the assistant adjutant general of having been ordered to send Barry's Company A and Hunt's Company M of the 2nd Artillery to New York. He is vexed at now having to accommodate "a regiment of undrilled New York City Volunteers, entirely undisciplined . . ."

At this time, Brown could have experienced few dealings with the 6th New York Zouaves, or Col. Billy Wilson, a former New York City alderman, their colorful commander. Nevertheless, Brown's reaction to these raucous "roughs" with their Billy Goat mascot and their politically appointed officers, mostly Democrats, was automatically negative.[123] They did not wear the gaudy uniform of other Zouave units; in fact, the only hint of Zouave about them was that Colonel Billy wore a sash. Rather, they had, figure 13, plain gray shirts and a brown felt hat.

Brown submitted a table of the armament at Fort Pickens, which he now regarded as complete, though the external water batteries were not. He repeats his complaint, however, that he has been stripped of sufficient officers and enlisted men to efficiently work all the guns.

FIGURE 13

---

120. O.R. Vol. 1, p. 430.
121. O.R. Vol. 1, pp. 432–435, 440.
122. Morris, p. 34.
123. Morris, p. 35; Figure 13 from August 31, 1861, *Harper's Weekly*, p. 552.

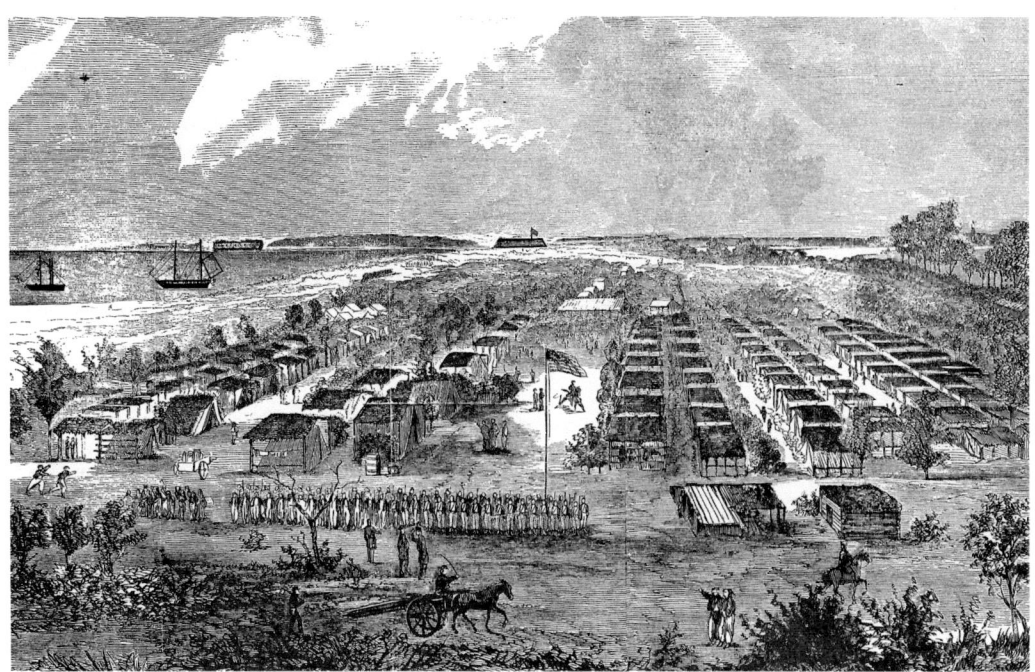

FIGURE 14. CAMP BROWN OF THE 6TH NEW YORK

Fort Pickens is in the distance, flying the flag. Fort McRee is beyond the ships off shore. Note the bowers built over the tents for protection from the intense sun. The troops in formation are still in their Zouave uniforms.
Harper's Weekly, October 26, 1861, p. 678.

Battery L to Pickens

To make up his perceived deficiency of men, on July 2nd, Brown ordered: "One of the artillery companies from Tortugas here.[124] I accordingly sent two companies of volunteers to that post. . . they may be expected in about a week, by which time I hope to have the guns and ammunition landed, and will be in a suitable state for offensive or defensive operations."

Thus, the rest of Battery L would join the dozen of its men who had been detached for service at Fort Pickens in April. Companies B and E of the 6th New York were the volunteers referred to that were traded off to Fort Jefferson. At the end of August Brown also sent Company A of the 6th to Key West,[125] thus reducing the total numbers of the 6th at Fort Pickens to seven companies, and presumably reducing his having to deal with their foibles.

---

124. O.R. Vol. 1, pp. 434, 435, 437.
125. Morris, p. 41.

"Record" 8/61
30 JUNE–31 AUGUST 1861, FORT PICKENS, FLORIDA

The Company left Fort Jefferson, Florida on the Steamship State of Georgia on the 5th July and arrived at Fort Pickens, Fla., on the 10th July 1861, per S.O. No. 38 Hdqrs. Dept of Florida, dated Fort Pickens July 2nd 1861.

Samuel K. Dawson Capt.
William Silvey   1st Lt. Reg. Adj. O. No. 7 Hdqrs. Fort Dallas, Fla. Aug. 13,'57. Left Co. Apr. 22,'54, S.O. No. 62 Hdqrs. N.Y. Apr. 18,'54.
Richard H. Jackson 1st Lt. Joined Co. at Fort Pickens, Fla. (from detached service) July 10th 1861.

Strength: 93. Sick: 19.

Sick Present: Lewis Keller, Isaac T. Cain, Edmund Anglin, Thomas Brook, William C. Brunskill, Patrick Cummings, Christopher Ebel, Charles A. Flint, James Harkins, Joseph Kutschor, Charles F. Mansfield, James McWaters, Morgan S. Shapley, Martin Stanners, William Schaffer, Andrew Stoll, Reuben Townsend, Henry A. Ward.

Sick Absent: Charles Riley Sick at Ft. McIntosh, Texas. Absent from Company since Feb. 25th 1861.

Joined:
Woodruff, George A. 2nd Lt. By Promotion June 24th 1861 (never joined Company).

Joined by Assignment at Fort Pickens, July 15th 1861:

Christopher Ebel	17 Jan.'61 New York,	Franklin W. Richards	6 Feb.'61 New York
Morris Galvan	11 Jan.'61 Boston	Morgan Shapley	10 Dec.'60 Buffalo, NY
James Hanney	8 Jan.'61 New York	Martin Stanners	19 Jan.'61 Boston
John Lanahan	9 Jan.'61 Rochester, NY	William Schaffer	18 Jan.'61 New York
Charles F. Mansfield	16 Jan.'61 New York	John Tomson	9 Jan.'61 Boston
Daniel McSweeney	11 Feb.'61 New York		

Joined from Detached Service at Fort Pickens, July 10th 1861:

James Ahern	Patrick Cummings	Daniel McCoy	Michael O'Sullivan
Thomas Brook	Christopher Foley	John Meyer	William Parketton
William F. Brown	James McCarthy	David Meyers	Thomas Poole

Discharged:
Golden, James Pvt. 2 Feb. '60 New York At Fort Pickens, Fla. By Surgeon's Certificate of Disability, July 11th 1861.

Extra Duty Assignments:

Isaac T. Cain	Carpenter	Robert Curran	Teamster	Miles McDonough	Boatman
George Hadley	Blacksmith	Edward McLaughlin	Teamster	James Harkins	Ord. Dept. Laborer
James McWalters	Teamster	Henry William	Teamster	William Wynn	Ord. Dept. Laborer
Wallace D. Wright	Teamster	William Creed	Boatman		

The large number of those listed as sick is not explained by any special note in the muster roll. Only two of the twelve men detached from the company, Thomas

Brook and Patrick Cummings, who had been at Fort Pickens since April 17th, were sick. This might be some indication that the conditions at Pickens were better than at Jefferson, with its mosquito-filled cisterns. Five of the eleven that were joined by assignment were sick.

Events Elsewhere

July 3: The Western Department is created, consisting of the state of Illinois and "all of the states west of the Mississippi River and on this side of the Rocky Mountains, including New Mexico" with the colorful "Pathfinder," the trailblazer of the West, and the first Republican candidate for President (1856), John C. Fremont in command, at St. Louis.[126]

July 21: The First Battle of Manassas (Bull Run), Virginia, is fought[127] and is regarded as a great victory for the South. However, this setback did not seem to cause panic in Washington. Instead, determination ruled; and Congress continued in session, conducting business in a manner that indicates a certain sense of ultimately prevailing in what was now clearly to be a great struggle.

July 22: Act of Congress authorizes 500,000 volunteers to be raised by the states to serve for three years; to be disbanded at the end of the war.[128] On May 3rd, Lincoln had called for 42,034 volunteers for three years' service, for 18,000 seamen for one-to-three years' service, and 22,714 more men for the regular army, but without any real legal basis. Here, Congress confirmed his action.

July 24: Act of Congress: "The navy may hire, purchase, or contract for, and furnish and arm in the most efficient manner, such vessels as may be necessary for the temporary increase of the navy . . ." No limitation of any kind stated, except the funds appropriated were $3 million.[129]

July 28: The Confederate center of attention is now focused even more on the eastern theater, and a request for more troops by Bragg is deflected by asking him how soon "and at what point you wish it to be assembled?"[130]

July 29: Act of Congress authorizes an increase in the regular army to 25,000.[131]

August 6: (1) The first Confiscation Act is signed by the President, confiscating property used for insurrectionary purposes. Thus, it set free any slaves employed by the consent of their masters against the government and lawful authority of the United States.[132]

---

126. O.R. Vol. 3, pp. 390, 406.
127. Nicolay & Hay, Vol. IV, pp. 358, 360, 365–368.
128. Appendix to *Congressional Globe*, July 18th, 1861, p. 27; *Harper's Weekly*, May 18th, p. 307.
129. Childs, p. 240.
130. O.R. Vol. 1, p. 469.
131. O.R. Ser. III, Vol. 1, p. 700.
132. *Congressional Globe*, August 6th, 1861, p. 455.

(2) Lest there be any confusion about any act of the President since his inauguration, another act on this date declared, "all acts, proclamations and orders of the President since March 4, 1861, respecting the army and navy of the United States . . . are hereby approved. . ."[133]

August 10: As the Union "ramped up" to the war, notably after Lincoln's first call for 75,000 volunteers the day after Fort Sumter fell (77,875 volunteered) and his second call on May 3rd, many former officers of the "old army" offered their services. After the Mexican War, many had found careers in civilian life more attractive than in the underfunded army and had resigned their commissions. One such was Ulysses S. Grant, West Point class of 1843. However, his resignation from the army in 1854 had been clouded by allegations that it was to avoid facing disciplinary action by his commander for alcoholism.[134] Another prejudice against Grant was that he was not successful in his time in civilian life, unlike other West Pointers who had cut grand swaths as lawyers, politicians, or business leaders.

Captain Grant offered his services in a letter to the adjutant general on May 24th, 1861. He never received an answer. In fact, the letter was not found until after the war, and then by accident. It has since been copied into the Official Records.[135] Undaunted, he began by assisting in organizing the Illinois State Militia. Having reintroduced himself to the profession, he was supported by Governor Yates and Congressman Elihu B. Washburne of Illinois when recommendations for appointments for brigadier generals of volunteers were requested by the Lincoln administration. He was confirmed by the Senate on July 31st, 1861, and assigned to the Western Department[136] as commander of the district of Ironton, Missouri.[137]

August 17: A seemingly final blow is handed to Bragg in a letter from the adjutant and inspector general's office at Richmond:

> General: Your requisition for ordnance and ordnance stores, inclosed with letter of 7th August, 1861, has been referred to the Bureau of Ordnance, and returned with report that there are no guns. These guns have been specially in demand for Manassas . . .
>
> R.H. Chilton, *Assistant Adjutant General*

August 19: McClellan is appointed commander of the Army of the Potomac.[138] George B. McClellan, West Point class of 1846, had resigned his commission in 1857 and had become vice president of the Illinois Central Railroad and then president of the Ohio and Mississippi Railroad. From this prominent position

---

133. Childs, p. 240.
134. *Encyclopedia Britannica*, 1911; *Harper's Weekly*, August 31sth, 1861, p. 559.
135. O.R. Ser. III, Vol. 1, p. 234, Grant Vol. I, pp. 239-240.
136. O.R. Vol. 3, pp. 390, 406, Grant Vol. I, p. 241.
137. O.R. Vol. 3, p. 430, Grant Letters, pp. vii-ix, 1-2.
138. O.R. Vol. 5, pp. 567, 575; Cullum, Vol. II, pp. 250–254; McClellan was Cullum no. 1273.

he was appointed a major general to the command of the Ohio militia in April of 1861 and then a major general in the regular army in the occupation of West Virginia. After the Union defeat at First Manassas, he became commander of the Army of the Potomac by order of General Scott. The "young [age 35] Napoleon" soon clashed with Scott,[139] as predicted by an earlier fundamental disagreement over the strategy for the prosecution of the war. Three months later, General Scott asked to be retired.[140] He was replaced by McClellan on November 1st.[141]

August 29th: Butler's force takes the Hatteras Inlet Batteries.

August 30th: Fremont issued an order declaring martial law in Missouri and the confiscation of the property "of all persons who shall take up arms against the United States . . . and their slaves, if any they have, are hereby declared freemen."[142] The fine point was that the slave need not be employed against the Union cause, which was the limitation of the First Confiscation Act. Fremont refused to back away from his sweeping order, and Lincoln then overrode him on September 11th,[143] ordering his command to conform to the provisions of the First Confiscation Act. Lincoln's concern was that Fremont's provision would "alarm our Southern Union friends and turn them against us; perhaps ruin our fair prospect for Kentucky."[144]

The "Truce" Ends

On September 2nd, Brown took the first action of an aggressive nature since his arrival. In May, Bragg's forces had towed a floating dry dock out from the navy yard. The tow line had allegedly parted, and it had drifted to a position off the western shore of Santa Rosa Island near Batteries Lincoln and Cameron, where it grounded. For weeks, the Federals watched as the Confederates attempted to refloat it. Concerned that these efforts might be a ruse to construct a floating battery, Brown ordered it destroyed.[145] Lieutenant Shipley of Company C, 3rd Infantry and 11 picked men rowed out to it shortly after 9:00 p.m. Finding no sentries on board, they put in place three "large" Columbiad shells and "combustible material," lit a fuse, and skedaddled. They had rowed "scarcely . . . twenty yards" before flames broke out, followed by the explosion of the shells. No one was injured, and the dry dock was destroyed.

In its turn, the navy conducted a raid on a Confederate schooner, the *Judah*,

---

139. Nicolay & Hay, Vol. IV, pp. 298–303, 322, 323, 351, 352, 368, 440, 445; *Harper's Weekly*, August 31st, 1861, p. 559.
140. O.R. Ser. III, Vol. 1, pp. 611, 613, 614.
141. O.R. Vol. 5, p. 639.
142. O.R. Vol. 3, pp. 466, 467.
143. O.R. Vol. 3, pp. 485, 486.
144. O.R. Vol. 3, pp. 469, 470.
145. O.R. Vol. 6, p. 665; *Harper's Weekly*, October 12th, 1861, p. 655.

moored at the navy yard, which was being fitted out as a privateer.[146] Approved by Adm. William Mervine, commander of the Gulf Blockading Squadron, the raid was led by Lt. John H. Russell of the flagship USS *Colorado* in the early morning hours of September 14th. About 100 sailors and marines of the squadron rowed with muffled oars to the schooner. Though they were fired upon by the alert Confederates, they boarded the *Judah* and set it afire. Others of the attackers spiked the only gun in the navy yard, a 10-inch Columbiad.[147]

Bragg later blamed the success of the attack on the conspiracy of nine Confederate marines who had deserted to the Federal side and brought intelligence of the Confederate preparations. For his part, Brown fully expected retaliation, but none quickly came. Brown was later criticized in a report to the secretary of war, noting that Fort Pickens did not provide covering fire at any time before or during the *Judah* raid and that none of Brown's troops had participated.[148] The operation was entirely to the credit of the more aggressive folks in the navy, who suffered three killed and eight wounded, including Lieutenant Russell. The view of the navy was that its duty was the enforcement of the blockade and that it could not countenance the outfitting of a pirate ship.

"Record" 10/61
31 AUGUST–31 OCTOBER 1861, FORT PICKENS, FLORIDA

[No entry in the "Record of Events"]

Samuel K. Dawson   Capt. Sick since Oct. 1, 1861
Edmund Kirby   1st Lt. Joined by promotion May 14,'61 S.O. No. 64 W. D. Wash. Aug. 22,'61. Abs. on det. ser. Since May 14,'61.
George A. Woodruff   2nd Lt. Joined by promotion June 24,'61 S.O. No. 64 W. D. Wash. Aug. 22,'61. Abs. on det. ser. Since June 24,'61.

Strength: 91. Sick: 7.

Sick Present: Samuel K. Dawson, Patrick Cummings, Michael Olvany, William V. Thompson.

Sick Absent: Thomas Brook, Ft. Hamilton, NY. Left Company Sept. 17, 1861
    William C. Brunskill   do.
    Charles Riley, Ft. McIntosh, TX Abs. since Feb. 25, 1861. Reduced from Sgt. to Pvt. July 12,'61 O. No. 2, Co. L 1st Arty Ft. Pickens, Fla. Subject to the approval of the Commander of the Regiment.

Died: Christopher Ebel, Pvt. Jan. 17,'61 New York, at Ft. Hospital Sept. 27, 1861, of scourbutics.
    James Harkins, Pvt. Oct. 1,'60 New York, at Ft. Pickens Hospital Oct. 8, 1861, of chronic diarrhea.

Transferred:
    William Silvey 1st Lt. Reg. Adj. O. No. 7 1st Arty. Ft. Dallas, Fla. Aug. 13,'57. Left Co. Apr. 22,'54 S.O. 62 Hdqrs. New York April 18,'54. Trans. By prom. As Captain to Co.

---

146. O.R. Vol. 6, pp. 437, 438; *New York Times*, September 30th, 1861.
147. *Harper's Weekly*, August 3rd, 1861, p. 495.
148. O.R. Vol. 6, p. 666.

"A" 1st Arty. To date from Aug. 14,'61. S.O. No. 64, W.D. Adj. Gen'l Office, Wash. Aug. 22,'61.

Richard H. Jackson 2nd Lt. Trans. By prom. as 1st Lieut. To Co. "B" 1st Arty. To date from Aug.14,'61. S. O. No. 64 Adj. Gen'ls. Office Wash. Aug. 22, '61.

Attached:
Richard H. Jackson 1st Lt. Since May 14th 1861. In Command of Company since Oct. 1st 1861.

The reduction in the number of men sick from 19 on the last roll to the seven listed here is interpreted to mean that the diet and sanitary conditions had improved, not to mention the weather. Note in the roll that Christopher Ebel died of scurvy and James Harkins of dysentery. Ebel, one of those newly assigned to Battery L, was apparently a healthy young man only six months before when recruited in New York, yet he is listed as sick when he was assigned to Battery L on July 15th. Harkins was with the second group of recruits into Battery L in 1860; he appears in table 2, Chapter 1. His service was cut short after only a year.

The quick recovery of the rest of the sick is perhaps an indication that the sickness was not due to one of the more serious lingering diseases like malaria. Health matters were originally drawn to Brown's attention by the surgeon on his staff, Dr. John Campbell, when, as early as May 9th, he noted scurvy "some cases of considerable severity"[149] in Company G [Slemmer's] and had recommended that they be shipped out, as was earlier noted. Brown had agreed, but now seemed to be convinced that the bulk of his command was healthy. Since the surgeon had intervened, it can only be speculated that he had invoked Army Regulation, paragraph 1,079, which states that when the surgeon considers it "necessary for the health of the troops, the commanding officer . . . may order issues of fresh vegetables, pickled onions, sauer-kraut, or molasses, with an extra quantity of rice and vinegar."

The considerable proportions of the sick on the Confederate side was also noted. In June,[150] news had been received from an escaped slave that the smallpox was raging.

Though Brown's theme song was always that he needed more troops in order to properly man the fort and deal with his adversary, he could well have considered the health effects of the numbers of men crowded into this corner of a narrow island with no water supply. There were also almost 100 horses and mules present at this time. With them came flies and sanitation problems, not to mention their requirement for four gallons of water each, per day. More crowding would exacerbate the problem, and another volunteer regiment, the 75th New York, would arrive on December 13th.

---

149. O.R. Vol. 1, pp. 408, 410.
150. O.R. Vol. 1, p. 432.

Chapter 4

Raid on Santa Rosa; "Record" 12/61; Bombardment of November 22nd & 23rd; Confederate Concerns; "Record" 2/62; Bombardment of January 1st, 1862; Confederate Withdrawals; "Record" 4/62; Reconnaissance on Santa Rosa

Raid on Santa Rosa

Brown's abstract for October lists 985 officers and men present for duty,[151] though his "aggregate" strength is listed as 1,203. The difference is those who are sick, on leave, in confinement, on other duty such as hospital attendant, detached[152] for duty elsewhere, sick elsewhere, or not there because of permanent assignment to regimental headquarters. By comparison, an abstract of the field return of the forces under Bragg for October 1st, 1861, totaled 6,533, with 5,247 present for duty.[153] Thus, Brown had about one-fifth the number of effectives as did Bragg. One might therefore ask why Bragg committed only 1,100 men to his raid when he could have potentially overwhelmed Brown.

Bragg's retaliation for the destruction of the dry dock and the raid on the *Judah* came as a raid on Santa Rosa Island. On the night of October 8th, the selected troops were detached from Bragg's forces and gathered at the navy yard. Under the cover of darkness, they were boarded on the steamer *Time*. Under the overall command of Brig. Gen. Richard H. Anderson,[154] the troops were organized into three battalions during the passage. The first, under Col. James R. Chalmers of the 9th Mississippi Regiment, was assigned to advance from the landing place along the north (bay) side of the island; the second, under Col. J. Patton Anderson of the 1st Regiment of Florida Volunteers, was to move along the south (outer or Gulf side) beach; and the third battalion, under Col. John K. Jackson of the 5th Regiment of Georgia Volunteers, was to follow Chalmers and push through the thickets to the middle of the island and attack Camp Brown, the site of the 6th New York, at

---

151. O.R. Vol. 6, p. 672. Brown mentions 1,300 for November, on page 469, and slightly exaggerates Bragg's strength at 8,000.
152. It is likely that there were none on detached duty at this time, Brown having concern about his strength. Soldiers were given a one-month leave for every year of service, so this would reduce some of the regulars present to something over 1,100. The primary alternate categories remaining, sick and in confinement, could vary widely.
153. O.R. Vol. 6, p. 750.
154. O.R. Vol. 6, p. 460.

the sound of gunfire. A special group of 53 men under Lt. James Hallonquist were lightly armed and carried materials to spike cannon and set fire to buildings and equipment. With a medical staff of 25, the raiding party totaled 1,089 men.

After difficulties with the transfer of some of the men to the steamer *Ewing* and boats and barges to be towed by the smaller steamer *Neaffie*, they got underway at midnight. They had not formed up at the landing site, a little more than four miles east of the fort, near the swampy ground indicated at the extreme right of figure 1, and about three miles from the pickets of Camp Brown, until 2:00 a.m. The dotted lines indicate the tracks of the advance of the three groups.

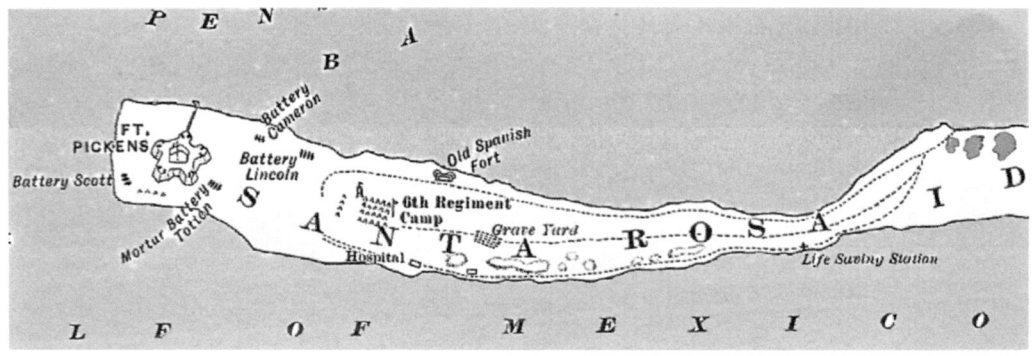

FIGURE 1

After some stumbling through the swampy ground, Chalmers' force of 350 men reached the 6th New York's picket posts at 3:30. Some of the pickets were shot down after a stiff resistance at the site of the old Spanish fort and others at the Gulf beach side, and all were soon driven in by Anderson's force.

The alarm having been given by the firing of the pickets, the 6th New York were rousted from their tents and formed up in what seems to have been in good order.[155] They then fell back toward water batteries Totten and Lincoln and avoided being surrounded. John K. Jackson's 260 Georgia volunteers were then able to advance into the deserted 6th New York camp. They did not push on immediately, using to their advantage the surprise and confusion of the moment. Instead, they stopped to loot. Hallonquist's men then went to work, made easy by the brush construction of the lean-tos and bowers that had been crafted for protection from the sun. By 4:00 a.m. the flames were visible from the fort. The delay took the initiative away from any planned Confederate advance.

The fort had long since been aroused, the east ramparts manned, batteries Lincoln and Totten prepared, and a force of 93 men sent out under the command of Maj. Israel Vodges. They proceeded up the north beach with Company G of the 6th New York deployed as skirmishers on the right flank to a point "some distance above Camp Brown."[156] A group of troops on their right and rear were

---

155. O.R. Vol. 6, p. 447; Figure 1, Morris, p. 70 altered.
156. O.R. Vol. 6, p. 449; report of J. McL. Hildt. Later analysis put the place at 3,331 yards from

encountered, which were at first confused with their own skirmishers, though it was a major part of the Confederate force returning to their boats. Major Vodges, on horseback, rode into them and was captured. His second in command, John Hildt, then took over. Being greatly outnumbered and almost cut off, they withdrew diagonally toward the south side of the island, and to quote Hildt, "as soon as his front was clear, the enemy proceeded along the north beach."

With word that Vodges had been captured, and his command in trouble, a second relatively meager group, under the command of Maj. Lewis G. Arnold, was sent out at about 5:00 a.m. They followed in the footsteps of Hildt's men and found them still engaged with the retreating rebels. Hidden behind the sand ridges along the beach, they fired upon one departing barge at about 1,200 yards. Making a further reconnaissance, they discovered the bulk of the rebel force embarking close to their original landing point about four miles up the beach. At the double-quick, Robertson's company was able to advance to within 200 to 250 yards of the steamer *Time,* in no small measure due to the fact that the flat she had in tow was aground.[157] From behind a tree-covered sand hill, they opened fire on the Confederate flotilla for 15 minutes. It is likely that at this time the greatest number of Confederate casualties were taken. Anderson listed 17 killed, 39 wounded, and 30 missing. Brown, revealing his status consciousness, reported: 4 regulars and 9 volunteers missing, with 7 volunteers wounded, and 10 regulars and 11 volunteers missing. The Confederate side saw it differently. A newspaper seen at Pensacola announced: "It is now certain that 175 in killed, wounded, and missing will more than cover our entire loss, while 250 will barely cover that of the Federalists."[158]

The reports Bragg submitted to the adjutant general[159] were exaggerated in almost every statistic, e.g., "600 or 700" was quoted as the strength of the 6th New York (vs. 250), and downright controversial in another. He publicly claimed in the *Pensacola Observer* that Confederate wounded had been "massacred" after being captured, something that an offended Brown hotly denied.

The Confederate raid failed by its own admission because Anderson allowed his men to loot and set fire to the camp of the 6th New York. If the camp had been deserted by the 6th, then it offered no obstacle to the overwhelming numbers of Anderson's troops to rush to attack the batteries and disable them. The few men sent piecemeal out from the fort (at first 93, then 112) could hardly have scared away 1,000 Confederates if they were determined to assault the outer batteries.

Incidents of poorly equipped Confederate troops stopping to loot were to be a part of Battery L's experience throughout the war. Though this minor incident

---

the fort, or about a mile beyond Camp Brown. Ibid., p. 442.
157. This per Robertson's report, O.R. Vol. 6, p. 451. Anderson has it as the steamer *Neaffie*; ibid., p. 462.
158. O.R. Vol. 6, pp. 442, 443, 462.
159. O.R. Vol. 6, pp. 458, 459, 670, 671.

at a remote outpost has been overlooked, other occasions later in the war caused a loss of momentum which may have contributed to serious Confederate defeats.

On October 7th, Bragg's command was expanded to include the "coast and State of Alabama," to be known as the Department of Alabama and West Florida.[160]

Bragg left for an inspection of the defensive preparations in and around Mobile on October 22nd. He met with General Withers on the 23rd and toured the area on the 24th. He found Fort Gaines to be "rapidly approaching a condition for strong defense, but is almost destitute of guns and ammunition." He found Fort Morgan to be in better condition, though not half armed, and with a limited supply of ammunition. Declaring that it was the key to the defenses of Mobile Bay, it must have heavy armament and ample supplies. "I shall at once *reduce my position at Pensacola to one of defense strictly*, and send what can be spared to this point, *though it will be totally inadequate to the wants here*."

The authorities in Richmond must have begun to wonder whether the substantial presence of the Yankees in Pickens and on the sea had trumped their ability to ever take Fort Pickens – even if this was still a worthwhile endeavor. Stalemate, in this remote place, perhaps would be sufficient if the war could be prosecuted with daring elsewhere. In addition to Bull Run (Manassas), success had been bestowed upon the Confederates at Wilson's Creek, Missouri,[161] in August; and at Lexington, which gave the Confederates control of southwestern Missouri; and again at Ball's Bluff, Virginia,[162] in October, which was a rout of the Union forces that killed Col. Edward D. Baker, a U.S. senator, and had shaken the Washington establishment.

In all, the raid was a story of ineptitude and timidity on both sides. The raid having been beaten back on October 9th, no significant events took place until a general bombardment of the Confederate positions on the mainland would be initiated by Brown on November 22nd, using the combined forces of the army and navy.

"Record" 12/61
31 OCTOBER–31 DECEMBER 1861, FORT PICKENS, FLORIDA

The Company was engaged in the bombardment on the 22nd and 23rd Nov.'61 at Fort Pickens, Fla. Corp'l Andrew J. Beeler was wounded in action on the 22nd Nov.'61. Lost his right arm.

1. Henry W. Closson   Capt. Promoted to Co. by letter War Dept. Nov. 21,'61. In Command since Dec. 8, 1861.
2. Edward L. Appleton   1st Lt. Trans. To Co. S.O. no. 321 Hdqtrs. Dept. of the Army, Washington, Dec. 5,'61. Abs. without leave.

---

160. O.R. Vol. 6, p. 751.
161. O.R. Vol. 3, pp. 2, 71–124. Italics added by the author.
162. O.R. Vol. 5, pp. 3, 32, 34, 290, 299, et. seq.

3. J.S. Gibbs	1st Lt.	Trans. To Co. S.O. no. 321 Hdqtrs. Dept. of the Army, Washington, Nov. 27,'61. Abs. without leave.
4. T. K. Gibbs	2nd Lt.	do.

Transferred:

1. S.K. Dawson	Capt.	To 19th Inf. [by promotion to Major] G.O. no. 65, War Dept. A.G.O. Washington, Aug. 23,'61. Left Co. Nov. 3,'61.
2. Edmund Kirby	1st Lt.	To Co. "L" 1st Arty. Sp.O. no. 321 Hdqtrs. Dept. of the Army, Washington, Dec. 5,'61. Never joined Co. "L".
3. George A. Woodruff	2nd Lt.	Absent on det. svc. since June 24,'61 Person never joined Co. L.

Detached: Richard H. Jackson 1st Lt. Left Co. Dec. 8, 1861.

Strength: 88. Sick: 12.

Sick Present: Andrew J. Beeler, John Buckley, Robert Curran, Michael Hanney, Michael Kenny, William Schaffer, Warren P. Shaw, William V. Thompson, Henry H. Ward.

Sick Absent: Thomas Brook, Ft. Hamilton, NY. Left Co. Sept, 17, 1861.
 William Brunskill do.
 Charles Riley Ft. McIntosh, TX. Abs. since Feb. 21, 1861. Reduced to Pvt. July 12, 1861.

Died:

1. Thomas Conroy	Sgt. 1 Jan.'59 San Antonio,	Killed at Ft. Pickens, Fla. by the accidental explosion of a shell while in the line of his duty.
2. Louis Hey	Pvt. 3 Sept.'58 New York	do.
3. Thomas Poole	Pvt. 4 Nov.'57 Detroit	do.
4. Michael Reedy	Pvt. 1 Jan.'57 San Antonio	do.

Capt. Henry Closson was transferred to the company by his promotion, and the transfer of Samuel Dawson, by promotion to major. Closson had previously been commanding Battery F, which had accompanied Battery L ever since they had arrived at Fort Duncan in 1860. Closson, as was previously noted, would remain in command of Battery L for the rest of the war and beyond; however, his brevet promotion to major on July 8, 1863, would place him in a position to be called away from Battery L for long periods of time. Battery L never saw him from October of 1863 to July of 1864, except as his duties as chief of artillery of the 19th Army Corps may have impacted Battery L. Again, during August of 1864, he was detached for service as chief of artillery in the Mobile expedition. Having been promoted to brevet lieutenant colonel, he was again detached in November and December of 1864 as chief of artillery and ordnance of the Cavalry Corps of the Middle Military Division in the Shenandoah Valley.

This meant that the direct daily command of the battery often fell to the second most senior officer or even the third most senior officer. Since two lieutenants, Kirby and Woodruff, who had been assigned but never joined, a third officer, Edward L. Appleton, was assigned on December 5th, but had not yet arrived. Appleton had been appointed a lieutenant on August 5th, directly from civilian life at Bangor, Maine. Edward never had any connection with West Point.[163]

The muster roll notation "Abs. without leave" is hardly just. Any individual would have to find a navy ship destined for the area – which could be weeks.

Note that Charles Riley, absent since February, has been reduced to private. This was to make way for the promotion of Alexander J. Baby to corporal and Julius Becker from corporal to sergeant.

Bombardment of November 22 & 23

One might ask: Why was there such a long time between the Confederate raid and any retaliation by Colonel Brown? Ever since April 16th, when he first took command, Brown seemed to have interminable excuses why he hesitated to initiate any action even after the "truce" was broken. By now, however, Brown was ready to respond to Bragg's raid of October 9th. He quickly consulted[164] with Admiral McKean, and the two agreed to begin a general bombardment of the Confederate positions on the 16th. Unfortunately, events at the "Head of the Passes," the location at the mouth of the Mississippi where it divides into three branches, caused McKean to advise Brown to postpone the plan. McKean's letter[165] to the Secretary of the Navy Gideon Welles explains:

> U.S. Flagship Niagara
> *Off Fort Pickens, October 15, 1861*
>
> Sir: The steamer *McClellan* has just arrived with disastrous news from the Mississippi. Our ships have been driven off from the Head of the Passes. . .
>
> The *Richmond* had several of her planks stove in by the steam ram[166] and all the guns of the *Vincennes* except four have been thrown overboard.
>
> A large increase of force and a supply of heavy rifled guns is essential . . .
>
> I shall leave the *Colorado* here to blockade the harbor and to prevent the landing of the enemy upon Santa Rosa . . .
>
> Wm. W. McKean,
> *Flag-Officer, Commanding Gulf Blockading Squadron*

---

163. Henry, Vol. 2, p. 36; U.S. Army Register, 1864, p. 23; Heitman, p. 168.
164. O.R. Vol. 6, p. 669.
165. ORN Ser. I, Vol. 16, p. 703. William McKean succeeded William Mervine on September 22nd, 1861.
166. The reference to the *Vincennes* throwing its guns overboard was to lighten it. It had become grounded, but this action allowed it to escape. Figure 2, from US Naval History and Heritage Command, photo no. NH608.

FIGURE 2

The "Battle at Southwest Pass," as it was called in the December 7th issue of *Harper's Weekly*, p. 779, described how, on October 12th, the ships of the Gulf Blockading Squadron were attacked by Confederate ships, including the iron-hulled ram *Manassas*, figure 2.

Having cleared up affairs off the mouth of the Mississippi, McKean returned to Pensacola on the morning of November 17th.[167] Now, plans could go forward for the opening of the guns on the Confederate installations. Fort Pickens, all its outer batteries, and the guns of the two powerful ships of the blockading squadron, the USS *Niagara* and the USS *Richmond*, would now pound Bragg into oblivion, or at least that was the plan.

The delay had been used to advantage, in that, mounted picket outposts were established, which extended far beyond the fort's former picket positions; and to add some strength and flexibility to this outer warning network, Battery L was supplied with four mountain howitzers and ordered to drill as a light battery. No horses being available, mules were substituted.[168] Thus, Battery L was modestly initiated into its eventual official conversion to a light battery a year later.

The mixed organization of the seven batteries of the fort is shown in figure 3. Note that Battery L people are in two locations, both under Lt. R. H. Jackson. The future officers of Battery L are all here. Franck Taylor, who would not be assigned to Battery L until January 15th, is now serving with Battery A, 1st Artillery. Outside the fort, Henry Closson is assigned to Battery Scott, one of the five outer sand batteries: Lincoln, Totten, Cameron, Scott, and the Spanish fort.

---

167. ORN Ser. I, Vol. 16, p. 772.
168. Haskin, p. 186.

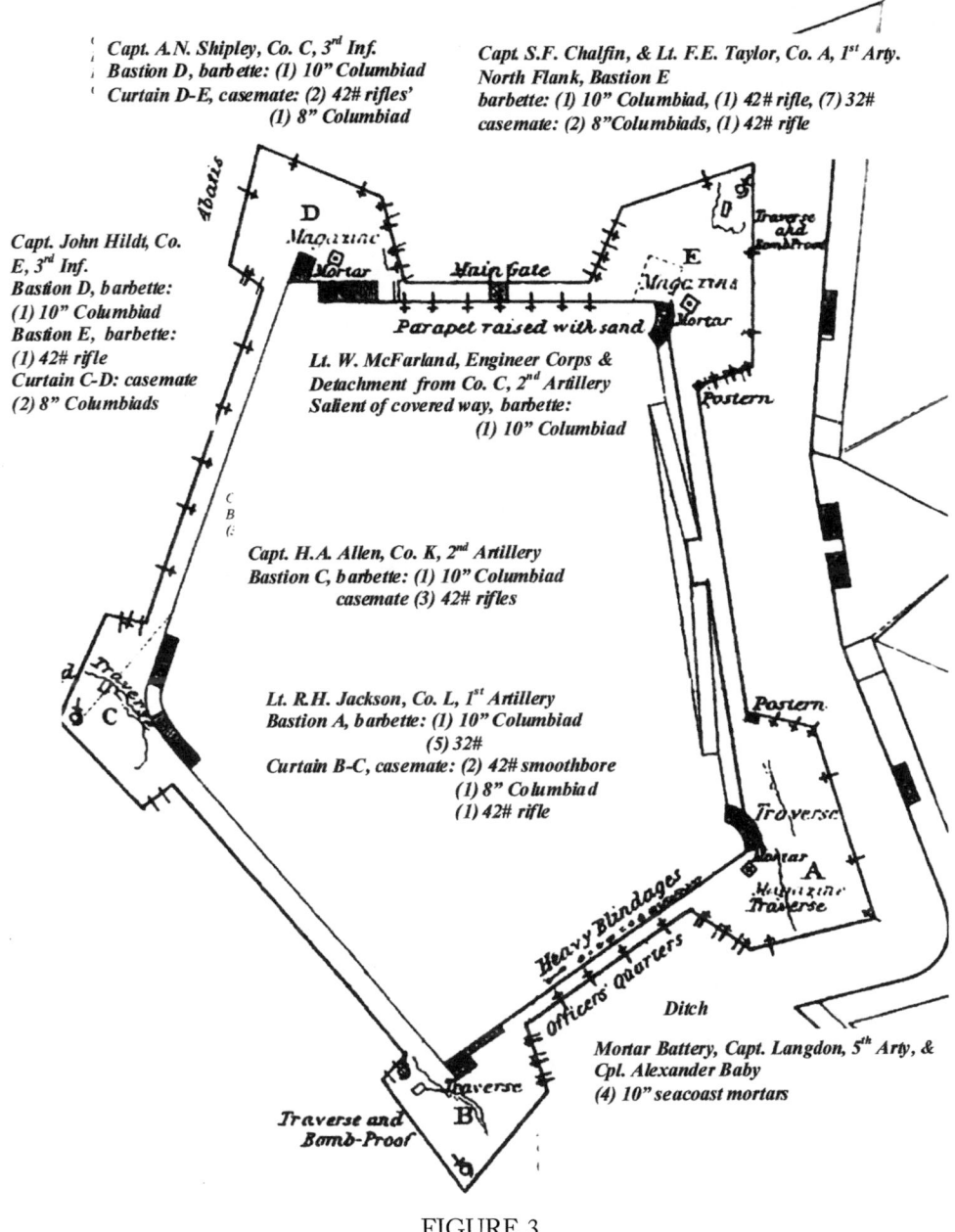

FIGURE 3

Figure 3 only generally indicates the gun locations at the various positions. Taken from *Atlas*, plate 5, it is altered to show the various troop deployments for the defense of the fort. Since the defensive arrangement is in reverse, the enemy being in the rear, none of the gun positions on the seaside are manned, nor were they able to be; Brown now had less than half of the guns that the fort could accommodate. The 7 batteries of the fort consist only of a total of 33 guns and 4 mortars.[169] Only 18 are barbette mounted, reaching over the parapet.

---

169. O.R. Vol. 6, p. 667; O.R. Vol. 1, p. 434.

Fifteen guns were mounted in the casemates under the parapet, figure 4.[170] The casemates on the outer curtain AB, now facing away from the enemy, are of little practical importance and are being used as officer's quarters. Their inner walls had been protected from plunging enemy fire by large timbers leaned against them and on which were layered sandbags. This is indicated by the note "heavy blindages."

FIGURE 4

The southeast bastion was known as "Hell Row," or "Long Hall," and it was, as Henry Closson remembered,[171] ". . .occupied at this time by a swarm of the younger officers of the garrison, who, when the toils of the day were over, used to congregate in this vicinity and console themselves for their fatigue and isolation with temperate appeals to pipe and cup."

FIGURE 5

---

170. Bastion E, Fort Pickens, 2008, photo by the author.
171. Haskin, *Closson's Memoirs*, pp. 356, 358; *Langdon's Memoirs*, p. 451. The 24-pounders were useless, as noted earlier, having insufficient range for the current circumstances. Note that the officers smoked while seated on the ammunition barrels. A budge barrel has a leather top, which provided a cushion.

Captain Closson's reminiscences confirm that, "Totten contained the Goliath of the armament, in the shape of a thirteen-inch seacoast mortar,[172] figure 5.

This monster required a crew of seven, three being required to load the 197-pound shell.[173]

The bombardment began at 10:00 a.m. by the firing of a signal gun located at the flagstaff, bastion A. Those batteries that could be directed at the navy yard, where the *Neaffie*, *Time*, and the gunboat *Nelms* were tied up, opened first, the *Time* being a favorite target.[174] Soon, most of the batteries of the fort were firing; at either the navy yard or Fort McCree, and the sand battery to the west of it.[175] The outer batteries variously opened on McCree, the lighthouse, the navy yard, the steamers, Fort Barrancas, the marine barracks, or Confederate masked [hidden] batteries as they were discovered by sighting their fire. The *Nelms*, though seen to have taken two hits, was backed out from the wharf and made its escape to Pensacola.[176]

The barbette guns of Fort McCree were silenced almost immediately after the initial firing from the ships and fort. Per Admiral McKean's report,[177] the sand battery south of McCree was silenced at 3:15. The fire of the casemate guns of McCree had slowed gradually, and they were silenced at about the same time. Brown reported that the fire from Barrancas was "reduced very perceptively," and the fire from the navy yard became silent by dark when the bombardment of the first day tapered off, rain interfering.

No hits from the Confederate guns were sufficient to silence any of Pickens' batteries. Private Cooper of the 6th Regiment of New York Volunteers was mortally wounded by a fragment of a shell that exploded at "about" the center of the fort, and Corporal Andrew Beeler of Battery L was severely wounded in the forearm by a shell fragment while serving one of the barbette mounted 10-inch Columbiads on curtain BC. His forearm was later amputated, and he was eventually discharged on a surgeon's certificate of disability. One private near the barbette 10-inch Columbiad of Company A, 1st Artillery, was slightly wounded in the head by a shell that landed and exploded on the parapet in front of the gun.[178]

On the second day, the bombardment was opened by Fort Pickens at "about

---

172. Figure 5, Miller, Vol. 3, p. 186. Referred to as "The Dictator."
173. Gibbon, p. 443.
174. *Harper's Weekly*, December 28th, 1861, p. 827.
175. O.R. Vol. 6, pp. 469–487. Capt. John Hildt, commanding bastion D and curtain CD of the fort, reports it as 9:30; Lt. Francis Seeley, at the old Spanish fort, has it as 8:00; and Capt. James Robertson in command of Battery Lincoln reports the time as 9:30. Problems with variable reports of time plague the historian here, and as will be seen, in future battle actions.
176. O.R. Vol. 6, p. 481, Lieutenant Seeley has three vessels at the wharf: the *Time*, *Neaffie*, "and another armed gunboat." Confederate general Anderson's report, p. 494, only mentions the *Time* removed after dark and the *Nelms*, which "made her escape at once."
177. ORN Ser. I, Vol. 16, p. 775.
178. O.R. Vol. 6, pp. 470, 474, 483.

the same hour." The USS *Niagara* was not able to get within range as a change in the wind had reduced the water depth. Though it was "careened" (tilted by moving guns and ballast) to increase the elevation of its guns, the efforts proved futile. It finally signaled "too great a range" and disengaged. The *Richmond* did not participate. Its damage was being surveyed, and it would likely go to Key West for repairs.[179]

The navy yard at Warrington was "much"[180] damaged according to the report of Brown's executive officer Maj. Lewis G. Arnold, (promoted after Vodges' capture) who had commanded the batteries. The church and parts of the town of Warrington were set afire by midafternoon. Later the fire extended into the town of Woolsey to the northeast. Though the general firing ended at dark, at 7:00 p.m. it was resumed by Battery Cameron, then Battery Lincoln, with mortar fire on the navy yard and surroundings to hamper Confederate efforts to extinguish the flames. Battery Lincoln's 10-inch mortar shells (weighing 87.5 pounds each) had been loaded with a mix of port fire (an incendiary) and regular powder.

All firing ceased at 2:00 a.m. on the 24th. Fort Pickens' damage was minimal. A 10-inch shell entered an embrasure of curtain CD under the command of Capt. John Hildt but did not explode. A shower of brick from the embrasure wounded six of Company E, 3rd Infantry. There were many close calls, but luck prevailed, in that many shells that landed near or in places that could have done serious injury did not explode. One 32-pounder throwing hot shot burst and threw fragments into bastions D and E but did no injury. The mortar battery commanded by Capt. Loomis Langdon, in the ditch in front of curtain AB and protected only with a parapet, had an 8-inch shell drop into its midst. It exploded with no one injured. A 10-inch shell struck the bombproof of Lieutenant Shipley's barbette battery, bastion D, and exploded, throwing sand bags around but without injuring anyone.

The companies of Captains Duffy, Heuberer, and Bailey of the 6th New York were praised for meritorious service. These volunteers, if not all of the 6th, had finally earned Brown's respect.

Both 1st Lt. Henry W. Closson and 2nd Lt. Franck E. Taylor, soon to be promoted to Battery L, were mentioned for good service, and Sergeants Keller, Becker, Conroy, and Newton; corporals Beeler, Wicks, and Spangler; and privates Jackel and Hanney, all from L, were mentioned for distinguished service.[181]

Bragg[182] lost "21 wounded – 1 mortally" all on the first day. A "defective" magazine at McCree collapsed and killed six others. Col. John B. Villepigue, Fort McCree's commander (West Point class of 1854 and classmate of Henry W. Closson), was one of the wounded.

Fort McCree suffered from the combined effects of fire from Fort Pickens

---

179. ORN Ser. I, Vol. 1, p. 780; also Vol. 16, p. 806.
180. O.R. Vol. 6, p. 475.
181. O.R. Vol. 6, pp. 475, 477, 483, 485.
182. O.R. Vol. 6, pp. 489–493.

and the ships, which "fired with much greater accuracy." It was so badly damaged on the first day that Bragg considered abandoning it and blowing it up.

Bragg complains of Brown turning Pickens' guns on the hospital, which was still flying its yellow flag on the second day, even though it had been evacuated. He does not admit to serious damage to the navy yard, though many houses were struck and damaged. Hot shot and shells landing in Warrington and Woolsey, however, burned "considerable portions of each" (50 buildings were quoted in General Anderson's report).[183]

Bragg waxed eloquent: "For the number and caliber of guns and weight of metal brought into action it will rank with the heaviest bombardment in the world. It was grand and sublime. The houses in Pensacola, 10 miles off, trembled from the effect, and immense quantities of dead fish floated to the surface in the bay and lagoon, stunned by concussion . . ."

He states that he fired "about 1,000 shots, and the enemy not less than 5,000."[184] He notes that he fired the last shot at 4:00 a.m. on Sunday "as a warning that we were on the alert . . ."

Bragg makes two curious statements in the context of the closing of the bombardment: "As they did not renew the action, and drew off with their ships in a crippled condition, our fire was not reopened on Fort Pickens, *to damage which is not our object*."[185] In a preliminary report,[186] he had stated: "Yesterday and to-day the enemy has not renewed the contest, and, for reasons which the [War] Department will appreciate, *it is not my policy to do so*."

This is entirely consistent with his declaration of the 24th of October of "*defense strictly*," if he would need to detach troops to reinforce Mobile. Though the Confederate command very realistically expected an assault on Mobile, it would never be made until August of 1864, the Union army and navy being distracted by calls for action elsewhere.

On November 25th, while cleaning up after the bombardment, a fatigue party was emptying the powder from unexploded rebel shells. Careless handling caused an explosion that killed Sergeant Conroy, privates Hey, Poole, and Reedy, and wounded two others of Battery L, plus one of Battery A, and others of the 3rd Infantry.[187]

---

183. O.R. Vol. 6, p. 495.
184. Not too much of an exaggeration. Calculating 55 total guns and mortars, for nine hours, for two days, averaging four shots per hour for the guns, and two for the mortars, results in 3,528. The naval bombardment did not begin until one-half hour later than the fire from the fort, was suspended briefly, and resumed on the afternoon of the first day. It did not continue on the second day but might have, nevertheless, gotten off 1,000 rounds.
185. O.R. Vol. 6, p. 491.
186. O.R. Vol. 6, p. 489.
187. Haskin, pp. 187, 361, 377. Those wounded in the explosion are not specified in the "Record."

Confederate Concerns

Sufficient Confederate spies roamed the North to alert the Confederate War Department to the preparations for another Union navy-army expedition[188] but not its destination. Of course, everyone on the Confederate-held coast was put on alert. On November 7th, Flag Officer S. F. DuPont's South Atlantic Squadron struck at Fort Walker, on Hilton Head Island, and Fort Beauregard on Bay Point, on the opposite shore. Known as the Battle of Port Royal Sound, the forts fell after five hours of bombardment. Gen. Thomas W. Sherman then landed some 12,000 men to occupy the area[189] and thus threaten Savannah and Charleston. Another post in Scott's Anaconda Plan had fallen to the Union.

On November 28th, Mississippi sent Bragg a regiment of new recruits.[190] He noted that he now had four new regiments, or about 3,000 men, though only 600 of them were armed. Bragg's responsibilities were expanded on December 10th to include Pascagoula Bay and "that portion of Mississippi east of the Pascagoula River." Just previous, Wood's 7th Alabama, one of Bragg's best, had been ordered to east Tennessee. In August, Wood had complained about his troops being demoralized from inaction and evidence that the "field of fight" was to be elsewhere.[191] Having just discussed his weakened situation, Bragg notes:[192] "The enemy landed about 1,000 men on Santa Rosa Island, and they are now encamped near the fort." This was the 75th New York Regiment of Volunteers, which began landing from the steamer *Baltic* on December 14th.[193]

On December 27th, Bragg received a letter from Secretary of War Judah P. Benjamin, requesting him to consider the position of commander of the Department of the Trans-Mississippi.[194] The explanation was circuitous, but the sum of it was that the Confederacy was in danger of losing Missouri, and that Gen. A. S. Johnston, whom it had been planned to send there, was detained in Kentucky by the threat of a Union invasion. Benjamin felt that defeat in Missouri was more of a risk than "misfortune" at Pensacola. The Trans-Mississippi Department would encompass Arkansas, Missouri, Northern Texas, and Indian Territory (Oklahoma).

On December 31st, forces from the Gulf Blockading Squadron landed at Biloxi, Mississippi. A detachment of marines and a gun's crew, a total of only about 60 men, then demanded its surrender, which was "acceded" to.[195] Though the Union forces left within hours, the Confederates were alarmed, and Benjamin wavered

---

188. O.R. Vol. 6, p. 758.
189. ORN, Ser. I, Vol. 12, pp. 261–319.
190. O.R. Vol. 6, p. 768.
191. O.R. Vol. 1, p. 470.
192. O.R. Vol. 6, pp. 771, 777, 782.
193. Babcock, pp. 21, 66.
194. O.R. Vol. 6, p. 788.
195. ORN Ser. I, Vol. 17, pp. 34, 40; O.R. Vol. 6, p. 809.

in his thoughts about reassigning Bragg.[196] The push-pull of where to defend the Confederacy was becoming more frantic. The Union navy had landed on Ship Island, 100 miles off the mouth of the Mississippi, on December 3rd and the scare at Biloxi "puts an end to all idea of assigning you to other duty."[197] This left Bragg's Department of Alabama and West Florida intact for the moment. Earl Van Dorn was assigned to the Department of the Trans-Mississippi.

By mid-January 1862, the scare seemed to pass, Confederate recruitment and reenlistment was picking up, and with the aid of Governor Shorter of Alabama, some 1,000 armed militia could be expected to reinforce Bragg's men. Gen. Jones M. Withers was to be assigned the immediate command of the forces at Mobile, thus freeing Bragg for the broader affairs of the department.

"Record" 2/62
31 DECEMBER–28 FEBRUARY 1862, FORT PICKENS, FLORIDA

The Company was involved in the Bombardment of Jan. 1st, 1862.

1. Henry W. Closson — Capt. In Command
2. Franck E. Taylor — 1st Lt. Joined Company by trans. S.O. no. 5 Hdqrts. Dept. of Fla. Jan. 15,'62
3. Edward L. Appleton — 1st Lt. Joined Co. from absent without leave, Jan. 6,'62
4. T.K. Gibbs — 2nd Lt. " " " " " Jan. 6,'62

Transferred:
1. J.S. Gibbs — 1st Lt. Left Co. by transfer to Co. "D" 1st Arty. Sp. O. no. 5 Hdqtrs. Dept of Fla., Ft. Pickens, Dec. 15,'62

Discharged:
1. Andrew J. Beeler Cpl. Feb. 9,'60 Boston, Ma by Surgeon's Certificate of Disability (of wounds received in action) Apr. 5, 1862.

Strength: 88, Sick: 9

Sick Present:
Julius Becker, Andrew J. Beeler, Michael Olvaney, William Schaffer, Warren P. Shaw, William V. Thompson

Sick Absent:
Thomas Brook, Ft. Hamilton, NY Left Co. Sept. 17, 1861
William Brunskill, do.
Charles Riley Sick at Ft. McIntosh, Texas. Abs. from Co. since Feb. 25, 1861. Reduced from Sgt. to Pvt. to date from July 12,'61. O. no. 2, Co. "L" 1st Arty, Fort Pickens, Fla.

On January 15th, Lt. Franck Eveliegh Taylor is recorded as being assigned to Battery L, essentially completing its officer makeup, which would survive almost unaltered for the rest of the war. Like Appleton, Taylor had been appointed directly from civilian life on the same exact date: August 5th. He was born in Washington

---
196. O.R. Vol. 6, p. 794.
197. O.R. Vol. 6, p. 803. Benjamin to Bragg.

D.C., the son of the prominent bookseller, Franck Taylor.[198] We shall see that for the rest of the war, Taylor and Appleton will be the mainstays in command of the battery, always there while Closson is detached on other duty.

Bombardment of January 1, 1862

At about 3:00 p.m., a steam tug approached along the sound from the east, and made a landing at the navy yard wharf[199]. This was the first ship to "come within the range of my guns," as Brown put it, since the previous bombardment. As it departed, it did not run straight toward Pensacola but veered into the bay, a man on deck waving the Stars and Bars and attracting the attention of Battery Lincoln. Regarding it as an act of bravado, calculated to draw his fire, Brown obliged. Three shots were directed at the steamer. One shot was returned from a rebel battery in the vicinity which was directed at the gun that fired into the steamer, and one shot was returned from Pickens. There the matter lay for about three-quarters of an hour, when all of the Confederate batteries opened up in a general bombardment. A slow rate of fire was selectively returned from Pickens, only using its heaviest guns, until sundown, and was continued after dark with mortars. In the previous bombardment, the incendiary "port-fire," was noted as being used to put to flame several of the buildings in Warrington. Brown had since ordered and received "rock-fire and carcasses," special shells filled with incendiary material, designed to set fire to the target.[200]

At about nine o'clock a bright light was seen in the navy yard. By 10:00 p.m "the whole firmament was illuminated, several of the largest buildings being set on fire." Fort Pickens maintained firing until about 2:00 a.m. The Confederates kept it up, despite all their troubles, until 4:00 a.m.

While Fort Pickens probably thought that the dense fog that had rolled in, followed by a drizzling rain,[201] was what had dampened the Confederate enthusiasm; it was actually the precipitate arrival of a fuming Braxton Bragg. Departing for his visit to Mobile, he had left Gen. Richard H. Anderson in command at Pensacola. The wild bombardment was the result of his drunken orders, "so much intoxicated as to be entirely unfit for duty," while Bragg was absent.

Fort Pickens suffered minor damage, few shots even hitting the walls. Brown estimated that the rebels used even more ammunition than they had in the two-day bombardment in November. Two men, one of the 6th New York and one regular

---

198. Henry, Vol. I, p. 226; Heitman, p. 632; *Name Authority File*, Library of Congress, *Taylor, Franck*, http://id.loc.gov/.
199. O.R. Vol. 6, pp. 495–498, ORN Ser. I, Vol. 17, p. 33; Hall, H. & J., p. 28. Brown's account differs. The above narrative is an amalgam.
200. Gibbon, p. 329.
201. Hall, H. & J., p. 29.

in the fort, suffered minor injuries. The entire 75th New York was ordered to leave their camp near the fort and move some two miles east, out of the fire. Bragg admitted to no casualties but fretted over the loss of ". . . a large and valuable storehouse" and the "criminal waste of means so necessary for our defense . . ." General Anderson would be relieved as soon as a replacement could be found. "We are sadly pressed for competent officers . . ."[202]

After the distractions of January 1st and 2nd, Bragg had a chance to respond to the adjutant general about the conditions in the Mobile area, saying that the forts were in a better condition than when appraised in October and that *with ammunition*[203] they would prevent any entrance to the bay.

A crushing reality check was received from the secretary of war on January 5th, 1862.[204] "I regret the total impossibility of supplying you with arms for your unarmed regiments." Also enclosed was a copy of the regulations "devised" for carrying out the new system of universal suffrage.

Confederate Withdrawals from Pensacola

In a letter to the Confederate adjutant general dated February 1st, Bragg noted that the Union forces on Ship Island were now estimated at 8,000 or 10,000 "but making no preparation for a decent on us at this time . . . We continue to receive supplies from Havana by small vessels running the blockade, both here and at Mobile." In the space of a month, Bragg had become relatively transformed in his estimate of the situation, saying: "Should no move be made against us in the next four weeks, I shall look upon their force as merely intended to hold us in check by threatening our positions, and would recommend a withdrawal of a portion of the oldest and best forces in this department for service elsewhere."

Only five days later, he informed Gen. Samuel Jones at Pensacola: "A large naval expedition left Hampton Roads on the 4th with additional land forces for the Gulf; supposed destination Mobile and Pensacola. Suspend all furloughs, and prepare to receive them."[205] This could have mistakenly referred to activity in support of the North Carolina Expedition of Gen. Ambrose Burnside, which had sailed from Fort Monroe on January 11th, 1862, and which attacked Roanoke Island on February 7th. It surrendered the next day.[206] Bragg's intelligence that the expedition was headed for the Gulf had been in error, or his spies at Hampton Roads premature in their judgment; the real danger, the Butler expedition to New Orleans, did not leave until February 25th.

---

202. O.R. Vol. 6, p. 802. Benjamin to Bragg.
203. O.R. Vol. 6, p. 793. Italics by the author.
204. O.R. Vol. 6, p. 794.
205. O.R. Vol. 6, p. 822.
206. O.R. Vol. 9, pp. 72, 74; Butler, p. 336.

A major blow – one that stunned the Confederacy – came with the news of the fall of Fort Henry, on the Tennessee River, on February 6th, 1862, to a joint operation between Gen. Ulysses Grant and the gunboats of Flag Officer A. H. Foote. First mentioned in the February 22nd issue of *Harper's Weekly*, anyone in Battery L who may have read the story would never have known of the existence of General Grant. He is not mentioned, perhaps correctly, since the fort surrendered to Foote. Bragg was now ordered to "as soon as possible send to Knoxville all the troops you can spare from your command..." The number the secretary of war was looking for was "at least" four regiments. Four regiments had already been ordered from Virginia[207] and 5,000 men from New Orleans.[208] Gen. A. S. Johnston was outnumbered, and the next and larger fort, Donelson, was now threatened. Bragg notified Gen. Withers at Pensacola that he would likely have to give up the 9th Mississippi Regiment as a part of what was to be sent to Tennessee.

On February 15th, Bragg offered the secretary of war some advice, constituting a remarkable reversal of his earlier sentiments. All coastal points except New Orleans, Mobile, and Pensacola should be abandoned. The protection of persons and property should, as such, be abandoned. Only important strategic points should be held. The whole of Texas and Florida should be let go and the enemy allowed to occupy and disperse himself there. As a result: "We could beat him in detail, instead of the reverse." The Confederacy could then concentrate its forces for the delivery of a heavy blow. The point of concentration now had defined itself as Kentucky. He goes on to wistfully note the need for modern rifled guns so as to protect those coastal, or river, "assailable points" from the enemy's shallow draft gunboats.

On February 18th, Bragg notified Benjamin[209] that he was sending the 5th Georgia, 9th Mississippi, 20th Alabama, and 23rd Alabama to Knoxville. That day, he received a telegram in response to his advice of the 15th: "Commence immediately the movement you suggest for aiding General Johnston." It also indicated he would get a policy decision sent by messenger *tomorrow*.[210] Dated the 18th, it arrived on the 27th, the messenger stating that he did not *leave* until the 21st!

The critical policy statement finally having arrived, it said, in part:

War Department, C. S. A.,
Richmond, Va., February 18, 1862

Maj. Gen Braxton Bragg,
Mobile, Ala.:

Sir: The heavy blow which has been inflicted on us by the recent operations in Kentucky and Tennessee renders necessary a change in our whole plan of

---

207. O.R. Vol. 6, p. 823.
208. O.R. Vol. 6, p. 825.
209. O.R. Vol. 6, pp. 826, 827, 894.
210. O.R. Vol. 6, pp. 827, 828, 830, 834.

campaign, as suggested in your dispatch of this date, just received.

We had had in contemplation the necessity of abandoning the seaboard in order to defend the Tennessee line, which is vital to our safety; but I am still without any satisfactory information from General A. S. Johnston. I know not the nature nor extent of the disaster at Fort Donelson…and am only aware of the very large loss we have suffered in prisoners through the dispatches in the Northern papers.

However, all this is beside the question. The decision is made, and the President desires that you proceed as promptly as possible to withdraw your forces from Pensacola and Mobile and hasten to the defense of the Tennessee line. …by the time you reach the Memphis and Charleston Railroad we will be able to determine towards what point they are to move…

J. P. BENJAMIN,
Secretary of War.[211]

The Confederate War Department also sent other orders to move troops out of Texas to supplement those in the Trans-Mississippi Department of Gen. Earl Van Dorn, and made a long string of demands on the Confederate general in command at New Orleans, Mansfield Lovell, who by now had replaced Twiggs, and on Governor Moore of Louisiana.

Bragg immediately issued orders to Gen. Samuel Jones to remove the heavy guns, tear up the railroad to its junction with the Mobile railroad, and "burn all from Fort McCree to the junction with the Mobile road." This was to be done at night and in "all the secrecy possible." The next day, Bragg turned over the command of the whole Department of Alabama and West Florida to Jones, and, on his own initiative, diverted some of the troops slated for Johnston, at Nashville, to P. G. T. Beauregard, at Corinth, Mississippi.

After almost two weeks of back-and-forth between Bragg and Jones, who resisted the idea of a complete abandonment, Jones began to implement a revised plan. All sawmills, lumber and boats that could be found in Escambia and Blackwater Bays were put to the torch, yet a force of "something over a thousand" recent volunteers from Alabama, only 300 of whom were armed, would remain.

At Fort Pickens, the command changed hands. On January 29th, 1862, Col. Harvey Brown was directed to turn over the Department of Florida to Lewis G. Arnold, now promoted to brigadier general, U.S. Volunteers.[212] Brown, having been brevetted brigadier general in the regular army, was transferred to New York where he had been assigned to the command of the defenses of New York Harbor.

The (Union) Department of Florida ceased to exist on March 31st, 1862, when it became the western district of the Department of the South. Headquarters of

---

211. O.R. Vol. 6, pp. 835, 836, 842, 843. Fort Davidson fell to Grant on February 16th.
212. O.R. Vol. 6, p. 694. A temporary, or brevet, rank, as Arnold was a still a major in the regular army. Cullum, Vol. I, p. 189.

the new department was at Port Royal, South Carolina, and the new commander was Maj. Gen. David Hunter.[213]

Those at Fort Pickens may or may not have observed the smoke of the burning lumber and ships at Blackwater Bay, nearly forty miles distant, on March 11th. However, confirmation[214] of a partial evacuation of the Pensacola area came in mid-March from a party of two whites and two blacks, who came to Fort Pickens from the area of Milton. This prompted Arnold to write to Flag Officer David G. Farragut, commander of the Western Gulf Blockading Squadron, on March 15th. He also wrote to the adjutant general on March 22nd, requesting that his "estimates" to the quartermaster department for a shallow draft steamboat and surfboats be filled without delay. This was necessary before any offensive operations could be initiated, the navy support being now at a minimum (only the sloop *Vincennes*). He requested the support of two or three gunboats to attack Town Point, on the north side of the peninsula opposite Deer Point. The information the visitors had provided indicated that all the guns and most of the 2,000 men had been removed, leaving a force of 400.

The March 1st issue of *Harper's Weekly* was no doubt by this time in the hands of someone in the Fort Pickens command, since friends and relatives of the 6th New York and now the 75th New York regularly sent newspapers. The front page was entirely covered with a depiction of the surrender of Fort Donelson. Perhaps this was the first time that Battery L, or anyone outside of Illinois, Grant's home state, had ever heard of General Grant. The article that followed, entitled "THE BEGINNING OF THE END" announced that the "fate of Columbus, Memphis, and consequently New Orleans is now sealed."

It took until April 8th for Arnold's letter to Farragut to arrive at the Head of the Passes, and for Farragut[215] to respond, but the answer was no. The reason was that he and Butler's forces were "on the eve of attacking New Orleans."

"Record" 4/62
28 FEBRUARY–30 APRIL 1862, FORT PICKENS, FLORIDA

The Company formed portion of an expedition which left Fort Pickens 26 Mar. Went into camp on Santa Rosa Island, 40 miles up the beach, on the morning of the 31st and at dawn shelled and dispersed a rebel force on the mainland. Returned to Fort Pickens April 2, 1862.

1. Henry W. Closson	Capt. In Command
2. Franck E. Taylor	1st Lt. Temp. detached to Command of Co. C 3rd Inf. By virtue of special order no. 19, Hdqrts. Dept. of Fla. Fort Pickens, Feb. 16, 1862
3. E. L. Appleton	1st Lt.
4. T. K. Gibbs	2nd Lt.

---

213. O.R. Vol. 6, pp. 257, 258.
214. O.R. Vol. 6, pp. 704, 705, 711.
215. O.R. Vol. 6, p. 712.

Discharged:
>Andrew J. Beeler, Corpl. Feb. 9,'60 Boston, Ma. By Surgeon's Certificate of Disability (of wounds received in action) Apr. 5,'62

Strength: 87. Sick: 7.

Sick Present: Isaac T. Cain, William Henry, William Schaffer, William E. Scott

Sick Absent: Thomas Brook, Ft. Hamilton, NY. Left Co. Sept. 17, 1861
>William Brunskill   do.
>Charles Riley, Ft. McIntosh, TX. Abs. since Feb. 25, 1861. Reduced to Pvt. July 12, 1861.

Reconnaissance on Santa Rosa

Arnold's reading of the signs of the deterioration of the Confederate lines and his impatience to act was indicated by ordering a "reconnaissance in force" on March 27th –31st.[216] Its purpose was to capture or displace what was heard to be a force of rebels some three companies in size who had attacked sailors from the blockading squadron at the Southeast Pass, at the eastern end of Santa Rosa Island. Arnold's force would consist of a detachment from Battery L, table 1, and Company K of the 6th Regiment of New York Volunteers. They would be under the command of Capt. Henry W. Closson of Battery L.[217]

Capt. Henry W. Closson	John Casey	Joseph Kutschor	William E. Scott
1st Lt. Franck E. Taylor	Jeremiah Connell	Charles F. Mansfield	Warren P. Shaw
1st Lt. Edward L. Appleton	Owen Coyne	James McCarthy	Joseph Smith
2nd Lt. Theodore E. Gibbs	William Creed	Daniel McCoy	Andrew Stoll
Sgt. Lewis Keller	Robert Curran	Miles McDonagh	William V. Thompson
Sgt. Thomas Newton	Patrick Donnelly	Terence McGaley	Reuben Townsend
Sgt. Julius Becker	Prosper Ferrari	Edward McLaughlin	John Tomson
Sgt. Alexander J. Baby	Charles A. Flint	James McWaters	Michael White
Corp. David J. Wicks	James Flynn	Dennis Myers	Joseph Wilkinson
Corp. Charles Spangler	Christopher Foley	John Murphy	Henry Wilkson
Corp. William Demarest	George Friedman	John G. Nitschke	Henry William
Privates:	Morris Galavan	Michael O'Sullivan	Owen A. Wren
James Ahern	George F. Hadley	William Parketton	Wallace D. Wright
Edward Anglin	John Holland	Franklin W. Richards	William Wynne
James Beglan	George Howard	Ludwig Rupprecht	
William F. Brown	Charles Jackel	Heinrick Schmidt	
John Burke	Michael Kenny	Phillip H. Schneider	

TABLE 1. BATTERY L MEMBERS OF THE EXPEDITION

Carrying rations for five days and with a six-mule team towing a 10-pounder Parrott rifle, they set out on the 27th and camped 12 miles east on the island. They were joined on the 28th by Company D of the 6th New York, led by Lt. Richard H.

---

216. The O.R. has the date as the 27th; as can be seen, the muster roll does not agree.
217. O.R. Vol. 6, p. 500.

Jackson (now Arnold's aide and assistant adjutant general) who was accompanied by one of the four rebel "refugees"[218] who had come in to the fort on the 27th. The refugee was to act as a guide. The rebel camp was indicated to be on the mainland, about 40 miles from the fort. Orders were to make a three-pronged approach, one company passing east past the rebel position, crossing the East Pass onto the mainland, and doubling back. The second group and the rifled gun were to shell the enemy from a position opposite, on the island, and the third was to cross the sound below the enemy camp and approach east along the mainland. Boats were to be furnished by the blockading schooner *Maria Wood*, stationed off Santa Rosa on a line opposite the enemy position.

The expedition went into camp on March 31st, some 36 miles from the fort and four from the supposed location of the rebel camp. The first part of the plan of attack was abandoned because the East Pass was too far from the position for the attack, some ten miles. The schooner was notified to deliver the boats to the point of a signal fire that night, determined by a close reconnaissance of the island by Lieutenants Jackson and Appleton. At sunset, the attacking force of 170 men moved to a point about two miles below the position of the rebel camp, the place where the crossing would be made. The signal fire was lit, and the first two boats arrived at the outer beach at 11:00 p.m. They were hauled the 800 yards across the island to the inner beach. The third boat did not arrive until 1:00 a.m. The late arrival and the sighting of, and unsuccessful chase after two rebel spies now "precluded all possibility of surprise," and the plans were changed. The Parrott rifle was brought into a position some 250 yards from the rebel camp, and at early dawn when the huts of the enemy camp could be seen, Lieutenant Jackson was directed to open fire. Quoting Closson: "The shells burst right in their midst." Though a "scattering" volley was fired from the rebel guard, they disappeared along with the rebels in camp, who could be seen running "through the brush in their shirt-tails making rapidly into the back country. After shelling the area thoroughly, I returned to my camp. My supply of rations and forage was nearly exhausted, the mules nearly broken down by a very severe pull of forty miles through the heavy sand of the beach. I therefore sent all my sick men [some six, two from Battery L: Scott and William] . . . back on the schooner." The command was back at the fort by April 2nd.

A search of the records indicates that the rebel camp was Camp Walton, which was occupied by McPherson's Walton Guards, about 60 men.[219]

The reconnaissance may have satisfied Arnold in that he obtained information on "the character of the upper end of the island," but he made no further comment. Certainly, there were no Confederates on the island; they were only camped on

---

218. Babcock. Babcock says "deserters," p. 55.
219. O.R. Vol. 6, pp. 762, 869.

what is today known as Fort Walton Beach. It is noted that no prisoners were taken, which precluded that source of information. However, the lack was soon made up by deserters. On April 4th, Col. Thomas M. Jones at Pensacola sent this message to Gen. Samuel Jones at Mobile: "I am confident the enemy now know my true condition, two men having escaped to Fort Pickens."[220]

---

220. O.R. Vol. 6, p. 870.

Chapter 5

The Secret Navy Plan; Troops to Ship Island; A Promising Outlook; Louisiana; Farragut and Butler; New Orleans; Vicksburg I; Failure at Corinth; "Record" 6/62; Occupation of Pensacola; "Record" 8/62; Reassignments; "Record" 10/62

The Secret Navy Plan

The plan that the navy Secret Board had put together for an attack on New Orleans was still underway, and its secrecy was intact. It was not known to anyone at the highest level, save the president, Secretary Welles, Assistant Secretary Fox, and General McClellan. Most importantly, the press never knew, and hence the Confederacy was never warned.

Of necessity, it was now revealed to General Butler and Edwin M. Stanton, appointed Secretary of War on January 13th, 1862.[221] From Welles:

> . . . Mr. Fox made known . . . the great object which had occupied the Navy Department for several months . . . Mr. Stanton seized hold of the information with avidity, and gave a hearty support to the movement – the more acceptable because General McClellan, who had known our object and was by express direction of President Lincoln to cooperate with the Navy, appeared indifferent and had little confidence in our success.

McClellan had been briefed[222] on the plan in mid-November by its three most serious proponents: Navy Secretary Welles, Assistant Navy Secretary Fox, and Cdr. David Porter. Initially, the plan to attack New Orleans was to support the blockade. It was one of the "points" in the Anaconda Plan. It had been proposed because "the capture of New Orleans itself would be less difficult, less expensive, less exhausting, would be attended with less loss of life, and be a more fatal blow to the rebels, than the most extensive, stringent, and protracted blockade that could possibly be established." McClellan had approved of it when he heard that the 10,000 supporting troops required were only to be there to "garrison the captured forts and hold the city, after the navy had obtained possession . . ."

---

221. O.R. Ser. III, Vol. 1, p. 964; Welles, Vol. I, pp. 60, 61.
222. Welles, *Galaxy*, November 1871, p. 677.

On January 25th, 1862, McClellan was again asked, this time by Stanton, what he thought of all of this secret navy planning. Giving Stanton a history of the murky origins of the expedition,[223] McClellan outlined his view that there were three "great points" from which operations should be conducted: the Departments of the Potomac, Ohio, and Missouri. The Potomac would face Richmond, the Ohio would oppose the rebels in Kentucky, and the Missouri would clear that state, and then have "for a prime object the control of the Mississippi River and operations against New Orleans." Operations on the seacoast were only to distract the enemy. McClellan estimated that 30,000 to 50,000 men would be required if Butler were to attack from the Gulf – that is, to meet the Missouri "Northern Column." This "new extension of the plan," as he called it, had insufficient troops, and it was his recommendation that General Butler's expedition should be suspended.

Brushing McClellan's objections aside, Stanton created the Department of the Gulf on February 23rd, 1862,[224] to "be occupied by the forces of Maj. General B. F. Butler." McClellan now had no choice other than to confirm to Butler his staff assignments and the troops involved. The objective, in support of Admiral Farragut of the navy, was initially to secure New Orleans, but then Baton Rouge and to "open your communication with the northern column [Halleck's Department of the Missouri] by the Mississippi . . ."

By February 17th, 1862, the force authorized had grown to 17 regiments of infantry, 3 companies of cavalry, and 6 batteries of light artillery; for a total of 17,945.[225]

Troops to Ship Island

On October 1st, the temporary Department of New England was created for Butler, at Boston, as a device for his recruiting efforts.[226] Planning for the southern expedition now began,[227] and an intermediate location for the troop buildup was chosen as Ship Island, which had been occupied by a token force of 170 Union sailors and marines since September 17th, four days after it had been abandoned by General Twiggs.

On December 3rd, 1861, the first of Butler's Massachusetts and Connecticut

---

223. O.R. Vol. 6, pp. 677–678.
224. O.R. Vol. 6, pp. 685, 687, 694, 695, 704. Only Chief of Staff Strong and Chief Engineer Weitzel knew the destination. Weitzel was recommended by the board of officers as an expert on the lower Mississippi.
225. O.R. Vol. 6, p. 688.
226. O.R. Ser. III, Vol. 1, p. 822; O.R. Vol. 51, p. 491.
227. Welles, *Galaxy*, November 1871, p. 673.

troops[228] arrived off Ship Island to begin its real occupation in any numbers. Of course, we have shown that Braxton Bragg was informed and there is no doubt that navy scuttlebutt at Pensacola made Battery L aware of it as well.

A Promising Outlook

Judah Benjamin's continued urgent calls for troops saw a massive buildup near the strategic point of Corinth, Mississippi, the junction of two important railroads in the Confederate defensive line.[229] Grant's Army of the Tennessee was directed by Halleck to follow, but not to engage, until Buell's Army of the Ohio should arrive.[230] The Union buildup took place about 20 miles north of Corinth, at Pittsburg Landing, on the west bank of the Tennessee River.

The Confederates were presenting a golden opportunity to the Union. They were concentrating in one place where they could be surrounded and beaten. However, General Johnston, aware of the imminent arrival of Buell's 40,000-man army,[231] decided to take the initiative, and launched a preemptive strike on Grant.

Johnston's 44,000-strong Army of Mississippi[232] – organized under Polk, Bragg, Hardee, and Breckinridge – left their entrenchments at Corinth and marched to engage Grant.

The Battle of Pittsburg Landing – or Shiloh, from the name of a log church that stood nearby[233] – was the "severest" according to Grant, of any fought in the west.[234] However, it was not decisive, because at a critical time, General Johnston was killed, causing confusion in the ranks, and a loss of initiative, which saved Grant. The Confederates withdrew back to Corinth. Now, the outlook was that Halleck's combined force would have little trouble in defeating Johnston's successor, Beauregard.

From the perspective at Washington, nothing looked more promising. Since January, there were the victories at Mill Springs, Kentucky, forts Heiman and Henry, Roanoke Island, and Fort Donelson; the occupation of Bowling Green, Nashville, and Columbus; the victory at Pea Ridge, Arkansas; the occupation of New Madrid, Missouri, and New Bern, North Carolina; victories at Kernstown, Virginia; Glorieta, New Mexico; the fall of Island No. 10; and victory at Shiloh.

---

228. O.R. Vol. 6, p. 466.
229. O.R. Vol. 10/II, pp. 33, 34.
230. O.R. Vol. 10/II, p. 41.
231. Grant, Vol. I, p. 332.
232. O.R. Vol. 10/II, p. 433. Derived from the Confederate return for April 15th. O.R. Vol. 10/II, p. 421, after the battle, but which shows 32,388 effectives. Assuming that 11,000 lost in the battle is correct, (Grant, Vol. I, p. 367) results in the 44,000 quoted here.
233. Grant, Vol. I, p. 338.
234. Grant, Vol. II, p. 356.

Farragut and Butler were poised to attack New Orleans and McClellan had stepped off on his vaunted Peninsular Campaign. It looked as though May 1862 would undoubtedly be the turning point of the war. In fact, the War Department issued General Orders No. 33 of April 3rd, 1862, directing that all volunteer enlistments be discontinued,[235] and on April 10th, Lincoln issued a Proclamation of Thanks to Almighty God for these "inestimable blessings."[236]

Unfortunately, Halleck would let a substantial strategic victory at Corinth slip from the Union's grasp, and McClellan's Peninsular Campaign would be a total failure. As a result, the country would be doomed to almost exactly three more years of warfare.

Louisiana

Early in April 1861, the governor of Louisiana had made note of the fact that no action to organize a protective force at New Orleans had been undertaken by the Confederate government. What was to happen if the city at the mouth of the strategic Mississippi was threatened? New Orleans was worth less than Pensacola?

Louisiana Governor Thomas O. Moore to L. P. Walker, the Confederate secretary of war, April 10th:

> The news from Washington creates considerable anxiety here. The ships sent from New York are believed to be destined for this place. [237]

The ships Moore had mentioned "from New York" were, of course, Brown's expedition to Fort Pickens, and an explanation to that effect was sent out by then Secretary of War Walker. It was also used to again remind Moore of the urgency of forwarding the remainder of the 1,700 Louisiana troops that had been requested for Confederate service in March. "Either send troops at once to Pensacola or call out volunteers for that service to fill requisition."[238]

The governor would have a year of anxiety before Farragut was, indeed, opposite the customs house; but in April 1861, a desperate Moore took the extraordinary action of enlisting men whom he knew the Confederate government would not demand be sent away. They were the men of the free colored population of New Orleans who were accepted into a militia unit known as the 1st Louisiana Native Guards.[239]

In May of 1861 Jefferson Davis offered somewhat of a palliative to Governor Moore and the citizens of New Orleans in the form of a "senior general,"

---

235. O.R. Ser. III, Vol. 2, p. 2, O.R. Vol. 8, p. 665.
236. O.R. Ser. III, Vol. 2, p. 14.
237. O.R. Vol. 53, p. 669.
238. O.R. Ser. IV, Vol. 1, pp. 134, 135, 175; 700 of the total were to garrison forts Jackson and St. Philip; O.R. Vol. 53, pp. 669, 670.
239. O.R. Vol. 15, p. 556; O.R. Vol. 53, pp. 669, 670.

David Twiggs, to supervise the construction of the city's defenses. Twiggs, at age 71 the oldest man in the Confederate service, was soon found wanting and was replaced in October by the respected Maj. Gen. Mansfield Lovell.

For the next four months Lovell's herculean efforts continued among repeated demands from Richmond for more troops and supplies. Now, in February, 1862 the Confederate War Department revealed that 5,000 more troops were needed for A.S. Johnston at Columbus, Kentucky, and that the policy decision had been made to abandon the coast. The extraordinary explanation was that: "New Orleans is to be defended from *above* by defeating them at Columbus; the forces now withdrawn from you are for the defense of your own command . . ."[240]

At the end of the month, Lovell reported that he had managed to comply. A total of eight regiments, two artillery batteries, a supply of 500 shotguns and 1,000,000 cartridges had been sent. He commented: "People are beginning to complain that I have stripped the department so completely, but I have called upon Governor Moore for 10,000 volunteers and militia for State service." Moore's task would be compounded by the fact that in January the state legislature had revised the Louisiana militia law. Going into effect on February 15th, it limited the militia to whites only, so the 1st Louisiana Native Guards had to be disbanded.[241]

A clearer picture of the Butler expedition having emerged after it left Hampton Roads on February 25th, Lovell took the same delusionary stance as had Bragg, i.e., that this was not a threat. Lovell's bravado bordered on hallucination: "A Black Republican dynasty will never give an old Breckinridge Democrat like Butler command of any expedition which they had any idea would result in such a glorious success as the capture of New Orleans." Governor Moore and the recently formed Citizen Defense Committee were, however, not hearing any of this nonsense and were distressed. "I had many unofficial conversations with General Lovell, and none of them inspired me with confidence in the safety of New Orleans, if vigorously attacked by the enemy" (testimony of A. D. Kelley, member of the New Orleans Citizen Defense Committee at the Confederate Court of Inquiry on the capture of New Orleans).[242]

On March 6th, a demand for 20,000 pounds of powder for Richmond brought Lovell's response that: "We have filled requisitions for arms, men and munitions until New Orleans is about defenseless."[243]

In early April, Lovell had to explain his situation to a new secretary of war, Randolph. The accumulation of drift on the river obstruction installed below Fort Jackson caused it to be carried away at the end of February, and a second one was put in place in March. This, consisting of schooners lashed together, was also

---

240. O.R. Vol. 6, pp. 823, 832, 847; author's italics.
241. Bynum, pp. 1, 14.
242. O.R. Vol. 6, p. 645.
243. O.R. Vol. 6, p. 841.

materially weakened and scattered by a storm on the night of April 11th.[244] They had a round of discussion about Lovell's need for large guns as well as small arms, which accomplished little. Next came the question of the potential saviors of New Orleans: the ram *Manassas*, and the ironclads *Mississippi* and *Louisiana*. The *Louisiana* would retain (they would not be made available to Lovell) all of its 16 guns, and when ready, soon after mid-April, it was ordered to *leave* New Orleans for Fort Pillow in Tennessee, to engage Foote's Western Gunboat Flotilla.[245]

Now with Farragut's ships in sight below forts Jackson and St. Philip, and his bombardment having begun on April 13th, Governor Moore sent a telegram to Jefferson Davis saying that it was "suicidal" to send the *Louisiana* up the river. Lovell also protested. Little did Navy Secretary Mallory realize that the deficient *Louisiana*, with its propulsion problems, would have to be towed up the river if it was ever to leave.[246]

By this time, New Orleans had an additional threat: the specter of starvation. A shortage of flour and basic provisions had been caused by the disruption of commerce due to the blockade, low water in the Red River, and "the want of communication by rail with Texas."

Farragut and Butler

David Glasgow Farragut, born in 1801 near Knoxville, Tennessee, joined the navy at the age of nine and, after more than 50 years of service, rising to captain, found himself at Norfolk at the outbreak of the war. Though a Southerner by birth, he remained loyal and, after serving briefly on the Naval Retiring Board, then specially constituted to rid the navy of unfit or disloyal officers, he was given the command of the West Gulf Blockading Squadron on December 23rd, 1861. Ordered to proceed to the Gulf, the expedition to capture New Orleans was his, and General Butler would join him with the supporting army troops.[247]

On the 25th of February 1862, Butler and a contingent of about 1,400 of his troops left Hampton Roads on the steamer *Mississippi*.[248] After passing through a severe winter storm and running aground off the Cape Fear Lighthouse, which created a serious leak in the ship's bow, a fuming Butler was not able to assume command at Ship Island until March 20th.[249] By the 30th, even though four

---

244. O.R. Vol. 6, pp. 512, 513, 523, 564.
245. O.R. Vol. 6, pp. 646, 873.
246. O.R. Vol. 6, pp. 877, 878, 879; ORN Ser. I, Vol. 18, pp. 346, 844, 845.
247. *Encyclopedia Britannica*, eleventh edition, Vol. X (1911), p.187; Welles, *Galaxy*, November 1871, pp. 679–683.
248. O.R. Vol. 6, p. 699.
249. O.R. Vol. 6, p. 704. He landed on the 21st; p. 708.

regiments and two artillery batteries of his force had not arrived,[250] he announced to Farragut that he could be at the mouth of the Mississippi[251] in 12 hours once notified.

Finally, by April 7th, Farragut notified Butler that he had his ships across the bar at the Southwest Pass and was loading supplies at Pilot Town. Butler was to come ahead at his own discretion.[252] He arrived with eight regiments and three artillery batteries on April 21st.

Though some of Farragut's ships had shelled the forts beginning on the 13th, Admiral Porter's mortar flotilla, consisting of 21 schooners and 6 supporting steamers, began a continuous bombardment of Fort Jackson on the 18th.[253] After four days there was no sign of surrender, and Farragut had had enough. On the night of the 20th, a demolition party was sent up to destroy the schooner obstructions across the river. Though the petards[254] set to blow it failed to detonate, in an act of great courage, the sailors manually cut loose enough of the chain to allow the current to open a passage sufficiently wide for the squadron to pass.[255]

New Orleans

On April 24th, at 3:30 a.m., 17 steamers of the squadron were seen to be moving, and by 4:40, 13 had passed the forts. Farragut described the passage as "one of the most awful sights and events I ever saw or expect to experience,"[256] a statement he would have to revise before the war was over.

A highlight was when the CSS *Manassas*, the ram that had so concerned Flag-Officer McKean back in October, was observed to be making a run at Farragut's venerable *Mississippi* (the oldest in the squadron, and the former flagship of Commodore Perry at Tokyo Bay in 1853). Undismayed, Farragut signaled Captain Smith to "turn and run her down." To avoid the *Mississippi*, the ram turned sharply and ran into the riverbank, and the *Mississippi* was able to pour two broadsides into her. Abandoned by her crew, she slid off the bank and drifted downstream in a sinking condition.

Later, during the surrender negotiations for the forts the few remaining Confederate naval personnel towed the *Louisiana* into the river from its refuge under the guns of Fort St. Philip and set her afire and adrift. Though the commander of the mortar flotilla, Porter, claimed that the burning hulk was aimed to come

---

250. O.R. Vol. 6, pp. 706, 707.
251. Leslie, Mrs. F., p. 444.
252. ORN Ser. I, Vol. 18, p. 521; O.R. Vol. 6, pp. 710, 876, 877.
253. ORN Ser. I, Vol. 18, p. 357.
254. Demolition charges, electrically ignited.
255. ORN Ser. I, Vol. 18, p. 156.
256. ORN Ser. I, Vol. 18, pp. 144, 152, 154.

down "upon" him, it blew up before it reached the *Harriet Lane*[257] – so much for another of Confederate Secretary of the Navy Stephen Mallory's ironclad saviors of New Orleans.

Lovell's evacuation of the city began on April 24th. He ordered that all of the steamers at the landing be detained until ordnance crews could load them with government stores.[258] However, the pilots and crews of nine or ten fled, leaving their vessels at the levee.

After an initial demand for surrender, Gen. Lovell declined, but restored the city authorities to power by revoking the martial law in effect and evacuated his 3,000 beleaguered troops, only 1,200 of whom were properly armed.

Farragut arrived off New Orleans soon after noon on the 25th. As soon as his ships were in sight, the last of Stephen Mallory's saviors, the *Mississippi*, unfinished on the ways at Algiers, was put to the torch.

New Orleans was occupied by Butler on May 1st 1862. Notwithstanding the complications of occupying a chaotic city of more than 160,000 people,[259] more than 20 percent of whom were not U.S. nationals, Butler took hold, with a sweeping proclamation of martial law.[260]

Vicksburg I

After the occupation of New Orleans, the next objective for Farragut and Butler's expedition was to "reach the invading forces from the upper Mississippi, under the command of Flag Officer Foote and General Halleck."[261] The naïve reasoning was that the only strong point on the river was Vicksburg and that it would be taken quickly.

To General Lovell's credit, he had taken the step of directing the remnants of his force to Vicksburg. Even before the surrender, he had telegraphed the following to Gen. Samuel Jones at Pensacola: "The enemy has passed our forts. It is too late to send any guns here; they had better go to Vicksburg."[262] He recognized that Vicksburg was the appropriate strategic defense point for the Mississippi Valley, and by May 5th, the 8th Louisiana Battalion and the 27th Louisiana Volunteers had arrived, and on May 12th, M. L. Smith, one of the engineer officers originally assigned to aid Twiggs in the construction of the defenses of New Orleans, arrived to take command. Already, three batteries of its defenses had been completed, and

---

257. ORN Ser. I, Vol. 18, pp. 250, 287, 288, 433, 434.
258. O.R. Vol. 6, pp. 568, 609, 717.
259. ORN Ser. I, Vol. 18, p. 234; U.S. Census Bureau.
260. O.R. Vol. 6, p. 717.
261. ORN Ser. I, Vol. 18, p. 132.
262. O.R. Vol. 6, pp. 515, 609, 624, 883.

a fourth begun.[263]

Of course, none of this was known to Butler or Farragut, and the plans to take Vicksburg went ahead. On May 2nd, Captain Craven, in the *Brooklyn*, and three other gunboats got underway. Without a pilot (the pilots were afraid of retribution if they aided the invaders), the *Brooklyn* ran aground some 60 or 70 miles up the river. Finally getting free, she was called back to Baton Rouge, which she had earlier bypassed.[264] Farragut had become worried about the navigation of such large ships on the river, an accurate assessment considering the difficulties to follow.

Some eight captured river steamers having been turned over to General Butler on May 7th, General Thomas Williams with the 6th Michigan Regiment, and six companies of the 4th Wisconsin Regiment got underway at New Orleans on the 8th.[265] Assigned by Butler to secure the Jackson Railroad as far as Manchac Pass,[266] Williams did not arrive at Baton Rouge until the 13th.

It was now learned that the Confederates had begun to fortify Vicksburg.[267] On May 16th, Commander Lee, in the *Oneida*, accompanied by five gunboats and two steam transports, with General Williams and his meager force of 1400 troops, departed Natchez for Vicksburg, arriving there on the 18th. Demanding its surrender, they were completely rebuffed by M.L. Smith.[268]

Finally arriving on the 20th, Farragut was disturbed by the situation. First, he blamed Lee for being slow to arrive. He was not happy that Williams' troops had run low on rations. The enemy had received Columbiads from Pensacola or Mobile and had placed three batteries below Vicksburg, and one above; 13 guns in all.[269] (There were six batteries completed by the 18th, per the report of M.L. Smith).[270] The bluffs being 200 or more feet high, Farragut's guns could not be elevated to fire on them. Further, the rebel tactic of drawing their guns to the edge of the bluff, firing, then drawing back out of view, made direct fire from the river useless. The mortar fleet would be required.

All of the ships having finally arrived in the vicinity by May 24th, a close reconnaissance of Vicksburg was made, which disclosed more guns and works. A final reconnaissance, made by Williams on the 25th, concluded that the only suitable place for a landing was at Warrenton, some eight miles below. He could

---

263. O.R. Vol. 15, p. 6.
264. ORN Ser. I, Vol. 18, pp. 467, 698.
265. Stanyan, p. 95.
266. O.R. Vol. 6, p. 507.
267. ORN Ser. I, Vol. 18, p. 478; O.R. Vol. 15, p. 7, Here it is indicated that they had begun as early as May 2.
268. ORN Ser. I, Vol. 18, p. 492. In three separate communications, by M. L. Smith, Commanding, J. L. Autry, Military Governor, and L. Lindsay, Mayor.
269. ORN Ser. I, Vol. 18, pp. 519, 703.
270. O.R. Vol. 15, p. 6.

not land and make his way to the batteries on the hill some 200 to 300 yards back from the river with his small force, now reduced by sickness, against an estimated 8,000 rebels.[271] To the question of an assault, there was a final unanimous "no" vote of the gathered army and navy officers. Even if they were aware of the position of the Northern Column at this time – some miles outside of Corinth – or of Admiral Foote, still above Memphis, they would have realized that there was no hope of gaining any support.

Farragut then decided to keep the *Iroquois, Oneida, Wissahickon, Sciota, Winona*, and *Itasca* on station as a blockading force and to harass the enemy by occasional bombardment. The river falling, the large ships and the troop transports would have to return downstream. Williams and the troops were back at Baton Rouge on the 29th.[272] The *Hartford*, with Farragut and the remainder of the flotilla, arrived at New Orleans on May 30th.[273] The six gunboats left behind began their bombardment on May 26th.

Failure at Corinth

The unconscionable performance of Halleck, in taking a month to arrive at Corinth and then letting the Confederate army escape, changed the momentum of the war in the west.

On closer scrutiny, Cullum's biography[274] of Halleck reveals that though he had served in the Mexican War, he did not serve in General Winfield Scott's command or in any major battles. At the outset, he was shipped from New York to California, and that alone took seven months. He did not arrive there until late in the year 1847 and was not involved in any fighting until November. As a 1st Lieutenant, he was involved in four skirmishes as the aide-de-camp of Navy Cdr. W. B. Shubrick, during operations on the Pacific coast of Mexico. Subsequent assignments were all as staff positions, most notably under the military governorship of California.

This man had never been in a field of battle and had never proven himself as a field commander in a war setting. His subsequent management of the advance to Corinth highlighted his unrealistic caution and lack of tactical sense.

Upon his arrival[275] at Pittsburg Landing from St. Louis on April 12th, which would be his first time at any frontline location, Halleck declared that Grant's

---

271. ORN Ser. I, Vol. 18, p. 706. An exaggeration, unless the total included those available on short notice, by rail, from Jackson.
272. O.R. Vol. 15, p. 22.
273. ORN Ser. I, Vol. 18, pp. 519, 707.
274. Cullum, Vol. I, no. 988, pp. 733–740.
275. O.R. Vol. 10/II, p. 99.

army, after the battle of Shiloh, was incapable of *resisting an attack*[276] and that it must be reorganized to make up deficiencies in men and equipment.

Grant describes the period up until April 30th as one of extensive preparation: "The roads toward Corinth were corduroyed and new ones made; lateral roads were constructed . . . . All commanders were cautioned against bringing on an engagement, and informed in so many words that it would be better to retreat than to fight."[277] Finally, the advance of the three armies – Grant's 50,554 strong Army of the Tennessee, Buell's 48,108 Army of the Ohio, and Pope's 21,510 Army of the Mississippi[278] – began on April 30th.[279]

From Grant: "The movement was a siege from start to close. The National troops were always behind entrenchments, except of course the small reconnoitering parties sent to the front to clear the way for an advance. Even the commanders of these parties were cautioned, 'not to bring on an engagement.' The enemy were constantly watching our advance, but as they were simply observers, there were but few engagements that even threatened to become battles. Roads were again made in our front and again corduroyed; a line was entrenched, and the troops were advanced to the new position."[280]

The investment of Corinth was not begun until May 28th. This vast Federal army had taken a month to advance the twenty-two miles from Pittsburg Landing to Corinth.

Beauregard had decided, however, to evacuate on the 26th. It was begun on the 29th and was completed on the 30th.[281] The Confederate army left "not a sick or wounded man . . . nor stores of any kind." Notwithstanding, on that day, Halleck announced in orders that there was the danger of a Confederate attack on Pope's front.

Beauregard's retreating army was chased by a detachment from Pope's forces, which consisted of the 2nd Iowa Cavalry and the 2nd Michigan Cavalry under the command of Col. W. L. Elliott, which briefly occupied Boonville, Mississippi, about 22 miles south of Corinth, on May 29th. They destroyed a portion of the Mobile and Ohio Railroad track which passed through Boonville, and burned some of its cars, which contained 10,000 stand of arms. They took some 2,000 "convalescent and sick of the enemy" into brief custody and "500 to 700" infantry prisoners, plus some 30 to 40 of their officers.[282] Learning that a large force of the enemy from Corinth was approaching Baldwyn and Guntown, they departed in haste without

---

276. O.R. Vol. 10/II, p. 106.
277. Grant, Vol. I, pp. 372, 373.
278. O.R. Vol. 10/II, pp. 146, 148, 151 (aggregate present).
279. Grant, Vol. I, p. 376.
280. Grant, Vol. I, pp. 376, 377.
281. Grant, Vol. I, p. 380.
282. O.R. Vol. 10/I, Elliott's and Sheridan's reports, pp. 862–865; Sheridan, Vol. I, pp. 147–149.

paroling the infantry prisoners and took away only the mounted officers.

Grant's comments on operations at Corinth: "For myself I am satisfied that Corinth could have been captured in a two day's campaign commenced promptly on the arrival of reinforcements after the battle of Shiloh." There was a pursuit of the retreating rebels, but it did not result in the capture of any numbers. It was back at Corinth by June 10th.[283]

Instead of the surrender of Corinth with the capture of upward of 50,000 prisoners, Halleck had burdened his army with occupying some real estate in a hostile territory, a fact little understood in Washington.

Beauregard having become ill, Braxton Bragg succeeded to the command of the Confederate Army of Mississippi, i.e. the bulk of the force that escaped from Corinth. He soon hatched a plan to try to gain control of Kentucky, and headed to Chattanooga. He left Van Dorn behind to oppose Grant.[284]

"Record" 6/62
30 APRIL–30 JUNE, 1862, PENSACOLA, FLORIDA

The Company was engaged in the bombardment of the rebel lines on the night of May 9th. Left Fort Pickens on May 10th, 1862 with the command, which recaptured Fort Barrancas. Marched on Pensacola – 9 miles – May 12th and went into quarters in the city.

1. Henry W. Closson   Capt. In Command of Company
2. Franck E. Taylor   1st Lt.
3. E. L. Appleton   1st Lt.

Detached:
> T.K. Gibbs 1st Lt. Prom. By virtue of letter A.G.O. Wash. April 21, '62. Assigned to duty with Co. A 1st Arty. Left Co. May 30, '62
> George Friedman Pvt. On det.svc. (as artillerist) at N.O. S.O. no. 11, West. Dist. Dept. of the South, Pensacola, Fla. May 24, 1862.

Strength: 86. Sick: 13.

Sick Present:
> Julius Becker, William Demarest, Robert Curran, Christopher Foley, Thomas Gilroyd, Dennis Myers, Michael Olvaney, John Roper, William Schaffer, Joseph Smith.

Sick Absent:
> Thomas Brook, Ft. Hamilton, NY, Left Company Sept. 17, 1861.
> William C. Brunskill   do.
> Charles Riley, Ft. McIntosh, TX Abs. since Feb. 21, 1861. Reduced to Pvt. July 12, 1861.

It is noted that the roll seldom indicates what the sickness was. To this point in time, we only know of "scourbutics" and dysentery as a cause of death, one in

---

283. O.R. Vol. 16/II, p. 33, Grant, Vol. I, pp. 381–383.
284. O.R. Vol. 17/II, pp. 614, 897.

each case. However, the increase in the number of sick from the previous roll would indicate the onset of warm weather related problems or perhaps indications that the water supply was becoming tainted on overcrowded Santa Rosa Island. Soon, the men would be relieved to be able to move to the healthier environs of Pensacola.

Occupation of Pensacola

On the night of May 9th, Fort McCree, buildings in the navy yard, the hospital, the barracks, and two steamboats, the *Mary* and *Helen*, were observed to be on fire.[285] Lumber had been piled in the casemates of McCree and set on fire rather than use precious gunpowder to blow it up. The same was allegedly done to Fort Barrancas, figure 1, but the fort suffered little, and survives to this day.[286]

FIGURE 1

On the morning of May 12th, a column consisting of the 6th New York led by their "Billy" goat mascot, labeled in red paint, five companies of the regulars, and the 75th New York was formed up at the outskirts of Pensacola. With the 75th Regimental Band at the head of the column, "the thrilling strains of Yankee Doodle"[287] were struck up, and a full dress marching entrance, white gloves and all, was made.

The city was described by Arnold as orderly and quiet and acting mayor Brosenham[288] "zealous and apparently loyal." About half of the buildings were burned,

---

285. O.R. Vol. 6, pp. 658, 660, 661; ORN Ser. II, Vol. 1, p. 89. The old side-wheeler, the *Fulton*, laid up at the navy yard at the time, was also destroyed. She had been considered for use by the Confederate navy but never fitted out.
286. Figure 1, photo by the author, 2008.
287. Hall, H. & J., pp. 41, 48. Henry Closson, (Haskin, p. 362).
288. O.R. Vol. 6, p. 659. From the diary of Mary E. Caro (display at the Wentworth Museum at

and many dwellings were deserted, the more militant of the inhabitants having fled along with the Confederate troops. "A good portion of the population that remained were loyally disposed Dutch."

The troops were delighted to be able to stand in the shade of pleasant flowered trees for the first time in months and to drink from a fresh water stream. Tents were pitched on fine wooden floors made from the "immense quantities of oak and pine lumber"[289] not destroyed by the retreating rebels. Battery L was located at the plaza at the city center, Captain Closson soon to be in charge of the defenses.

The 91st New York arrived on May 20th, having been transferred from Fort Taylor.

The rebel troops had evacuated to the railroad, at Oakfield, six miles north. Five companies of Confederate cavalry were detached there to stay in the area and "watch the enemy's movements."[290] This band of Alabama volunteers, or a part of it, remained on the outskirts of Pensacola for months. Seldom seen, they never attacked, but a number of reconnaissances were sent out to determine their strength and to engage them if possible. An expedition to Bagdad, that lasted from August 7th to August 10th, met four families who were delighted to escape the "tyranny of their oppressors," and were apparently taken back to Pensacola.[291] The tyranny alluded to might have consisted of demands for, or the theft of, food and forage by roaming "marauders" or "bushwhackers" pretending to be affiliated with the Confederate army. A considerable factor in this regard was the first Confederate Conscription Act of April 1862. Under this act, all white men 18 to 35 were required to serve. The age limit was soon raised to 45. In desperation for able-bodied men, the Confederate government allowed the conscription methods used to deteriorate into on-the-spot impressment. From the start, there had been rather lukewarm sentiment for the war among poor backwoods whites who had no slaves, and who had little connection with official government, whether State or the Confederacy. The families of these subsistence farmers would suffer if the men were dragged away, by the "gangs that . . . infest the country for the purpose of plunder and . . . enforcing . . . the Conscription Act."

Other reconnaissances accomplished little militarily, but what was discovered delighted the troops, such as "unguarded" watermelons, peaches, figs, and cattle.[292]

Pensacola), it is noted that Mayor Bobe had fled to Alabama on the 10th, and Union officers came as early as the 9th to raise the Union Flag. Porter, in his *Anecdotes and Incidents of the Civil War*, pp. 41–42, describes the "Mayor" as a cynical opportunist, earlier a close friend of Bragg's and today a true loyalist.

289. Hall, H. & J., pp. 42, 48.
290. O.R. Vol. 6, pp. 459, 662.
291. O.R. Vol. 15, pp. 126, 503; O.R. Vol. 16, p. 789.
292. Hall, H. & J., pp. 44–47.

"Record" 8/62
30 JUNE–31 AUGUST, 1862, PENSACOLA, FLORIDA

[No Entry in the Record of Events]

1. Henry W. Closson   Capt. In Command.
2. F.E. Taylor   1st Lt.
3. E. L. Appleton   1st Lt. On special duty as Bat. Adj. by virtue of O. no. 31 Hdqrts. Reg. Bat., Pensacola, Fla. Aug. 22, 1862

Detached:
> George Friedman Pvt. On det. svc. (as artillerist) at N.O. S.O. no. 11, West. Dist. Dept. of the South Pensacola, Fla. May 24, 1862.

Died:
> Patrick Carroll Pvt. Dec. 17,'60 Ft. Duncan, Tx. At Ft. Pickens, Fla. Aug. 19th 1862, of consumption.

Discharged:
> James McWaters Pvt. July 16,'57 Newport, Ky. By exp. of svc.; at Pensacola, July 16, 1862.
> Thomas Gilroyd Pvt. Aug. 26,'57   do.   Aug. 26, 1862.

Strength: 82. Sick: 6.

Sick Present: John Tomson

Sick Absent: Thomas Brook Ft. Hamilton, NY. Left Company Sept. 17, 1861, discharged on Surgeon's Certificate of Disability, July 25, 1862 (thus, he is counted as both sick, and discharged.)
> William C. Brunskill Ft. Hamilton, NY. Left Co. Sept. 17, 1862.
> Joseph Kutschor Ft. Pickens, FL. Left Company Aug. 3, 1862.
> Charles F. Mansfield Ft. Pickens, FL. Left Company Aug. 12, 1862.
> Charles Riley Ft. McIntosh, TX. Abs. since Feb. 21, 1861. Reduced to Pvt. July 12, 1861.

The drop in sickness since relocating to Pensacola is noted, but we must add another kind of disease to the list: tuberculosis. The list of those who have been transferred out due to serious long-term sickness and needing treatment in the north keeps edging up.

Reassignments

The occupation of New Orleans by General Butler's forces soon affected one of Battery L's men, Pvt. George J. Friedman. The 23-year-old goldsmith – an emigrant from Baden to New York, who joined Battery L in 1860 – was detached on May 24th, 1862. Sent to New Orleans as an "artillerist," he undoubtedly helped to fill Butler's need for people to train his green New England volunteers and additional units that would soon be recruited in New Orleans. The 1st Regiment of Louisiana Infantry (white) would begin recruitment in July, the Native Guards (free black) in August, and recruiting for the 1st Regiment of Louisiana Heavy Artillery (former slaves) had begun by the fall.

Friedman left Pensacola and never returned. He was not formally transferred out of Battery L until July 3rd, 1864,[293] when arrangements had been made to assign him as a second lieutenant in Company F of the 10th Regiment, U.S. Colored Heavy Artillery. He was mustered in on August 19th, 1864. He was promoted to first lieutenant on November 19th, after the 77th Colored Infantry was consolidated into the 10th Regiment. He had achieved the rank of brevet captain before being mustered out when the regiment was decommissioned on February 22nd, 1867. He and his regiment had served entirely in the defenses of New Orleans.

Arnold's requests for reinforcements were initially honored by the Department of the South. General Brannan at Key West had been ordered to transfer the 91st New York Infantry Regiment to Pensacola in May, as previously noted. However, the Army of the Potomac had better use for the regulars of the 3rd Infantry. Both Companies C and E were shipped out in June.[294] Arnold's request for a small steamer for transport between his supply base at Pickens and the mainland was honored when the *Creole* arrived on July 19th.[295] It came from the Department of the Gulf: General Butler. However, the flow of requests would quickly reverse. Butler was searching for reinforcements. He feared an attack on Baton Rouge. As it was, Butler's anticipation was correct. It came on August 5th.

On August 8th, the Western District of the Department of the South became a part of Butler's Department of the Gulf.[296] Halleck, general-in-chief since July 11th, as noted in the following chapter, had given Butler Pensacola in lieu of any additional reinforcements, which he claimed were not available. Butler immediately called upon Arnold to consult with him as to what troops he could spare. Needless to say, as a result, the 75th New York Infantry embarked on the *Ocean Queen* at Pensacola on August 30th, and arrived at New Orleans on September 2nd.[297] As somewhat in recompense, the 15th Maine Infantry Regiment was transferred from Camp Parapet, near Carrollton, above New Orleans, to Pensacola on September 11th.[298]

In the days soon after the occupation of Pensacola, Battery L's commander, Henry Closson, was named chief of artillery and given charge of planning and supervising the construction of the defenses of the city. However, as the work of construction ground on through the summer, the troops must have seen that

---

293. Pension record application no. 836550, National Archives; S.O. 200, A.G.O. Department of the Gulf. He is listed as being from Bavaria, according to his wife, who spelled her married name as Freeman.
294. They appear at White Post, Virginia, on June 19th, 1862. O.R. Vol. 11/I, p. 1,032.
295. O.R. Vol. 15, p. 526.
296. O.R. Vol. 15, p. 544.
297. Hall, H. & J., p. 56.
298. Shorey, p. 28.

Pensacola was to become a backwater in the strategic plan of the war. Pensacola was now a mere outpost of the Department of the Gulf rather than being the jumping-off point for a land thrust at Mobile, or a link-up with Union forces thrusting south. It was to be occupied initially simply to deny the Confederates the use of the harbor and to retain the Union hold on the navy base. However, after inspecting the base on June 2nd, the commander of the mortar flotilla, David D. Porter, wrote to the Secretary of the Navy Gideon Welles: "This place is so far superior to Ship Island that I would respectfully recommend removal of all naval property to this place."[299] The navy soon had ships and supplies back at Warrington. By mid-June, 33 schooners of the mortar flotilla were counted moored in the bay.[300] If Battery L had only recently regarded itself as stationed in a remote backwater of the war, it now found itself in a town bustling with sailors on shore leave with news from everywhere.

"Record" 10/62
31 AUGUST–31 OCTOBER 1862, PENSACOLA, FLORIDA

[No entry in "Record of Events"]
1. Henry W. Closson Capt. Sick
2. Franck E. Taylor 1st Lt. In temp. Command of Co.
3. Edward L. Appleton 1st Lt. Absent with leave at N.O. Left Co. Oct. 21,'62.
4. James A. Sanderson 2nd Lt. App. To Co. by prom. vice Gibbs G.O. no. 73, War Dept. A.G.O. Washington, July 4, 1862 (never joined Co.)

Detached: George Friedman Pvt. On det. svc. (as artillerist) at N.O. S.O. no. 11. West. Dist. Dept. of the South Pensacola, Fla. May 24, 1862.

Discharged:
Thomas Brook Pvt. Feb. 9,'60 Boston. Ma. by surgeon's list of disability at Fort Hamilton, New York, July 25, 1862.
William Schaffer Pvt. Jan. 8,'61 New York by surgeon's list of disability at Fort Hamilton, New York, Nov. 6, 1862.
Robert Curran Pvt. Oct. 6,'57 Newport, Ky. By exper. of service, at Pensacola, Fla. Oct. 6, 1862.
John Holland Pvt. Oct. 13,'57 Detroit, Mich. By exper. of service, at Pensacola, Fla. Oct. 13, 1862.

Temp. Attached:
1. William Bruce Pvt. June 30,'58 Ft. Independence, Ma. S.O. no. 73, Hdqrts. Troops in West Fla. Dept. of the Gulf, Pensacola, Fla. Aug. 30,'62
2. Sigmund Loeb Pvt. Dec. 5,'57 Ft. Snelling do.
3. Christian Allendorf Pvt. Sept. 29,'58 New York, NY do.

Strength: 80. Sick: 13.

Sick Present: Henry W. Closson, James Ahern, William Brown, Bernard Farrell, Charles F.

---

299. ORN Ser. I, Vol. 18, pp. 481–482.
300. Hall, H. & J., p. 51.

Mansfield, James McCarthy, Daniel McCoy, Thomas Newton, William E. Scott, Reuben Townsend, Michael White.

Sick Absent: William C. Brunskill, Ft. Hamilton, NY. Left Co. Sept. 17, 1861
Charles Riley Ft. McIntosh, TX Absent since Feb. 12, 1861. Reduced to Pvt. July 12, 1861

The increase in the number of sick may indicate seasonal related sickness, e.g., malaria, and sanitation problems with a now more crowded Pensacola. Regarding discharges, note that Robert Curran was the last of the three who had been recruited in 1857 at Newport, Kentucky. McWaters and Gilroyd left in July and August, respectively. Holland was another from the midwest who was apparently not interested in this war. The three may have missed the notification of the operation of General Orders No. 74 issued on July 7, 1862, and General Orders No. 154, issued on October 9th, offering a $100 bounty for new recruits as well as reenlisting soldiers.[301]

General Arnold was transferred out of Pensacola at the end of September to be the military administrator of New Orleans and Algiers. He was briefly replaced by the next highest-ranking officer, Col. William "Billy" Wilson of the 6th New York, who duly moved into the commander's quarters, the mansion of the ex-senator from Florida, Stephen R. Mallory.[302] Wilson was soon displaced by another commander, Gen. Neal Dow, who arrived in October.[303] The 7th Vermont Infantry arrived on November 14th. It had sweated and suffered in Louisiana the previous summer.[304] Another unit for the Pensacola rest camp.

On November 16th, the 6th New York boarded the steamers *Nassau* and *Creole*, bound for New Orleans. They were initially assigned to Thomas W. Sherman's division at Carrollton.

---

301. O.R. Ser. III. Vol. 2, pp. 187, 206, 654, 687.
302. Morris, p. 82. All references to the 6th here on pp. 80–84. There is reason to question this source. Babcock states, p. 63, that the commander's quarters were at the plantation of Major Chase.
303. Shorey, p. 28.
304. Benedict, Vol. 2, p. 46.

Chapter 6

Vicksburg II; Wavering in Washington; Halleck General-in-Chief; Baton Rouge; The Department of the Gulf; Banks; "Record" 12/62; The Close of 1862

Vicksburg II

Upon Farragut's May 30th return to New Orleans from the Vicksburg I episode, to his chagrin, he was met by a message of such urgency and import that the secretary of the navy had sent copies of it on three separate ships to ensure that it reached him. Dated the 19th of May, it said: "The President of the United States requires you to use your utmost exertions (without a moment's delay) to open the river Mississippi and effect a junction with Flag Officer Davis, commanding pro tem, the Western Flotilla."

Injured in the attack on Fort Donelson, Admiral Foote's wound continued to fester, and he was forced to relinquish command to Davis on May 9th.[305]

Next day, an apprehensive Farragut notified Porter to bring from six to ten of his mortar boats up from their anchorage at Pensacola. After all, the river was falling. His view was that his ships had barely made it out from the last trip. Porter responded on June 3rd that he was underway,[306] though, in his remarkably candid way, he offered the opinion that the whole expedition was better off being given up.

The June 7th entry of the diary of navy commander H. H. Bell[307] notes: "Troops went up the river to join those at Baton Rouge, for the combined attack upon Vicksburg. Talk of General Butler going up. Considered rash to take troops from city while those sent, 5,000 in all, are considered insufficient for the work. Mortar fleet arriving."

If Butler realized that the troops being sent were insufficient to capture and hold Vicksburg, a canal would bypass it; thus, the Mississippi would be opened anyway. General Williams' task upon arrival at Vicksburg was: "You will send up a regiment or two at once and cut off the neck of land beyond [opposite] Vicksburg

---

305. ORN Ser. I, Vol. 18, p. 502. Foote was replaced permanently by Davis on June 17th, ORN Ser. I, Vol. 23, pp. 86, 213.
306. ORN Ser. I, Vol. 18, p. 576.
307. ORN Ser. I, Vol. 18, p. 708.

by means of a trench across, thus . . ." He included a sketch in his written order. A later drawing appears as figure 1. The first mention of the Vicksburg canal appears in this directive, dated June 6th.[308]

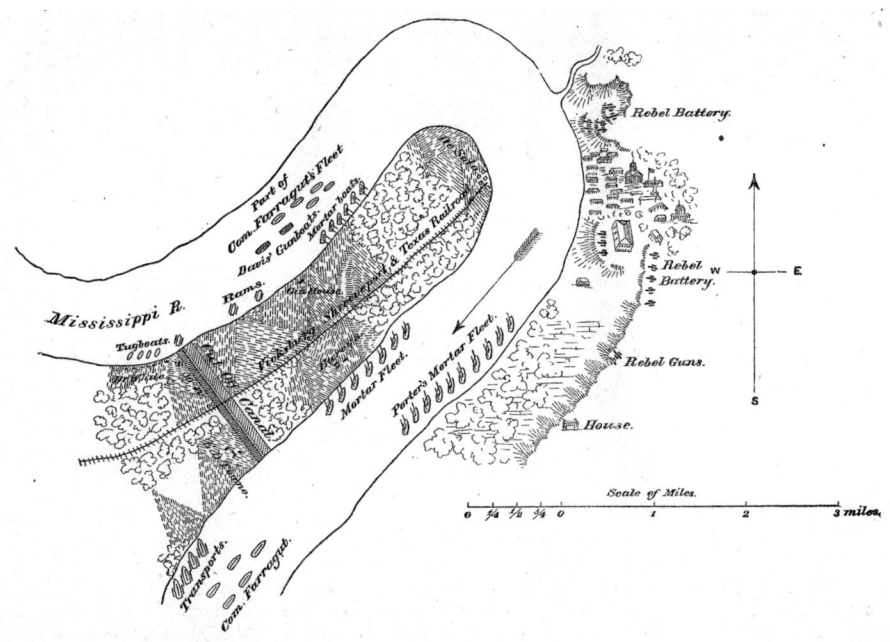

FIGURE 1

By June 25th, almost the entire expedition, including 16 of Porter's mortar schooners, had anchored seven miles below Vicksburg.[309] The troops were landed at Burey's Point, where the proposed trench was to be dug, with Nims', figure 2, and Everett's Batteries placed further east to support Farragut's run past to link up with Davis' Northern Column, proposed for the 28th.[310]

At 3:00 a.m. on the 28th, the squadron weighed anchor and began its run past Vicksburg. Supported by the mortar boats and fire from the two batteries, eight of the 11 ships had made it past by 5:30 a.m. Above, Farragut met the army's new ram fleet under the command of Col. Charles Ellet, who had been hunting the Confederate ram *Arkansas*. Dispatching letters to Davis and Halleck, Farragut reported his having run past Vicksburg, noting that the enemy now had "Van Dorn's division of Beauregard's army" there, "some eight or ten thousand troops,"[311] making any landing on the east side of the river impossible.

---

308. O.R. Vol. 15, pp. 25, 29.
309. ORN Ser. I, Vol. 18, p. 710. Figure 2, Miller, Vol. 2, Pt. II, pp. 180, 181. The author identifies Nims' Battery by their bronze guns.
310. ORN Ser. I, Vol. 18, pp. 711, 712.
311. ORN Ser. I, Vol. 18, Farragut's Report, p. 590; Davis' Report, p. 592. Beauregard had been replaced by Bragg on June 20th.

FIGURE 2. NIMS' BATTERY AT THE PENTAGON BARRACKS, BATON ROUGE
No similar "portrait" of Battery L has been found, though it would be nearly identical.

On June 28th, Davis, at Memphis, received the request from Farragut to aid him.[312] Davis would leave "at the earliest possible moment," which was the next day.

Davis hove into Farragut's sight at 8:00 a.m. on July 1st. The difference between the two flagships, Farragut's *Hartford*, figure 3, and Davis' *Benton*, figure 4,[313] pointed up Farragut's risk of a falling river. Though the two vessels were of similar length, the *Hartford* at 225 feet and the *Benton* at 202 feet, all similarity ended there. The draft of the *Hartford* was 17 feet and that of the *Benton* was only 9. Farragut mentioned that it was the furthest away from the ocean his ship had ever been.

FIGURE 3. USS *HARTFORD*

Spirits were high for a few moments until it was disclosed that Davis brought no troops – had none to bring – and that Farragut's strength, with Butler's force of 5,000, was insufficient even if it were not already depleted by sickness.

On June 30th, Farragut had sent another message, directed through Memphis[314] to Halleck, which said that it was "absolutely necessary to have additional troops to occupy Vicksburg

FIGURE 4. USS *BENTON*

---

312. ORN Ser. I, Vol. 23, pp. 231, 232.
313. Figure 3, *Hartford*; Leslie, Mrs. F., p. 440; Figure 4, *Benton*, ORN Ser. II, Vol. 1, p. 44a.
314. Grant, Vol. I, pp. 385, 386.

after the batteries are silenced."[315] On July 3rd, Halleck responded that: "The scattered and weakened condition of my forces renders it impossible for me at the present to detach any troops to cooperate with you at Vicksburg."

Halleck's paranoid caution spoiled the show. The curtain fell when word was received from Butler on July 16th ordering the troops to return. The enemy was reforming and an attack on Baton Rouge was likely.

Wavering in Washington

As far back as May 22nd, Farragut had received an inquiry as to whether the mortar fleet could better be used at Mobile.[316] He had to patiently explain, in his answer of June 30th, the importance of having it at Vicksburg. Now, on July 5th, he was ordered to send 12 of Porter's schooners to Hampton Roads.[317] Those in Washington (particularly Stanton) had become alarmed for the safety of the city after the impending failure of the Peninsular Campaign and the successes of now-famous Confederate Gen. Stonewall Jackson in the Shenandoah Valley. "How strange . . ." Farragut noted, sending some two thousand miles for mortar boats.[318] Further, he was still on the north side of Vicksburg and was worried about again running the gauntlet and extricating his ships from a river that had been falling for the past six weeks. On July 13th, he sent the following curt telegram (via Cairo on the 17th) to Gideon Welles: "In 10 days the river will be too low for the ships to go down. Shall they go down, or remain up the rest of the year?"[319] Apparently, Welles had anticipated his concern, and Farragut had been ordered down on July 14th, four days before his telegram left Cairo. Davis' flotilla was to remain.

So much time and opportunity had passed. Washington had now forgotten about the 19th of May order to Farragut: "The President of the United States requires you to use your utmost exertions (without a moment's delay) to open the river Mississippi and effect a junction with Flag Officer Davis . . ." By the end of June, it was clear that events elsewhere would cause the President and the cabinet to refocus their attention. "Elsewhere" was a few miles outside of Washington itself.

This was the period (March 11, 1862 to July 23, 1862) when Lincoln and Stanton were operating without a general-in-chief. It is an example of a period of anxiety and a lack of balance; Lincoln subjected to the full force of Stanton's volatility and often exaggerated fears.

The Army of the Potomac had moved from Fort Monroe in late March, first appearing before Yorktown. Fearful of making a direct attack on what he conceived

---

315. ORN Ser. I, Vol. 18, p. 593.
316. ORN Ser. I, Vol. 18, p. 591.
317. ORN Ser. I, Vol. 18, p. 629.
318. ORN Ser. I, Vol. 18, p. 632.
319. ORN Ser. I, Vol. 18, p. 594.

as a powerful Confederate force there (untrue), McClellan laid siege. This lasted from April 5th to May 4th when the Confederates slipped away toward Williamsburg. The Peninsular Campaign,[320] McClellan's planned advance to Richmond, continued with engagements at:

2. Williamsburg	3. Eltham's Landing	4. Drewery's Bluff	5. Hanover Court House
6. Seven Pines	7. Oak Grove	8. Beaver Dam Creek	9. Gaines' Mill
10. Garnett's Farm	11. Savage's Station	12. Glendale	13. Malvern Hill

Many of these were inconclusive, including the large battle (84,000 troops, total) at Seven Pines. The Confederate victory at Gaines' Mill on June 27th turned the tide, convincing McClellan to discontinue his advance. The final engagement, Malvern Hill, was fought on July 1st, 1862; and though technically a Union victory, it really only served to protect McClellan's withdrawal.

Halleck General-in-Chief

The failure of McClellan's campaign had become evident by the end of June, and concern for the safety of Washington only increased. At that time, Lincoln personally initiated correspondence with Halleck, asking for more troops. In the series of letters between them that followed,[321] Halleck refused to send any, and lectured Lincoln on the value of his army to the holding of west and southwest Tennessee, and that it should not be weakened . . . etc., etc. Rather than becoming enraged, Lincoln accepted the advice, and, in the words of his secretaries Nicolay and Hay:[322]

> . . . McClellan's reverses had been unduly exaggerated, and that by straining resources in the East the Western armies might be left undiminished. But with this conviction President Lincoln also reached the decision that the Richmond campaign must be remedied by radical measures. To devise new plans, to elaborate and initiate new movements, he needed the help of the highest attainable professional skill. None seemed at the moment so available as that of Halleck.

After all, at this time, Halleck had successes to his name like no other. He had gone to the west and made order out of the chaos of Fremont's old department, and had victories at forts Henry and Donelson, Pea Ridge, Shiloh, New Madrid, Island No. 10, and Corinth. Nevermind that the bulk of these victories were attributable to the initiative of subordinates, and that Corinth, under Halleck's direct control, was a disastrous non-event.

---

320. Williamsburg, May 5th – Malvern Hill reoccupied August 8th; O.R. Vol. 11/I, pp. 1–3. Seven Pines CWSAC VA 014, Gaines' Mill CWSAC VA 017, Malvern Hill CWSAC VA 021.
321. O.R. Vol. 17/II, pp. 53, 70, 71, 72.
322. Nicolay & Hay, Vol. 5, p. 355.

The president was influenced, and finally visited West Point, where he saw General Scott, who was known to favor Halleck.[323] As a result, Halleck was named to the command on July 11th, 1862.

Baton Rouge

On June 19th, Earl Van Dorn was assigned to the Confederate Department of Southern Mississippi and East Louisiana.[324]

Van Dorn had resolved to attack Baton Rouge as part of a plan to secure the Red River and recapture New Orleans. He sent Gen. John C. Breckinridge and his Kentuckians to unite with a small force under Gen. Daniel Ruggles at Tangipahoa, and then proceed, on July 30th, to attack Baton Rouge.

Due to "appalling" sickness, the actual strength of Breckinridge's force was reduced from about 4,000 to 2,600 on the early morning of August 5th, when the attack began.[325] It was repulsed by the hard, face-to-face, often at 40 yards or less, fighting of the units of the garrison, some 2,500 in number. They held off the attackers despite their sickness (many left the hospital to fight) and General Williams' poor defensive preparations (regiments beyond supporting distance from one another and few preparations such as earthworks).[326] The sharpshooters of the 6th Michigan, though limited in their traditional role by the morning fog, may have made a surprise impression on the close action by bringing to bear their Henry repeating rifles, able to fire 15 shots in as many seconds.

The *Arkansas*, an ironclad ram, recently constructed nearby, and in which great faith was placed by the Confederates, was ordered from Vicksburg to support the attack. According to Van Dorn's report, "Broke machinery five miles above Baton Rouge. On way down was attacked by enemy. In this condition fought well, inflicting great damage to gunboats, and was then blown up by crew, all of whom escaped."[327] Failure of the *Arkansas* to support the attack was concluded by the Confederate side as the reason for its failure.

Butler's fear of another attack at first drove frantic efforts to improve the defenses of the town, but fear of an attack on New Orleans prompted his order for a withdrawal on August 16th.

---

323. Nicolay & Hay, Vol. 5, p. 86; O.R. Vol. 17/II, p. 90.
324. O.R. Vol. 17/II, pp. 612, 613, 616, 617.
325. O.R. Vol. 15, p. 77, Pirtle, SHS, Vol. 8, pp. 327, 328. The quote "appalling" is Breckinridge's comment. The force began with 4,000 men, about one-third of them with no shoes. Water on the march from Tangipahoa was scarce and the heat intense. Irwin, *Military Operations in Louisiana*, Vol. III, pp. 582–583.
326. McGregor, p. 422.
327. O.R. Vol. 15, p. 14; ORN Ser. I, Vol. 19, pp. 137, 138.

The Department of the Gulf

Only ten days before Arnold and Battery L moved into Pensacola, Butler had occupied New Orleans and found it to be in desperate condition. The blockade had reduced commerce and food was scarce. On May 3rd, he authorized the safe conduct of food from various sources within the Confederacy[328], and requested of the administration that the blockade be lifted, which was done on May 12th.[329]

Butler's administration was going to be humane, but tough and efficient, as promised in his declaration of martial law on May 1st.[330] Due to disorder and neglect, the city was filthy and in danger of disease. Plans were put forward to employ the starving to clean up.

The initial response to all of this sweeping efficiency was positive.[331] The New Orleans *Bee* noting on May 8th: "The federal soldiers have done nothing . . . to provoke the difficulty . . . . The city is as tranquil and peaceable as in the most quiet times."

However, the initial shock of the Yankee occupation of the largest city in the Confederacy having worn off, and with full stomachs, diehards began to make trouble. The Union soldiers were cursed and spat upon, mostly by women. On May 15th, after Admiral Farragut had a chamber pot emptied upon him, Butler issued his famous woman order, in which if any woman showed contempt for any soldier of the U.S. ". . . she shall be regarded and held liable to be treated as a woman of the town plying her avocation." It had no specified punishment other than the stigma. It worked.

That settled, the Mumford incident had to be dealt with. Farragut had raised the Stars and Stripes on the staff of the U.S. Mint on April 27th, and Mumford had subsequently torn it down. He was placed on trial and found guilty by a military commission on June 5th.[332] Butler refused clemency, and Mumford was hanged on June 7th.

The outside world, particularly those British merchants who were poised to make fortunes in running the blockade, all reacted with feigned outrage. Eventually, Jefferson Davis joined the crowd, and in a proclamation, declared Butler a felon, and reserved him to be executed if captured.

The only part of this nonsense that was detrimental to Butler was in how members of Lincoln's administration reacted to it. A fatal move involved the investigation of the activities of the foreign consuls, whom Butler would soon require to take an oath of neutrality and surrender their arms. In one case, his men discovered

---

328. O.R. Vol. 7, pp. 720–722.
329. O.R. Vol. 7, p. 723; O.R. Ser. III, Vol. 2, p. 31.
330. O.R. Vol. 7, p. 717.
331. Butler, pp. 395, 417, 418.
332. Nicolay & Hay, Vol. 5, pp. 278, 282.

$800,000 in Mexican coin in the liquor store of the Netherlands consul, suspected to be ready for shipment to the Confederacy.[333] In another case, it was discovered that the "British Guard" had sent their arms and uniforms to Beauregard after New Orleans fell.

All of this concerned the State Department and Secretary Seward, who, fearful of offending the foreign powers, promptly sent a representative, Reverdy Johnson, "that Baltimore secessionist" to look over Butler's shoulder. He, "in every case" reversed Butler.[334] Others sent by Treasury Secretary Chase reported regularly. Their suspicions of the activities of Butler's brother (referred to as Col. Butler) gradually grew, and two of the last from George Denison are quoted here:[335]

December 10:
> An occurrence has just taken place which causes me to feel much indignation and some chagrin. Col. Butler has three or four men in his employ who manage his business for him. The principal one is a Mr. Wyer. Some days ago Wyer loaded a vessel for Matamoras. She was loaded in the New Basin, and when she got into the Lake, ran into Ponchitoula . . . I am satisfied that it was a predetermined plan to take the cargo to . . . the rebels . . . I am also satisfied that Col. Butler was the sole owner . . .

December 23:
> I do not know your opinion and wishes concerning Gen. Butler, but it is certain that his removal gives great satisfaction to all classes including officers, soldiers and citizens.

Banks

Banks was one of the first to offer his services to the President and was appointed a major general on May 16th, 1861. First in command at Annapolis, he was credited with keeping Maryland in the Union, though we know from Chapter 3 that Butler had also taken significant and decisive actions there. Banks later commanded two divisions in the Shenandoah Valley in the spring and summer of 1862.[336]

When Pope's Army of Virginia was consolidated with McClellan's Army of the Potomac on September 5th, Banks was given the command of the defenses of Washington. He was also named as commander of the 12th Corps, though Brig. Gen. Alpheus Williams is listed as serving in Banks's place.[337] Under Banks's brief tenure, order was restored in Washington with thousands of wounded, convalescents, and stragglers placed appropriately, in a full army corps, which was sent to

---
333. O.R. Vol. 7, p. 72; Vol. 15, p. 422; O.R. Ser. III, Vol. 2, pp. 116-125.
334. Butler, p. 522.
335. Chase, Vol. II, pp. 310-342.
336. O.R. Vol. 12/II, p. 133; CWSAC, VA 022.
337. O.R. Vol. 19/I, pp. 1,157.

the field. The fortifications were completed, etc. This all attributed to a man still recovering from injury.

On October 27th, he was relieved and ordered on a confidential assignment to New York, where he made his headquarters. The pattern of recruitment for a special expedition, much as that of Butler, was followed. Banks was to lead a "southern expedition"[338] to renew the effort to open the Mississippi from that quarter.

On November 8th,[339] he was assigned to replace Butler in command of the Department of the Gulf. Banks' first instructions from Halleck, the day after he was assigned,[340] were to open the Mississippi River: "A military and naval expedition is organizing at Memphis and Cairo to move down the Mississippi and cooperate with you against Vicksburg and any other points which the enemy may occupy on that river." After the capture of Vicksburg, Banks was to, first, destroy the railroads east of there to cut off all connection by rail to Mobile and Atlanta and, second, ascend the Red River as far as it was navigable and "thus open an outlet for the sugar and cotton of Northern Louisiana." This would be a good base for operations in Texas, for Banks was to command not only the Department of the Gulf, but Texas as well.

By November 1st, Banks reported that he now had three Maine regiments, that Governor Andrew of Massachusetts had assigned the 41st Massachusetts and seven other nine-month militia regiments to the expedition, and that three New York regiments had been assigned.[341] On the 7th of the month, 10,000 troops assembling at Fort Monroe were assigned to Banks by Halleck. By November 16th, Halleck told Banks to send the troops from Fort Monroe without waiting for those from New York. Banks sailed on Thursday, December 4th.

Underway, sealed orders were opened, and the destination of his immense fleet was revealed as Ship Island. Banks carried with him[342] Gen. Andrew J. Hamilton who had been appointed as the military governor of Texas, whose task was to "re-establish the authority of the Federal Government in the State of Texas" and who was authorized to raise two regiments of Texas Volunteers. Banks was to support him "by a sufficient military force, to be detailed by you from your command . . ."

At Ship Island, Banks gave instructions to the fleet to proceed to New Orleans. Banks and the transports left on Sunday, December 14th, arriving that evening, figure 5.[343] He immediately met with Butler and delivered Halleck's order relieving him of command. Butler issued his gracious farewell address the next day.

---

338. O.R. Vol. 19/I, p. 3; O.R. Ser. III, Vol. 2, p. 691.
339. O.R. Vol. 15, pp. 3, 610–613; Irwin, pp. 59, 60.
340. O.R. Vol. 15, p. 590.
341. O.R. Vol. 15, pp. 705, 706, 712, 713, 784, 879, 913, 955.
342. O.R. Ser. III, Vol. 2, pp. 782, 783, 913.
343. O.R. Vol. 15, pp. 613, 191, 609. Figure 5 from *Harper's Weekly*, January 10th, 1863, p. 21.

ARRIVAL OF THE TRANSPORT "NORTH STAR," WITH MAJOR-GENERAL BANKS AND STAFF, AT THE LEVEE AT NEW ORLEANS.—[SEE PAGE 27.]

FIGURE 5. ARRIVAL OF BANKS AND STAFF AT NEW ORLEANS ON THE *NORTH STAR*

Banks delivered an introductory letter from President Lincoln to Farragut on the morning of the 15th, noting that he was delighted with Farragut's cordial promise of cooperation in the enterprise that had been outlined by Halleck's orders.[344] Farragut's report of the meeting notes that it was he who had recommended the occupation of Baton Rouge, which was agreed to by Banks, who assigned the 4,500 effectives of Grover's command to the task.

Arriving at daybreak on the 17th, four companies of the 26th Maine Regiment were the first to land "and drive the rebs out of the city."[345]

Thus, Banks' first military initiative was a complete success. The second, to Galveston, spoiled the initiation. Never mind that Hamilton was impatient to get to Galveston; Farragut was also interested. He had requested troops from Butler to reinforce it soon after its capture;[346] and with the threat of it being attacked by the aggressive new Confederate commander of the district of Texas, New Mexico, and Arizona, John B. Magruder, he was more concerned than ever. After consulting with both Farragut and Butler, Banks agreed to send troops. The lead group, consisting of three companies, a total of 15 officers and 249 enlisted men of the 42nd Massachusetts Infantry, had already arrived at New Orleans and encamped at

---

344. ORN Ser. I, Vol. 19, pp. 342, 409, 415, 417, 418.
345. Maddocks, pp. 18, 19; Tiemann, p. 17.
346. ORN Ser. I, Vol. 19, pp. 300, 319. Magruder: assigned October 10th; O.R. Vol. 15, p. 826. Assumed command November 29th, p. 880.

Carrollton, but were ordered away and landed at Galveston on the 24th.[347]

Early on the morning of the first of January, they were attacked by a force commanded by Magruder that was estimated at 3,000 or more. The supporting troops for the 42nd, arriving on another steamer the next day, were too late. The entire detachment of the 42nd had surrendered. Being hurried out from New Orleans, they were ill-equipped and too few. Of the gunboats of the squadron, the *Harriet Lane* had been boarded and captured, and the *Westfield* had been grounded and then abandoned and blown up, killing Commodore Renshaw and four others.[348]

Banks did not officially announce taking command of the Department of the Gulf until the 17th, when his staff assignments were announced. The organization of the units in the department as of the end of December 1862 is shown in table 1.[349] Those shaded were newly arrived as a part of the expedition. The remainder were Butler's old units or those recently raised by him. The 1st Texas Cavalry and Louisiana Artillery are listed, raised by Butler in New Orleans too late in November to be involved in his Lafourche campaign or in the reinforcement of Galveston.

GROVER'S DIVISION

13th Connecticut	52nd Massachusetts
24th Connecticut	6th New York
25th Connecticut	91st New York
2nd Louisiana	131st New York
12th Maine	133rd New York
22nd Maine	159th New York
26th Maine	161st New York
41st Massachusetts	173rd New York
174th New York	2nd Massachusetts Battery
4th Wisconsin	1st U.S. Artillery, Battery L
1st Louisiana Cavalry, Co. C	2nd U.S. Artillery, Battery C
2nd Mass. Unattached Cavalry, Co. B	

W. T. SHERMAN'S DIVISION

26th Connecticut	114th New York
14th Maine	156th New York
31st Massachusetts	160th New York
42nd Massachusetts	162nd New York
6th Michigan	2nd Massachusetts Cavalry, Co. A
15th New Hampshire	18th New York Battery
16th New Hampshire	1st Vermont Battery
110th New York	

---

347. O.R. Vol. 15, pp. 200–211; O.R. Vol. 26/I, pp. 5–7; Bosson, pp. 63, 72, 74, 86, 109.
348. ORN Ser. I, Vol. 19, pp. 458, 460.
349. O.R. Vol. 15, pp. 627, 628.

RESERVE BRIGADE

12th Connecticut
1st Louisiana
75th New York
1st Louisiana Cavalry, Co's A & B

2nd Massachusetts Cavalry, Co. B
1st Maine Battery
6th Massachusetts Battery

DEFENSES OF NEW ORLEANS

9th Connecticut
26th Massachusetts
30th Massachusetts

1st U.S. Artillery, Batteries A & F
5th U.S. Artillery, Battery G

DISTRICT OF PENSACOLA

28th Connecticut
15th Maine

7th Vermont
2nd U.S. Artillery, Batteries H & K

INDEPENDENT COMMANDS

23rd Connecticut
21st Indiana
1st Louisiana Native Guards
2nd Louisiana Native Guards
3rd Louisiana Native Guards
13th Maine
8th New Hampshire

128th New York
165th New York
177th New York
8th Vermont
1st Texas Cavalry
Louisiana Artillery (one company)
4th Massachusetts Battery

TABLE 1

Counting regiments only, Banks had brought, up to this date, 25, more than doubling the strength of Butler's old command. More units, both veteran and freshly recruited, would continue to arrive in the next few weeks. The final tally would total 39 infantry regiments, six artillery batteries, and three cavalry units.[350] Note that Battery L appears as assigned to General Grover.

"Record" 12/62
31 OCTOBER–31 DECEMBER 1862, BATON ROUGE, LOUISIANA

Embarked Battery on board steamer "Che Kiang," left Pensacola Dec. 24th – reached New Orleans Dec. 25th and received orders to proceed to Baton Rouge. Reached that place Dec. 27th and took up quarters in the State Penitentiary. Company serving as Mounted Artillery with Battery of: 4 light twelves, 2 ten lb. Parrot rifles

1. Henry W. Closson   Capt. In Command
2. Franck E. Taylor   1st Lt.
3. Edward L. Appleton   1st Lt. Absent on det. svc. At Pensacola – left Co. Dec. 24,'62

---

350. O.R. Vol. 15, pp. 712–714.

4. James A Sanderson   2nd Lt. App. to Co. by prom. vice Gibbs G.O. no. 73 War Dept. A.G.O. Washington July 4, 1862 (never joined co.)

Detached:

Edmond Cotterill	Pvt. Abs. on det. svc. at Pensacola. Fla. As clerk in A.G.O. Abs. from Co. since Dec. 24, 1862.
George Friedman	Pvt. Abs. on det. svc. at N.O. (as artillerist). Left Co. May 24, 1862.
Franklin W. Richards	Pvt. Absent, confined at Ft. Pickens, Fla. Since Nov.10, 1862.

Transferred:[351] Charles Riley, Pvt. Trans. (from abs. sick) to Co. F 1st Art. U.S.A New Orleans, La.

Discharged: William Wynne, Pvt. Oct. 9, 1862 By Expiration of Service
Michael Harrington Pvt. Dec. 23,'62 Pensacola, Fla. Discharged as a minor and returned to his Company "I" 15th Me. Vols.

Strength: 126. Sick: 8.

Sick Present: Charles A. Flint

Sick Absent: William C. Brunskill Ft. Hamilton, NY. Left Company Sept. 17, 1861.
Charles F. Mansfield Left sick at Pensacola since Dec. 24, 1862.
Wallace D. Wright Left sick at Pensacola since Dec. 24, 1862.

The ratio of those sick to total strength is noted as being the lowest since leaving Texas.

Horses: Serviceable: 88, Unserviceable: 0

Temporarily Attached: not included in Strength:

William Bruce,	Extra Duty – Ord. Dept.	Co. C, 2nd Arty.
Sigmund Loeb,	" " " "	Co. A, 1st Arty.
Christian Allendorf	" " " "	Co. F, 1st. Arty.

Joined:

47 Volunteer unit Privates enlisted in Battery L at Pensacola, 30 in November and 18 in December. $100.00 Bounty Due.

1. Allen, James H.	A 91st New York	25. McEnearny, Cornelius	A 91st New York
2. Baker, John	A 91st New York	26. Mahoney, Thomas	F 15th Maine
3. Card Rowland	A 91st New York	27. Moore, Churchill	H 91st New York
4. Campbell James	F 15th Maine	28. Meese, Christian	A 6th New York
5. Champion, Henry	H 91st New York	29. Miller, John	A 6th New York
6. Comfort, James	B 6th New York	30. Montgomery, Solomon V	I 91st New York
7. Crowley, William	I 15th Maine	31. Morgan, Frank	H 91st New York
8. Chase, George	A 91st New York	32. Orcutt, Ephraim	H 91st New York
9. Deering, John	F 15th Maine	33. Parks, William	F 15th Maine
10. Deal, Charles	C 91st New York	34. Pelky, Henry	C 91st New York
11. Flynn, Arthur	A 6th New York	35. Parslow, Joseph	I 15th Maine
12. Fudge, William	C 15th Maine	36. Ranahan, Michael	F 15th Maine
13. Gibbon, Patrick	I 91st New York	37. Stewart, William	I 15th Maine
14. Hubbard, Hiram	H 91st New York	38. Smith, Hiram	A 15th Maine
15. Hall, Benjamin O.	H 15th Maine	39. Smith, James H.	C 15th Maine
16. Harrison, George	C 91st New York	40. Smith, William H	A 15th Maine

---

351. The muster roll record is unreadable, though the "Strength" section lists one private transferred.

17. Jessop, Francis	B 6th New York	41. Walton, Charles E.	6th New York
18. Kilburne, Sirenus T.	I 7th Vermont	42. Welsch, Peter	H 91st New York
19. Kelly, George	A 91st New York	43. Wilder, Joshua E.	H 7th Vermont
20. Kelly, John	A 91st New York	44. Wilson, Thomas W.	C 91st New York
21. Lewery, John	A 15th Maine	45. Winn, Abram F.	A 91st New York
22. Lashner, Joseph	H 91st New York	46. Winn, Joel T.	A 91st New York
23. McKenney, John	I 15th Maine	47. Woods, John C.	A 91st New York
24. McGinnis, Angus	F 15th Maine		

Having absorbed 47 new recruits since the end of October, Battery L had reached its highest strength ever, achieving close to "Full War Organization," with six guns (two 10 pounder Parrott's and four 12 pounder Napoleons), and 86 horses. The "Record of Events" section duly notes that they were now serving as a "mounted" battery – supporting infantry. The addition of the horses required new "Extra Duty" assignments. Luckily, the skills required were found in the newly recruited members, who were from mostly rural locations:

1. Henry Champion, blacksmith since Feb. 5th, 1863, recruited from the 91st New York Infantry, Company H, which he had joined at Hillsdale, New York, on October 5th, 1861.

2. John Deering, artificer since Dec. 24th, 1862, recruited from the 15th Maine Infantry, Company F, whose men hailed from either Cumberland or Aroostook counties. Deering's position as artificer was additional to that of Isaac T. Cain, who remained separately listed as artificer on the muster roll.

3. George F. Hadley, blacksmith since Dec. 24th, 1862, the only original "Regular."

4. Sirenus T. Kilburne, blacksmith since Jan.1st, 1863, recruited from the 7th Vermont Infantry, Company I, which he had joined at the picturesque south-central town of Manchester, on Feb. 7th, 1862.

5. Frank Morgan, saddler since Nov. 14th, 1862, named the same day that he was recruited from Company H of the 91st New York Infantry; which he had joined at Albany. His quick appointment indicates that his skills, and at age 32, his maturity, were desirable assets. (Referred to as Harness Maker in the December 1862 Monthly Report.)

When Enos Deal, of C company, 91st New York, stepped up to Lieutenant Appleton, Battery L's recruiting officer, he may have been hoping for a new beginning. He gave his name as Charles, no. 10 in the list. The "new" Deal did not last long; he deserted on September 15th, 1863, just as Battery L was leaving New Orleans.

The recruits from the volunteer units all were due a $100 bounty, which had been originally planned to be paid upon their honorable discharge.[352] Recognizing that this incentive was too distant in time to be effective, many, including most of the state governors, proposed that one-fourth of the bounty be paid up front. This was so ordered on July 1st, 1862, but additional recruits taken into Battery L at Baton Rouge in February 1863 were not paid up front, despite this order.[353]

Regular army recruiting of those already serving in volunteer units was specifically authorized by the issuance of General Orders No. 154 dated October 9th, 1862, by the adjutant general's office.[354] It ordered that all batteries, battalions, and regiments appoint "one or more recruiting officers, who are hereby authorized to enlist, with their own consent, the requisite number of volunteers to fill the ranks of their command to the legal standard." The day previous to the day the volunteer enlisted in the regular army, he was to have been considered as honorably discharged from his volunteer unit. The volunteer unit commander had no choice in the matter, which was objected to by the state governors.[355]

The mysterious case of Sgt. Charles Riley must now be closed. Carried on the rolls as "sick at Fort McIntosh, Texas; Absent from Company since Feb. 25, 1861" for almost two years, and in the interval reduced to a private, he now appears as having been transferred to Company F of the 1st Artillery at New Orleans.

Fort McIntosh was abandoned on March 12th, 1861, according to the report of the commander, Maj. C. C. Sibley. The report refers to a post return, which had been "herewith, inclosed,"[356] but is missing from the official records. The post return is, however, available from the National Archives.[357] Examination of the February return shows under "Those Casually at Post" Riley, Charles, Sgt. 1st Arty L 26 Feb'y 1861. He does not appear on the final, or March return, cut short, as it was, to the 12th. This means that he was shipped out before the fort was abandoned and was therefore not a part of Sibley's command when it left.

Riley's pension record[358] has supplied some answers. Riley did not apply for a pension until 1882. Because the time interval from the event to the time the pension application was made had been 22 years, memories had faded and people had died. Affidavits were required by the Pension Bureau for proof of service. The adjutant general's office confirmed his enlistment and service in Batteries L and F and his prisoner-of-war record. Col. Richard C. Jackson, who commanded Battery

---

352. O.R. Ser. III, Vol. 1, p. 382, Act of Congress, July 22nd, 1861. Worth $13,000 if calculated for a private's monthly pay in 2019.
353. O.R. Ser. III, Vol. 2, p. 187.
354. O.R. Ser. III, Vol. 2, p. 654.
355. O.R. Ser. III, vol. 2, pp. 694, 737.
356. O.R. Vol. 1, pp. 560–561.
357. National Archives Microfilm Publication M 617, Returns from U.S. Military Posts, 1800–1916, M 671D, roll 681.
358. Case no. 476558 National Archives, Civil War Pension Files.

L at the time of the march out from Fort Duncan, couldn't remember Riley at all, much less the incident at Willow Pond, Texas, where Sergeant Riley was thrown from his horse, resulting in a severe multiple fracture of his right leg and injury to his ankle. Fortunately, the Fort McIntosh post surgeon, Lt. Charles C. Byrne, remembered Riley's arrival there and that he had set his leg. "Flat on his back," Riley was not abandoned at the fort; but after a private collection from members of the battery, he was left with provisions and money for his support, in the care of an apparently loyal private family in Laredo. All this was to no avail; he was taken prisoner on April 15th, 1861.

As was described in Chapter 2, those enlisted men taken prisoner were sent to Camp Verde and then dispersed to different posts. Riley mentions being at Ringgold Barracks and Camp Cooper, as well as Verde. He remained a prisoner until his "escape" on October 19th, 1862. He gave no details of his return to New Orleans, though a prisoner exchange took place at Vicksburg in December. He was assigned to Battery F at the end of October, and Battery L records finally caught up with that fact in December. He was described as "lame" by one witness. He had been partially disabled since his accident. Nevertheless, he was accepted into service in Battery F upon his return from prisoner of war status, as he wrote, "for the reason that any man who could mount a horse was not sent to hospital."

A subsequent report of the involvement of Battery F in the engagements at Fort Bisland and at Jeanerette, Louisiana,[359] in April 1863, gives mention of those who were distinguished in battle. Sgt. Charles Riley is among those listed.

For those at McIntosh who had been solicitous of Riley, judging it best to take up a collection to provide for his care and to leave him with a "loyal" family, in hindsight, was a mistake. Remember from Chapter 2 that the Fort McIntosh command and others under Major Sibley were taken prisoner, but paroled, and left Texas at the end of April. If they could have only predicted the deprivations and exposure that the still-injured Riley would suffer as a prisoner for 19 months, they might have caused him to suffer the jolts and jounces of their travel to be rid of Texas promptly.

The Close of 1862

The "record of events" section statement closes the history of Battery L at Pensacola. The *Che Kiang*, a relatively new side-wheeler built for the China trade,[360] had arrived from New Orleans on December 22nd, bringing the 28th Connecticut Regiment to replace the 91st New York. Embarking on the 24th, Battery L, along with the 91st New York, would have arrived at Balize, on the passage up the Mississippi, on

---

359. O.R. Vol. 15, p. 337.
360. *New York Times*, November 9th, 1862. Compiled Service records, 91st New York, National Archives, M594, roll 128, p. 21.

Christmas morning. As it happened, this was the same day that the 15th New Hampshire Regiment, one of Banks' nine-month militia units, passed up the river. Their observations and those of another volunteer unit, the 49th Massachusetts, provide us with impressions of the scene:[361] "... dawn found us on deck, gazing at Secessia" ten feet below them from the height of the levee. As they passed forts Jackson and St. Philip, they were "greeted with songs by the colored people on the shores and the waving of bandannas." They marveled at the rice, cane and cotton fields, the magnolias, and the "prolific orange ... groves, bending with their golden fruit" that they saw near Quarantine. "Now, green fields ... palatial mansions, and slave cabins fully occupy our attention." It was a pleasant time of the year. The weather made things upbeat, except for "the sight of the contrast between the homes of the planters and of their laborers is suggestive of the fact that, in the everduring struggle between the privileges of the few and the rights of the many, victory had here crowned the wrong."

FIGURE 6. CAPITOL, 1862

Captain Closson remembers that upon arriving at New Orleans at 4:00 p.m. on the 26th that they had lain on board at the foot of Canal Street "looking up its long vista of lights."[362] In contrast, a member of the 49th Massachusetts saw nothing but "deserted rotting levees, frowned on by numerous gunboats, everything to denote war and its destructiveness, nothing to convince us that we were before the great metropolis of the Southwest. Divers pieces of calico and a few carriages reminded us of the friends and comforts of the North. Peddlers brought us pies, apples and oranges: of the latter you could get three as large as pound apples for 10 cents. You had better believe that those who had any money left rapidly invested it in this inviting stock."

L and the 91st soon received word to proceed to Baton Rouge, the state capital. Approaching from miles away, on December 28th, they could see this "beautiful Southern capitol ... situated on ... a plateau, the first bluff above the Head of the Passes." The capitol building[363] presented "a fine appearance ... its massy snow white-walls and towers remaining intact ..." figure 6.

The 110 members of Battery L aboard the steamer that day were among the

---

361. McGregor, p. 197; Johns, H. T., p. 135.
362. Haskin, *Closson's Memoirs*, p. 363; Johns, H. T., p. 136.
363. Leslie, Mrs. Frank, p. 476.

last to see it in its nearly original condition for 20 years.[364] Late in the afternoon, it caught fire and was gutted, only the walls remaining. An investigation revealed that the cause was a defective flue, not the action of the Union troops.

Disembarking across the derelict steamer *Natchez*, which was used as a wharf, Battery L took up residence in the state penitentiary, figure 7.[365]

FIGURE 7

When General Banks took over the command of the Department of the Gulf, he initiated a review of the troop dispositions at Pensacola. A recommendation was made to abandon the city and withdraw to the peninsula on which stood the small communities of Woolsey and Warrington, the navy yard, and Fort Barrancas. Bounded by Bayou Grande to the north, and Pensacola Bay to the east and south, the only access to it was a road bridge from Pensacola or the land approach from the west.[366] If the bridge was destroyed, a defensive line from Fort Barrancas to the advanced redoubt was considered adequate.

After Halleck had approved, the plan was carried out. The evacuation took place between March 17th, and March 23rd, 1863.[367] Of the 1,100 or so residents, about 1,000 Union sympathizers who feared Confederate retribution, or conscription, were transported out. The Spanish consul, Francesco Moreno[368] remained, and under his "flag of neutrality," the 70 or so remaining citizens hoped to be afforded protection from the "marauders" alluded to in Chapter 5.[369]

---

364. O.R. Vol. 15, pp. 630–633. Maddocks, p. 19.
365. Tiemann, p. 17; Figure 7, G. H. Suydam Collection, Mss. 1394.
366. O.R. Vol. 15, pp. 677, 1,109.
367. O.R. Vol. 15, p. 699.
368. O.R. Vol. 15, p. 1,036. Moreno was the father-in-law of the Confederate Secretary of the Navy Stephen A. Mallory.
369. Shorey, pp. 33, 34.

Chapter 7

*"Record" 2/63; The Nineteenth Army Corps;
Engagement at Bayou Teche; Grant; The Mississippi Marine Brigade;
Return of the 42nd; Indianola Lost; Banks' Plan;
Farragut Passes Port Hudson*

"Record" 2/63
31 DECEMBER 1862–28 FEBRUARY 1863, BATON ROUGE, LOUISIANA

Record of Events: [This portion of page missing].
1. Henry W. Closson   Capt. Commanding
2. Franck E. Taylor   1st Lt. sick
3. Edward L. Appleton 1st Lt. Joined from det. svc. Feb. 22, 1863 (Acting Ordnance Officer, Pensacola since Dec.24, 1862)
4. James A. Sanderson 2nd Lt. Appointed to Co. by promotion/ Gibbs prom. G.O. no. 73, War Dept. Washington July 4, '62 (never joined Co.)

Joined: Edmond Cotterill Pvt. From det. svc., Jan.13, 1863 (as clerk in A.G.O. Pensacola, since Dec. 24, 1862)

Detached: George Freidman Pvt. On det. svc. (as Artillerist) at N.O. S.O. no.11, Hdqtrs. West. Dist. Dept. of the South, Pensacola, Fla. May 24, 1862

Strength: 149. Sick 14

Sick Present: Franck E. Taylor, Charles A. Flint, Benjamin O. Hall, George Harrison, Daniel McCoy, Denis Myers, Angus McGuiness, John G. Nitschke, Ephraim Orcutt, Joshua E. Wilder, Thomas M. Willcox.

Sick Absent:
William C. Brunskill, Fort Hamilton, NY Left Company Sept. 17, 1861
Charles F. Mansfield, Sick at Pensacola, Fla. since Dec. 24, 1862
Wallace D. Wright   do.

Horses: Serviceable: 86 Unserviceable: 8

Volunteer unit Privates enlisted at Baton Rouge in February, 1863; $100. Bounty due:

1. O'Brien, Sholto   Co. B, 13th Conn. Vols.   13. Kenny, Theodore W   Co. C, 131st New York Vols.
2. Breen, Michael   Co. G, 174th New York Vols.   14. Kastenbader, John M. Co. F, 13th Conn. Vols.
3. Bieber, Peter   Co. K, 13th Conn. Vols.   15. Leonard, George F.   Co. D, 174th New York Vols.
4. Brooks, William   Co. G 174th New York Vols.   16. Lowry, John   Co. F,   "   "   "
5. Clinton, Thomas   Co. K, 13th Conn. Vols.   17. Mansfield, Herbert B. Co. H, 13th Conn. Vols.
6. Cook, Charles   Co. B,   "   "   "   18. Moore, Daniel   Co. B, 174th New York Vols.
7. Dickson, Clark   Co. B,   "   "   "   19. Mint, William   Co. F, 13th Conn. Vols.

8. Eisle, Joseph	Co. K,	"	"	"	20. Moran, John H.	Co. H, 2nd Louisiana Vols.	
9. Foote, Edward A.	Co. B,	"	"	"	21. Pfiffer, George	Co. F, 13th Conn. Vols.	
10. Harrington, Jas. R.	Co. E,	"	"	"	22. Tieghe, Michael	Co. F,	" " "
11. Howard, Daniel	Co. C,	"	"	"	23. Woodruff, Lyman	Co. E	" " "
12. Hughs, Benjamin	Co. B,	"	"	"			

As to their health while at Baton Rouge, both Battery L and Nims' 2nd Massachusetts Battery could look to the services of Assistant Surgeon William Y. Provost of the 159th New York Regiment.[370]

The Nineteenth Army Corps

On January 7th, Banks reported to Halleck that correcting matters regarding the abuses of trade by private parties under Butler had occupied much time, which had interfered with military matters. He noted that the troops were "not in condition for immediate service." Banks also noted that he had no heavy guns, as would be required to assault Port Hudson, "whose works . . . have been in progress many months and are formidable." On that date, there were reported to be 10,700 Confederate effectives at Port Hudson, counting Gregg's Brigade, which had been ordered to report to General Gardner. Gardner had replaced Ruggles in command there on December 29th.[371]

A substantial difficulty was a lack of sufficient transport steamers to take advantage of the flooded condition of the country north and west of New Orleans and pursue any plan to approach Port Hudson from the river systems on the west side of the Mississippi. Only seven steamers were available by February 12th, when plans for a campaign were outlined to Halleck.[372]

Banks' command was not completely manned until February 11th. Having taken so much time to be mustered, equipped, and transported to the scene of operations, some of the nine-month militia regiments would have only until May 1863 before the expiration of their service. All of those 22 regiments whose service term was nine months would have expired by August. Thus, Banks was under pressure to accomplish his mandate to open the Mississippi in short order. He appreciated the dilemma, and as early as April, he began to request replacements.

Examples of administrative problems in the Department surfaced by mid-January; horses and even food, shoes, and clothing, were scarce; and Banks felt that he needed more cavalry.[373]

Back on November 22nd, Lincoln had warned Banks about horses and of the danger of employing too much "*impedimenta*," which

---

370. Tiemann, p. 19.
371. O.R. Vol. 15, pp. 640, 641, 913, 933.
372. O.R. Vol. 15, pp. 24, 241.
373. Tiemann, p. 21. O.R. Vol. 15, pp. 618, 619, 647, 671, 688, 691, 1099–1105.

. . . has been, so far almost our ruin, and will be our final ruin if it is not abandoned . . . You would be better off anywhere, and especially where you are going, for not having a thousand wagons to feed the animals that draw them, and taking at least two thousand men to care for the wagons and animals who otherwise might be two thousand good soldiers.[374]

This was a rare example of an angry Lincoln. Before he left, Banks had made requisitions for such a large amount of supplies that Lincoln felt it could not be filled "for an hour short of two months. I enclose you a copy . . . in some hope that it is not genuine – that you have never seen it."

Banks' whining seems to have prompted Lincoln to wonder if he had made a mistake in relieving Butler. Butler's commission as a major general had not been revoked, and the issue of his replacement was still apparently a topic at Washington. On January 8th, Senator Sumner, the Republican from Massachusetts, wrote to Butler[375] that he had seen the President, who had said that "he hoped very soon to return you to New Orleans." A letter from Lincoln to Stanton on the 28th suggested it.

All this was followed by a formal offer on February 11th, to return to New Orleans "at my request, for observation." Butler did not have to refuse, he alleges,[376] because Seward had tendered his resignation over the matter, and Lincoln had to back down.

The military situation did not cease to need attention. On January 6th, Maj. Godfrey Weitzel described that Confederate Gen. Richard Taylor's (son of former President Zachary Taylor) forces had fallen back across Berwick Bay, a wide spot in the Atchafalaya River, 80 miles west of New Orleans, to an earthwork called Fort Bisland. They had beaten the naval force sent there to trap them and had taken the steamer *J. A. Cotton* up Bayou Teche.[377] The navy had remained in Berwick Bay ever since, protecting "my left flank," but if Admiral Farragut removed them, New Orleans would be in danger. Farragut was indeed planning to remove them to serve as replacements for the *Harriet Lane* and *Westfield*, lost at Galveston. Control of the gunboats had fallen to the navy by a July 16th Act of Congress, the news of which had just filtered down.[378]

Weitzel, West Point class of 1855, was well known to Battery L as an engineer at Fort Pickens. He had arrived with the Brown expedition in April, but had moved on in September having been called into the planning for the Butler expedition. Having worked as assistant engineer on the defenses of New Orleans from

---

374. Nicolay & Hay, Vol. VII, Ch. 11, pp. 312, 313.
375. O.R. Vol. 53, pp. 546–548.
376. Butler, pp. 569, 570.
377. Taylor, p. 120; Cullum Vol. II No. 1676.
378. O.R. Ser. III, Vol. 2, pp. 227, 644; O.R. Vol. 15, pp. 637, 638, Congressional Globe, 37th Congress, 2nd series, p. 411.

1855 to 1859, he was qualified to help plan the attack on the city. He had remained on Butler's staff since, and had planned and commanded Butler's campaigns in the LaFourche district, west of New Orleans.

Eight months of Butler's rule had made little impression on the greater part of the state, only occupying 13 of 48 parishes, what with the small force he could bring to bear. Though he had to give up Baton Rouge, he had extended Union control as far west as Berwick Bay, despite resistance by Taylor.

Taylor had been appointed to the command of the District of Louisiana on August 20th, only days after Baton Rouge was abandoned. Initially given the assignment of recruiting troops for the eastern campaigns, the Confederate War Department soon expanded his role, with orders to "confine the enemy to the narrowest limits and recover lost ground, if possible."

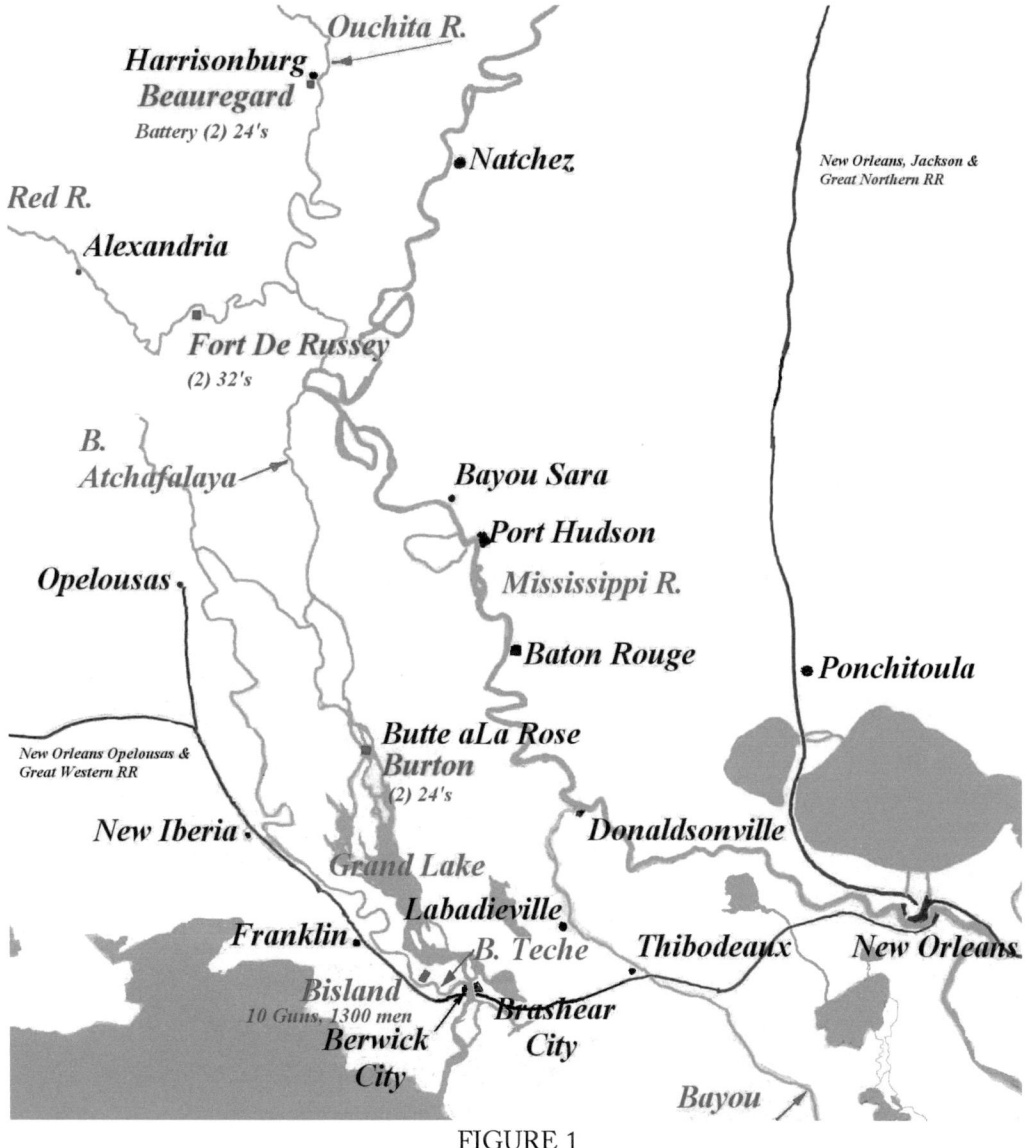

FIGURE 1

Writing to the Confederate adjutant general at the end of December, Taylor reports having enrolled 3,000 men, 1,000 of whom were now serving at Port Hudson. The number of conscripts brought to his camps of instruction were described as "small, and these have to be hunted down by detachments from the small command which I have at my disposal and brought in tied and sometimes ironed" – an embarrassing revelation about the lack of enthusiasm of the people of the Attakapas country for the war.

As of January 1863, he reports his strength as 7,233.[379]

Taylor had been constructing a series of forts, Bisland, Burton, DeRussy, and Beauregard, figure 1 (prepared by the author), for the purpose of denying the navigable watercourses west of the Mississippi to the Federals. These "forts" were earthen water batteries, manned by small crews. There were 60 men at Butte-la-Rose, and at Beauregard and DeRussy there were from 50 to 100. The largest was Fort Bisland, on Bayou Teche, which was supported by the gunboat *Cotton*.

At all but Bisland, which had ten, there were but two guns, all salvaged from the waters of Barataria Bay and Berwick Bay, where they had been thrown after the fall of New Orleans. All of this now confronted Banks.

Engagement at Bayou Teche

Weitzel's description of the situation resulted in Banks' assigning more reinforcements to him. A strengthened Lafourche District, coupled with the completion of fortifications at Brashear City on the Atchafalaya and Donaldsonville at the mouth of the Lafourche,[380] would ensure the safety of New Orleans' approaches while Banks' main force at Baton Rouge focused on attempts to open the Mississippi.

A fresh campaign to weaken Taylor and maintain firm control in the west was necessary. On January 13th, at 3:00 a.m., Weitzel's troops, consisting of six of his volunteer regiments, elements of four batteries of his artillery, and Barrett's Company B of the 2nd Louisiana Cavalry, began being ferried across Berwick Bay by the navy. The whole force was disembarked and formed up at Pattersonville. After waiting for the gunboats to make a reconnaissance, they advanced upstream about two miles to Lynch's point and went into camp under the cover of the gunboats. They were just within sight of the *Cotton*, lying a short distance above the earthworks of Fort Bisland and above a floating bridge, where obstructions and torpedoes had been placed in the Teche near the Cornay residence, figure 2.[381]

---

379. O.R. Ser. IV, Vol. 2, p. 380; O.R. Vol. 15, pp. 874, 919.
380. Fort Butler was not completed until January 31st, 1863, O.R. Vol. 15, pp. 163, 240.
381. Figure 3, Atlas, portion of plate 156. O.R. Vol. 15, pp. 234–237; Taylor, R., pp. 120, 121; ORN Ser. I, Vol. 19, pp. 337, 516.

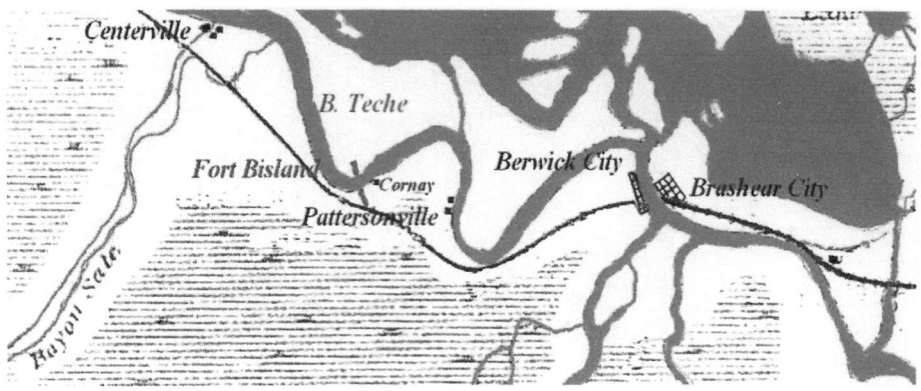

FIGURE 2

Next morning,[382] 60 volunteers were assigned from each of the 8th Vermont and the 75th New York to act as sharpshooters, who were to move up to the *Cotton* and shoot down her gunners. The regiments were then to follow. Navy Lieutenant Commander Buchanan's supporting gunboats, led by the *Kinsman*, moved upstream with the advance of the regiments; and with enfilading fire from Bainbridge's, Carruth's, and Bradbury's artillery, the enemy was driven from his positions, first on the west side and then on the east. The gunboats continually fired into the *Cotton*, which withstood the onslaught "for some time" until it backed out of range. It returned, only to back out of range once again.

The *Kinsman* had a torpedo explode under her stern, causing her to back out of the action. Buchanan was killed by a rifle bullet from shore as the *Calhoun* advanced forward to replace the *Kinsman*. All told, there were 3 killed and 6 wounded on the *Calhoun* and 1 on the *Kinsman*, none on the *Estrella* or the *Diana*.[383] Weitzel reported 6 killed and 27 wounded.

Next morning, the *Cotton* was seen to be on fire, swung across the bayou as to form an obstruction. At this, Weitzel ordered the entire expedition returned to Camp Stevens near Thibodeaux. A successful result might have fairly been predicted considering that Weitzel brought some 6,000 men, 18 field guns,[384] and a company of cavalry – all supported by gunboats – against 1,300 men behind the Confederate works. However, the Confederates did not retire, and the *Cotton*'s two guns were salvaged and mounted on the west side of the Teche.[385]

Banks quickly wrote to Halleck "of the complete success of the expedition,"[386] and it was gratifying news to receive, following on the heels of the Galveston

---

382. O.R. Vol. 15, pp. 234, 235.
383. ORN Ser. I, Vol. 19, pp. 517–520.
384. O.R. Vol. 15, p. 234: (4) of Battery A, 1st Arty, (2) of the 4th Massachusetts, all of the 1st Maine, and all of the 6th Massachusetts (each having 6 guns if properly equipped).
385. Taylor, pp. 120, 121.
386. O.R. Vol. 15. pp. 233, 234. Dated January 16th.

episode, though in reality it failed to weaken Taylor's forces or his resolve.

Farragut learned of the action and the death of Lieutenant Commander Buchanan on January 14th, from a telegram from the captain of the *Estrella*.[387] Impressed by the severe action, he dropped plans to redeploy the gunboats.

Grant

Now would begin an exploration of the water routes in the country west of Port Hudson as a way to skirt around it to unite with Grant above and then turn to its reduction.[388]

Grant's situation was a mystery, and a message just received from Halleck, dated January 4th added nothing: "You will learn from the newspapers that our last advices General W.T. Sherman has his hands full at Vicksburg." Banks was already aware that Sherman had attacked Vicksburg on December 29th and by his own admission had "failed." Further details of that attack might indicate some idea of Grant's future plans, which would likely not repeat that bloody frontal assault, but there was no communication from him.

Communications were poor. Any from Grant could not come by telegraph. Two hundred miles of enemy territory intervened. Communication by courier, whether by land or riverboat, could hardly be contemplated for much the same reason. Grant could telegraph from Memphis to Washington, but Banks' communications with Washington were by steamship, at least ten days to two weeks.

The Mississippi Marine Brigade

The takeover of Ellet's army riverboats by the navy in July may have been revealed by Farragut in the context of his need for replacements for those ships lost at Galveston, but certainly other navy developments were not of concern so far south as here – or were they?

The Western Flotilla, which had been under the command of Admiral Davis since just before the futile meeting with Farragut above Vicksburg on July 1st, was regarded by Welles as needing a more aggressive leader. There was no senior officer that seemed to fit the bill, save one junior officer with "great energy" but who was "reckless, improvident…too presuming and assuming…" It was David Dixon Porter.

Porter, revealed to us by the opinion he gave to Farragut at the time of the second Vicksburg expedition in chapter 6, may be indicative of Welles' estimate of Porter's "…presuming and assuming…" Nevertheless, Welles' promoted him over the heads of many, to vice rear-admiral. It was an "experiment."[389]

---

387. ORN Ser. I, Vol. 19, pp. 515, 518, 519, 525, 526.
388. O.R. Vol. 15, p. 655.
389. Welles, Vol. 1, pp. 167, 180; Society of Survivors, pp. 129, 138.

Only a few weeks passed before Porter demanded control of the Ellet Ram Fleet, the last vestige of the army gunboats on the upper Mississippi. In a heated cabinet meeting in November, Stanton, who had resisted the July Act of Congress, lost and Porter won. He immediately renamed and strengthened the fleet, directing that 1,500 men be added as a support unit. It was to be known as the Mississippi Marine Brigade.

Though little of the strengthened force had materialized in early February, Porter ordered strikes by Ellet on enemy shipping. A successful expedition had captured two ships bringing supplies to Port Hudson as well as having destroyed the Confederate steamer *City of Vicksburg*. A second strike, in which Ellet's powerful ram, *Queen of the West*, was to join with the ironclad *Indianola*, figure 3, was ordered on February 10th. Without waiting for the *Indianola*, the young colonel in command, Charles Ellet's son – and as equally rash as Porter – entered the Red River, looking for the many steamers known to use that critical Confederate supply route. He ran up to Gordon's landing about one hundred miles above the mouth of the Red.[390] On rounding a bend, as the *Queen* approached Taylor's Fort DeRussy, she grounded. A sitting duck, she was shot up and disabled, with many of Ellet's men taken prisoner.

The fatal result was that the abandoned *Queen* was soon repaired, and now flying the Confederate flag, was ready for its own hunting expeditions. The first victim was Porter's *Indianola*, but all of this was up in Grant's command, unknown to Banks, and apparently of no consequence to the operations in the Department of the Gulf.

FIGURE 3

---

390. Fort De Russy is the "Fort Taylor" referred to by both Ellet, Porter, and in some Confederate correspondence. To add to the confusion, it was also referred to as Norman's Landing. ORN Ser. I, Vol. 24, p. 378.

Return of the 42nd

There was some good news. The prisoners taken at Galveston, plus a contingent of 278 men of the 8th Infantry with their wives and children, who had been surrendered in Texas in 1861, plus 21 of the crew of the *Queen of the West*, arrived at Baton Rouge on the Confederate steamer *General Quitman* on February 24th.

Those of the 42nd, along with their navy comrades from the *Harriet Lane*, had been taken to Houston and confined. Though ordered out on January 22nd, their route through Beaumont and up the Sabine River to Burr's Landing, Louisiana, and then Alexandria, had taken until February 13th. News that the *Queen of the West* had entered the Red River briefly delayed their journey, until it became known that she had been captured. Paroled on the 18th and 19th, the prisoners boarded the *Quitman* on the 23rd for the final leg. Sadly, the black crewmembers of the *Lane* remained held in the Texas state prison at Huntsville.[391]

Indianola Lost

The fearsome Union ironclad was lost on February 24th[392] having been rammed seven or eight times by a repaired *Queen of the West* and another Confederate gunboat, the *Webb*. Though an ironclad, and not vulnerable to the light guns of the Confederate ships, she was towing two coal barges which slowed her and thus made her vulnerable to ramming. She was in "almost powerless condition" when she ran into the bank below Warrenton and surrendered.[393]

By the afternoon of March 2nd, Porter's initial telegram, sent on the 27th from Memphis, had reached Secretary Welles. It mentioned only that the *Indianola* "had fallen into the hands of the enemy."[394] The reaction was sober:

> The disastrous loss of the Indianola may, if she has not been disabled, involve the most serious results to the fleet below. Without due knowledge of all the circumstances under which you are placed at Vicksburg, the Department is not prepared to give a positive order, but rather suggests that a sufficient number of ironclads be sent to destroy her or ascertain her fate. The Department has no means of notifying the fleet at New Orleans.

The "no means of notifying" is a clue as to how upset Welles was. There was, of course, the same means as always, by ship – ten days or more – clearly unthinkable.

---

391. Bosson, pp. 173–193; O.R. Ser. II, Vol. 5, p. 397.
392. ORN Ser. I, Vol. 24, pp. 390, 393.
393. Figure 3, *Harper's Weekly*, February 7th, 1863, p. 84.
394. ORN Ser. I, Vol. 24, p. 388.

Banks' Plan

Having completed his organization, the next day Banks wrote to Halleck giving the details of his broad plan of attack. It would consist of a large force sent north through the watercourses to the west of the Mississippi, the Atchafalaya, and Bayou Plaquemine to attack Taylor's fort at Butte la Rose and gain control of the country opposite Port Hudson. Then, Admiral Farragut "will attack the works on the river, and will probably run the batteries with one or more vessels, placing us in communication with the forces above." The land force at Baton Rouge would then attack Port Hudson from the rear.

Unfortunately, reconnaissance found that the proposed watercourses west of Plaquemine were blocked by miles of drift, and a revised plan, described in a letter to Halleck dated February 21st, used an altered route "via Berwick Bay and Grand Lake."[395] All was set, and the troops that had been drilling for weeks were ready to step out on the grand mission for which they were sent to Louisiana.

But wait – only six days later, all the preparation for Banks' plan was on hold. After hearing the news of the *Indianola*,[396] Banks again succumbed to Farragut's entreaties and postponed his own plans.

Farragut Passes Port Hudson

Admiral Farragut had learned of the *Indianola* disaster from "secesh" newspapers. He had been moored off New Orleans since returning from Pensacola on November 13th,[397] at that time, concerned about Butler fulfilling his promise to hold Galveston. His presence at New Orleans[398] had allowed close consultation with Butler and now with Banks, and he was therefore on top of the local news.

The West Gulf Blockading Squadron would once again step into the Mississippi above Port Hudson. In a report to Welles dated March 3rd, Farragut announced that he was "all ready to make an attack on or run the batteries at Port Hudson, so as to form a junction with the army and navy below Vicksburg.[399] The army of General Banks will attack by land or make a reconnaissance in force at the same time that we run the batteries." Banks' force would divert the attention of the Port Hudson garrison and reduce – or eliminate – its fire on Farragut.

Though Farragut had been under orders to guard the lower part of the river, "…

---

395. O.R. Vol. 15, pp. 242, 243, 248, 249, 1104; Taylor, p. 127.
396. O.R. Vol. 15, pp. 1104–1106. Banks had little choice. The navy was a critical element in the security of New Orleans, and without close and friendly cooperation – recall Weitzel's earlier concern about Farragut removing the gunboats from Berwick Bay – life would have been infinitely more difficult.
397. ORN Ser. I, Vol. 20, p. 6.
398. ORN Ser. I, Vol. 19, pp. 255–260, 344; O.R. Vol. 15, p. 201.
399. ORN Ser. I, Vol. 19, p. 644.

especially where it is joined by the Red River, the source of many of the important supplies of the enemy,"⁴⁰⁰ the *Indianola* was an emotional thing for Farragut. As described in the *History of the 19th Army Corps*,⁴⁰¹ "Farragut took fire." Porter was his brother by adoption, and "the impudent little Confederate flotilla had laid low the hopes and plans of his brother admiral."

Farragut was, of course, unaware of the content of Porter's telegram to Welles, saying "fallen into the hands of the enemy" or the balanced wording of Welles' response "… destroy her or ascertain her fate." No action at all was necessary if the *Indianola* was a hulk, beyond repair, whether sunken or not. If repairable, the question was: How long before she was back in action?

Grant had telegraphed to Halleck on the same date as Porter's telegram to Welles, adding that the *Indianola* had sunk, so at least Washington knew that the ironclad would not be running wild immediately. There was less than extreme urgency, and by 5:00 p.m. on March 15th the end of the story was known in Washington. The note was from Porter, who sent a transcript of an article in the *Vicksburg Whig* of March 5th, which said in part: ". . . Indianola . . . had been blown to atoms; not even a gun was saved."⁴⁰²

As it turned out, Porter had built a huge (300 feet long) fake gunboat complete with wheelhouse, burning smokestacks and fake guns, which he set adrift. As the threatening monster approached the site of the wreck of the *Indianola*, those Confederates working to remove her guns decided to blow her up. Left to his own devices, Porter had solved the problem.

In complete ignorance of all of this, Farragut and Banks went ahead with their plans.

Baton Rouge and the surrounding country was now one vast camp; for days the troops of Augur's division had been arriving from New Orleans. The journal⁴⁰³ of the USS *Richmond*, one of the four major warships Farragut had been able to gather for the planned run, notes: "February 25 – New Orleans. Two regiments and one light battery went up to Baton Rouge to-day. Steamer loads of provisions and stores are going up every hour in the day . . ." The journal continued with similar commentary ("mules," "wagons," "troops") every day, right through to the last minute.

Farragut and the elements of the blockading squadron that could be spared for his exploit had arrived off Baton Rouge on March 11th, but last-minute preparations were still being made. On the *Richmond*, figure 4,⁴⁰⁴ for example, the work involved

---

400. ORN Ser. I, Vol. 19, pp. 245, 318.
401. Irwin, pp. 75, 76.
402. ORN Ser. I, Vol. 24, p. 397; Porter, pp. 134, 135.
403. ORN Ser. I, Vol. 19, pp. 767, 768.
404. Figure 4, Miller, Vol. I, p. 227; Figure 5, Persec map, portion. Note Judah P. Benjamin's plantation below Port Hudson.

"sending down the running rigging and putting up splinter netting on the starboard side" and mounting two more guns, which consumed all that day, the next, and into the 13th. While underway to their anchorage for that night, about fifteen miles above Baton Rouge, below Profit Island, figure 5, one of the final preparations was the whitewashing of the decks and gun carriages. This increased the visibility for the gun crews for night action and is an indication that Farragut had already altered his thinking about the original plan, which was that the passage would be made in "the grey of the morning."

FIGURE 4 USS *RICHMOND*

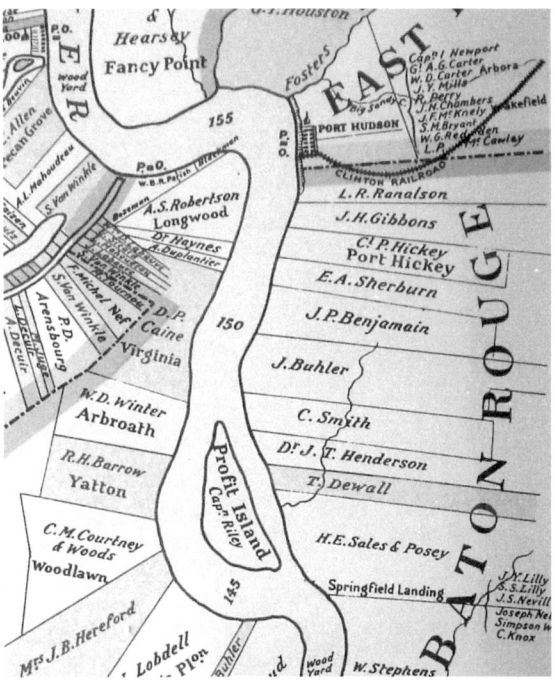

FIGURE 5

The land force,[405] on the afternoon of Friday the 13th, with Grover's 4th Division as the advance guard, flying its red flag with the white star in the center, marched out on the Bayou Sara Road. The brigades followed by number, baggage wagons sandwiched between. First rode the yellow-trimmed cavalry. The artillery, including Battery L, was sprinkled through the column, brightening the scene with their red trim. Included was a small detachment of signal corps officers and flagmen, assigned to open communications with the fleet. The weather was "grand." The road was "just what it should be, not muddy, not dry enough to be dusty; but smooth and soft enough for the foot to feel like a cushion . . ." They were shielded from the sun by a tall Magnolia forest.

A provisional brigade consisting of a section of Nim's battery, Company E of the 1st Louisiana Cavalry, three companies of the 26th Maine, and the 159th New York, all under the command of Col. E. L. Molineaux, marched up the Clinton Plank Road toward the rear of Port Hudson. Emory, with the 3rd Division, followed Grover's route the same day. Augur's 1st Division brought up the rear on the 14th.

---

405. O.R. Vol. 53, pp. 548–551. Irwin, pp. 6, 77, 78, 79; Tiemann, p. 23; O.R. Vol. 15, pp. 252, 253.

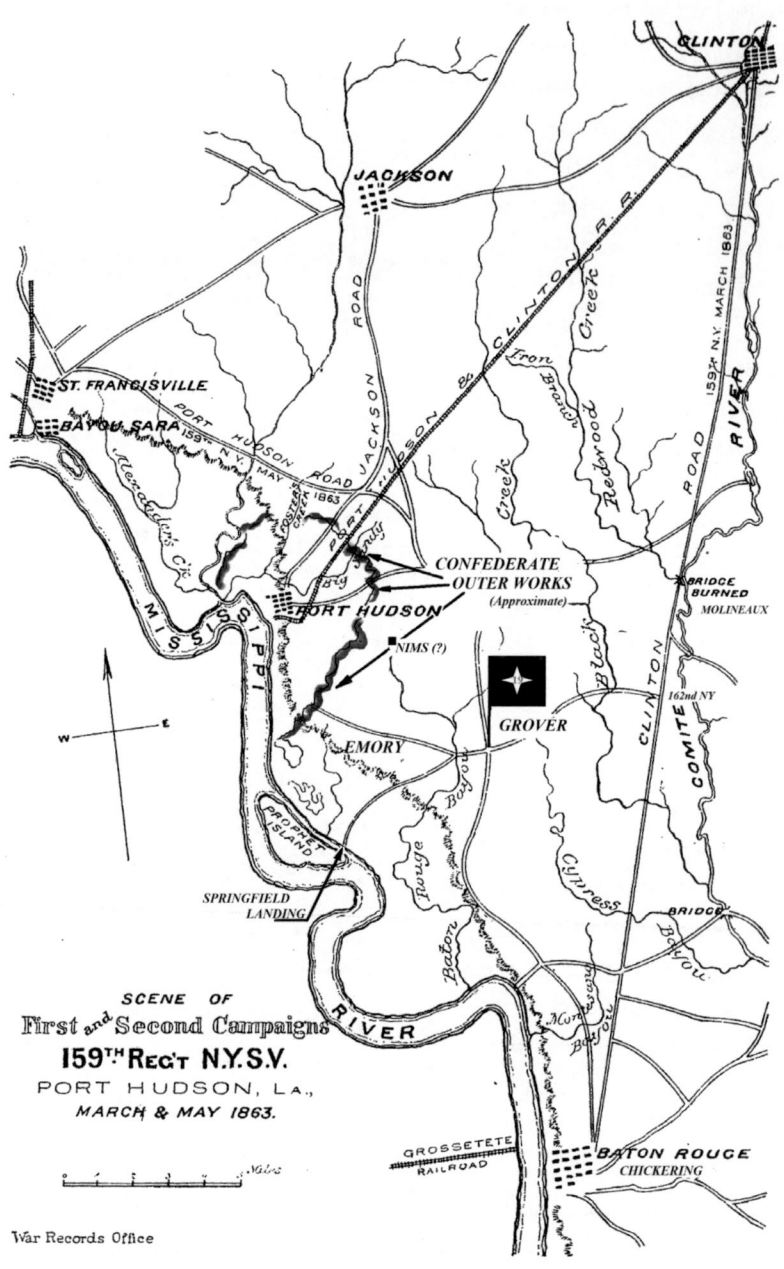

FIGURE 6

By 8:00 p.m. Grover had reached Green's Plantation, about eight miles from Baton Rouge, and halted for the night. Resuming the march on the next day, described as "beautiful," the column was now out of the shade. Soon the sun began to burn, and the road had become dusty. Men began to fall out. The sweaty 4th Division reached Barnes' Crossroads, about four miles from Port Hudson, at noon. Here, the division went into camp, and the signal detachment was sent off to Springfield Landing to open communications with the fleet. Nearby, Grover's headquarters was established – note the flag in figure 6. Emory's division formed on the left of Bayou Baton Rouge. Augur formed in the rear. The total mobile force was about 12,000.[406]

At 1:30 p.m. on the 14th, a message was "sent off" according to the signal officer Lt. Joseph L. Hallett[407] from Banks to Farragut: "My command is at Barnes' Cross-roads, and occupies the road to Ross Landing, on the flank and rear of the rebel batteries. When will you open fire? We shall be ready this evening."

At 5:00 a.m. that day, the flagship, the *Hartford*, had made signal to get underway from its anchorage off Springfield Landing, it being cloudy with a "heavy mist" hanging over the river.

By 8:00 a.m., the fleet had come to anchor within sight of Port Hudson, and at 10:00, Farragut signaled for all of the commanders to "repair on board the flagship" for a council of war. The decision was to make the attempt that night, not as planned, in the grey of the following morning. Thus, Farragut answered Banks, by a dispatch received at 5:00 p.m., that though he had earlier been delayed,[408] he would move at eight o'clock that evening and expected to be past the batteries by midnight.

Farragut's arbitrary decision to move that night was too quick for Banks to finalize the positions of his forces. He took it rather blandly: "I immediately directed the best disposition of our forces that circumstances would admit of, in view of the fact that our maps were in many important respects very unreliable. Of the enemy's position we had not the information necessary to enable us to approach it with confidence, and had no time to obtain this information." Add to this the fact that Banks was made cautious in that he believed that Gardner had "not less than 20,000 men."[409] This was quite accurate. General Gardner reported 15,572 aggregate present for the month of February, and Rust arrived on March 7th, with 2,771 men. The 6th Mississippi Regiment was expected.[410] Gardner's March return shows 20,388 aggregate present.

In a later report, Banks noted: "Had the original purpose been carried out my batteries would have been in position before morning." This may have been true, but it is of note that the guns of his heavy artillery would likely have better gained the attention of the Port Hudson garrison, and the 1st Indiana Heavy Artillery does

---

406. Figure 6, Tiemann, insert, pp. 22–23, modified to add captions showing approx. troop locations. A more detailed map is available: "Baton Rouge to Port Hudson; Showing Position of 19th Army Corps, Maj. Gen. N.P. Banks Com'd'g. On the 14th March 1863," a copy of which was obtained, courtesy of the Port Hudson State Historic Site, Jackson, Louisiana. It was used as a reference for the preparation of the much less detailed map shown here. See also O.R. Vol. 15, p. 1113.
407. O.R. Vol. 15, pp. 260–262.
408. O.R. Vol. 15, pp. 251–256, 1112–1114; Tiemann, p. 23; O.R., Vol. 19, p. 768.
409. O.R. Vol. 15, pp. 255, 678.
410. O.R. Vol. 15, pp. 1000, 1005, 1032.

not appear in any of the marching orders. They, in fact, remained at Baton Rouge.[411] As it happened, the fire from the light batteries accompanying the 3rd and 4th Divisions, and the cavalry skirmishing that had taken place as it had advanced, would be the full extent of any engagements on the land. The whole affair would prove to be ineffective – in fact, a useless waste of time and energy.

Only those sections of Nims' battery not with Molineaux had been sent forward to shell the fort.[412] It is presented in their history as a farce; they did not know where they were, and hardly knew in which direction to fire.

> We arrived at a certain point on the road and having passed through the woods, were ordered to halt, unlimber and go into battery. Said the lieutenant to the guide: 'Where is Port Hudson?' 'Right ober dar.' Was the reply. 'Which way is that?' 'Right ober dat away.' 'How far is it?' 'Oh right smart aways.' The gunners then elevated their guns, and each fired a few shots, after which "all was still and dark as before . . . Then we limbered up, thinking of our tents and stole away back to camp and turned in.

Regardless of the forward positions of the few detachments mentioned, the vast body of Banks' forces were never much closer than five miles from the Port Hudson outer works. The monthly report of Battery L mentions merely: "Marched toward Port Hudson."

Farragut got underway, and reached a position at 11:20 p.m. where the Port Hudson lower batteries opened upon the *Hartford*. The river depth was such that the fleet had to pass close to the Port Hudson shore – so close, in fact, that the voices of General Gardner's gunners could be distinguished above the clank and hiss of the *Hartford*'s steam engines.

The numerous guns on the bluff opened upon the fleet as they moved further upriver at their painfully slow pace, less than three miles per hour, their net forward speed being reduced by the speed of the current. Soon, the smoke from the Confederate guns, the ship's guns, and their smokestacks – their coal-fired boilers straining at maximum power – had, in this still, damp evening, settled upon the river to a degree that made it almost impossible to distinguish anything of the action. The *Hartford* had the advantage of being in the lead, running clear, but leaving those to follow in difficulty.

Continuing the narrative, now from the log of the *Richmond*:

> The rebel sharpshooters opened on us, but they were soon silenced. The engagement now became terrible. The rebel's guns raked us as we came up to the point. The lower batteries were all silenced as we passed, but misfortune now befell us: as we were turning the point almost past the upper batteries we

---

411. Ewer, p. 64.
412. Hosmer, p. 96; Moors, p. 76; Whitcomb, p. 44. Unfortunately, no specific mention of Battery L has been found.

received a shot in our boilers . . . Our steam was all gone; we could not steam up against the current with one boiler. . . . We turned her head downstream.

The admiral and the Albatross were the only vessels that succeeded in passing up the river. . . . The Mississippi got aground and could not be got off, and the rebel shell were tearing her to pieces. The crew left her in boats, some coming down the river and some going ashore. After they deserted her they set fire to her . . . The Monongahela and Kineo were pretty badly cut up; they got down safe again.

FIGURE 7. FARRAGUT PASSING PORT HUDSON

An engraving from *Harper's Weekly* of April 18th, 1863, pp. 248–249. The artist has forgotten that the ships were forced to pass close to the Port Hudson side where there was adequate water depth. Where shown, all would have run aground, as has the *Mississippi*, the last in line and now in flames.

At about 4:00 a.m. the sky was lit, and a dull roar, "resembling distant thunder," shook the ground where the 26th Maine Regiment was standing, four miles away from the river.[413] It was the *Mississippi*'s magazine, which demolished the venerable veteran.

At 6:00 a.m., Signal Officer Hallett sent off a message to Banks, indicating details of the disaster and that only the *Hartford* and the *Albatross* had passed Port Hudson. The action was ended. Orders to Banks' troops were soon sent out that the object of the expedition had been achieved, and that they should be ready to march. "Much wondering at this Delphic announcement, not yet knowing that Farragut had successfully passed the batteries . . . we marched at daybreak."[414] Given these

---

413. Maddocks, p. 25.
414. Sprague, p. 104. The 52nd Massachusetts (Hosmer); 26th Maine Regiments (Maddocks) list

orders to retreat in apparent disgrace, without firing a shot, and in ignorance of the purpose of the plan, the troops were sullen.[415]

In the afternoon it began to rain in torrents, the dry pleasant weather of the previous week forgotten. Then, there was straggling and complaining about the insufferable heat and the casting aside of various pieces of their equipment. Now, the complaint was where to find shelter and cook supper. An area previously chosen for their bivouac was now a swamp, but they must obey orders. Many spent the night standing in water from a few inches to a foot deep. No supper, no breakfast.

Despite the spectacular nature of the "disaster" as Farragut very correctly referred to the affair,[416] the numbers of navy casualties were small. The preliminary reports of the casualties were (*K* [killed], *W* [wounded]):[417]

	HARTFORD	ALBATROSS	RICHMOND	GENESEE	MONONGAHELA	KINEO	MISSISSIPPI
K	1	1	3	0	6	0	25
W	2	2	12	3	21	0	9

As a measure of the activity on land, Confederate losses were reported as one killed and nineteen wounded.[418] The only reference Banks makes to casualties is to those of the navy; General Dwight, 1st Brigade, Grover's Division, reported no losses.

Some of the regiments were ordered directly back to Baton Rouge. Others were detained for several days to escort wagon trains carrying cotton confiscated from the area.[419]

The historian of the 6th New York commented: "Altogether this raid was valuable training, as accustoming everybody to be cheerfully uncomfortable, but it affected little in the way of suppressing the rebellion, and when the Sixth settled back into its camp, cleaned its trousers, and got itself into shape, it had an internal feeling that perhaps the high military authorities of the Department of the Gulf were not so much wiser than the rest of the world . . ."

The soldier's view about not suppressing the rebellion was a misunderstanding, as Irwin, in his *History of the Nineteenth Army Corps*, insists. Irwin's optimistic view (also Farragut's) was that as the result of Farragut's presence above Port Hudson, the Confederacy was denied the use of the Red River to carry supplies across the Mississippi, "save in skiff-loads." The Confederate supply road from Galveston and

---

noon (p. 99) and 3:00 p.m. (p. 27), respectively. The 26th was the rear guard. O.R. Vol. 15, p. 262.
415. Hosmer, p. 100.
416. ORN Ser. I, Vol. 9, pp. 667, 670, 676, 680, 683, 688, 689, 693–695.
417. The number of missing for the *Mississippi* was listed on March 20th as 76, with an initial count of 233 saved. The numbers shown are those reported by the surgeon of the *Richmond*.
418. O.R. Vol. 15, pp. 252, 265, 278.
419. Moors, p. 92; Morris, 6th New York, p. 89.

Matamoros was closed. Also, of course, the operation resulted in the army gaining "some facility of movement, some knowledge of its deficiencies, and some information of great future value as to the topography of the unknown country about Port Hudson . . ."[420]

Irwin seems to have forgotten that only two of Farragut's ships had managed to pass Port Hudson and that to accomplish anything in terms of future patrolling of the Red River, Farragut would have to ask for assistance.

Banks' letter to Grant describing the diversion, dated March 13th, had been carried on the *Hartford* and therefore was able to be delivered to Grant on March 20th. Grant made his reply on the 23rd, and gave it to Farragut. Apparently feeling that his note to Banks was unclear, he wrote another to Farragut, whom he requested to "inform the general of the contents of this." Unfortunately, Farragut appears to have disregarded the matter, and went off on his own fruitless expedition. Consequently, Banks heard nothing from him, or the content of General Grant's reply, until April 10th, which is discussed in the next chapter.

In the interval, Banks returned to New Orleans, where, on March 25th, he issued marching orders for his much delayed campaign. It was designed to wrest control of the river systems west of the Mississippi from Taylor, as had been attempted in January, and open a path to unite with Grant. It would become known as the Teche Campaign.

---

420. Irwin, pp. 81, 84.

Chapter 8

"Record" 4/63; The Teche Campaign; Chasing the Fox; Opelousas; Cooperation with Grant; Grierson's Raid; Alexandria; Farragut Departs; Turn to Port Hudson

"Record" 4/63
28 FEBRUARY–30 APRIL 1863, BARRE'S LANDING, BAYOU CORTABLEAU

Battery with Gen. Grover's Division, left Baton Rouge the 13th. Marched toward Port Hudson. Remained in the field until the 20th and returned. Left Baton Rouge on Steamer "Laurel Hill" & arrived at Donaldsonville, La., on the 27th. Took up line of march and camped at Assumption Parish, La., 31st, 15 miles. Took up line of march April 1st for Brashear City, La.,[421] and arrived April 9th, distance 80 miles. Embarked on Steamers and Flats, night of the 11th, disembarked, morning of the 13th on Madame Porter's plantation / Grand Lake. The Battery was thrown to the front at 6 a.m. Appleton's sections detached each to secure and hold bridges across the Teche Bayou. This was accomplished, and at the left bridge, by Appleton's section, under a very annoying fire from 4 guns of the Rebel artillery and his Sharpshooters. Casualties two horses wounded. During the fight of the 14th the Battery was held in reserve. On the 15th, took up line of march and reached Vermillion River the 17th, estimated distance 45 miles. At Vermillion Bridge, Taylor's section of the Battery was opened upon by Sharpshooters and the Rebel artillery of 4 guns from the cover of the opposite bank. Casualties 2 horses killed, the remainder of the Battery in action soon shelled the Rebels out. On the 20th marched into Opelousas, an estimated distance from Grand Lake landing 60 miles. On the 26th inst. Marched from Opelousas and encamped same day at Barre's Landing,[422] Bayou Courtableau, La. distance from Opelousas, 8 miles.

Henry W. Closson	Capt. Commanding Battery and Chief of Artillery Grover's Div. 19th Army Corps.
Franck E. Taylor	1st Lt. Ass't. Comm'y Of Musters S.O. no. 16 Hdqrts Dept of the Gulf 19th Army Corps Apr. 9, 1863
Edward L. Appleton	1st Lt.
J.A. Sanderson	2nd Lt. Appointed to the Co. by promotion: vice Gibbs promoted War Dept. A.G.O. Wash. July 4, 1862 (never joined co.)

---

421. Now Morgan City.
422. Now Port Barre.

Detached: George Friedman	Pvt. On Det. Svc. at N.O. as Artillerist. Left Co. May 24, '62.
Amelius Straub	Pvt. Absent on Ex. Duty as Cook in Gen. Hosp. Baton Rouge since Mar. 14, '63.
Martin Stanners	Pvt. Absent on detached serv. At Bayou Boef, La., since April 7, 1863.

Strength: 143. Sick: 15

Sick Present: Julius Becker, John Baker, John Burke, Edward A. Foote, Edward McLaughlin

Sick Absent: William C. Brunskill, Fort Hamilton, NY, Left Company Sept. 17, 1861
Isaac T. Cain Brashear City, La. since Apr. 22, 1863
George Chase Brashear City La. since Apr. 22, 1863
William Crowley Absent sick in General Hospital, Baton Rouge, La. since Mar. 27, '63.
Joseph Eisele In Hosp. at Franklin, La. since April 15 ,'63
Benjamin O. Hall Abs. sick at Baton Rouge, La. since Mar. 27, '63
Michael Kenny In Hosp. at Franklin, La. since April 15, '63.
Charles F. Mansfield sick at Pensacola, Fla. since Dec. 24, '62
Joseph Smith Brashear City, La. since April 22, '63
Wallace D. Wright sick at Pensacola, Fla. Since Dec. 24, '62

Horses:	Serviceable: 96 Unserviceable 8.
Died:	Daniel McCoy Pvt. 22 Oct. '60 New York, at Gen. Hosp. Baton Rouge, La. Mar.4,'63 Charles A. Flint Pvt. 22 Sept. '60 Boston,　　"　　"　　"　April 13,'63
Deserted:	James A. Allen Pvt. 17 Nov. '62 Pensacola. From Baton Rouge, La., on Mar. 12, '63 James R. Harrington Pvt. 27 Feb. '63 Baton Rouge. From Camp at Thibodeaux, La., April 17, '63 Abram P. Winn Pvt. 16 Dec. '62 Pensacola　　　"　　"　　" Joel T. Winn Pvt. 26 Dec '62 Pensacola　　　　　"　　"　　"

The sick list is long, and two deaths due to it appear. The cause of death is not listed, and that portion of Danield McCoy's service record is unreadable. Substantially more information has been found in Sarah Flint's application for a mother's pension. She cites the surgeon's report that lists Charles as having died of "Typho Malarial Fever." Her file also contains a personal letter from Captain Closson, which is reproduced here as figure 1.

It is the only one that has been found from the many records examined relating to deaths in Battery L. There was apparently no official requirement that the company commander write such a letter or that it pass through channels. The only official requirement was that "final papers" be made out as for a typical discharge. Enumerated were pay adjustments that we have seen before; those that may have been owed to the sutler, laundress, or for "camp and garrison equipment."

It is believed, however, that either Henry Closson or Franck Taylor wrote a letter in every case of a Battery L death, because an affidavit in yet another pension application, that of Clarissa Parslow, makes reference to such a letter: "Said Joseph

H. Parslow died in battle as his colonel wrote my husband . . ." Parslow died at Smithfield, West Virginia, chapter 15.[423]

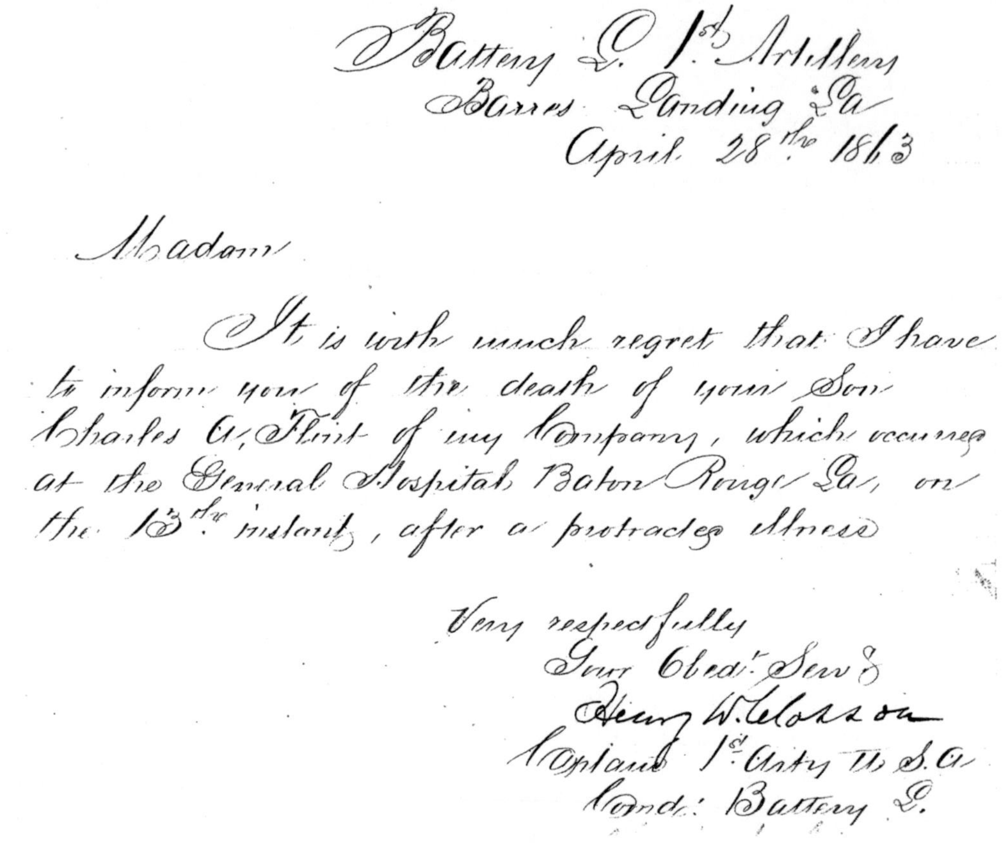

Battery L 1st Artillery
Berwick Landing La
April 28th 1863

Madam,

It is with much regret that I have to inform you of the death of your Son Charles A. Flint of my Company, which occurred at the General Hospital Baton Rouge La, on the 13th instant, after a protracted illness

Very respectfully
Your Obedt. Servt.
Henry W. Closson
Captain 1st Arty U.S.A.
Comd: Battery L.

FIGURE 1

The muster rolls, as earlier mentioned, were used to verify service for pensions granted long after the end of the war. This repeated usage resulted in many of the rolls being in poor condition. Corrections and status changes as may have been effected by laws subsequently enacted by Congress were simply overwritten into the original document. This roll is an example. Above Allen's name, there appears an overwrite: "Also, under prov. of Act of Congress approved March 2, 1889, charge of desertion is removed, vide AG 1741499."[424] His story explains why – he reenlisted sooner than the required four months limit stated in the law.

There are numerous overwrites on the names of the two Winns, which are nearly illegible. Below the original entry, there appears this new entry:

---

423. Flint file no. 43709, Parslow file no. 147787; National Archives.
424. The number is the adjutant general's file number. This decision was made on April 13, 1911. The law referred to is recorded in the 1890 Annals of Congress, 50th Congress, ch. 390, pp. 692–694.

#Abram P. Winn} See Prisoner of War records for report of capture May 16/63, subsequent parole and desertion.
Joel T. Winn}   do.   do.

Though the subject of prisoner parole and exchange had had a checkered history right from the beginning, it was at least operating up to 1863, as we have seen regarding the troops in Texas, and the exchange of Sergeant Riley. The Dix-Hill Cartel was agreed to on July 22nd, 1862. The following February, noting that the Confederacy was violating the terms of the cartel, the War Department was caused to issue new instructions[425] in the form of General Orders No. 49, dated February 28th, 1863, to guide commanders on how to handle prisoners. The final rules for the parole and exchange of prisoners were not worked out until April of 1863 and were issued as General Orders No. 100[426] on May 20th, 1863. However, this strict construction was suspended on May 25th, by Halleck in the case of Confederate officers who, henceforth, were not to be paroled and were to be confined until further orders. The case of the Winn brothers tested the system with a rather strange twist.

Having deserted at Thibodeaux, on April 17th, as Battery L was marching to Berwick City, the Winns were taken prisoner on May 16th at "Jackson." Though the record doesn't specify it, we assume that it was the small town of Jackson, Louisiana in Confederate territory. Sent to Richmond, they were paroled, with the remark "deserter," at City Point, Virginia, on June 13th. Sent to College Green Barracks, Maryland, they were examined at hospital and then sent to Camp Parole, Maryland, on the 22nd, arriving there on the 23rd. There they again deserted and never returned to Battery L.

As the criteria for an invalid pension was eased by a sucession of laws after the war, both applied for pensions. Both were denied. After Joel's death in 1877, his wife Mary repeatedly applied for a widow's pension, a process she carried on for some forty years. Finally, after a Special Act of Congress, dated December 23rd, 1923, she was granted $30 a month.

The Teche Campaign

The campaign would begin at the lower waters of the Atchafalaya and follow where Longfellow's Father Felician had acted as guide to Evangeline.[427] Heading north along the Teche, figure 2, it would clear Taylor out of the Attakapas (which today comprises St. Martin, St. Mary, Lafayette, Vermilion, and Iberia parishes), and the western

---

425. O.R. Ser. II, Vol. 5, pp. 256, 306, 307.
426. O.R. Ser. II, Vol. 5, pp. 671–682, 696; Abram Winn, pension file no. 766906; Joel Winn, file no. 957542; National Archives.
427. Ewer, p. 68.

bank of the Mississippi opposite Port Hudson, as far north as Opelousas.[428] At that point, the plan, as Banks outlined to Grant in a letter dated April 10th, 1863, was to return to Baton Rouge to "cooperate with you against Port Hudson."

The plan would step off in a move essentially identical to what Weitzel had made the previous January. Weitzel's reinforced brigade, separated from Augur's division, would be transported from Brashear City across Berwick Bay to be landed near Pattersonville to again attack a revived Fort Bisland, which once again had a gunboat, the recently captured *Diana*, to support it.[429] This time, however, Weitzel would

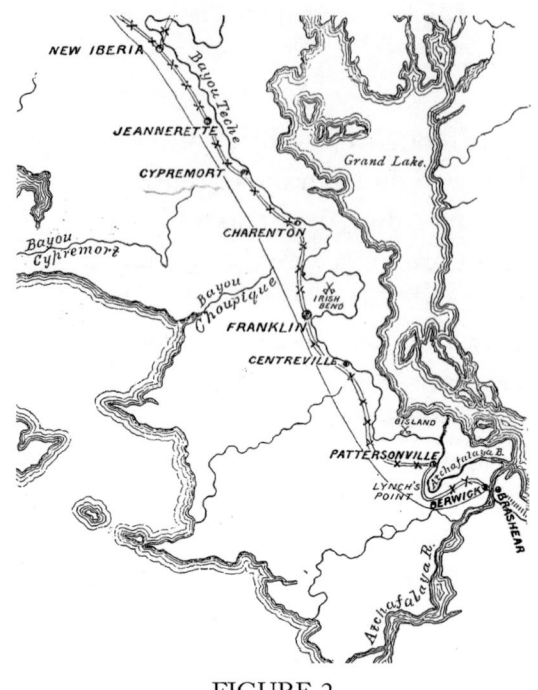

FIGURE 2

be followed and supported by Emory's division; and as part of a classic flanking movement, Grover's division would be transported up Grand Lake to a point above Franklin, where the shore of Grand Lake lies close to one of the bends in the Teche. There, he would land in the rear of Taylor's forces, seize the road, and cut off any possible avenue of retreat.

Augur was left to defend Baton Rouge with his 1st and 3rd brigades, and T. W. Sherman was in command of the defenses of New Orleans and the Lafourche.[430]

Grover's division was ordered to move on transports to Donaldsonville and to march from there to points on the Great Western Railroad. Emory's division would be transported to Algiers as soon as the transports were released from Grover to move to Brashear City by rail. Battery L boarded the transport steamer *Laurel Hill*, figure 3, on the 27th and arrived at Donaldsonville, some 90 miles south of Baton Rouge the next day.[431] Some of Battery L's colleagues from the days at Pensacola, Billy Wilson's 6th New York, found whiskey on board their steamer, the *Morning Light*, got very drunk, and a part of the resulting shenanigans was an attempt

---

428. ORN Ser. I, Vol. 20, p. 50; Sprague, pp. 108, 109; Irwin, pp. 80 (Figure 2), 104, 105.

429. Taylor mentions having brought the USS *Diana* to the fort, where "her 'Parrott' became a valuable adjunct to our line of defense." It had been captured on the Teche on March 28th. Taylor, R., p. 128; ORN Ser. I, Vol. 20, pp. 109, 110, 113.

430. Irwin, p. 88; Sprague, p. 106; O.R. Vol. 15, p. 258.

431. Figure 3, March 21st, 1863, *Harper's Weekly*, p. 21; the steamer was rated at a capacity of 1,000 troops. O.R. Vol. 15, pp. 1102; Haskin, p. 189; O.R. Vol. 15, p. 365, Closson's report; pp. 380, 381, Day's report.

to throw General Dwight overboard.⁴³² Woe unto those officers, Billy Wilson among them, and the 24 enlisted men who were involved. Dwight saw no humor in the incident. They were all disarmed and arrested.

Who was this Dwight?

Born in Massachusetts, he attended West Point from 1849 to 1853 but resigned to enter private business in Boston. He entered the army as a captain of the 13th Infantry on May 4th, 1861, but in June was appointed a lieutenant colonel of the 70th New York Volunteer Infantry. At the Battle of Williamsburg – where the 70th lost half of its men killed, wounded, or missing – Dwight was twice wounded and left for dead on the field. He was taken prisoner and was eventually exchanged and promoted to brigadier general on November 29th, 1862.⁴³³

FIGURE 3

Though most of Grover's infantry marched south from Donaldsonville to take the railroad at the Lafourche or Terrebonne stations, Battery L, being fully mounted, and towing heavy cannon, marched all the way to Brashear City. On Wednesday, April 1st, with a cool northeast wind blowing, Battery L set out. The weather remained clear and cool, only warming on Sunday the 5th. There had been no rain, and the ground was dry. It was an easy pace of about ten miles per day – nothing like the urgency of the march out of Texas or the pull in the sand during the reconnaissance on Santa Rosa. Memories were accumulating and were attaining the usual rosy glow.

They reached Bayou Boeuf, just east of Brashear City on April 9th, where Grover's division had been encamped since the 4th and the whole division then moved on to Brashear City.

Weitzel began to cross the mile-wide Berwick Bay at about 10 o'clock on the morning of the 9th, the navy gunboats *Estrella*, *Clifton*, *Calhoun*, and *Arizona* adding to the ferrying capacity of the transports, the *St. Mary's*, *Laurel Hill*, and *Quinnebaug*.⁴³⁴ Emory followed, and the four brigades – some 12,000 officers and

---

432. Bissell, p. 26.
433. Johnson, Brown, eds., *Biographical Directory*, Vol. III.
434. Sprague, p. 108. A fourth small steamer, the *Sykes*, was in the area but was retained by Weitzel to transport troops across the Teche as part of his advance on Fort Bisland; O.R. Vol. 15, p. 355.

men – were bivouacked near their landing place that night. The short distance across Berwick Bay had allowed many round trips, the landing points were clear, and the vessels could be loaded without regard to crowding, the criterion being merely that the vessel remained afloat for the short trip. Banks and staff arrived at Brashear City that evening.

For Grover's force, consisting of 13 regiments, three artillery batteries, and one cavalry company – more than 4,000 officers and men – the situation was different. The only available vessels[435] had to be loaded for a one-way 20-mile trip up to the planned landing on Grand Lake. There could be no return for supplies if this was to be a surprise operation. The situation taxed the vessels and the constitution of the troops to the extreme.[436] For example, the *St. Mary's* – formerly a New York and Galveston ocean liner, "a beautiful vessel" built to carry 500 passengers at a pinch was loaded with the 52nd Massachusetts, the 25th Connecticut, the 12th Maine, the 24th Connecticut, and a portion of Nims' battery – some 2,500 men. Conditions were described by J. F. Moors of the 52nd Massachusetts as "full to overflowing, so that the men appear to lie two to three deep on deck; while some hang up in the shrouds, and others stand leaning against or crouching on the bulwarks. We are literally as thick on board as three in a bed." Of course, there was no room for cooking; and when the planned day trip had dragged on to 40 hours – one day, two nights, and part of another day – the sanitary conditions had become terrible, the troops had had little sleep or food, and they were hardly ready to be thrown forward into battle.

The little flotilla had to be supplemented in its capacity by towing flatboats,[437] or scows, as they were labeled by the navy,[438] which were "picked up along the bay."[439] Portions of Rodger's and Nims' batteries, as well as Battery L, their guns, caissons, and some of their horses, were assigned to the flats. Add in the cavalry and various stores, and the loading process could not be completed until the night of the 11th, though the original plan was to have Grover at his landing point that morning.

To compound Grover's problems, a fog developed on the night of the 11th, which did not clear until 8:00 the next morning. Only then did they finally get underway.

The operation was soon spotted by Taylor's scouts, and Taylor ordered Vincent, with the 2nd Louisiana Cavalry and a section of Cornay's Battery, upstream to observe and oppose the movement. A second section of Cornay's battery was sent

---

435. ORN Ser. I, Vol. 20, p. 106; O.R. Vol. 15, p. 383.
436. Irwin, p. 90, 104; Ewer, p. 69; Tiemann, p. 27; Moors, pp. 111, 112; Hosmer, p. 125; Bissell, McManus, pp. 33, 34; O.R. Vol. 15, pp. 294, 326, 364, 365, 382, 712; Taylor, p. 129.
437. O.R. Vol. 15, p. 294; Irwin, p. 91.
438. ORN Ser. I, Vol. 20, p. 134.
439. O.R. Vol. 15, pp. 358, 359.

later in the day, and on the morning of the 13th, Reily's 4th Texas was sent out to oppose any landing, though there is no evidence that any of these troops fired upon the flotilla while it progressed up Grand Lake.

Taylor had long realized that his little fort at the Bisland Plantation was vulnerable to this sort of flanking movement[440] and with the "Increased activity of the enemy at Berwick's Bay" had ordered Fuller, the former captain of the *J. A. Cotton*, to Alexandria to repair the *Queen of the West* and to prepare "one or two other steamers as gunboats." Fuller had then brought the *Queen* to Butte la Rose, where he now looked for the arrival of his two other steamers, the *Grand Duke* and *Mary T*, delayed awaiting their guns. If Fuller could arrive in Grand Lake before Grover's troops could be disembarked from their overloaded transports, there was potential for hundreds of drowned men.

Arriving at Hutchin's Point,[441] opposite the Porter plantation, in darkness, a detachment of Holcomb's 1st Louisiana under Fiske was sent out on a reconnaissance. It was 9:30 p.m. before they returned with the news that the shell road planned as the landing point, which gave access to Irish Bend, was under water. The flotilla again got underway and anchored about six miles farther north, opposite the McWilliams Plantation at Indian Bend, where a second road west was known. At 1:00 a.m., a detachment of the 6th New York, and two of Dwight's staff, were sent out on a reconnaissance which revealed that there was a "practicable" plantation road leading to the Teche, and debarkation began at daybreak, figure 4.[442]

The position of the flotilla having been reported to Taylor on the evening of the 12th, Vincent's 2nd Louisiana Cavalry, and the two sections of Cornay's battery were ordered there. Vincent failed to obey Taylor's orders, assigning only pickets at the landing, with his main force encamped west of the Teche. Thus, the small force at the landing was only able to establish "a scattering fire" on the landing. The 1st Louisiana, on the *Clifton*, was the first to land and was "at once opened upon by . . . the enemy's artillery, supported by . . . sharpshooters."[443] Colonel Fiske and four men were wounded.

The heavy guns of the *Clifton*, a converted Staten Island ferry,[444] guided by signal flags on the *Laurel Hill* "sent out, now a solitary puff, now three or four

---

440. Taylor, pp. 127–129.
441. Irwin calls it Miller's Point. We defer to Taylor, p. 130, Tiemann, p. 27, calls it Hudgin's point. Present-day Grand Lake has filled in and been altered to the point that these landmarks are no longer recognizable.
442. O.R. Vol. 15, p. 371; Figure 4, Irwin, portion of p. 112a.
443. O.R. Vol. 15, p. 363. Report of the signal officer on board the *Clifton*. He also states that the fire direction was from his colleague Lieutenant Hall on board the *Laurel Hill*. Hosmer's quote has been retained, though we remain wary of many observations in volunteer unit histories, written long after. However, Hosmer wrote this in 1863.
444. Bosson, p. 70.

nearly together; while in the pauses of these heavier firings, came from the shore the fainter fusillade ...At length it ceased."[445] Cornay's field pieces were no match for the IX-inch bow gun of the *Clifton*, its fire directed by signal corps officers assigned to it, and the *Estrella*. The 1st Louisiana advanced, and Vincent's pickets retreated.

The second and third units to land were the 13th Connecticut and the 159th New York. They were both on the same transport, the *Laurel Hill*, which had shallow draft, allowing it to come in close to shore. Grover, "vigorously puffing" on his cigar, was there to greet them; and by 9:00 a.m., they were detached from Birge's 3rd Brigade and loaned to Dwight's 1st Brigade to reinforce the 1st Louisiana. By 10:30, Battery L and Barrett's cavalry, on flats, had disembarked and were assigned by Grover to Dwight. Dwight now ordered an advance.[446] Grover had granted him permission to detach this force to "prevent the destruction of the bridges across the Teche, which would be almost indispensible to our crossing." The Teche is too deep for troops to easily ford and would have been impossible for the artillery. The bayou runs for miles, remarkably like a man-made canal, figure 5. It has remained relatively unchanged by mother nature, industrial development, or flood control, and is estimated to be about one hundred fifty feet wide here.

The rest of the artillery had to wait for the assembly of a bridge of flatboats to disembark its heavy guns and caissons, and it was not until 4:00 p.m. that the whole division had disembarked, the *Saint Mary's* not being able to approach closer than a half mile from shore.

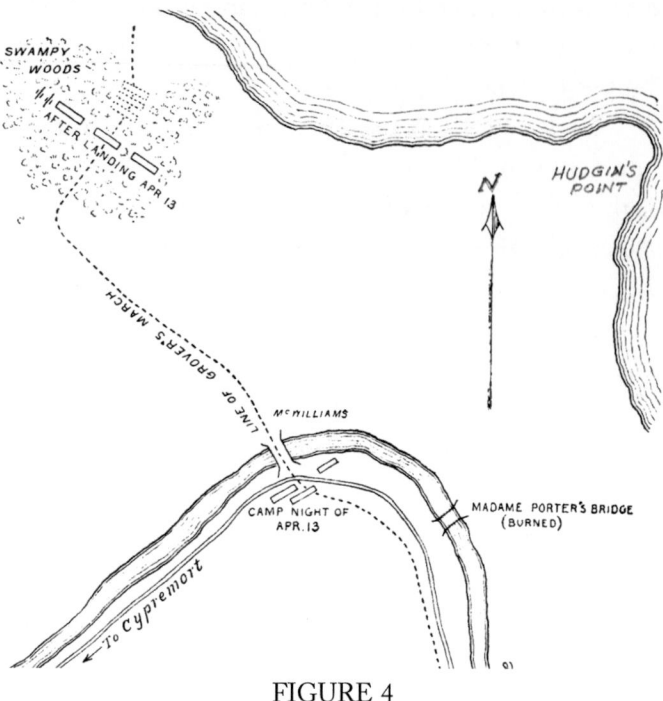

FIGURE 4

---

445. Hosmer, p. 127; Irwin, p. 107; O.R. Vol. 15, pp. 363, 390, Bissel/McManus, p. 35.
446. O.R. Vol. 15, pp. 359, 371; Tiemann, p. 27; Sprague, p. 108; Figure 5 photo by the author.

FIGURE 5

The enemy cavalry had withdrawn behind a sugarhouse on the McWilliams plantation, but when Holcomb's 1st Louisiana skirmishers pushed forward and Barrett's cavalry advanced along the road, the Confederate cavalry, now reinforced with "some" infantry sharpshooters, reappeared and opened with their long-range rifles. Supported by the 159th New York and 13th Connecticut, the 1st Louisiana arrived at the McWilliams bridge, about five miles distant from the landing, by noon. By now, the Confederates had withdrawn out of range, simply observing. The McWilliams bridge was found to be partially destroyed by fire, which was extinguished, and slaves from the nearby Porter plantation were used to repair it.[447]

While those sections of Battery L that remained under Lieutenant Taylor (Captain Closson was division chief of artillery and Lieutenant Sanderson had not yet reported) were detached to help secure the upper bridge, Appleton's section, along with Barrett's cavalry and two sections of Rodger's battery, galloped south to Madame Porter's bridge, about a mile or so below. Under the "very annoying fire of 4 guns of the Rebel artillery, and his Sharpshooters," it was, nevertheless, seized. The 131st New York, the 22nd Maine, and a detachment of the 6th New York were then sent to support the 1st Louisiana in holding it. Casualties were two men of the 1st Louisiana killed, four men of the 131st New York wounded, one mortally, and one of Battery L's horses wounded.[448]

Colonel Day of the 131st included an echo of a past problem in his report of the action: "We found by this day's experience that our pieces were very defective and exceedingly short of range." He does not state what type of weapon had been issued to his men, but apparently, the inspector general's examination in January and the report of his findings of many defective and "worn"[449] weapons had not been fully acted upon. Here it is April of 1863, after two full years of war, and the Union army still has defective weapons.

Having approved of the detachment of the force to seize the bridges, Grover

---

447. O.R. Vol. 15, pp. 366, 367, 379–381.
448. O.R. Vol. 15, p. 381. The record of events lists two horses wounded.
449. O.R. Vol. 15, pp. 648, 649.

became cautious. He reports that he was unsure enough of the strength of the enemy in his front that he did not wish to "expose a large force to the enemy until the whole division was in readiness to cut its connection with our point of landing."[450] That said, he had seized the road – that is, the road along the Teche, leading north to Cypremort and New Iberia – and the trap was closed, at least as far as he understood.

The slow pace of the landings allowed the 13th Connecticut to relax a bit.[451] "In perfect silence we passed a mile through the dense woods in pursuit of the enemy, who retreated to Madame Porter's plantation. Emerging from the woods, we saw a few of their cavalry a mile distant . . ." Following Grover's orders not to engage, they rested. "While the other regiments and batteries were coming up, we improved the opportunity to take a lunch. Crossing the bayou bridge . . . near the beautiful mansion of Madam Porter," Oaklawn Manor, figure 6,[452] they helped themselves liberally to sugar from the large mill that stood next to the bridge.

FIGURE 6

Aboard the *Saint Mary's*, and consequently one of the later regiments to land, the 25th Connecticut noted that all the commotion had drawn the plantation inhabitants and its mistress outside to witness their passing. "This stately, handsome lady, surrounded by scores of fat, happy looking and well clad slaves, stood in front of her elegant home and sadly watched as we passed. No farm in Connecticut, however carefully supervised, could show better evidence of wise management . . ." Oaklawn Manor still survives, and its current picture fits well into the description

---

450. O.R. Vol. 15, p. 359.
451. Sprague, pp. 109, 110, 122; Bissell (McManus), p. 35.
452. Picture taken in 2012 by the author with the kind permission of Governor & Mrs. M. J. Foster.

given in 1863: "The elegant mansion, the delightful grounds, the fences, granaries, fields, slaves and everything, were in perfect order . . . and Madame Porter herself, a splendidly beautiful woman, all looked lovely as peace itself in contrast with the ugliness of war."

While halted at the bridge crossing, Grover was approached in haste, and in much agitation, "by a very stately lady . . . a matron of fine bearing, elegantly attired, her face full of character; she is bareheaded . . . She sweeps by us hastily . . . and stops at the stirrup of Gen. Grover. . . . She has come to intercede . . . for her son, [Alexander, age 17] who has just been taken prisoner – a fine fierce boy . . . who stands, haughty and tall, close by, among a group of captive rebels."[453] This was Mary Porter, 40, the second wife and widow of James Porter, and proprietress of one of the largest plantations in Louisiana – the 1860 federal slave census listing 297. She, her son, two daughters, and personal staff were well-known residents "in former seasons"[454] of the exclusive social colony at Newport, Rhode Island, had now been swept up into this maelstrom of war, her son a suspected partisan.

She pleaded, "Do let him go, general; he is all I have" many times. Fortunately, Capt. Homer Sprague of the 13th Connecticut Volunteers, himself a prisoner before the war was over,[455] ends the drama for us by noting in his diary: "Her young son was taken prisoner, but soon released."

There was no incident of pillage to the Porter plantation, perhaps for the reason that the troops were focused on the mission, under the threat, as they were, of Vincent's cavalry. Another factor may have been the close supervision of the officers of the command, most of whom were quite literally on the ground. All the other officer's horses and baggage had been left behind.[456] Another factor was that Mary Porter had not fled from her property, an act regarded as an almost sure sign that the owner was a rebel sympathizer. She was there to insist that she was loyal, a declaration, whether sincere or not, which would preserve her rights as a slave owner. *This, after all, was St. Mary's Parish, one of those exempt under the terms of the Emancipation Proclamation.* In any event, the troops were simply in awe of, and full of respect for her and her plantation.

Grover had feared that there was a strong enemy force in the woods to his front and concluded that the night was too dark to "dislodge or even find his position, or for our own skirmishers to keep up the connections." The delays of the day had taken their toll, and Grover's caution and, as noted earlier, his apparent

---

453. Hosmer, p. 130. Hosmer was a corporal in the 52nd Massachusetts Volunteers, also aboard the *St. Mary's* and passing in the same time frame. Additional details of the plantation at that time are given in the February 6th, 1864, issue of *Frank Leslie's Illustrated Newspaper*. Today the plantation remains open to visitors.
454. Duganne, p. 103. Data from 1860 federal census of Newport, Rhode Island.
455. Sprague, p. 122.
456. Bissell (McManus), p. 36; Duganne, pp. 103, 104.

satisfaction at having captured the road, decided the matter; they would bivouac for the night.[457] Grover's attitude is revealed by this statement in the history of the 25th Connecticut Regiment: "Our generals had believed, and we had hoped, that as soon as Taylor would find this large force . . . suddenly occupying the road in his rear, he would submit to the inevitable and surrender . . ."

Unfortunately, Grover was ignorant of the fact that there was a cut-off road on a causeway across the swamp about three miles to the west of his bivouac, figure 7 (from National Archives RG 77, map 111, altered by the author). It was the one critical to Taylor's escape. If Grover had sent out Barrett's cavalry, or a few of the staff on a reconnaissance along the northern arm of the bayou road, they would have discovered the cut-off in 15 minutes. Little did Grover know that it was only Vincent's cavalry that opposed him. Reily's troops were encamped at Franklin, and would soon be asleep; Clack's battalion, called by Taylor from the salt works at New Iberia, had not yet arrived.[458] Grover then called in Appleton, Rodgers, and Barrett, the detachment that had saved the lower bridge; and with that, the 1st Louisiana and its attached force of portions of the 131st New York, the 6th New York, and the 22nd Maine were ordered to burn it and retire.

Grover's orders were to advance to Franklin, and it was upon that direction only that he had remained focused.

Who was this Grover? Graduated from West Point in 1850, he had been in continuous service in the army ever since: at Fort Leavenworth, Kansas, involved with the Northern Pacific Railroad exploration; in garrison duty at forts Crawford and Snelling, in the Utah Expedition of 1857–1858; and, rising to captain in the 10th Infantry, at Fort Garland, Colorado, and forts Union and Marcy, New Mexico, through 1861.

Much like many others in the

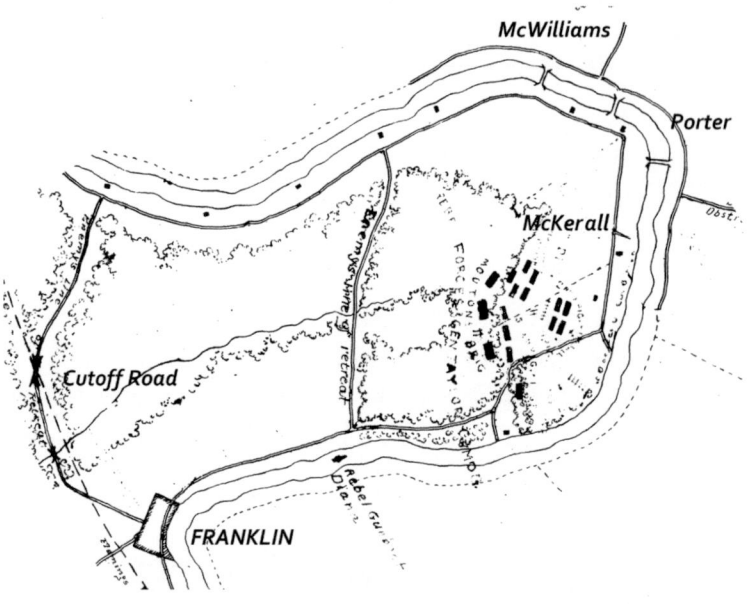

FIGURE 7

---

457. Sprague, p. 110; Grover's report, O.R. Vol. 15, p. 359; Bissell (McManus), p. 36.
458. Taylor, p. 132. Clack's "battalion" consisted of 90 men.

early months of the war, Grover, a captain in the regulars, was vaulted up to become brigadier general of volunteers in April of 1862. He served in the Army of the Potomac during the Peninsular Campaign, in the battles of Yorktown, Williamsburg, Seven Pines, Savage's Station, Glendale, Malvern Hill, and the skirmish at Harrison's Landing. In the Northern Virginia Campaign, he was at Bristoe Station and Second Manassas. He was brevetted lieutenant colonel for meritorious service at Williamsburg and colonel for meritorious service at Seven Pines.[459]

Weitzel's and Emory's troops had remained encamped near their landing on the 10th because it was Banks' intent not to attack in force until something of Grover's progress was learned.

At noon on the 11th, Weitzel had begun a halting advance on Bisland.[460] By six o'clock they had advanced ten miles and bivouacked about a quarter of a mile beyond Pattersonville, opposite Sibley's Texas men. Emory followed with the 3rd Division and went into bivouac on Weitzel's left, figure 8, from Irwin, facing p. 96.

Finally, early on the morning of the 12th, Banks gained the word that Grover was underway. It was now necessary to occupy the enemy, "yet not too strongly, lest he abandon his position too soon and suddenly spoil all."[461] Weitzel pushed forward, stiffly resisted by the rebel cavalry, which was supported by two regiments of infantry. By 4:00 p.m., the advance, consisting of a line of battle made up of the 12th Connecticut on the left; the 160th New York and 75th New York at the center, with the 114th New York and the 8th Vermont on the right, had traversed through three miles of rough cane fields coursed with muddy trenches. Contested all the way by the enemy skirmishers, they finally reached just short of the range of the guns of Brig. Gen. H. H. Sibley. Here was observed a row of cane shocks "arranged with careful negligence parallel with and about half to three quarters of a mile from the plainly visible works." To Weitzel's discerning eye, these had been placed to sight and range the rebel guns, and he halted the command. He then sent word to Banks that he was just outside of the enemy's range and asked to open his artillery before making any further advance on the fort.

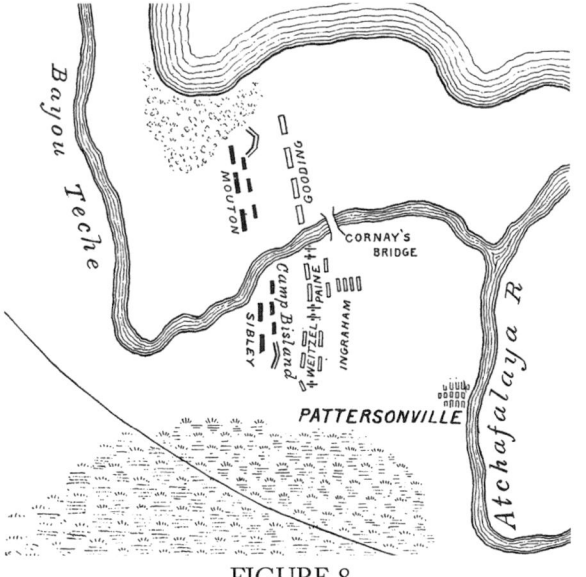

FIGURE 8

---

459. Cullum, Vol. 2, no. 1453, pp. 256, 257. Seven Pines was also known as Fair Oaks.
460. O.R. Vol. 15, pp. 294, 324–325; Hall, H & J, pp. 90–92.
461. Irwin, p. 91.

Banks' reply was to advance. The order was obeyed, and in less than five minutes, Weitzel's suspicions were verified. The rebels opened a "quite severe" fire. Two men of the 75th New York were killed and three wounded. The "political general" was thus taught his lesson, and he ordered a cease to the advance, with the troops to lie in the trenches of the cane field. Bainbridge's battery on the right and Carruth's battery on the left were brought up and opened a return fire, which lasted until sunset, when the troops were withdrawn about a mile; to pass a night "besieged with mosquitoes from the adjacent swamp."

Finding Mouton in his recently-thrown-up and only-partially-completed works on the opposite side of the Teche too strong for the cavalry, Banks ordered Emory to reinforce it with the 31st Massachusetts and the 175th New York, along with a section of the 1st Maine Battery. They were deployed, and with Gooding in command, they were ordered to observe and conform to the advance of the main line.[462]

Banks had waited for the sound of Grover's guns, and none having been heard, Banks gave orders to continue the assault the next day, thus risking the potential for Taylor withdrawing before Grover was ready.[463]

It was at this time that Taylor had received word of the position of Grover's flotilla, off Hutchin's Point, and he had sent Vincent's regiment to oppose it. A few hours later, Taylor left Fort Bisland to visit Vincent, leaving behind orders with General Sibley to prepare for a bold plan. Taylor had no thought of a withdrawal. Instead, Sibley was to prepare for an attack on Banks the next morning. Taylor had calculated that the result would be that Banks would recall Grover for reinforcement.[464] Returning at daylight, Taylor was angry to discover that Sibley had failed to organize for the attack, considering it impractical.[465]

The dense fog that so characterized that time of the year – which had delayed Farragut at Port Hudson and, yesterday, Grover – had prevailed during the evening, but it lifted on the morning of the 13th to reveal a bright sunny day. Banks' attack was resumed with the whole line returning to within musketry range.[466] Banks ordered the right strengthened, sending the remainder of Gooding's brigade – the 38th Massachusetts, the 53rd Massachusetts, the 165th New York – and the remaining sections of the 1st Maine Battery, to the opposite side of the Teche.

Taylor then brought the gunboat *Diana* well out in front of his earthworks to provide enfilading fire on the advancing Yankee line, but it was silenced by a fortunate shot from one of the 30-pounder Parrotts of the 1st Indiana Heavy Artillery. It was then observed to withdraw up the bayou and out of the fight. This risk

---

462. Irwin, p. 97.
463. O.R. Vol. 15, pp. 296, 297.
464. O.R. Vol. 15, p. 389, 390.
465. O.R. Vol. 15, pp. 1093–1095.
466. O.R. Vol. 15, p. 296; Irwin, pp. 93–103; Bissell (McManus), p. 35.

removed, the artillery engagement became general.

By afternoon, when the Union forces had reached advanced positions in front of the works, Banks began to believe that something had gone awry with Grover. He feared losing an opportunity to defeat Taylor then and there. Somewhat in desperation, he gave discretionary orders to Weitzel and Emory to form a plan to storm the Confederate works if a favorable opportunity presented itself. Weitzel and Halbert Paine of the 4th Wisconsin, commanding Emory's 2nd Brigade, conferred, and then concluded to attack. Gooding, as already ordered, would conform to whatever took place.

Every preparation having been made, the generals waited for Banks to give approval. It was already past 4:00 p.m., and the opportunity was passing. Banks was still anxiously waiting, weighing the cost of the assault versus the chance of news from Grover.

Suddenly, a shell was seen to travel high over their heads, which burst on the Confederate works. Then followed the deep roar of the *Clifton*'s IX-inch bow gun. Having covered Grover's landing, it had been dispatched back to the Teche, with the news so long awaited. Relieved, and assuming Grover to be in place, Banks cancelled any further operations for the day and withdrew his front line out of the range of the Confederate musketry. The trap was closed, and the full assault could begin the next day.

After dark, signal rockets sent up by Grover confirmed the news.

At one point in the day's attack, Taylor had worried that some of his green troops had been shaken sufficiently that Weitzel might have broken through. However, he claims, he had steadied his men toward the end of the day to a degree that he was looking forward to any assault "feeling confident of repulsing it."[467] The *Diana* was expected to be repaired, and behind the breastworks "we had suffered but little." We must question this as bluster, written as it was in 1879. According to a former member of Taylor's staff, Taylor "was the most anxious man in the Southern Confederacy, when . . . he learned that Grover's division had landed by way of Grand Lake . . ."[468]

However, at 9:00 p.m., all that quickly changed. Reily appeared with the news that, yes, Grover had landed and advanced to the Porter plantation but had not entered Franklin. The indication was that the cut-off road to New Iberia was open. From Taylor: "Here was pleasant intelligence! There was no time to ask questions." Mouton was directed to withdraw to Franklin, starting the artillery – one section to reinforce at Irish Bend – followed by the trains and the infantry.[469] Green, with the 5th Texas Regiment, Waller's battalion, and the rifle section of Semmes' battery

---

467. Taylor, pp. 132, 133.
468. McManus, *Battlefields of Louisiana*, p. 17.
469. O.R. Vol. 15, p. 1092.

constituted the rear guard. Semmes[470] was ordered to get the *Diana* to Franklin by dawn. Taylor then rode for Franklin with Reily and arrived at his camp above the village at about 2:00 a.m. The sleeping men were awakened and sent off to position themselves in Nerson's Woods, at Irish Bend.

Taylor now rode farther on and discovered Clack and his men camped along the cut-off road. Originally ordered by Taylor to reinforce him at Bisland, Clack had stopped for the night. They were near the Yokely Bridge, almost within sight of the Federal camp.[471]

The road and the causeway bridge were safe. Taylor's comment: "It was a wonderful chance. Grover had stopped just short of the prize. Thirty minutes would have given him the wood and bridge, closing the trap on my force." Taylor is quite right. Grover was camped about three miles away, about a half hour at the quickstep.

Back at Bisland, Halbert Paine, commanding Emory's 2nd Brigade, noted in his diary: "At 1:00 a.m., a messenger from the picket line reported the moving of artillery. A personal observation upon the picket line failed to satisfy me whether a general evacuation was going on or only a transfer of guns from one part of the fortification to another. . . . Soon Gen. Emory ordered me to go into the works if I could. A like order came from Gen. Banks . . . The 8th New Hampshire being deployed as skirmishers along my entire front, marched to the entrenchments. As I took and planted the flag of the 8th New Hampshire on the breastworks, they all bounded in with three loud cheers." [472]

An immediate advance was ordered. Quoting from the *History of the 114th New York Regiment* [Pellet]:

> At four o'clock a.m. of the 14th, we advanced on the works. Not a shot was fired. We moved almost breathlessly. Soon we saw one of our flags waving from the parapet of the works. The cause was apparent. The enemy had fled. Such was the battle of Fort Bisland.

Fifteen miles or so to the north, it was dawn at Grover's encampment at Irish Bend when the troops of Birge's 3rd Brigade fell into line and headed south along the bayou road toward Franklin. It having rained during the night, they had little sleep, lying on the freshly plowed ground of Mrs. Porter's cane field "drenched to the skin, and covered with the red soil of Louisiana, we looked like a moving brick yard."[473] Five companies of skirmishers of the 25th Connecticut led off, followed by five companies in reserve. Next, came two companies of the 26th Maine, Bradley's section of Rodger's battery, eight companies of the 26th Maine, the 159th New York, the other two sections of Rodger's battery, and, finally, Birge's own: the 13th

---

470. Son of the *Alabama* captain.
471. Bissell (McManus), p. 35.
472. Stanyan, p. 199; Pellet, p. 69; order of march: O.R. Vol. 15, p. 384.
473. Maddocks, p. 354.

Connecticut. Dwight's brigade and Battery L were next in the column, followed by Kimball, with Nims' battery.

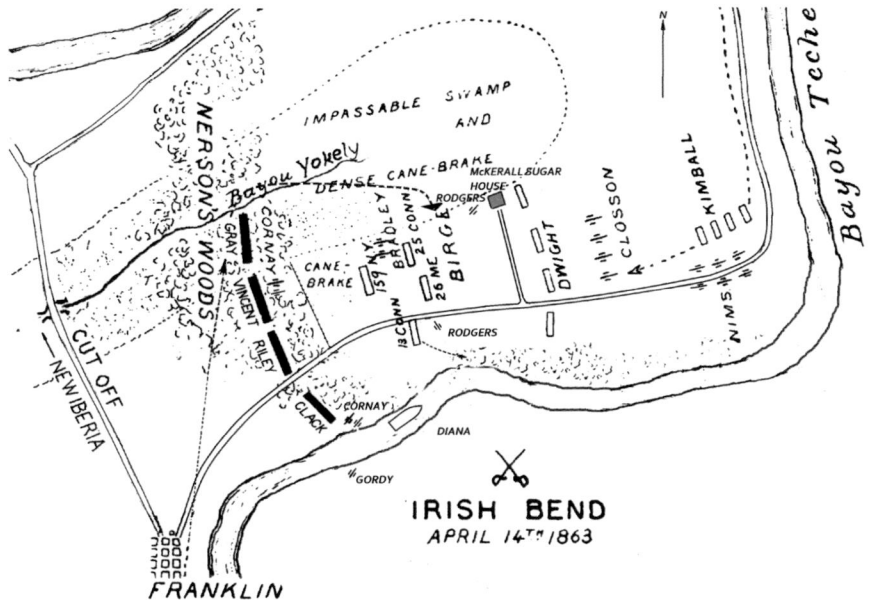

FIGURE 9

Figure 9 provides an introduction to the battlefield.[474] Note that the impassable swamp surrounding Yokely Bayou in the center, and Nerson's Woods south of it, confine the battlefield to the open cane field and the small section of woods to the west.

Marching south for about a half hour, they passed a sugarhouse on their right. They had left the Porter plantation and entered McKerall's,[475] a position where a sharp bend in the road makes its turn to the southwest.

The 25th Connecticut advanced into the cane field. Grover having expressed the thought that there was "nothing more than a picket" in the woods – quite a change in attitude from the night before.[476]

Of course, in place, and unseen at the edge of the woods, were the dismounted cavalry of Reily and Vincent, and Clack's battalion. Compare this situation with that of Weitzel at Bisland and Banks' order to advance without a preliminary bombardment. The result then was casualties; and Banks, realizing his mistake, allowed Weitzel to withdraw. Then, according to accepted tactics, Weitzel brought up the artillery.

---

474. Figure 9, portion from Irwin facing p. 112.
475. O.R. Vol. 15, pp. 391, 392.
476. Sprague, p. 111.

FIGURE 10

From Harper's Weekly, May 16, 1863 – From a sketch by Wm. Hall of the 22nd Maine, Dwight's Brigade. Assumed to show when Dwight was called forward following Gray's flank attack. Note the fence on the left, and the direction of the cane furrows – indicating that the lead regiments had turned to face the attack.

Leading to their right, or west, they entered "an immense cane field, its furrows in line with the road. On the west the field was bounded by a rail fence, beyond which arose a dense wood of magnolias, cottonwood and semi-tropical trees looking like a long green wall . . . near to the ground this forest was absolutely impenetrable to the sight, by reason of the suffocating briars, vines, palmettos and underbrush. We ought to have occupied these woods the night before . . ."[477]

As the 25th Connecticut's skirmishers marched forward, they were greeted by "a puff of smoke from the green wall in front of us and a second or two later the crack of a rifle." Next, it would be the shells from a section of Cornay's St. Mary's Cannoneers, placed as they were at the edge of the woods, by the road, and at the upper portion of the woods.[478] The skirmishers were now called in, and the regiment changed to the right "front forward on first company."

Those few cracks of Taylor's riflemen could have been effective when the 25th was some three hundred yards from the edge of the "green wall," but the majority of Taylor's force still carried smoothbore muskets. These would not be effective until the 25th marched much closer; to less than 50 yards, if they were to continue to advance.[479] They did, marching erect and in plain sight of the enemy

---

477. Bissell (McManus), p. 36.
478. O.R. Vol. 15, p. 391. Cornay had six guns. One was destroyed at Bisland (Taylor, p. 134), and the fifth was brought here by Cornay and the 4th Texas Regiment (Noel, *A Campaign...* p. 47).
479. McMorries, pp. 62–63. "Our arms were the old flintlock musket (but they were a sure fire) not effective over forty yards . . ."

– something out of the tactics of the wars of Napoleon.

By now, (it is estimated that the time was 6:45 a.m.) the 25th had advanced more than 800 yards, loading and firing as it went, and it had become evident that there were more than a few skirmishers in the woods. They were opened upon with buck-and-ball "four shots to our one in return . . ."[480] Soon shells began to burst overhead. The 25th was ordered to lie down in the furrows of the cane field, a difficult position from which to reload their Enfield rifles. They were initially alone on the field. "An unexpected force of the enemy having thus developed, Grover then ordered up the rest of Birge's brigade." Instead, it would have been the time to order his artillery to shell Taylor's positions before sending in more troops to the slaughter.

The 26th Maine was ordered forward and took position to the left of the 25th Connecticut. They were also ordered to lie down and fire "when the enemy could be seen." Bradley's section of Rodgers' Battery B, was ordered forward by Birge, and placed in the interval between the 25th Connecticut and the 26th Maine, about five hundred yards from the woods.[481]

At about 7:00 a.m., the 28th Louisiana, under Gray, reported to Taylor from Franklin, and unseen behind bushes and canebrake, had begun deploying (the dotted line and arrow in figure 9) along a ditch at the northern portion of the field. Grover would soon have a surprise flanking attack to deal with, Gray nearly doubling the strength of the force Taylor had initially deployed.

Then, the "peculiar whistle of a Parrott shell was heard, and Semmes appeared with the *Diana*[482] . . . Roger's battery was brought up to reply to the rebel artillery and also to the *Diana*, whose guns now swept the field."[483] One gun was placed eight hundred yards away and another, some four hundred yards in advance.

In the meantime, the 13th Connecticut, remaining south of the road, had advanced in the woods between the road and the bayou. Company A's skirmishers led the advance at the double-quick. The cover of the trees allowed them to quickly advance without firing until they reached a stretch of open ground close to where the guns of Cornay's St. Mary's Cannoneers were in battery. Crossing the 300-yard open stretch, they had the advantage of their skirmishers being equipped with Sharps breechloading rifles and like the 25th Connecticut, the remainder of the troops had received training to load and fire their Springfields while advancing. This steady barrage added up to their quickly overpowering Cornay's position.

At this point, the 13th was far in advance of the rest of the division, and they knew that a furious battle was taking place to their right and rear. They were withdrawn to a position where they posted sharpshooters to try to pick off the *Diana*'s gunners and stood ready to support the advancing 159th New York. When

---

480. Bissell (Ellis), p. 30; Bissell (McManus), p. 37.
481. O.R. Vol. 15, pp. 366, 384.
482. Taylor, p. 133.
483. Sprague, pp. 112, 117, 118.

the 13th withdrew, the 4th Texas Cavalry came to the aid of Cornay's Cannoneers, and they were able to make off with their most important possessions: their guns. Nevertheless, the 13th retained some 50 to 60 prisoners, 2 caissons, a limber, and the Cannoneer flag.[484]

The 26th Maine had originally formed to the left of the 25th Connecticut, but now had the 159th New York, marching at the double-quick, move through its left and then through the depleted 25th Connecticut, which had turned to meet Gray's threat. Firing as they had been, for more than an hour, the ammunition of the 25th was beginning to run out, and the number of men wounded had significantly depleted their ranks. Those wounded that were able to crawl, retreated.

It was fortunate that the 159th had arrived because the 26th Maine was having troubles of a different kind. Its old Harper's Ferry muskets, firing buck-and-ball, were fouling after becoming heated from repeated firing. Their weapons, one by one, were becoming useless.[485] They were in desperate straits, looking for available guns left by the wounded, all the while lying down under the increased fire from the 28th Louisiana. Philip Holmes remembers having found a Belgian rifle near one of the dead of the 159th, and, while reaching over to obtain the contents of his cartridge box, was wounded by three almost simultaneous shots. He was helped to the rear by a comrade.

The response of the 159th faltered when its colonel, Molineux, was struck in the mouth by a ball and fell, just as he had raised his sword to order a charge. Soon, Gray's whole line drove forward, led by a mounted Taylor and members of his staff.[486] This shattered the Yankee troops, who ran to the rear in disorder, abandoning Bradley's battery, which was forced to withdraw to a new position. In the process, Bradley had 8 men and 6 horses wounded.[487] Seeing this, Rodgers briefly waited for orders to turn his guns toward Gray and finally fired over the heads of the Union line to cover "the retrograde movement as well as possible."

From Taylor: "The enemy then displayed a much larger force . . ." This was Dwight, with the 1st Brigade, which had been next in the marching column, along with Battery L. Dwight's regiments had been ordered forward while Battery L and the remainder of the artillery, 14 guns, stood in reserve. With that, Gray fell back, and the general withdrawal that had been the core of Taylor's plan was set in motion.

Taylor reports that all his wagons and troops from Bisland had passed through Franklin by 9:30 a.m. Fear of Banks' force from Bisland arriving in his rear, Taylor had to wrap it up quickly.

Nims' battery had come up with the 2nd Brigade; initially posted by Captain

---

484. Noel, *Autobiography*... pp. 92, 93; Sprague, p. 118.
485. Maddocks, pp. 354, 355; O.R. Vol. 15, p. 384, Birge's report.
486. Tiemann, p. 29; Taylor, p. 133. Here, Taylor recounts his charge as happening earlier. Taylor's report in O.R. Vol. 15, p. 392, places it along with Gray's charge.
487. O.R. Vol.15, pp. 367, 392, 393. Subsequently altered to 1 killed and 7 wounded.

Closson near the road, it was shifted to the edge of the bayou to meet the *Diana*, should it move farther up. It did not, but it kept firing for some time while Dwight advanced. Semmes had been instructed to work the *Diana's* gun to the last moment, then get ashore with his crew and destroy the boat. Fortunately for Dwight's troops, many of its shells failed to explode.[488]

When Dwight's regiments arrived at the edge of Nerson's Woods, they were ordered to halt, Grover fearing that Taylor had massed his troops for a counter-attack – which is yet another example of Grover's misreading of the situation – Taylor had long since vanished.

An estimate of time, other than Taylor's mention of 7:00 a.m. when Gray arrived, and 9:30, when his main force had passed through Franklin, is found in Birge's report: "About 2 p.m., was ordered by General Grover to advance . . . to feel the enemy. Before this order could be executed he suddenly withdrew, and the Diana was discovered to be in flames."

Given that the *Diana*'s crew were captured, they being the last to retire, and allowing time for Semmes to back out of his position, abandon the *Diana*, and then set it on fire, it is concluded that the last of Taylor's force (Mouton's) had crossed the burning Yokely Bridge by about 11:30 a.m.[489]

Dwight heaped abuse on the 3rd Brigade for the "embarrassing" retreat of the 25th Connecticut, but the casualty figures show that by the time his brigade had made its advance it met only token resistance.[490] It is clear that by that time Taylor had turned the page of his playbook and had begun his skillful withdrawal. The respective lists of casualties tells the story: the 3rd Brigade, the 25th Connecticut had 86; the 26th Maine, 61; the 159th New York, 97; and the 13th Connecticut, 54; numbers approaching 20 percent. In contrast, Dwight's 1st Brigade had a total of 16, with none in the 6th New York or the 1st Louisiana.

McKerrall's sugarhouse, which was used as a hospital, was now filled with over 250 "Union and rebel" subjected to the medieval medical procedures of that era by only three or four overwhelmed surgeons. Passing "a pile of legs, feet, arms, hands beside the table" while searching for over an hour for his wounded men, Capt. Homer Sprague of the 13th Connecticut detailed a man to provide for them, and only then left; "sick at heart of war and all its surroundings."[491]

Taylor did not report any casualties, save commenting that Reily was killed and Vincent was wounded "… and many others went down."[492]

It is clear that the Battle of Irish Bend was mismanaged by Grover. Battery L; Nims' 2nd Massachusetts, and two guns of Rodgers' Battery C were held in reserve,

---

488. O.R. Vol. 15, pp. 365, 367; Taylor, p. 134.
489. Tiemann, p. 30. O.R. Vol. 15, pp. 393, 399: "midday."; O.R. Vol. 15, p. 393.
490. O.R. Vol. 15, p. 319.
491. Sprague, p. 120.
492. Taylor, p. 133.

despite their potential to have shelled Taylor into oblivion.

Additional points made by Irwin[493] in his 19th Corps history, are that Grover knew nothing of the surrounding country and that it was Weitzel who had done all of the preliminary planning. Weitzel had recommended a smaller force – two brigades of infantry with six guns. Thus, they would have had fewer problems with their transportation to the site, and it is implied that Weitzel would have known of the cutoff road and would have moved to block it.

It was Banks who ordered the larger force, innocently enough, and appointed the senior general to its command, innocently enough. Unfortunately, it was an example of a lack of practical military judgment and misplaced trust.

It is interesting to find current reports of Irish Bend being a Union victory. Nothing could be further from the truth. Taylor escaped as he had planned, and he would reorganize, be reinforced, and go on to ruin Banks' Red River Campaign, finally driving the Yankees out of northwestern Louisiana permanently.

Chasing the Fox

Banks' forces entered Franklin on the heels of Taylor's forces leaving via the cutoff. Grover having failed to trap Taylor, there was nothing left for Banks to do other than begin a fruitless month-long march toward Alexandria, figure 11, much like the proverbial fox and hounds. The fox was always ahead and just out of reach.

Taylor never published a casualty list, but bemoaned the fact that his force began to fall apart on the march north, a fact that he fails to mention in his postwar memoir. "Nearly the whole of . . . Fournet's battalion, passing through the country in which the men had lived before joining the army, deserted with their arms . . ."

Theophilus Noel,[494] a member of the 4th Texas Cavalry, remembers:

> We started for the rear; those under command toward Alexandria and those in desperate want of home comforts toward Niblett's Bluff, Texas; very few of whom ever afterward heard the sound of their own reveille toot-horns, much less Yankee cannon or musketry.

The ability of the Texans to disappear into the vastness of Texas was to be repeated in 1865. Then, rather than surrender, their numbers would be seen to have just melted away, some fleeing to Mexico.

Taylor's superior, E. Kirby Smith, commanding the Department of the Trans-Mississippi, offers the estimate of Taylor's loss as one-third from "straggling and desertions." Few were killed, and with their arms, they could – and they would – be *impressed* once again.

Duryea's Battery F of the 1st U.S. Artillery, which accompanied Weitzel's

---

493. Irwin, pp. 118, 119.
494. O.R. Vol. 15, pp. 393–396, 387; Figure 11, Irwin, pp 80a, 80b; Noel, *Autobiography*, p. 93. The road west to Niblett's Bluff left from Vermilionville, now Lafayette.

brigade, took the lead in chasing Taylor as he retreated. There was a skirmish at Jeanerette, where Taylor had encamped, as the artillery of Green's rear guard faced them. It was in this action that Sgt. Charles Riley, among others, was cited for coolness and accuracy in sighting their guns, which drove Green's guns back.[495]

Taylor's first prolonged stop was at his first defensible position, Vermilion River, which he reached on the afternoon of the 16th. After they crossed, the bridge was burned, and his troops were allowed to rest until midday on the 17th, when Grover's division caught sight of Green's Texans, posted so as to prevent its reconstruction. Dwight's skirmishers were then deployed on either side of the road in support of Battery L and Nims' battery, which "shelled the Rebels out."

Battery L's record of events gives an unusually detailed account of the action: "On the 15th took up line of march and reached Vermillion Bayou on the 17th estimated distance 45 miles. At Vermillion bridge, Taylor's section of the Battery was opened upon by Sharpshooters and the Rebel artillery of 4 guns from the cover of the opposite bank. Casualties 2 horses killed, the remainder of the Battery in action soon shelled the Rebels out." Battery L had remained with Grover's division as did two sections of Nims' battery, which also participated, under Closson's direction.[496] Grover had proceeded directly to that point while Banks, with

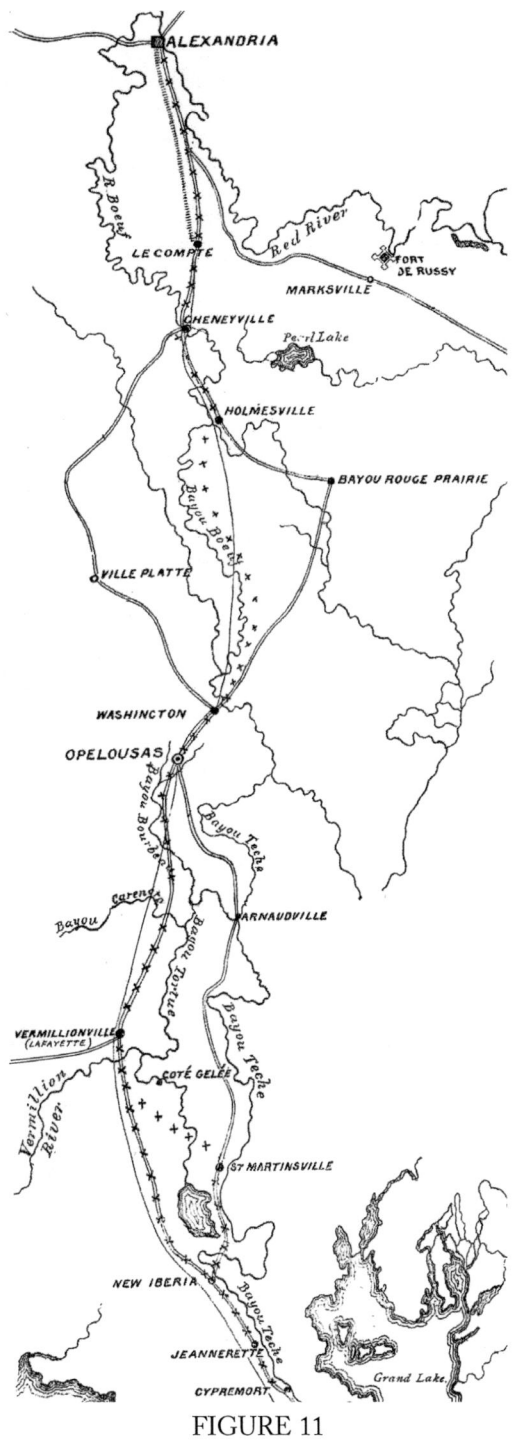

FIGURE 11

---

495. O.R. Vol. 15, pp. 335–337, 393.
496. O.R. Vol. 15, p. 365; Irwin, p. 122.

Weitzel and Emory, had branched off the main road at New Iberia and headed to Saint Martinsville. The purpose of the diversion was to capture or destroy five more of Taylor's rivercraft, but it was learned on the 18th that four had been destroyed by Taylor's men. Upon hearing of Grover being stopped at Vermilion River, Banks ordered the expedition to return.

The facility with which Taylor's forces, many mounted, had kept ahead of Banks' footsore troops had highlighted the issue of cavalry. True, many of Taylor's troops were mounted, but many were not, but by their background and long service were simply inured to these tough marches. In any event, Banks had been looking for horses to convert some of his infantry to cavalry ever since he had arrived in the department. We remember his having been rebuffed and lectured by Lincoln for having made too many requisitions. In March, Banks had written to Halleck about the deficiency of cavalry, especially regretting the fact that the 2nd Massachusetts, originally raised by Governor Andrew for the Banks expedition, had instead been sent to Virginia. On April 18th, Banks sent a telegram to Halleck again requesting that the 2nd Massachusetts be sent to him.

Banks finally took his own initiative in view of the large new prairie area he had just occupied. The 114th and the 159th New York, having arrived at Vermilion Bayou, were ordered to proceed back down the country and collect all the horses and cattle they could find, and deliver them to Brashear City. After a day's rest, which allowed the men to wash their clothing, which was "blackened with dust, and to care for their feet, which were in wretched condition," they set off on the 19th.[497] He also ordered Gen. T. W. Sherman at New Orleans "to collect all of the saddles and bridles in the city . . ."

Earlier, the locals had successfully hidden horses and supplies from the Yankee invaders; not so this time. The 114th arrived at Berwick City on the 28th with 3,000 head of cattle and 800 horses; the 159th arrived the next day, having spent the extra time to destroy the earthworks at Bisland. They had collected "about five thousand head, as well as a number of horses and mules . . ."

Opelousas

The bridge at Vermilion Bayou repaired, Grover's force crossed and proceeded in rain and mud to Carrion Crow Bayou. The next day, April 20th, they arrived at Opelousas, to where the government of Louisiana had initially fled after the fall of New Orleans[498] and from which it now had fled to Shreveport. Opelousas was as far as Banks had originally intended to go, and here his headquarters remained until May 4th, to sort out his situation regarding cooperating with Grant, and to make several other significant decisions:

---

497. Pellet, pp. 73, 74; Tiemann, p. 36; O.R. Vol. 15, p. 703.
498. O.R. Vol. 15, pp. 735, 307.

1. The number of horses already confiscated allowed the creation of more cavalry. The 41st Massachusetts, the 4th Wisconsin, and Company G of the 8th New Hampshire were designated as mounted. Eventually, the name of the 41st was changed to the 3rd Massachusetts Cavalry, though on May 17th, it had become the "Mounted Rifles."[499]
2. General orders no. 35, Department of the Gulf,[500] dated April 27th, 1863, required registered enemies of the United States to leave the department "before the 15th day of May proximo. The provost-marshal-general is charged with the peremptory execution of this order."

   Butler had first required the oath on June 10th, 1862. That those who had refused to take it were still there is telling. Banks had first tried reconciliation when he arrived, calculating that Butler's harsh measures had only alienated the citizenry. Not true, reconciliation had not worked. These declared enemies would cause trouble if the city's security was weakened by drawing off forces to Port Hudson; and they could no longer be tolerated, in view of the fact that there were as many, or more, hypocrites who were not willing to sacrifice their personal interests, and had declared themselves loyal.

   The June 6th issue of *Harper's Weekly,* p. 363, noted: "You can readily imagine what a flutter this has caused in the ranks of the Secesh . . . It is really quite amusing to spend an hour at the [provost marshall's office]. People – principally ladies – are constantly flocking in to try if there is no possible way of avoiding the dreadful alternative of starvation in Dixie or bowing to the horrible Yankee flag."
3. General Orders No. 40, Department of the Gulf, dated May 1st, 1863, proposed the "Corps d'Afrique."[501] It would consist of 18 regiments of former slaves. Officers for these units were to be obtained from "the nomination of fit men from the ranks and from the lists of non-commissioned, and commissioned officers . . . respectfully solicited from the generals commanding . . ." All officers were to be white, a blow to the faithful free blacks who had volunteered for the Native Guards, and a reversal of "Beast" Butler's promise. This, we shall see, opened up opportunities for the enlisted men of Battery L.

On April 20th, Lt. Commander Cooke, commanding the *Estrella*, and his three other gunboats – the *Arizona, Clifton,* and *Calhoun* attacked Taylor's Fort Burton at Butte-la-Rose, figure 1, chapter 7. A detachment of six companies of the 16th New Hampshire were on board as sharpshooters and were to occupy the fort if it fell.

---

499. Stanyan, pp. 209, 211; Ewer, p. 79.
500. O.R. Vol. 15, p. 710.
501. O.R. Vol. 15, pp. 716, 717.

The *Clifton*, with the heaviest armament, pushed ahead, followed by the *Arizona*, *Calhoun*, and *Estrella*. After "two or three" shots, "the fort immediately ceased firing and surrendered."

About 60 of the garrison were taken prisoner, the few remaining having escaped on the *Webb* and *Mary T*. A large quantity of ammunition and stores was captured.[502] Only four of the 16th New Hampshire men were casualties, with two dead and two wounded.

Stepping ashore, the conquerors were greeted by the Confederate commander, with the pregnant statement: "You will be doubtless glad to get here, but you will be gladder when you leave . . ." Next, we quote from the author of the *History of the Sixteenth New Hampshire Volunteers* (Townsend), who penned: "No prophet of early times ever has offered a truer prediction."

Fort DeRussy, further north on the Red River, would be the next target for the navy, so the New Hampshire men were left at Burton. They were basically forgotten, without sufficient supplies for their 600, in the center of a malarial swamp. Arrangements were finally made for the fort's abandonment on May 30th, but from privations due to a lack of rations, overcrowding, and disease, there were but 150 effectives that day.

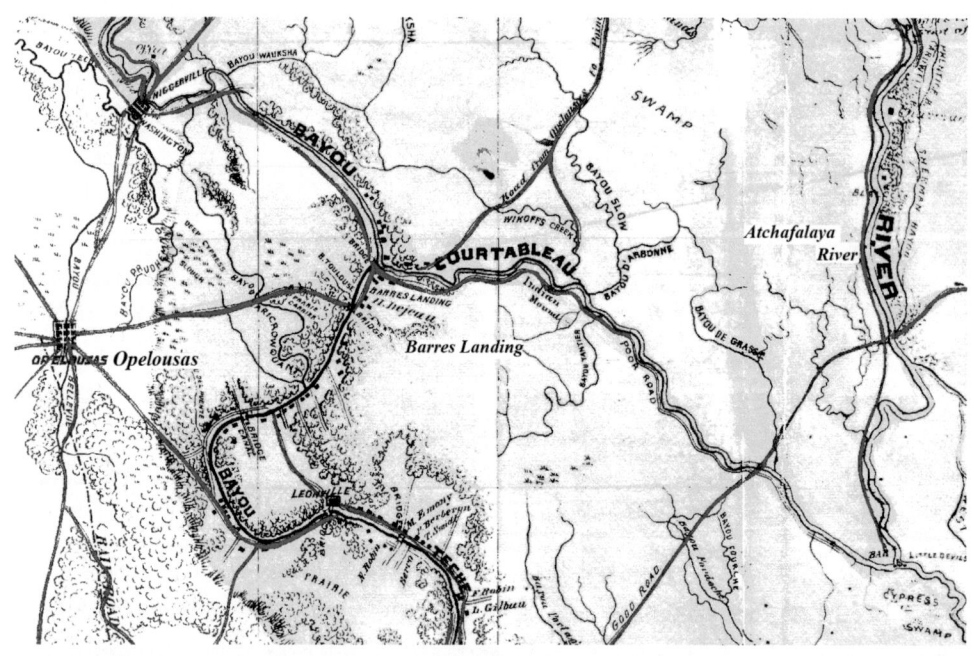

FIGURE 12

Barre's Landing, at the junction of the Teche and Bayou Courtableau, which entered the Atchafalaya to the east, figure 12,[503] was taken on the 22nd. This now

---

502. ORN Ser. I, Vol. 20, pp. 153–155; Townsend, pp. 152, 159, 166, 196; Irwin, pp. 126, 127.
503. O.R. Vol. 15, pp. 313, 338, 714; Ewer, p. 79; Figure 12, portion of NOAA Historical Map

opened water communication back to Brashear City by the Atchafalaya, as well as the Teche, and of course, it allowed direct passage to the Red River and the Mississippi. The alternate water passage past Port Hudson, so long sought, was now in Banks' hands, though falling water would soon render it useless.

Barre's Landing became a suburban camp to Opelousas, eight miles west. Battery L, still with Grover's division, arrived there on April 26th and did not leave until ordered to march to Alexandria on May 5th.

While here, it was regarded that the important work of seizing the products of the countryside be carried on, as well as "receiving and guarding [the slaves] Gen. Butler's historic 'contraband of war'."[504] From this, an unusual plantation story arose:

Colonel Greenleaf of the 52nd Massachusetts was named the commandant of the post, and six companies of the regiment were assigned as a part of the post command. Soon after taking command, the colonel had his attention drawn to a planter who professed to be a Union man. He had voluntarily delivered his cotton, sugar, and molasses, hauling it to the steamboat himself, and taking the quartermaster's receipt. It was also learned that this Mr. Gantt had on one occasion saved the quartermaster by warning him of an enemy ambush. All that Gantt had asked in return was that he be protected "from molestation."

Soon, the planter introduced himself to the colonel, invited him to visit the plantation, and once assured that he would be protected, Gantt departed. Quoting the colonel:

> A few days later a delegation of negroes – intelligent, healthy, hearty looking fellows – waited upon me, as they said to 'advise with me about the situation.' They had noticed . . . their people flocking to the post by the thousand, and that they were protected and fed . . . They wanted my advice with regard to their own coming, also, with their families. I asked them whence they came and to whom they belonged. They answered that they came from a plantation about two miles away, and that they belonged to 'Massa Gantt' – our Union friend and planter.
>
> I then asked them if . . . Gantt was kind to them. They said he was. If they fed and clothed them well. They said he did; that they had no fault whatever to find with his treatment of them, but they 'wanted to be free.' I answered them that they could come within our lines, with their families, and that, if they came they would be protected the same as the others; but that from what I had learned of 'Massa Gantt' before, and what they had just told me, my advice to them would be to remain, for the present, just where they were, explaining that 'I had little doubt' that they all would be free when the war was over, but that if they undertook to follow the fortunes of the army . . . I feared but few

---

CWAB, Atchafalaya Basin, 1863.
504. Moors, pp. 131, 132, 141, 142.

of them would live . . . The death rate from disease in the great congregations in the camps had proven to be large.

They thanked him and departed.

"A week or two later," the colonel accepted Mr. Gantt's invitation to visit the plantation. Seeing the conditions under which Gantt's slaves were living, Colonel Greenleaf was confirmed in the opinion that he had given them wise counsel.

On May 21st, the same slave group was seen in the refugee column that had evacuated Barre's Landing. Despite Colonel Greenleaf's advice, the promise of freedom had overcome all else.

Cooperation with Grant

On April 10th, Farragut's secretary, Gabaudan, finally arrived at Banks' headquarters at Brashear City with the memorized substance of the letter from Grant to Farragut dated March 23rd.[505]

In the original letter, Grant had noted that a Lake Providence experiment was expected to be successful and that: "This will give navigable water through by that route to the Red River." Further, assuming that transports would become available, and if "Admiral Porter gets his gunboats out of the Yazoo," Grant could send McClernand's 20,000 "effective men" to cooperate with Banks.[506] Though there were two big ifs in the original, Gabaudan may not have stressed them. In any case, Banks sent the secretary back with a letter, which he was also to memorize and destroy, with the answer that he, Banks, could be at Baton Rouge "to cooperate with you against Port Hudson. I can be there easily by May 10." Significantly, this was not received by Grant until May 2nd.[507]

Now, at Opelousas, on April 23rd, he again wrote to Grant,[508] giving a summary of his situation. Taylor's forces were dispersed. One portion, Sibley and Mouton, were on the road to Texas, and Taylor's artillery and some cavalry were on the road to Alexandria.

In the interval between March 20th and May 2nd, Grant had known nothing of Banks' situation, and on April 30th, Grant's forces had begun crossing the Mississippi.[509] His long-planned southern attack on Vicksburg was finally in motion. There was no turning back,[510] in the middle of an assault that had been impatiently

---

505. ORN Ser. I, Vol. 20, pp. 765, 788; O.R. Vol. 15, pp. 294, 295; O.R. Vol. 24/III, p. 225; Irwin, p. 135.
506. O.R. Vol. 24/III, pp. 131, 182, 183.
507. ORN Ser. I, Vol. 20, p. 71; Irwin, p. 139. This shows it in Farragut's hands on the 1st. Irwin noted Grant received it on the 2nd. In fact, Gabaudan did not destroy it but carried it in his mouth when chased by enemy pickets as he hiked across the peninsula opposite Port Hudson.
508. O.R. Vol. 15, pp. 304, 305; O.R. Vol. 24/III, p. 225.
509. O.R. Vol. 24/I, p. 34.
510. O.R. Vol. 24/I, p. 30.

anticipated by everyone looking over Grant's shoulder, from Lincoln on down, Grant did not feel that he could send troops to Banks.

Communications finally became faster since Farragut had established the protocol that he would descend the Mississippi with the *Hartford* every Thursday to send and receive flag signals across the point opposite Port Hudson, masthead to masthead, to Alden, in the *Richmond*, below.

Nevertheless, on the 10th of May, Grant put an end to the negotiations, for the lack of a better description. It may be useful to quote his note in some detail. After describing that his advance was far along, he said:

> It was my intention, on gaining a foothold at Grand Gulf, to have sent a sufficient force to Port Hudson to have insured the fall of that place with your co-operation, or rather to have co-operated with you to secure that end . . . Meeting the enemy, however, as I did, south of Port Gibson, I followed him to the Big Black, and could not afford to retrace my steps. I also learned, and believe the information to be reliable, that Port Hudson is almost entirely evacuated . . . I would urgently request, therefore that you join me or send all the force that you can spare . . . Grierson's cavalry would be of immense service to me now, and if at all practicable for him to join me, I would have him do it at once.[511]

Banks, now at Alexandria, received this early on the morning of May 12th. It had been hand carried. Finally, the two generals had a knowledge of each other's status that was something akin to current. Banks responded[512] that he had no water or land transportation to:

> . . . be of service to you in any immediate attack . . . The utmost I can accomplish is to cross for the purpose of operating with you against Port Hudson . . . Were it within the range of human power, I should join you, for I am dying with a kind of vanishing hope to see two armies acting together . . .
>
> The only course for me, failing in co-operation with you, is to regain the Mississippi, and attack Port Hudson, or to move against the enemy at Shreveport. Port Hudson is reduced in force, but not as you are informed. It has now 10,000 men and is very strongly fortified. I regret . . . my inability to join you. I have written Col. Grierson that you wish him to join you, and have added my own request to yours.

The famed Grierson is discussed below.

The decision on immediate cooperation had been made. The words "immediate cooperation" are used for the fact that if Banks turned toward Port Hudson and kept a considerable number of the enemy occupied by his attack, he would be de facto cooperating with Grant. Given that Banks had heard that 7,000 Confederate

---

511. O.R. Vol. 24/III, pp. 288, 289.
512. O.R. Vol. 24/III, pp. 298, 299.

troops had left Arkansas to join Kirby Smith at Shreveport, and that a Texan column was coming to join Taylor at Grand Ecore, Taylor might then succeed in reconstituting some of his army and potentially the whole effort of the Teche Campaign would be lost. Should he chase after them?[513] Clearly, this would be fruitless. Though Banks' reports were filled with the numbers of prisoners and the immense amount of confiscated goods taken, he mentions nothing of the basic failure of the Battle of Irish Bend and the consequent lost opportunity to trap Taylor.

On the same day, May 12th, Banks turned from being beaten in his attempt to have Grant join him, to a decision to retreat back to New Orleans, and start again for Baton Rouge and Port Hudson; something that should have been obvious long before. Orders to that effect were issued.[514] The next day, however, his engineers having reconnoitered the area, it was determined that it would be feasible to march to Simmesport and board any watercraft available there to be transported to Grand Gulf. Banks decided on this course. The orders of the previous day were countermanded, and messages then sent to both Farragut and Grant that his headquarters was being moved to Simmesport as the first stage of the move.

The plan was to enter the Mississippi from the Red, near Simmesport, and then head north to join Grant. Later in the day, however, unshakeable in his obsession, Banks sent Grant another message, trying to sell the idea of a combined attack on Port Hudson.[515] This was carried by Dwight, duly escorted by 28 of the few healthy men left from the 16th New Hampshire's recent trial, with instructions to urge Grant's consideration of Banks' views. (Compare Dwight's entourage with that of Farragut's lone man Gabaudan.) Dwight proceeded to Grand Gulf by steamboat, and riding forward toward Grant's location, he was just in time to witness the Battle of Champion's Hill on the 16th of May. Dwight sent back the promise "to secure the desired cooperation" but urged Banks not to wait for it. This was apparently only Dwight's opinion and was no real assurance whatever, and the offer was reduced to 5,000 troops.

On May 25th, Grant sent a staff member to Banks to explain that he now had all of his available force at Vicksburg, which was fully invested. He was concerned about a reported force of the enemy "now about 30 miles northeast of here." This was Joseph E. Johnston. He was asking for 8,000 to 10,000 men from Banks and again asked for Grierson.[516] At this point, the discussion of cooperation finally ended.

---

513. Irwin, pp. 149–151; O.R. Vol. 26/I, pp. 494, 495, 500; O.R. Vol. 26/I, pp. 6–21 (Banks' final report, April 6, 1865).
514. O.R. Vol. 15, pp. 729, 730.
515. O.R. Vol. 15, pp. 731, 732; Irwin, pp. 157, 158; O.R. Vol. 26/I, pp. 12, 489; Townsend, p. 172.
516. O.R. Vol. 24/III, p. 347; Irwin, p. 159.

Grierson's Raid

Grierson's Raid[517] and a demonstration by Gen. W. T. Sherman before Haynes Bluff, was planned by Grant as a diversion for his long-awaited crossing south of Vicksburg.

Grierson's force consisted of his own, the 6th Illinois Cavalry; the 7th Illinois Cavalry, under Prince; and a brigade of the 2nd Iowa Cavalry, under Hatch. Supplemented by a battery of artillery, the total force came to 1,700 men. It was intended to break up Vicksburg's (Pemberton's) rail and telegraph communications with Johnston to the east, and it caused Pemberton to order all his available cavalry, and even some infantry, chasing after it. An item in favor of Grierson was that Pemberton had few cavalry. It had been assigned to Van Dorn, and he had been sent to Bragg in Tennessee.

Leaving La Grange, Tennessee, on the morning of April 17th, figure 13[518] and heading south, they played havoc through Mississippi and a portion of Louisiana for sixteen days and 600 miles, before riding into Baton Rouge on May 2nd.

The last leg had been 28 hours without food or rest. The cloud of dust that announced them was spotted and Augur sent out Godfrey's cavalry to escort their long column in, figure 14.[519]

All told, losses were 3 killed, 7 wounded, 9 missing, with 5 left en route sick, including the sergeant major and the surgeon of the 7th Illinois who stayed behind to care for the wounded.

Regarding the raid, the laconic Grant allows one of his rarest and most extraordinary comments.[520] "Colonel Grierson's raid from La Grange through Mississippi has been the most successful thing of the kind since the breaking out of the rebellion. . . . The Southern papers and the Southern people

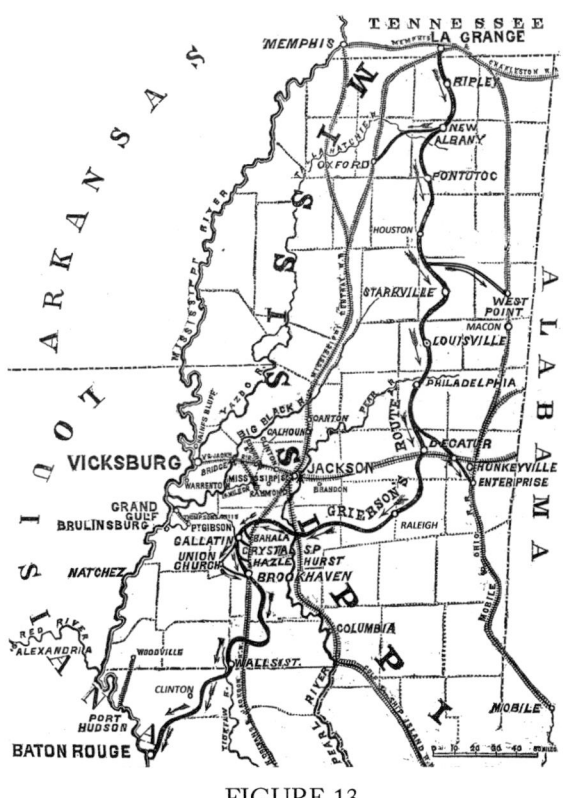

FIGURE 13

---

517. O.R. Vol. 24/III, p. 50; O.R. Vol. 24/I, pp. 522–531.
518. *Harper's Weekly*, June 6th, 1863, p. 358. Corrections/additions by the author.
519. Miller, Vol. 4, p. 130.
520. O.R. Vol. 24/I, pp. 33, 34.

regard it as one of the most daring exploits of the war. I am told the whole state is filled with men paroled by Grierson."

FIGURE 14. GRIERSON ENTERING BATON ROUGE.
The raid became the basis for the 1959 John Ford production, "The Horse Soldiers."

Farragut Departs

After passing up the Mississippi past Port Hudson, Farragut was embarrassed to find himself with only the *Hartford* and the *Albatross*, insufficient to patrol the river. Uninformed of Porter's whereabouts, he sent off a message to Porter on March 20th requesting help, and in a message to Grant, requesting coal, he mentioned: "I do not know what Admiral Porter would suggest if he was here, but I think he might spare one or more of his rams."[521]

Word of Farragut's request got out to Alfred Ellet of the Mississippi Marine Brigade, and on the 23rd, Farragut invited Ellet on board the *Hartford* for a consultation. He and Ellet then agreed that Ellet would run the Vicksburg batteries on the night of the 24th, to bring Ellet's two rams, the *Switzerland* and the *Lancaster*, down to Farragut. Leaving later than planned, the two boats were caught in dawn light and shot up to the extent that the *Lancaster* was sunk and the *Switzerland* was lucky to drift past the remaining batteries after having her center boiler shot through. Finally informed of Farragut's presence, Porter asked Farragut to keep the ram, which, it turned out, could be easily repaired.

Farragut promptly made use of what he had by attacking the Confederate guns at Grand Gulf and Warrenton, and then, on April 1st, proceeded to the mouth

---

521. ORN Ser. I, Vol. 20, pp. 6–23.

of the Red River to begin its blockade.

Finally, on May 4th Porter appeared off the mouth of the Red, bringing the news that Grant had secured his landing at Grand Gulf,[522] and had begun his march on Vicksburg. Porter was now free to patrol the Red and Farragut realized that it was time to leave. In fact, on April 15th, Welles had informed Porter that "The Department wishes you to occupy the river below Vicksburg, so that Admiral Farragut may return to his station."

The May 8th deck log of the Hartford reads: "At 4:40 a.m. the following officers left the ship, on board the *Sachem*, for passage to New Orleans…" Though Farragut left personally, he gave Captain Palmer of the *Hartford* the option to choose his time for running the gauntlet. At Banks' request, he remained to support the move against Port Hudson.

Alexandria

The rest at Opelousas had recuperated the troops, and on the march "there was not the least appearance of straggling or disorder in any portion of the command." No opposition was met.[523] Porter had destroyed the casemates at the unoccupied Fort DeRussy, removed the raft that had been placed across the river, and arrived off Alexandria on May 7th. Banks arrived that night, ahead of his infantry.

Hearing that Banks was in Alexandria and Grant was in Jackson, Mississippi, caused great dissatisfaction at Washington.[524] In a letter dated May 19th, Halleck reminded Banks that:

> . . . the Government is exceedingly disappointed that you and General Grant are not acting in conjunction. I thought to secure that object by authorizing you to *assume the entire command* [Note that even at this juncture, Banks was regarded as more fit to command than Grant.] . . . The opening of the Mississippi River has been continually presented as the first and most important object to be attained. Operations up the Red River, toward Texas, or toward Alabama, are only of secondary importance, to be undertaken after we get possession of the river . . .

Turn to Port Hudson

The tortuous decision regarding cooperation with Grant settled, Banks ordered a recall on May 13th. His decision was tortuous for all that had gone on before, but there was another reason as well – noted in Halleck's letter – he would have to turn

---

522. Irwin, pp. 144, 152.
523. O.R. Vol. 15, p. 313; ORN Ser. I, Vol. 20, p. 42.
524. O.R. Vol. 26/I, pp. 494, 495, 500; ORN Ser. I, Vol. 19, pp. 699, 700; ORN Ser. I, Vol. 20, p. 767.

away from the assumption of the command of the combined forces.

Dwight's and Weitzel's troops, who had gone farther beyond Alexandria toward Grand Ecore, would be the rear guard and would remain in Alexandria until May 17th.[525] A portion of the troops would march to Simmesport, cross the Atchafalaya, remain there briefly, and from there, be transported up the Atchafalaya to the Red, then down the Mississippi to Bayou Sara. Landed there, they would march on Port Hudson. However, on the 19th, the water level of the Mississippi having dropped, a land route to Morganza was found,[526] so some of the troops or trains would divert to Morganza on the Mississippi, the arrow in figure 15, and be ferried across to Bayou Sara by whatever water transport that could be made available.

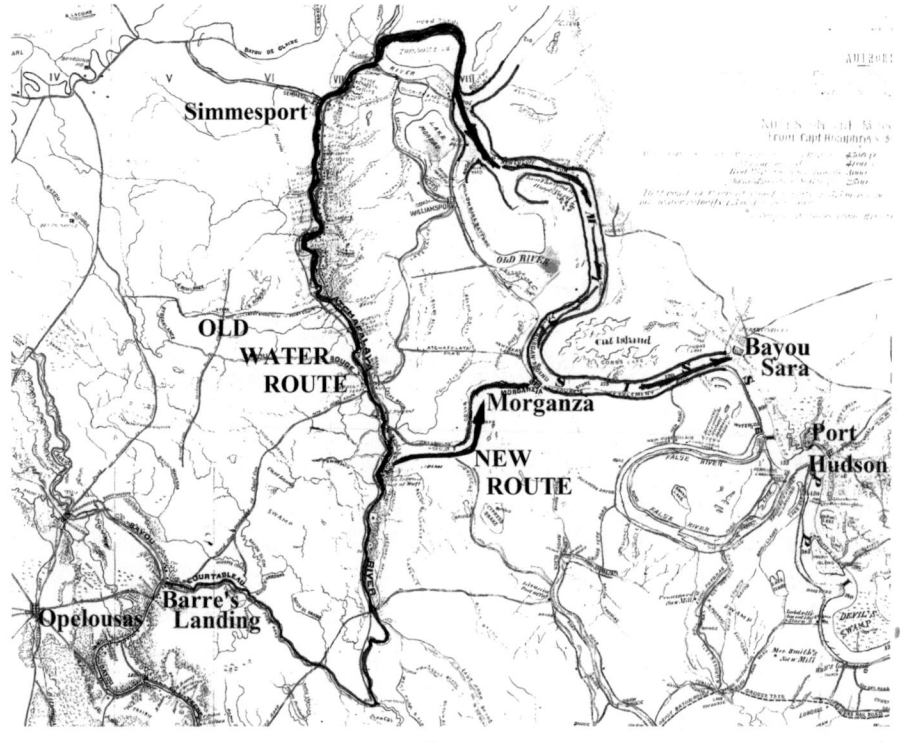

FIGURE 15

The decision to move to Port Hudson meant abandoning the Teche country. Banks' small force could not both hold this area and conduct operations against Port Hudson or whatever might develop on the east side of the Mississippi. One of the concerns Banks had was that: "The force west of the Mississippi, which I had dispersed, would reorganize by re-enforcements from Texas, and move directly upon the Lafourche and Algiers . . . both of which were nearly defenseless."[527]

---

525. O.R. Vol. 26/I, pp. 11, 12; O.R. Vol. 15, p. 325; Tiemann, p. 37; Irwin, p. 152.
526. O.R. Vol. 26/I, pp. 493, 495, 502. Figure 15, NOAA CWAB, portion, altered.
527. O.R Vol. 26/I, p. 12.

Retreating south and back across Berwick Bay to more defensible positions was a necessity. Essentially, this was an admission that the Teche Campaign was a failure. In fact, Taylor soon returned to Alexandria from Natchitoches, where he had stopped at the end of the chase. In a few days, he was informed that Walker's division was on its way from Arkansas to reinforce him, and that his first assignment was to attempt to relieve the pressure Grant was putting on Vicksburg.[528]

The finale of Banks' move was that of a provisional brigade,[529] to be commanded by Col. J. S. Morgan of the 90th New York, which would convoy a train of supplies and refugees back to Brashear City. The brigade included three companies of the 41st Massachusetts Mounted Rifles, the 90th New York, the 52nd Massachusetts, the 22nd Maine, the 26th Maine, Company E of the 13th Connecticut, and Snow's section of Nims' 2nd Massachusetts Battery. Some elements of the 110th and 175th New York regiments also participated.[530]

Under way from Barre's Landing, on the morning of the 21st, the column was joined at St. Martinsville on the 23rd by the 114th New York. The train consisted of 50 army wagons carrying ammunition and stores, 500 "emigrant" wagons carrying the household goods of the former slave families, 5,000 former slaves, 2,000 horses and mules, and 1,500 cattle. It was almost eight miles in length. Though harassed by guerillas, notably at Franklin, when the rear guard was attacked on the evening of the 25th, it arrived safely near Berwick City on the 26th.

---

528. Taylor, p. 137.
529. O.R. Vol. 26/I, pp. 40, 41; Irwin, p. 155, Ewer, pp. 81, 82; Moors, p. 145; Maddocks, p. 37; Pellet, pp. 80, 81.
530. Pellet, p. 79. Pellet mentions the 110th and 175th New York; Irwin does not. Colonel Morgan's report mentions individuals from these units. The statement above is the compromise.

Chapter 9

"Record" 6/63; Port Hudson; Demonstrations; Plains Store; Closing the Ring; May 23rd; May 24th; May 25th; May 26th; The Assault; Weitzel Advances; Sherman Advances; Augur Advances; Cease Fire

"Record" 6/63
1 MAY – 30 JUNE, 1863, PORT HUDSON, LOUISIANA

MAY 5th Left Barre's Landing Bayou Courtableu, La. & arrived at Stafford's Plantation, La., 70 miles, May 8th - May 11, moved to Webb's Plantation, 4 miles & 12 miles from Alexandria, Red River, La. May 14th broke up camp and took up line of march for Port Hudson via Simmesport, Atchafalaya River. May 24 reached Rebel pickets which were driven in and position taken within 750 yards of the rebel intrenchments on our right centre. Drove in skirmishers fired a storehouse and commenced fortifying. June 14th drawn up for the advance of that day, casualties 2 horses wounded. Held this position up to June 20th. At work upon Rebel artillery, camps, storehouses and other shelter. June 20, Sgt. Becker with one section detached on escort duty, drove the Rebel Artillery from the train & behaved with most Commendable coolness & ability. June 25th took up position on the extreme left in the Redoubt 300 yards from Rebel citadel, operating upon rebel sharpshooters, covering our working and assaulting parties and shelling out the Rebel defenders. June 27th Pvt. Casey mortally wounded by Sharpshooters while at his gun. Rebel colors three times shot away by 1st Sergt. Louis Keller. June 28, Lt. Taylor and two sections detached to Donaldsonville, La, engaged and dispersed the Enemy. June 30, two sections of Battery in Redoubt as before. One section under Sgt. Becker in Redoubt at Donaldsonville, La.

May 24th Pvt. Buckly fatally wounded while cannonading the rebel lines.

It is noted that though Battery L was present and participated in the shelling during the May 27th assault, not a word is mentioned here.

Henry W. Closson	Capt. Commanding Battery and Chief of Artillery, Grover's Div. 19th Army Corps.
Frank E. Taylor	1st Lt. Ass't Comm'y of Musters S.O. no. 16, Hdqrts Dept of the Gulf 19th A Corps, Apr. 9, 1863
Edward L. Appleton	1st Lt.

J. A. Sanderson	2nd Lt. App. to Co. by prom. Vice Gibbs prom. G.O. No. 73 War Dept. A.G.O. Wash, July 4, 1863 (never joined Co.)

Detached:
George Friedman	Pvt. On Det. Svc. at N.O. as Artillerist. Left Co. May 24,'62.
Amelius Straub	Pvt. Absent on Ex. Duty as Cook in Gen. Hosp. Baton Rouge since Mar. 14,'63
Martin Stanners	Pvt. Absent on detached service at Bayou Boeuf, since April 7,'63

At Donaldsonville:

Julius Becker	Sgt.	
William Demarest	Cpl	
William Wynne	Cpl.	
James Beglan	Miles McDonough	Ephriam Orcutt
James Comfort	Daniel Moore	Philip H. Schneider
Clark Dickson	Cornelius McEnearny	Andrew Stoll
Daniel Howard	Christian Meese	Michael Teigh
Benjamin Hughs	John H. Moran	Henry H. Ward
John Kelly	John A. Nitschke	Joseph Wilkison
John Kastenbader	Sholto O'Brien	Thomas M. Wilcox

Strength: 141, Sick: 11 [1 not accounted for]

Sick present: none

Sick Absent: William C. Brunskill, Ft. Hamilton, NY, Left Company Sept. 17, 1861.
John Casey absent, sick in 2nd Div. Hosp.
George Chase Brashear City, La., Since April 22,'63
William Crowley Baton Rouge, Since Mar. 27,'63
Benjamin O. Hall   "   "   "   "   "
Joseph Kutschor Brashear City, Since May 4,'63
Charles F. Mansfield Pensacola, Fla. Since Dec. 24,'62
Solomon J. Montgomery Baton Rouge, Since June 11,'63
Wallace D. Wright Pensacola, Fla. Since Dec. 24,'62
Peter Welsch Baton Rouge, Since June 11,'63

Died: John Buckley Pvt. 30 Sept.'58 New York Of wounds received before Port Hudson, May 24,'63.

Discharged: Franklin W. Richards Pvt. 6 Feb.'61 New York. At Pickens, FL April 28,'63 After serving 6 mos. confinement. Gen'l Court Martial O. No. 26.

Horses: [Not Listed.]

Deserted: None

There was no money noted as owed to laundresses, and it is therefore assumed that there were officially none, though many of the officers and groups of the enlisted men throughout the department had by this time hired servants from the population of escaped slaves. The August–October muster roll notes enlisting "colored" cooks, who had been "cooking in the Company since May 20, '63." This was brought about by an Act of the 37th Congress, Chapter LXXVII, Sections 8–10, dated March 3rd, 1863. It would finally modernize the way the army handled the soldier's ration. First, it directed the medical staff to supervise the cooking, "as an important sanitary measure." Also, "from the privates of each company" there were to be detailed cooks, one for a company of up to 30 men, and two for each company numbering over 30 men, for ten day assignments. Moreover, to each cook, two "under-cooks of African descent" were authorized to be enlisted. They were to be given one ration per day, and paid $10 – less $3 allowed for clothing.

Section 11 added pepper to the ration.

Port Hudson

The arrival of Banks' forces at Port Hudson would find it weakened but not abandoned, as had been alluded to in one of Grant's messages. On April 30th, after Grant's forces had landed at Bruinsburg, the critical moment had come, and the reinforcements for the Confederate force predicted as needed for the defense of Vicksburg and Port Hudson would ultimately not arrive, due to the direct intervention of none other than Jefferson Davis.

In November of 1862, Joseph E. Johnston had been appointed commander of the Confederate department embracing Tennessee, Mississippi, Alabama, and that portion of Louisiana east of the Mississippi; and to whom both Pemberton, E. Kirby Smith (Taylor's superior), and Bragg reported. Noting the Union buildup of Grant's army, Johnston had predicted the invasion of Mississippi, and had requested troops from Arkansas, which were never sent[531] though there were two corps in the Trans-Mississippi Dept. under Holmes at the end 1862.[532]

In February, before Grant began to pressure him, Pemberton sent Rust's brigade[533] to reinforce Gardner at Port Hudson. It consisted of two Arkansas regiments, the 12th Louisiana and the 6th and 15th Mississippi, plus three artillery batteries, all of whom had arrived in March. However, Grant's activity increased. Thus, Pemberton ordered Rust back out of Port Hudson on April 4th – a total of 3,400 troops. On May 1st, with Grant now having crossed the Mississippi, Pemberton ordered Gregg's brigade out – about 3,800 troops. Then again, on May 4th, Pemberton ordered Gardner to "come and bring with you 5,000 infantry," including Maxey's brigade.[534] This would leave only Beall to garrison Port Hudson

---

531. O.R. Vol. 17/II, pp. 784, 801, 811, 823, 828, 838.
532. O.R. Vol. 22/I, pp. 903, 904; Johnston, *Battles and Leaders...* pp. 473, 474. Randolph, the Confederate secretary of war, had directly ordered Holmes in Arkansas to send reinforcements to Pemberton, but it was countermanded by Jefferson Davis. A few days later, Randolph resigned, "unfortunately for the Confederacy," quoting Johnston.
533. O.R. Vol. 15, pp. 985, 1005, 1032, 1033, 1035, 1069.
534. O.R. Vol. 24/III, pp. 808, 828, 835, 839. On p. 839, a Gardner note dated May 6th appears: "Gregg's division ought to reach Brookhaven next Thursday or Friday." See also Vol. 15, pp.

with about 3,800 troops. The message was in cipher, and Gardner initially could not understand it, delaying any action on it until May 6th, when Gardner and Gregg's brigade left. Pemberton appeared to be on the brink of abandoning Port Hudson as had been implied in a message from Johnston to him on May 1st: "If Grant's army lands on this side of the river, the safety of Mississippi depends on beating it. For that object you should unite your whole force."

Jefferson Davis now entered the picture with a May 7th direct communication to Pemberton, bypassing Johnston. The message began with information, among other things, that the eastern reinforcements for him from Beauregard would be limited to the 5,000 already sent. He then closed with a message devastating to Johnston's plan of uniting the few Confederate forces: "To hold both Vicksburg and Port Hudson is necessary to a connection with Trans-Mississippi."[535] Naturally, on the next day, the 8th, Pemberton ordered Gardner to return with 2,000 troops to Port Hudson "and hold it to the last."

On May 17th, Pemberton reported a series of setbacks to Johnston, who replied,[536] "If you are invested in Vicksburg, you must ultimately surrender. Under such circumstances, instead of losing both troops and place, we must, if possible, save the troops. If it is not too late, evacuate Vicksburg and its dependencies, and march to the northeast" (Johnston's location at Vernon, Mississippi).

Johnston also sent a separate order to Gardner by courier on the 19th, apparently not trusting Pemberton to do so: "Your position is no longer valuable . . . Evacuate Port Hudson forthwith . . ."

Gardner may have received the order or heard of its contents verbally because he is reported[537] to have evacuated, and to have reached Clinton, yet upon hearing of Augur's advance from Baton Rouge, (see "Demonstrations" below) retreated. The actual copy of Johnston's order arrived at the commander of the Port Hudson outposts, Col. John L. Logan's headquarters at Clinton, on the 25th, but by that time it could not be sent through, Port Hudson having been surrounded. Logan reported that Gardner had intended to cut his way out on the night of the 24th, and had he succeeded, his line of retreat to Johnston would have been so long as to "have been attended with great loss."[538]

Gardner's "incomplete" return indicates 5,715 for the "aggregate present" at Port Hudson on May 19th, 1863, slightly more than half of Banks' estimate of 10,000 given to Grant on May 12th.[539]

Some few more souls would be added to the Port Hudson garrison by paroled

---

1071, 1074, 1076.
535. O.R. Vol. 24/III, pp. 842, 845, 846.
536. O.R. Vol. 24/III, pp. 887, 888, 896, 897; O.R. Vol. 24/I, pp. 222, 242.
537. Stanyan, p. 247; Freret was present within the fort during the siege.
538. O.R. Vol. 26/I, p. 180. Logan's report, written from Gardner's outpost at Clinton, LA.
539. O.R. Vol. 26/II, p. 10.

prisoners. An example of ". . . the useless fashion of the time," as Irwin referred to it, is found in the journal of the *Richmond*, which had been on station below Port Hudson since May 2nd. Its entry for May 11th reads: "At 10:00 a.m., the *Iberville* came up with a flag of truce. She had between three and four hundred rebel prisoners on board. The *Iberville* went up to the batteries at Port Hudson, where the paroled prisoners were landed." Described here as paroled, they would have to have been officially exchanged if they were to legally reenter active service. Regardless of the fine point, under Gardner's dire need for men, the suspicion was that they would be pressed into service. This suspicion is also evidenced by a comment regarding the 100 or so prisoners Grierson brought into Baton Rouge. Company G of the 49th Massachusetts was detailed to escort them to New Orleans, "where, I suppose, they will be paroled, and set loose to fight us at Port Hudson."[540]

Demonstrations

Long before the decision to turn to an attack on Port Hudson, Banks had insisted upon Augur, at Baton Rouge, making "offensive demonstrations every day in the direction of Port Hudson, with the view and effect of deterring him [Pemberton] from weakening his forces there."[541] Keeping Gardner bottled up at Port Hudson, he assumed he would quickly dispose of him and manage his own small force so as not to lose New Orleans.

A part of the planned program of demonstrations relied, once again, on Farragut and parts of his squadron. South of Port Hudson, a naval bombardment involving the Mortar Flotilla, the *Essex*, and the *Richmond*, would commence. Later, Farragut would add the *Kineo* and the *Genesee*, with the admiral himself occasionally visiting from New Orleans on the *Monongahela*. On Farragut's orders of May 5th, the Mortar Flotilla was brought into position near the *Essex* and began shelling the Port Hudson works on the 8th.[542]

After it was invested on May 23rd, the navy shelling would be continued nightly until the surrender on July 9th, 45 days by the reckoning of the commander of the Western Gunboat Flotilla, D. D. Porter. He failed to mention that Farragut had started it 16 days earlier.

Plains Store

The demonstrations were over. Augur would try to secure as much of the approaches to Port Hudson as possible and await the arrival of the forces from Simmesport via Bayou Sara, and Thomas W. Sherman's forces from New Orleans. The provisional

---

540. Irwin, p. 190; ORN Ser. I, Vol. 20, pp. 790–804. There had been earlier exchanges; Johns, p. 214; O.R. Ser. IV, Vol. 2, p. 615.
541. O.R. Vol. 15, p. 704.
542. ORN Ser. I, Vol. 20, pp. 178, 247, 218, 261; Defenders, SHS, Vol. 14, 1886, p. 313

brigade that had marched south to Brashear City would ultimately join them, coming upriver from New Orleans to Springfield Landing.

Plains Store was at the intersection of the old Zachary Highway and the Port Hudson – Plains road, about three and one-half miles due east of Port Hudson. Here stood a two-story white wood-framed structure, which, typical of the time, consisted of a "drug-store and post-office . . ." with a Masonic Lodge above.

Augur's First and Third Brigades, led by Grierson's cavalry, had advanced to here on May 21st, to discover a Confederate force commanded by Col. F. P. Powers.[543] The Confederate cavalry pickets were dispersed by Godfrey's cavalry, and the column of Augur's 3rd Brigade, led by the 30th Massachusetts,[544] continued on until within about three-quarters of a mile from the Plains store, when they were fired on by Abbay's Mississippi battery of light artillery. One section of Battery G of the 5th U.S. Artillery was brought up and joined one section of the 18th New York Battery. They engaged the enemy battery for a half hour and failed to silence it. General Augur then ordered up four pieces of the 2nd Vermont Battery, supported by the 174th New York Infantry. With an additional section of Battery G on the right, supported by the 30th Massachusetts, and a flanking move by the 2nd Louisiana, supported by the 161st New York, this overwhelming force drove Powers' force from their position. The 3rd Brigade then occupied a position near the store, the advance chasing Powers as far as the Clinton Railroad. The fighting tapered off by noon.

In the afternoon, on Gardner's orders, Col. W. R. Miles[545] with 400 of his Legion and Boone's battery, arrived on the road from Port Hudson to run into the 48th Massachusetts which was stationed in the open, supporting a section of Battery G, which had been placed to cover the road. The 48th, thrust into its first action ever, was surprised, overwhelmed, and consequently retreated into and through the 49th Massachusetts, which was coming to its aid. Battery G's guns were briefly taken. Miles then managed to flank and get into the rear of the 49th and the 116th New York, which had followed the 49th. The 116th and the 49th were both then "faced about," putting the 116th in a position to engage Miles. Augur now ordered a charge, to which the 116th responded. A second charge routed Miles and thus ended the Battle of Plains Store. The total casualties on the Union side were 102, the 116th taking the most: 11 killed and 44 wounded. Miles reported a total of 89, though it was incomplete.[546]

---

543. O.R. Vol. 26/II, p. 5. Powers was detached from the 14th Arkansas, at Port Hudson, on May 15th; Defenders, SHS Vol. 14, p. 314.
544. O.R. Vol. 26/I, pp. 120–122; Howe, p. 47; O.R. Vol. 26/I, pp. 137, 179.
545. O.R. Vol. 26/I, pp. 167, 168, 144; ibid., Pt. II, p. 10; Johns, p. 235; Clark, pp. 78–80; Irwin, p. 161.
546. Eighty-nine enlisted and 4 officers *killed*, Defenders, SHS Vol. 14, p. 314; Stanyan, J. M., p. 247

Augur was now in place, holding the ground southeast and east of Port Hudson. Col. John L. Logan, commanding the remainder of Gardner's "outside" troops, reported that his force was too weak to attempt to attack Augur, but would keep on Augur's right flank, after a move to Clinton.[547]

T. W. Sherman, now in command of the 2nd Division,[548] came upriver from Carrolton via Springfield Landing and reported to Augur late on the 22nd. He was directed to deploy his troops to the east and south and hold the road to Baton Rouge. He moved within sight of the rifle-pits the next day, his cavalry pickets exchanging shots. He then looked to the task of repairing the bridges on the road to Springfield Landing.

FIGURE 1

At Simmesport, Grover's division was left considerably reduced by the detachment of five regiments, and portions of two others, who had been assigned to the provisional brigade under Morgan, which was just getting under way. The remainder of the division was ordered to load "two artillery batteries on board the steamers Empire Parish and St. Maurice, with as large a part of his infantry as the two steamers can carry," destination Bayou Sara,[549] figure 1.

---

547. O.R. Vol. 26/I, pp. 179, 180.
548. Emory took over as the commander of the defenses of New Orleans after Sherman. Weitzel was temporarily assigned to the command of the 3rd Division for one day, May 13th, and was then returned to the command of the 2nd Brigade, 1st Division, plus that of Dwight, plus other cavalry and artillery detachments, to hold Alexandria while the rest of the army marched out on the 17th. Paine took over the command of the 3rd Division. O.R. Vol. 26/I, pp. 486,492, 493, 500, 501; Irwin, p. 159; Hanaburgh, p. 35; Stevens, W.B., p. 130; Johns, pp. 228–230.
549. O.R. Vol. 26/I, p. 499; Figure 1 from June 20th, 1863, *Harper's Weekly*, p. 389.

One regiment, however, did not go with Morgan. It was the 6th New York, whose term of service having already expired, had been assigned to the 3rd Division, now under Brig. Gen. Halbert E. Paine, and had been doing provost duty at Alexandria.[550] The 3rd marched from Alexandria on the 19th and arrived at Simmesport on the 21st. Since both Battery L and its old comrades-in-arms were together on the same day, there seems no doubt that there was a chance for some rather stunned goodbyes, the 6th, thanks to Dwight, no longer under the command of its colorful Billy Wilson. Had almost two years passed since the days at Pickens and the sands of Santa Rosa? The war was now a much wider one, and the attention the darlings of the City of New York had gotten while at Fort Pickens had long since faded. Many, many more units had been mustered and sent off. The attention of the city and the state had lately been focused on unpleasant news of events elsewhere. The political scene had changed as well, Seymour was now governor, and his positions against Lincoln on emancipation, military arrests, and conscription had wafted an antiwar scent into the New York ether.[551] The deplorable consequences of this, the draft riots, would not take long to appear.

Closing the Ring

Banks, with Grover's division, landed after midnight on the 22nd at the remains of the town of Bayou Sara, which "presented a very battered appearance," having been shelled and raided repeatedly by the navy ever since the Battle of Baton Rouge in August.[552]

Resting until morning, they began the march to the north side of Port Hudson. Climbing up the bluff on which stood St. Francisville, then turning south, they encamped near Thompson's Creek that night, figure 2.

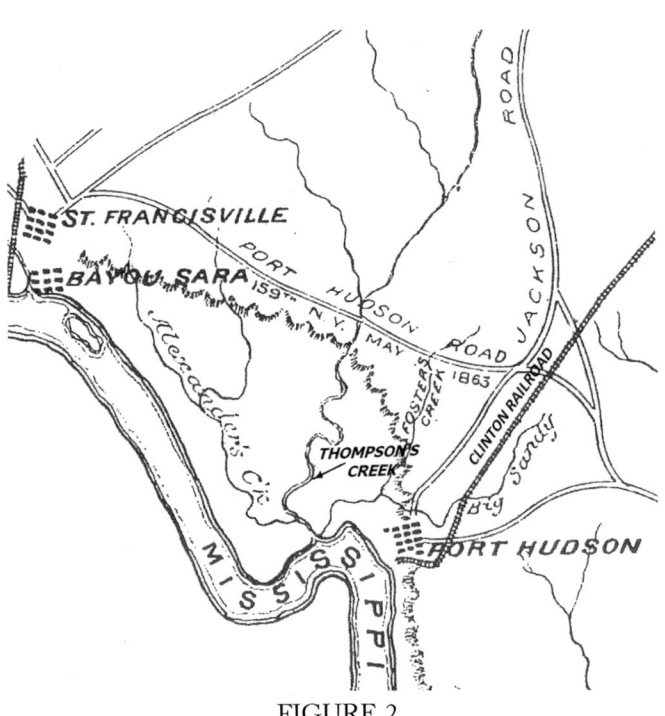

FIGURE 2

---

550. Morris, pp. 114–116.
551. *Encyclopedia Britannica*, 1911, Vol. 24, p. 755.
552. Sprague, pp. 131, 132, 134; Bissell (McManus), pp. 43, 44; Willis, p. 117; Figure 2, Tiemann, facing p. 23, altered.

Resuming the march on the 23rd, Grover's troops, though still annoyed by Wingfield's 9th Louisiana Rangers and skirmishers of the 1st Alabama under Locke, linked up with Grierson's cavalry,[553] who were the advance of Augur's troops. This was near the junction of the Jackson and Port Hudson roads, above the Clinton Railroad, to where Grierson had advanced after the Battle of Plains Store. It was about 9:30 a.m.

Gardner had originally felt that the north side of Port Hudson had too many natural obstacles to overcome and that no attack in force would come from that direction. Now that it was obvious that the Yankees were approaching, he sent a considerable force there, under the command of Col. I. G. W. Steedman of the 1st Alabama. He placed Steedman in command of the whole left wing of the defenses, essentially the area north of the Clinton Railroad, including that portion of the earthworks known as the Priest Cap and Fort Desperate, where Col. Ben Johnson's 15th Arkansas was deployed, figure 3.[554] The irregular line indicates the prepared Confederate fortifications, which extended from the river in the south to the advanced work, Fort Desperate. Beyond it, following a line curving around south of Sandy Creek, and extending to the swamp above the river, was undefended.

Gardner then ordered Steedman to "observe the enemy and to oppose his advance upon our works, but without risking a serious engagement..." This rather peculiar directive, in retrospect, may have been Gardner's idea to probe the Yankees for an escape route.[555]

May 23rd

Figure 3 roughly indicates the positions of both the Union and Confederate troops as of May 23rd, when the Union line closed around Port Hudson. Weitzel was still crossing the Atchafalaya at Simmesport. Paine's troops had crossed Thompson's Creek and there, acting with Prince's regiment of Grierson's cavalry, captured the steamers *Starlight* and *Red Chief*, which, it was feared, could have been used as part of a Gardner escape plan. Paine now covered the right of the Union line from the swampland above the river on the north side of Sandy Creek up to Grover's position.[556] Grover extended along from the north side of the Sandy to one half mile north of the Clinton Railroad, to Augur's right.

General Banks' Headquarters, near Grover, was on the Riley Plantation.[557] Sherman, who, as described earlier, had arrived at Springfield Landing on the 22nd, was deployed south of Augur and directed to move to "a position on the Western

---

553. Hanaburgh, p. 36; Sprague, pp. 136, 137; Stanyan, p. 220; O.R. Vol. 26/I, pp. 165–167.
554. Irwin, facing p. 192, altered. O.R. Vol. 26/I, pp. 84, 501, 504.
555. Defenders, SHS, Vol. 14, pp. 316–318; Stanyan, pp. 220, 247, 248.
556. Stanyan, p. 220; O.R. Vol. 26/I, p. 138, Report of Col. Edward Prince, 7th Illinois Cavalry.
557. Irwin, p. 166; *Harper's Weekly*, June 11th, 1863, p. 446.

Port Hudson Road" with pickets within sight of the enemy rifle pits, which he reported he would "carry . . . tomorrow." Others of his troops conducted limited reconnoitering of his flanks and front.

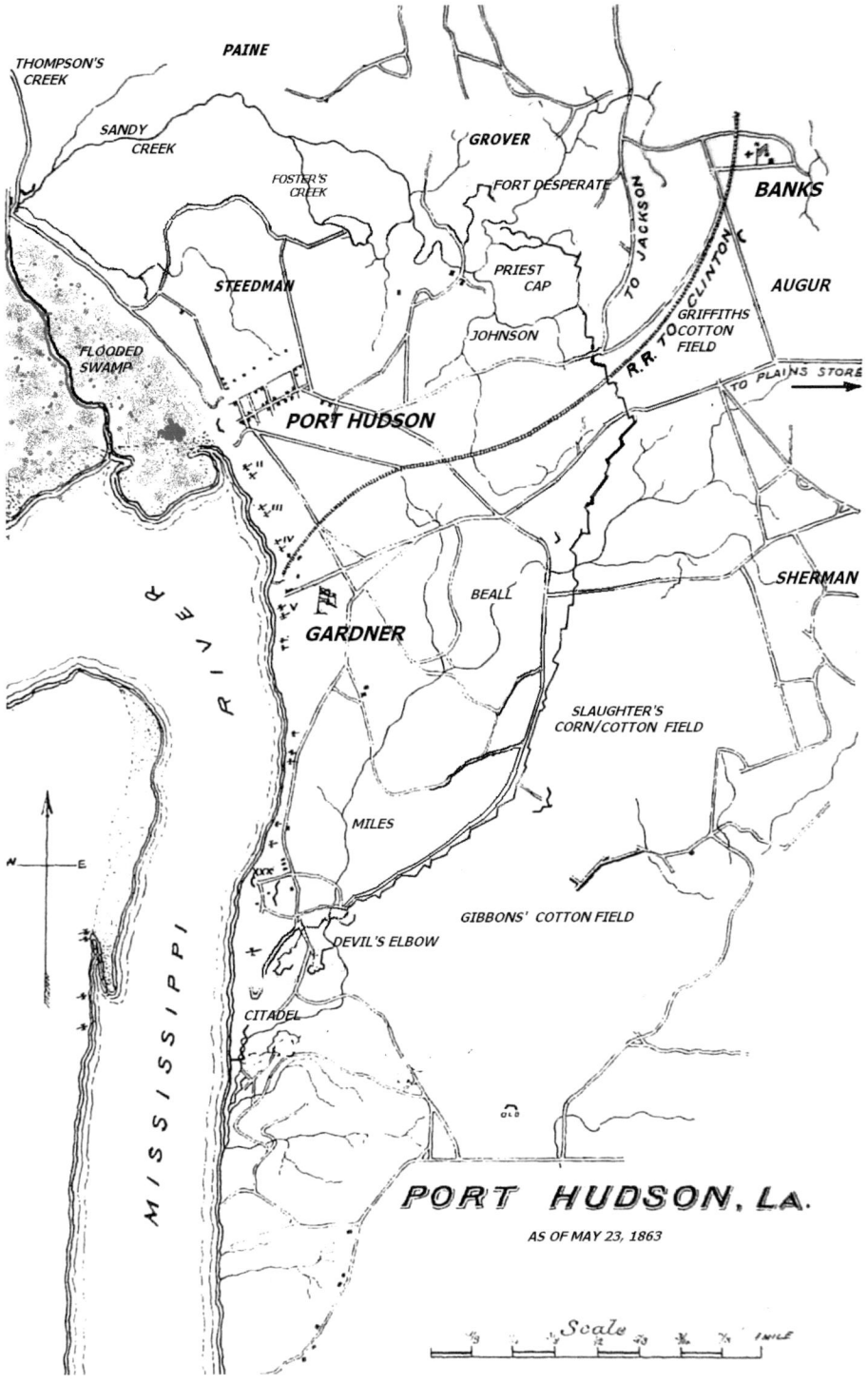

FIGURE 3

The Confederate defensive earthworks, four and a half-miles long[558], and lined with heavy guns, ran along a natural prominence overlooking the corn and cotton fields of the Gibbons, Slaughter, and Griffiths plantations. The tortured terrain in the north was described as "very peculiar, looking like the skeleton of a huge fish, the backbone representing the long ridge running from the woods towards the fortifications, and the ribs the short ridges . . ." with deep ravines in between.

Hilly and forested areas in front of the works had been cleared by the Confederates and the trees left in every conceivable position, making what was termed a "slashing." The slashing opposite Weitzel's and Augur's position appears in figure 4.[559] Weitzel (Paine's troops under Weitzel), were to face a similar front after Steedman completed his preparations. In areas where "the ground was open and favorable to assault," such as Gibbons' and Slaughter's fields, a ditch had been dug. Described as anywhere from three to four feet deep and five to six feet wide and rumored to be twelve feet deep and fifteen feet wide, it undoubtedly varied and was deeper and wider where natural features helped. From the bottom of the ditch to the top of the parapet was no more than seven or eight feet. About one hundred yards of cleared ground extended out from the ditch, all swept by the fire of the guns on the parapet.[560]

FIGURE 4

Sherman's front faced Slaugther's field on the edge of the wood opposite the site of the Slaughter house, which was said to be occupied by rebel sharpshooters.

---

558. Hanaburgh, p. 38.
559. Figure 4, Tiemann, p. 159 "Scene of Charge, May 27, 1863." The same picture appears in the "Album of the 2nd Battalion Duryee Zouaves" (pages not numbered but approx. p. 49). This was the 165th New York Regiment.
560. Tiemann, p. 41; Pellet, p. 85; Clark, p. 83; Bacon, p. 131; Johns, p. 243; Smith, p. 42; Woodward, p. 34. Figure 5, McGregor, p. 463.

Fire from the 1st Vermont Battery was concentrated upon it on the afternoon of the 26th, which drove out the sharpshooters, and it was then burned down by members of the 128th New York regiment.[561] Its chimney can be seen to the right in figure 5. Beyond, the "white face of their parapet could be seen zigzaging away to the right and left until it disappeared from view." The view is from the area of the Slaughter home at battery 16, in figure 9, which was not completed until June. No battery positions appear in figure 3 because as of May 23rd there were none.

FIGURE 5. VIEW FROM SHERMAN'S FRONT, CONFEDERATE PARAPET 1000 YARDS DISTANT
Today's undergrowth totally obscures this view.

May 24th

Having completed the encirclement, a consolidation began which would test and push in the Confederate outlying defensive lines and picket positions. Casualties began to mount, and the ferocity of the skirmishing on Steedman's front grew when Paine advanced. Finally, on the afternoon of the 24th, having observed Paine's advance of about a mile, Steedman was ordered by Gardner to "determine the enemy's strength, if possible, and drive him from my front." With Weitzel not having yet arrived, Gardner may have conceived of a weakness in Paine's 3rd Division line, whereby he could break out, which he was apparently still considering.[562] Breakout or not, it is clear that Steedman was placed there in urgent desperation because, in Steedman's words: "Up to Monday night, the 25th of May, no works of any description had been thrown up to defend this position, extending from Col. Johnson's advanced work, on the right . . . to a point within five hundred yards of the river on the left . . . about three-fourths of a mile."

As Grover advanced along the Jackson Road, the first shots of the day occurred

---

561. McGregor, pp. 318, 323, 424, 434, 463; Hanaburgh, p. 36.
562. O.R. Vol. 26/I, p. 180; SHS, Vol. 14, p. 318; Stanyan, p. 220.

at noon, when "sharp skirmishing ensued. A labyrinth of woods, lanes and ravines afforded cover for both offensive and defensive operations . . ." Here, the two sides were close, much closer than at Paine's, Augur's, or Sherman's fronts. Grover was facing a portion of Steedman's troops, the 15th Arkansas, under Johnson, whose line was now forced back to his main work on Commissary Hill, figure 6.[563]

A relevant part of Battery L's "Record of Events" notes:

> May 24 Reached Rebel pickets which were driven in and position taken within 750 yards of the rebel intrenchments on our right centre. Drove in skirmishers fired a storehouse and commenced fortifying.

Lieutenant Taylor's section of Battery L, with Grover's advance, was the first to come into action, at about 10:00 a.m. Skirmishers, sharpshooters, and artillery on both sides were active; the artillery positions had little or no protection at this time, and Confederate fire resulted in the death of Battery L's Pvt. John Buckley, "fatally wounded while cannonading the rebel lines."[564]

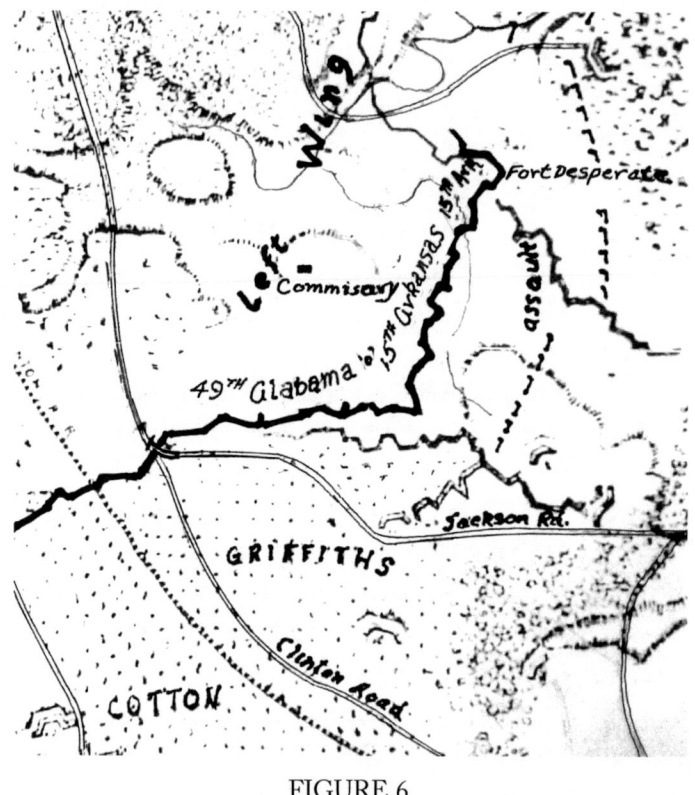

FIGURE 6

W. R. Miles, now back in place from the engagement at Plains Store, and sitting on the right wing of the Confederate defensive line opposite Sherman, reported "that no attack or menace has been made on my line today. The shelling

---

563. Sprague, p. 137; Figure 6, portion of map "Defense of Port Hudson," from McMorries, p. 48.
564. Haskin, pp. 191, 193; Whitcomb, p. 48.

from guns and mortars has, however, been extraordinarily furious" with the 12th Louisiana Heavy Artillery suffering 3 killed and 3 wounded.[565]

May 25th

Miles' turn came today. He was threatened by two Union advances resulting in a number of Union casualties, compared to none reported by him. Viewing the Yankees removing their dead and wounded under a white flag, Miles complained to Gardner of its "illicit" use, saving those who might have become prisoners.

The rear guard, Weitzel, landed at Bayou Sara and hurried "for it was feared that the Confederate garrison might attempt to evacuate the place and escape. At this time, Weitzel was named to command the right wing, which consisted of his own troops, the 3rd Division under Halbert Paine, Dwight's brigade of Grover's Division under Van Zandt, and the 7th Illinois Cavalry under Prince.[566]

Commodore Palmer, on the *Hartford*, signaled to Admiral Farragut on the *Monongahela* that General Banks had requested him to shell the west side of Thompson's Creek at midnight "to prevent any attempt of the enemy to escape over that side."[567]

General Paine wrote:[568] "25th Crossed the Big Sandy [figure 2], Dwight's brigade of Grover's division in advance. Drove the enemy steadily through the woods to within a half-mile of his fortifications."

A battle ensued along the whole line, and an attempt was made to overwhelm both Steedman's right and left flank. He was reinforced but notes: "Thus reinforced the right repelled every attack; but in consequence of my inability . . . to extend my line to Sandy Creek, the enemy marched a body of troops around the extreme left and seriously threatened our rear." This was the 38th Massachusetts Regiment, supporting Mack's 18th New York Battery, who drove back Wingfield's Partisan Rangers. The skirmish resulted in the 38th reporting two killed and one wounded. One of Wingfield's officers was killed.

In the afternoon, the 1st and 3rd Louisiana Native Guards arrived at Paine's camp and were ordered to join the 38th at its position, which was on the north side of the still flooded Sandy Creek, near where the telegraph road to Bayou Sara had crossed,[569] figure 7. Command of the Native Guards was transferred by Paine to Dwight, since his division occupied the remainder of Weitzel's right. The positions

---

565. O.R. Vol. 26/I, pp. 168, 169.
566. O.R. Vol. 26/I, pp. 505, 506; Stanyan, p. 224.
567. ORN Ser. I, Vol. 20, p. 209.
568. Stanyan, p. 220; Defenders, SHS, Vol. 14, pp. 316, 317.
569. Figure 7, McMorries, portion of map, p. 48; Powers, pp. 90, 91; Stanyan, p. 222; Bosson, pp. 364, 365; O.R. Vol. 26/I, pp. 128, 167. It is instructive to note that the 18th New York was at this time assigned to Sherman, yet they were deep in Paine's area; Defenders SHS, Vol. 14, p. 318.

of Dwight's troops are indicated by the small rectangles. Note the gap between the position of the Native Guards on the north side of Sandy Creek and the closest of the others – almost three-quarters of a mile. Dwight then assigned Col. John A. Nelson, commander of the 3rd Native Guards, to the command of the two units brigaded for the assault.

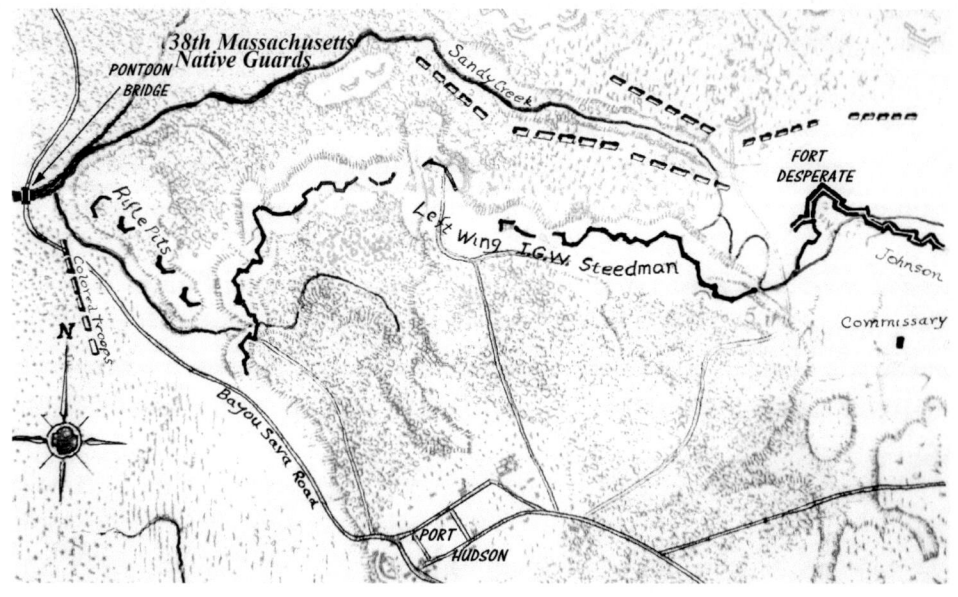

FIGURE 7

Darkness terminated the action, with Steedman withdrawing within his "fortifications." He noted:

> The enemy demonstrations on this day convinced me beyond a doubt that he determined to attack our lines in the vicinity of our commissary depot, arsenal, etc. I reported my convictions to the Major-General . . . and he ordered all the available tools, negroes, etc., to be placed at the disposal of the chief engineer. The work was promptly laid out . . . and ere the dawn of Tuesday, considerable progress had been made.

May 26th

Orton S. Clark of the 116th New York records:

> A council of war was held during the day by General Banks, in which all of the general officers participated, and at which it was determined to at once assault the works.[570] But, two of the Generals dissented from this determination – Gen'l Auger and Gen'l Weitzel – whose better judgment was, in the fatal result of that assault, recorded, as it were, in blood.

Though Clark has the meeting as "during the day," Irwin[571] says,

---

570. Clark, p. 84. Grant, Vol. 1, pp. 530, 531.
571. Irwin, pp. 166, 167.

> When night fell . . . the division commanders met at headquarters at Riley's on the Bayou Sara Road to consider the question of an assault. No minutes of this council were kept and to this day its conclusions are a matter of dispute.

The latter is certainly true, as little regarding the meeting has been found in any of the many histories of the regiments there at the time. It is reasonable to assume that preliminary orders to prepare for an assault on the 27th had already been issued, and an attack on the 27th was a foregone conclusion. Whatever the case, nothing explains the disarray that was to characterize the events of the morrow.

Though Banks knew of Grant's successes (which drove Pemberton back into Vicksburg) he was apparently unaware of Grant's rebuff in his first assault there, on the 19th (942 killed, wounded, and missing) or of his bloody failure on the 22nd (3,199 killed, wounded, and missing).[572] Grant had worried that Johnston, in his rear, was likely to come to Vicksburg's aid, and ". . . defeat my anticipation of capturing the garrison if, indeed, he did not prevent the capture of the city." If Grant had rushed because of Johnston, Banks was now doing so because of Taylor.

Grant later admitted that he regretted making the assault of the 22nd, and classed it with another failure at Cold Harbor a year later. Other frontal assaults under similar circumstances, were also bloody failures – the frontal assault of the 54th Massachusetts on Fort Wagner, in July 1863, being, perhaps, the most well known of the war.

However, other Civil War frontal assaults were dramatically successful – that of the Union assault on Fort Fisher, in January 1865 being an example. There, the *veteran* attackers had vastly superior firepower in the form of Spencer repeating rifles, which were used in a carefully orchestrated advance, and the attack was preceded by a naval bombardment of historic proportions. The devil is in the details.

If Augur and Weitzel had, in fact, dissented, as Clark alleged, it was likely within the context of a discussion that the character of the ground over which the attack was to be made was relatively unknown, what with Weitzel having been in position for less than a day. The fact that Weitzel's artillery was not in place was ignored. His front was covered in dense woods "where it was necessary to clear a path for the artillery with an axe." Another factor was that "roads" for direct communication between adjacent commands were "not open,"[573] a fatal flaw.

There is a chance that T. W. Sherman dissented, as well, given his opinion of Banks. Graduated from West Point, class of 1836, he was a veteran of the Seminole War and the Mexican War. He was the commander of the ground forces in the famously successful Port Royal Expedition (chapter 4, *Confederate Concerns*), and was subsequently a division commander in the Army of the Tennessee at Corinth,

---

572. Grant, Vol. I, pp. 529–531; Fox, p. 545; Grant, Vol. II, p. 276.
573. Irwin, p. 167; Hanaburgh, p. 39; Haskin p. 192.

until reassigned to the Department of the Gulf in September of 1862.[574] As an experienced professional, he looked down upon the volunteers and Banks. He is quoted as never having forgotten his glorious achievement at Port Royal and as having referred to Banks as "that d–d militia colonel . . ." Sherman was too big a man to be kept "in penal service here under Banks much longer."

Regardless of his sour attitude, there seems to be no hint of mutiny. It was merely a fact, in his professional view, that a hurried frontal assault against Confederate cannon and sharpshooters, entrenched on a parapet with a ditch, fronted by an abatis, was essentially suicide. His front was the most open, and the advance across it was more than a thousand yards, figure 5. In this situation, the attacker is always at the disadvantage. He began preparations for the worst. A forlorn hope and fascine bearers were called for, as is noted in the history of the 15th New Hampshire regiment.

The situation was similar on Augur's front, though the ground was more broken, and he reacted as had Sherman. Henry T. Johns[575] quotes from a letter he wrote at 9:00 a.m. on the 27th:

> Yesterday morning we were aroused to the solemnity of a soldier's life. Volunteers to constitute a 'forlorn hope' were called for. One field officer, four captains, eight lieutenants, and two hundred men were desired from each brigade. We were expected to furnish five from each company. As nearly as we could learn, a part were expected to run from the woods and bridge the ditch in front of the enemy's parapet or breastworks with fascines and then return; the other part to cross the bridge thus made and assault the enemy at the point of a bayonet.

The brigades of the right wing, commanded by Weitzel, with the exception of Kimball and Birge, who were under Grover, are shown in figure 8.[576] The organization was temporary, and the positions are approximate. Dwight commanded the extreme right, i.e., Nelson, Van Zandt, and Thomas. Paine commanded what were formerly Emory's troops: Fearing and Gooding. The regiments of Grover, Paine and Dwight near the Confederate Fort Desperate were intermixed on the morning of the 27th as final movements were made in preparation for the assault, rendering any brigade listing nearly meaningless. The cavalry, and particularly the artillery, were assigned where needed regardless of their division affiliation.[577]

---

574. Cullum, Vol. I, no. 859; Bacon, p. 101.
575. Harding, p. 334. Fascines are bundles of sticks variously five to eight feet long and a foot in diameter. Johns, p. 243; Plummer, p. 36; Woodward, pp. 31, 32; Clark, p. 84; Howe, p. 48; McGregor, p. 376.
576. Grover was at the right center. Weitzel, in command of the assault on the "right wing," included Grover's command, though it seems Grover took over command in the afternoon, after Weitzel's assault had failed. Stevens, W. B., p. 146.
577. Organization chart of the Department of the Gulf as of April 30th, 1863, O.R. Vol. 15, pp.

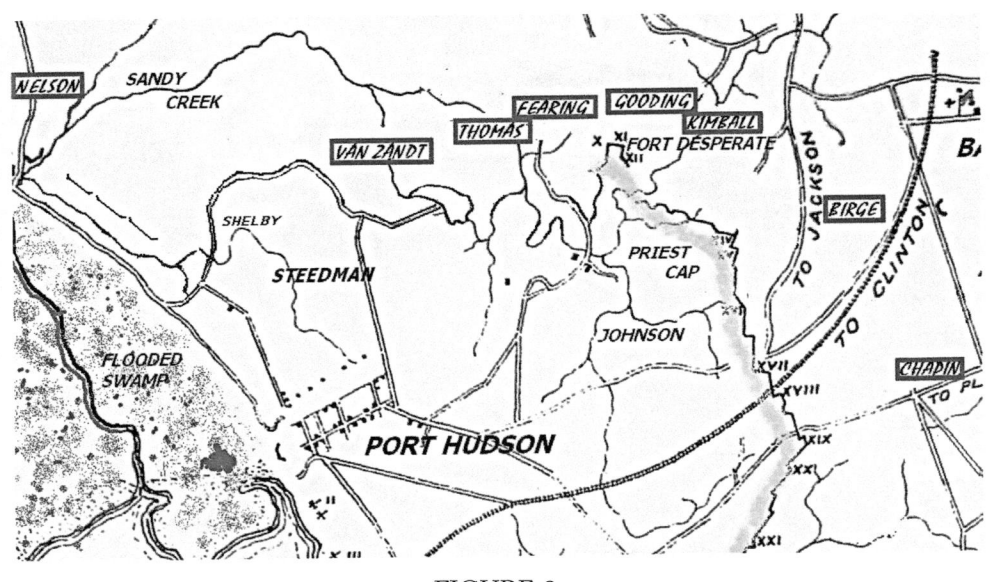

FIGURE 8

The connection of Grover's troops with Augur, who was south of the Clinton Railroad, is indicated by the position of Chapin, a part of Augur's command.

ORGANIZATION FOR THE ASSAULT

DWIGHT'S DIVISION

NELSON - 1st and 3rd Louisiana Native Guards; 1st Louisiana Cavalry, Company A; 8th New Hampshire, Company G, mounted.

VAN ZANDT - 1st Louisiana (Holcomb); 131st New York; 91st New York.

THOMAS - (Weitzel's old Reserve Brigade)[578] 75th New York; 12th Connecticut; 160th New York; 8th Vermont.

PAINE'S DIVISION

FEARING - 8th New Hampshire; (298)[579] 133rd New York; 173rd New York; 4th Wisconsin, mounted; 162nd New York, nine companies.

GOODING - 31st Massachusetts (300); 38th Massachusetts; 53rd Massachusetts (377).

GROVER'S DIVISION

KIMBALL - 12th Maine, eight companies; 24th Connecticut.

BIRGE - 13th Connecticut; 25th Connecticut; 159th New York.

The remainder of the dispositions for the Union "left center" and left are indicated

---

712, 713. The 1st Indiana Heavy Artillery, the Native Guards, and some cavalry, were unattached. Figure 8 portion of Figure 3 altered.

578. The 114th New York had not yet arrived.

579. The numbers in parentheses are the number of men fit for duty, if available in the unit's history. Stanyan, pp. 245, 209; Willis, p. 158; Johns, p. 267; 162nd New York, p. 15; Irwin, p. 203. Figure 9, Irwin, facing p. 192 altered.

in figure 9. There was no occupation of the ground south of the Gibbons house, all the way to the river. Security there was left to Grierson.

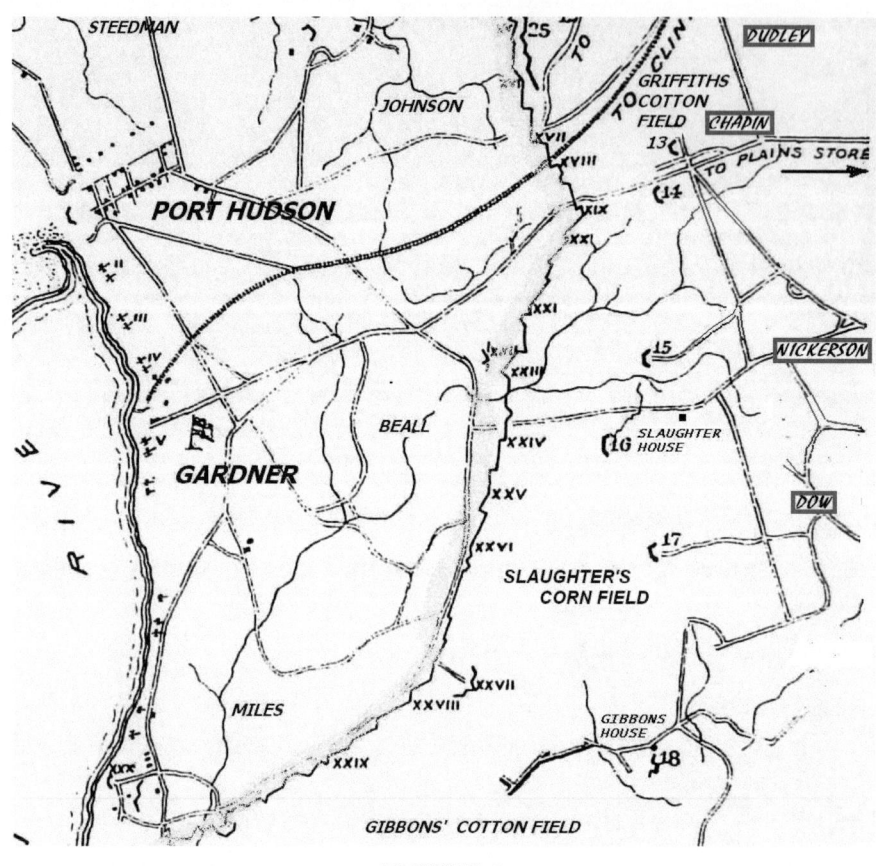

FIGURE 9

AUGUR'S DIVISION

CHAPIN - 48th Massachusetts; 49th Massachusetts (233); 21st Maine; 116th New York; 2nd Louisiana; 2nd Vermont Battery.

DUDLEY - 30th Massachusetts; 50th Massachusetts, four companies; 161st New York; 174th New York.

SHERMAN'S DIVISION

DOW- 6th Michigan; 15th New Hampshire (250); 26th Connecticut; 128th New York.

NICKERSON – 14th Maine; 165th New York (350); 177th New York.

RIGHT WING ARTILLERY

WEITZEL - 1st Maine, Morton; 1st US Battery A, Bainbridge.

PAINE - 4th Massachusetts, Briggs; 1st US Battery F, Duryea; 18th New York, Mack.

DWIGHT - 6th Massachusetts, Carruth; 2nd Massachusetts, Nims.

CENTER ARTILLERY

GROVER - 1st US Battery L, Closson; 2nd US Battery C, Rodgers.

AUGUR – 2nd Vermont, Holcomb; 5th US Battery G, Rawles; 1st Indiana, (7 batteries).

SHERMAN – 21st New York, Barnes; 1st Vermont, Hebard; 1st Indiana, (1 battery).

There were 39 regiments actually in place ready for the assault. The few figures available for the numbers of fit men (the numbers in parentheses above) show an average of less than 300 per regiment, which fits with Banks' quote of a total of 13,000.[580] This number, vs. 6,000 Confederates, nevertheless, met the test of the then current theory that a weight of two to one for an assault on a fortified position was required. However, the devilish details were far more complicated by reasons that were soon revealed.

The Assault

Special Orders No. 123[581] were issued on the night of May 26th. Significant portions are reproduced here:

> III. Generals Augur and Sherman will open fire with their artillery upon the enemy's works at daybreak. They will dispose their troops so as to annoy the enemy as much as possible during the cannonade, by advancing skirmishers to kill the enemy's cannoneers and to cover the advance of the assaulting column. They will place their troops in position to take instant advantage of any favorable opportunity, and will, if possible, force the enemy's works at the earliest possible moment.
>
> V. General Weitzel will, according to verbal directions already given him, *take advantage of the attacks* on the other parts of the line to endeavor to force his way into the enemy's works on our right.
>
> VI. General Grover will hold himself in readiness to re-enforce within the right or left, if necessary, or to force his own way into the enemy's works. He will also protect the right flank of the heavy artillery, should it become necessary.
>
> VII. General's Augur, Sherman, Grover, and Weitzel will constantly keep up their connection with the commands next to them, so as to afford mutual aid and avoid mistakes.
>
> VIII. The fire of the heavy artillery will be opened by General Arnold at as early an hour as practicable, say 6 a.m.
>
> IX. Commanders of the divisions will provide the necessary means for passing the ditch on their respective points of attack.
>
> X. All of the operations herein directed must commence at the earliest hour practicable.
>
> XI. Port Hudson must be taken to-morrow.
>
> By Command of Major-General Banks:
>
> Richard B. Irwin
> *Assistant Adjutant-General.*

---

580. O.R. Vol. 26/I, p. 44; Johns, pp. 245, 250.
581. O.R. Vol. 26/I, pp. 508, 509; author's italics.

Regarding part X, one must make the observation that: "the earliest hour practicable" is not definite, and if given to different commanders not equally situated, the assault will begin at their convenience. In plain language, it allows for confusion. That said, what actually happened was inexcusable; there was an absence of any "connection" as ordered in part VII, since no connecting paths had been cleared, and as a result, there was little or no knowledge of affairs in the other parts of the line. *No attacks*, as ordered in part V, were initially made by Augur or Sherman, with the result that Weitzel had none to "take advantage of."

Previously having informed Admiral Farragut of the planned assault, Banks sent a final message at midnight on the 26th: "I have ordered the light artillery to open fire on the enemy's works at daybreak tomorrow morning, and the heavy batteries concentrated on the left center to open at 6 a.m. . . . Your fire should cease as soon as you observe our artillery cease its fire, which will probably be about 10 o'clock, though the time is dependent upon circumstances."[582]

From the *History of the 8th Vermont Regiment*:[583] "In accordance with orders, the fleet in the river opened with their guns on the morning of the 27th of May, and rained shot and shell upon the garrison; the land batteries began firing with great spirit and determination; and the ground fairly shook . . ."

Weitzel Advances

Weitzel's troops, awakened at 3:00 a.m., were formed at 5:00 a.m., and finally, at 6:00 a.m., *assuming* that Augur and Sherman had done so, Weitzel ordered Dwight to advance on the Confederate left center, i.e., that portion of I. G. W. Steedman's defensive line commanded by Col. M. B. Locke, which extended west from Fort Desperate for approximately 1,000 yards. Paine was to follow. Before reaching the objective, the troops would have to carry an advanced line, Locke's, about a half mile in front of the fortifications, which occupied a hill beyond the ridge on the north side of Sandy Creek. It was held by elements of the 15th Arkansas, the 10th Arkansas, the 1st Alabama, and the 1st Mississippi regiments, about 500 troops in total.[584] Figure 10 shows the area of the attack.

---

582. O.R Vol. 26/I, pp. 506, 507.
583. Carpenter, p. 114.
584. Irwin, pp. 169–184; Hanaburgh, pp. 40, 41; Carpenter, pp. 114, 115; Willis, p. 122; Hall, H. & J., pp. 112–117; Defenders, SHS, Vol. 14, Official Report of I. G. W. Steedman, pp. 319–324; Stanyan, pp. 224–226, 229, 230, 248–250, 261; O.R. Vol. 26/I, pp. 157, 163, 166, 509, 510, 551, 626; museum.dmna.ny.gov *160th Infantry Regiment, Clippings*; McMorries, pp. 62–64; Smith, D. P., p. 63; Haskin, p. 192; Johns, pp. 281, 282.

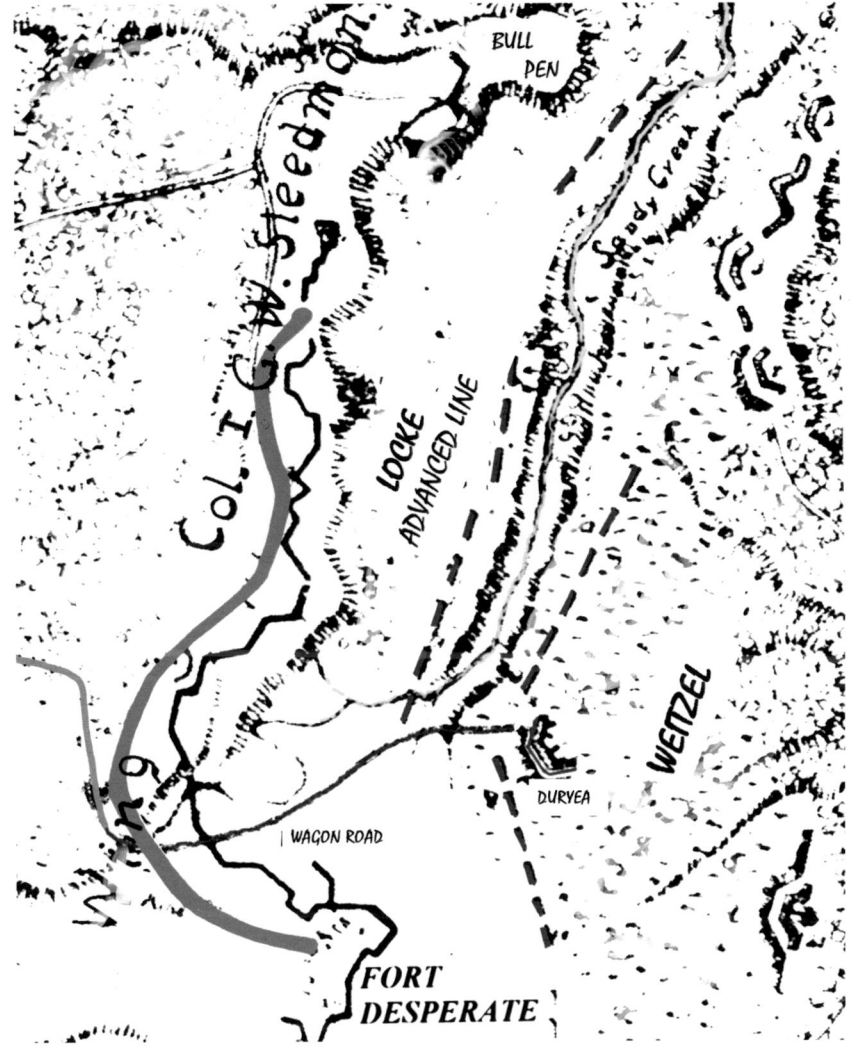

FIGURE 10

A portion of the "Defense of Port Hudson" from McMorries, obtained courtesy of the Port Hudson State Historic Site. The heavy double curved line is the mark of Col. IGW Steedman, indicating his defensive line on May 27th. Additional captioning by the author. Courtesy of the Port Hudson State Historic Site.

Van Zandt's 91st New York led Dwight's brigade, in column of regiments, with Weitzel's, under Thomas, of the 8th Vermont, following. Paine's division followed in support, Fearing's 8th New Hampshire in front, and Gooding in reserve. Pushing through the dense forest, the men of the 91st New York, with the 131st New York on their left, and the 1st Louisiana on their right, soon ran up against the outer rifle pits that Locke had prepared. They were taking casualties from the raking fire, and they were ordered to lie down.

Thomas had deployed the 12th Connecticut, the 75th New York, the 8th Vermont, and the 160th New York behind Van Zandt, who was now stalled before

Locke's riflemen. Thomas was ordered by Weitzel to drive through Van Zandt's line.

> His men responded nobly, led by the gallant colonel on foot, for orders had been given for field and staff officers to leave their horses in the rear.

Charging "into the fearful storm of missiles," their overwhelming weight of numbers caused Locke to turn to his prearranged line of retreat, slowly and deliberately into the Confederate works.

As mentioned previously, having had little time to reconnoiter the area, the nature of the ground ahead was unknown. Paine noted: "We had not definite knowledge of the ground over which we were to fight, for the enemy occupied it. The forest was so dense that glasses were useless." Thomas came out on the edge of a hill which overlooked a deep ravine, with Sandy Creek at its bottom, and the Confederate fortifications on the opposite side. The whole area, once covered with pine and magnolia, had been haphazardly "slashed," figure 4. Locke's sharpshooters still lingered, hidden behind some of the few trees still standing, and sheltered behind a succession of gullies and the fallen timber. Thomas' open position on the hill was swept by fire from the fort, and a decision had to be made whether to retreat back under the cover of the forest, or plunge into the purgatory before them. A single rough road on the left, the wagon road in figure 10, which led into the fort, was covered by a battery of four guns on Commissary Hill, and two guns of Battery B of the 1st Mississippi, in Johnson's 15th Arkansas line at Fort Desperate.

Plunging ahead into the slashing, Thomas' line of battle disintegrated, both from the rebel fire and the confused maze that confronted them. The men were now advancing nearly one by one, and only maintaining cohesion by what little that could be seen of each other in the smoke, or from behind logs or gullies. Their organization had broken down into groups or squads. Their training, which had consisted of a few weeks of close order drill around the fields of Baton Rouge, had not contemplated this. Colonel Thomas and a group broke off to the right, and another group followed the 75th's Colonel Babcock, who was on the wagon road to the left. In a desperate charge, they made it to the protection of a gully about 50–80 yards from the parapet. In Babcock's words:

> I reached the most advanced position which we have yet occupied, and saw the rebs running up the hill beyond into their inner line of rifle pits and found myself here with only five or six men, one of whom was Johnny Mathews and another, a boy of the 91st Regt., who was already hit twice.

That Babcock had managed to do so, we must hark back to Irish Bend. The majority of the Confederate defenders there were equipped with old flintlock muskets, with an effective range of 40 yards, which was also the case here at Port

Hudson. Thus, in the initial part of the assault, the danger had come mainly from sharpshooters and the enemy's artillery.

Van Zandt, with elements of the 91st New York, the 131st New York, and the 1st Louisiana, had since flanked off to the right toward the field called the Bull Pen, the Confederate slaughter pen, where there may have been an opportunity to advance, since it looked to be lightly defended, and it was further from the guns on Commissary Hill. From his position on Commissary Hill,[585] Steedman saw the move, and requested reinforcements from Gardner. The 23rd Arkansas, under Col. O. P. Lyles, arrived in time to check any Federal advance.

The position of Battery L, which had not changed since it first arrived at Port Hudson, but not definitely shown on any map, can only be inferred from the fact that it could pour fire into Steedman's line and cover at least a part of Thomas' advance. They had opened fire since early in the morning, and the battery "was warmly engaged until the coming into action of batteries A and F."

Bainbridge's Battery A, Duryea's Battery F, and Morton's 1st Maine Battery had been on the move since 5:30, following Weitzel's advance in the woods. So little advance preparation had been made that it was necessary for pioneers to chop trees to open their way. They arrived in the rear area into which Paine had driven, with Paine personally directing the placement of Duryea's battery near the wagon road. The 38th Massachusetts, previously ordered to return from its duty with the 42nd Massachusetts and the Native Guards, had "been hunting in the woods for the brigade," when General Paine rode up, and sent it forward to support Duryea. The other two batteries, supported by the 53rd Massachusetts, were deployed several hundred yards farther to the left. "The four batteries now in position soon silenced the guns in this part of the field."[586] However, they were not in place in time to have covered Babcock's advance, and it is noted that the "heavy batteries," those of the 1st Indiana, were not in place until 10:00 and never were a factor in the action.

With the helpful fire of the newly arrived batteries, and the distraction created by Paine's advance, more and more troops were able to pick their way forward and join their comrades on the slopes. Yet they were stalled, and now their own artillery had to cease firing, "lest its fire should prove fatal to friend as to foe." Thomas fell back to better cover in the ravine, and reported his position to Weitzel. He was told to hold in place if possible. As to Babcock, he never moved, and though "the 75th asked leave to charge the works above it on the slope, and penetrate the citadel" – no order came. Weitzel reported his situation to Grover, and sought the desired permission.

---

585. McMorries, pp. 63, 64; Smith, D. P., p. 71.
586. Powers, pp. 92–94; Haskin, p. 192.

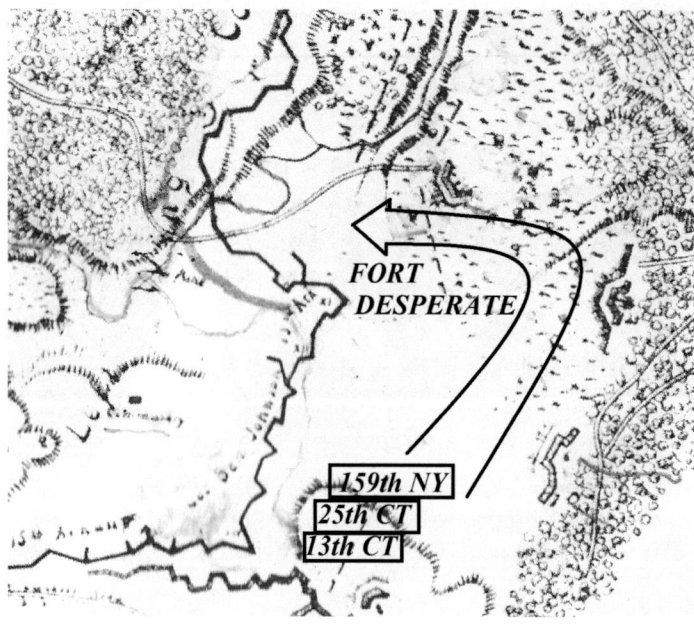

FIGURE 11

Grover now moved to Weitzel's support. Two of Birge's three regiments, the 159th New York and the 25th Connecticut, were marched out from the slightly advanced positions they had secured on the southeast side of Fort Desperate, figure 11.[587]

Under the command of Maj. Charles A. Burt of the 159th New York, they attempted to maintain the double-quick, but their path (the large arrow) was made difficult by brush and fallen timber.[588] In spite of every exertion, they took over an hour to struggle the half mile to the base of the hill on the north side of Fort Desperate, which was the point of the planned assault. They were met by Grover and ordered to charge on the earthworks. Here, Grover gave out the same deranged opinion that he had given the 25th Connecticut at Irish Bend – that there were "hardly any rebels there. Major Burt . . . was told that his regiment alone was able to carry the works and to send back our regiment if it wasn't needed."

A too familiar scenario then played itself out; with the fallen and hidden of Weitzel's command in their path, they advanced out into a "valley . . . filled with felled trees, and heavy underbrush, while thick and black rolled the battle-smoke." Fortunately, the 24-pounder in Fort Desperate was kept quiet by "a handful of brave men firing a stream of bullets at that piece. For six long hours the gunners did not dare approach to load . . . and that wicked looking piece was kept silent."

Finally, Burt's troops reached close to the base of the Confederate works.[589] "The nature of the ground was such it was impossible to form in battle line . . ." so the attack was made in three columns. They waited "for a few moments with beating hearts . . . for the forward charge. The word came, and with a terrifying yell, we rose to our feet and rushed forward. It was a terrible time, when bounding over the last tree and crashing through some brush, we came out within a short

---

587. McMorries, p. 48, portion, altered.
588. Irwin, p. 171; Tiemann, p. 40; Bissell (Ellis), pp. 46–48; Sprague pp. 139–140.
589. Irwin, p. 172; Bissell, (Ellis), p. 47.

distance of the enemy's entrenchments, and it seemed as though a thousand rifles were cracking our doom. This fire was too deadly for men to stand against. Our brave fellows, shot down as fast as they came up, were beaten back." The Confederate line had been given adequate time to observe the Union advance, and had massed four deep. The front man would fire, turn to the rear to reload, allow the next man to fire, and so on.

A "short time" later, a second charge was made, with the same deadly results. Between the two regiments, there were 80 men killed and wounded. There, within 30 yards of the parapet, the charge was stalled, and they were forced to seek shelter. Grover then tried a diversion to allow the attack to regain the initiative, ordering his 12th Maine, supported by the 13th Connecticut, to attack the west face of Fort Desperate. It did not succeed, even though some men were equipped with Sharps breechloaders.

Irwin relates:

> After the first attack on the right had wellnigh spent itself, and when its renewal, in *conjunction with an advance on the centre and the left, was momentarily expected*, Dwight thought to create a diversion and at the same time to develop the strength and position of the Confederates toward their extreme left,[590] where their lines bent back to rest on the river, and to this end he ordered Nelson to put in his two colored regiments.

Irwin, of course, here refers to the expected attacks by Augur (the center) and Sherman (the left), which, painful to relate, had not yet begun and would not until 2:00 p.m.

Absent Augur and Sherman, Dwight's plan seemed like a good alternative. The order was given. Colonel Nelson and the Native Guards, accompanied by their artillery and cavalry, were observed to be advancing across Sandy Creek on the pontoon bridge just after 7:00. The 1st Regiment led, with the 3rd following, their total number being 1,080.

They were opposed by Shelby's 39th Mississippi Regiment and elements of Wingfield's Battalion,

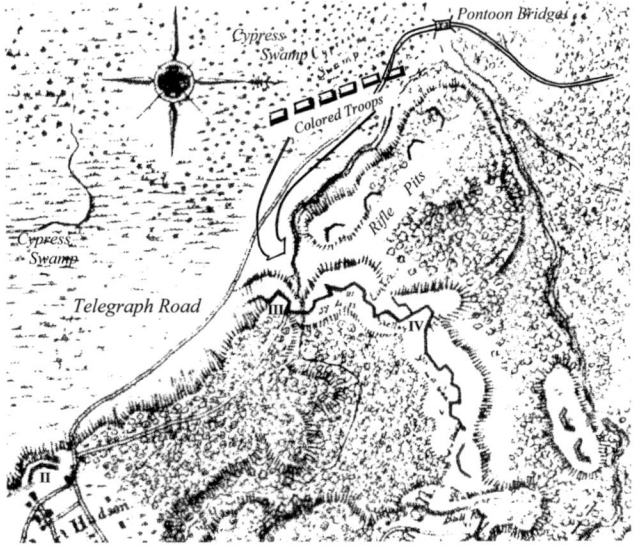

FIGURE 12

---

590. Irwin, p. 175; author's italics.

who were arrayed along a detached ridge, which ran for four hundred yards along Telegraph Road and up to within two hundred yards of the bridge, figure 12. Along its sharp crest ran a line of rifle pits. In Shelby's words: "This ridge was a strong position and easily held . . . it was abrupt and inaccessible." Still farther back, along the natural prominence which was Steedman's line, were emplaced Shelby's light gun batteries III, and IV. To the south, a detachment of Company K of the 1st Alabama manned a 24 pounder in battery II.[591] Backwater from the still-flooded Mississippi filled this end of Sandy Creek, which left a narrow stretch along the road as the only avenue for Nelson's troops to advance. The 6th Massachusetts battery had just crossed the pontoon bridge and begun to unlimber when they were hit by "rapid and effective fire" from a gun in the Confederate line. Shelby's report: "The enemy's artillery, after firing one gun retreated across the creek." The Native Guards crossed under the same fire and filed to their right "and under cover of the willows formed in line of battle and advanced." They continued to advance at the double-quick, shown by the arrow in figure 12, ". . . until they reached to within about two hundred yards of the extreme left, when ". . . two pieces of light artillery on Col. Shelby's line opened ". . . on them with canister, and at the same time, the infantry (in their anxiety to fire – fired without orders) opened on them, driving them back in confusion and disorder, with terrible slaughter. Several efforts were made to rally them, but all were unsuccessful, and no effort was afterward made to charge the works during the entire day."

The Native Guards, like the 6th Massachusetts Battery, were able to fire but once before falling back. They were raked from the rifle pits as they advanced, and when they had just made it to where they should turn to an assault, they were hit by the three separate batteries of artillery.

After the bloody repulse, one of Nelson's aides was sent to Dwight's headquarters for permission to retire, they being "all cut up." Dwight would hear none of this and ordered, "Charge again, and let the impetuosity of the charge counterbalance the paucity of numbers." The aide left. Dwight had been skeptical about the use of black troops, and allegedly having been drinking since before breakfast, he was showing no sympathy for their fate. Dwight's rant was ignored. No further charges were attempted after the regiments had retreated to cover, where they remained under heavy fire for the rest of the day.[592]

The Federal attack on the Confederate left was now entirely stalled. We turn to Irwin, Banks' adjutant general, for comment:

> The morning was drawing out when these movements were well spent, and

---

591. Defenders SHS, Vol. 14, pp. 321, 322; Stanyan, pp. 249, 250; Bosson, p. 364; Smith, D. P., p. 63; McMorries, pp. 61, 62.
592. Irwin, pp. 173–175; Stanyan, p. 229. Grover was senior to Weitzel and replaced him as the commander of the right wing that afternoon.

the advanced positions were simply held without further effort to go forward. The hour may have been about 10 o'clock . . . Grover had been ordered to support either the right or the left, or attempt to make his way into the works, as circumstances might suggest. This last he had tried, and failed . . . On his left there was no attack to support.

Riding to meet Weitzel, Grover gave him the "counsel of prudence" and either ordered or convinced him to ask for "fresh orders" before continuing the attack. Steedman confirms when the attack on the Confederate left had petered out:

> The battle on the left wing . . . was an assault or series of assaults for the first two hours: at the end of that time the enemy had been repulsed signally at every point, and he had withdrawn a short distance and concealed his men under the cover of the trees, logs, ravines, etc., and from this hour, about 11 o'clock, until five o'clock, the firing relaxed and could only be called sharpshooting.

No attack to support? Nothing had been heard from either Augur or Sherman, and it was Banks who now addressed a note to Weitzel. Dated 1:45 p.m. It read:

> General Weitzel:[593]
>
> General Sherman has failed utterly and criminally to bring his men into the field. At 12 m. I found him at dinner, his staff officers all with their horses unsaddled, and none knowing where to find their command. I have placed General Andrews in command, and hope every moment that he is ready to advance with Augur, who waits for him. . . .
>
> N.P. BANKS
> *Major-General, Commanding.*

Of course, at this point, Banks should have had sense enough to call it off. The attack had lost any element of surprise, coordination, or concentration.

Sherman Advances

Andrews, Banks' chief of staff, rode immediately for Sherman's campsite, in the woods southeast of the entrance road to the Slaughter plantation, figure 13, another portion of the McMorries map. When he arrived, he found that the whole division had been deployed, with Sherman on horseback, ready to lead the advance.[594] Andrews handed Sherman a message. Sherman was seen to throw his hat to the ground, "and after some words and excited gestures, he turned to the troops and cried 'Forward! Double Quick! Double Quick.'" It was 2:10 p.m.[595] Andrews deferred from taking over the command but remained to observe the action.

---

593. O.R. Vol. 26/I, pp. 509, 510; Irwin, p. 177.
594. Hanaburgh, p. 41.
595. O.R. Vol. 26/I, p. 125; Bacon, p. 138.

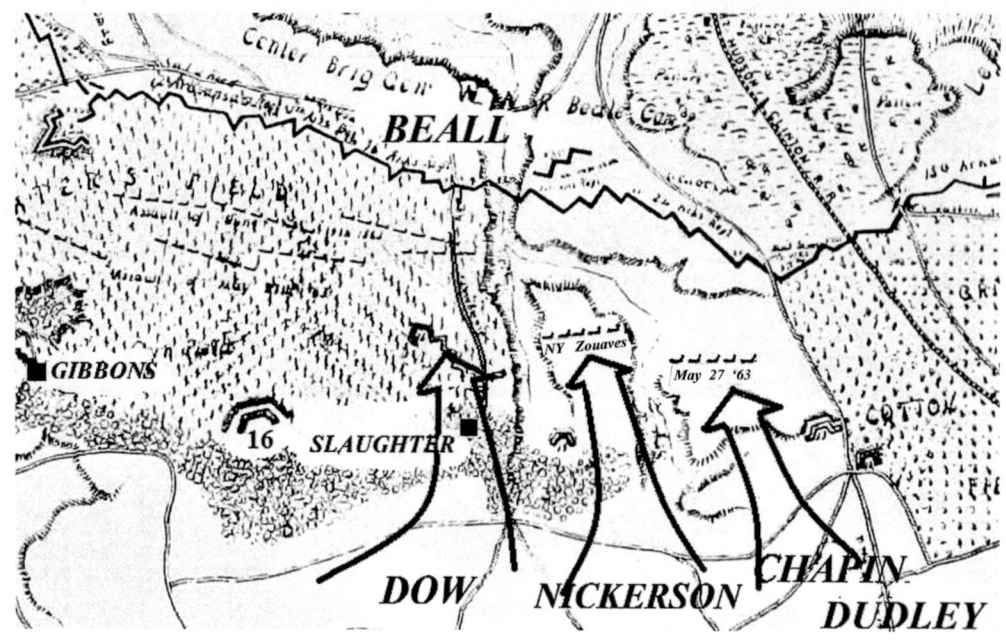

FIGURE 13

Sherman was able to react so quickly because, regardless of his seemingly careless attitude about the timing of the assault, he had roused his troops, as had Augur, early in the morning, and they had been active in preparations ever since. In the case of the 15th New Hampshire, advanced picket positions on the front had been established since the evening of the 26th, and Companies A and K, along with Companies A and C of the 128th New York, had been skirmishing near the site of the Slaughter home since daybreak. Companies D and E of the 15th were further south, part of a line that extended across the front.[596] Companies B, G, I, and H were ordered to join the 26th Connecticut in support of one of the batteries, which since early morning had been firing "twenty shots per minute; their shells sound like a distant train of cars." At 9:00, the regiment was drawn up in line of battle, and at 10:30, it was advanced to "half way through the intervening woods" to the site of the Slaughter house. At 12:15, Dow's attack column was formed up. It consisted of the 6th Michigan, the 15th New Hampshire, the 26th Connecticut, and the 128th New York. Nickerson had done the same, his pickets advancing at 9:00, with the 14th Maine acting as skirmishers. His attack column consisted of the 165th New York, the 14th Maine, 24th Maine, and the 177th New York.[597] Sherman and his entire staff, all "splendidly mounted," were positioned in between the two brigades.

An account from the History of the 15th New Hampshire Regiment:

---

596. McGregor, pp. 319, 320, 331, 337, 338.
597. Irwin, p. 177; McGregor, pp. 334–339.

... Some teams now drive up loaded with heavy poles; negroes shoulder them, two to each, and are placed in front of the skirmish line. Those who volunteered from our regiment are each provided with a 2-inch plank a foot wide and about four feet long, the design being to force the negroes up to the face of the enemy's parapet, and compel them to lay the poles across the ditch in front, the plank carriers then to lay on their planks, and so bridge over.

Now on the Slaughter grounds, they broke into two wings, one passing to the right and one to the left. They were slowed and their organization disrupted by fences around flower gardens,[598] and the right fell into a ravine beyond, the left remaining on the cornfield, directly exposed to the "murderous musketry fire that mows all in winrows, and thickly covers with the dead and dying." The advance was stopped, then rallied, with Sherman[599] directing Lieutenant Colonel Blair of the 15th New Hampshire to "lead them ahead – straight ahead – dead upon the enemy's works." Shortly after, Blair, now at the head of the column, looked back "and hardly anybody but the dead and wounded were in sight. General Sherman was dismounted when I had left him, but pushing forward bareheaded encouraging the men around him . . . I ran back . . . near the edge of the ravine and found him . . . with his left leg shattered."

The ravine can be seen in figure 13,[600] running toward the Confederate defense line to the right of and parallel to the road past the Slaughter plantation. It afforded some cover, and those in the groups that advanced through it were able to reach the most advanced position "but from which all efforts to scale the enemy's works proved futile."

Dow was felled next, Clark of the 6th Michigan was wounded,[601] and Cowles of the 128th New York, to whom the brigade command fell after Dow was wounded, was killed. The brigade command then devolved to Nickerson, who was not notified. Kingsley of the 26th Connecticut was then wounded while trying to rally the men. Leaderless, for a period of time in which no direction was given, Dow's columns ground to a halt.

Nickerson had stepped off on the right of, and at the same time as Dow, with the 165th New York leading the column. Remarkably, the faint original inscription "NY Zouaves" can be seen in figure 13, at the head of the arrow indicating Nickerson's advance. The description of what happened next is left to an extract from

---

598. O.R. Vol. 26/I, pp. 123–125.
599. McGregor, pp. 347–350.
600. Figure 13, a portion of the McMorries map, labels and arrows added, and altered to conform to a definition of where the edge of the forest extended, (five hundred yards from the works, near the Plains Store road) per p. 32 of Woodward.
601. Bacon, pp. 126, 139. Clark was allegedly stunned by the concussion of a cannon ball, which passed near him. He was unmarked when carried off the field. Bacon accuses him of cowardice; Irwin, pp. 178, 179.

the Southern Historical Society Papers:[602]

> While the battle was raging on this [Beall's] part of the line, a New York regiment of Zouaves came dashing out of the swamp on the extreme right of the field, making, with their red breeches and caps [figure 14], a magnificent spectacle. To meet this new danger our troops were thrown rapidly to the right and opened a hot fire on the advancing Zouaves, who, nevertheless, came dashing on, deploying from column into line with the precision of veterans, as they neared our works, we mowing them down by the scores, when they were ordered by their colonel to lie down, who, himself walked back and forward with as much apparent coolness, as if he were giving orders on parade.

FIGURE 14

In a moment more, the ranking officers of the 165th, Col. Abel Smith, and Maj. Gouverneur Carr, had fallen, and many of the men then broke and fled for cover. A few remained behind stumps and logs as sharpshooters, the assault of the Zouaves had been stopped.

Augur Advances

As soon as Banks heard the rattle of musketry on the left, and saw from the smoke of the Confederate guns that Sherman was engaged, he ordered Augur forward.[603]

Augur had also been prepared. On the 26th he had asked for volunteers for a 200 man storming party, and had begun making fascines, which would be used to fill the ditch along the base of the Confederate parapet. Col. James O'Brien of the 48th Massachusetts had volunteered to lead it.

Companies B and G of the 116th New York had been thrown out as skirmishers on either side of the Port Hudson – Plains Store road to a place "only less dangerous than the open field," where they formed a part of the support for Holcomb's 2nd Vermont Battery.[604] Colonel Chapin, the 1st Brigade commander, had been up since daylight, directing its firing. Colonel Dudley, the commander of the 2nd Brigade, was concerned about this being the first time that the 50th Massachusetts[605] would be under fire and had "routed" them out at 3:00 a.m. for "advice and instruction" before breakfast. The cooks later brought bread up to their position near Hadden's Co. D of the 1st Indiana Battery on the north side of the Plains Store road, where they were sent to protect it.

Also before breakfast, Chapin's men received their orders to assault, and they

---

602. *Defenders*, SHS, Vol. 14, p. 324, Figure 14, Duryée Zouaves, frontispiece.
603. Irwin, p. 179.
604. Johns, p. 249; Plummer, p. 36; Clark, p. 87.
605. Stevens, W. B., pp. 137, 138, 141.

were moved to their assault positions in the woods along the Plains Store road, figure 9. Chapin's brigade was to lead, and Dudley's was to support.

The 21st Maine was to act as skirmishers to be followed by Lt. Col. James O'Brien's, forlorn hope, with 15 officers and 370 men,[606] in the advance of the regiments.

At 1:00 p.m., the forlorn hope had formed up in the woods across the Plains Store road, about a mile from the enemy's works, some only with muskets, others with muskets slung and carrying fascines, and some who had stacked their muskets, and carried only fascines.[607] The fascine bearers were to be in the vanguard. Chapin had addressed the storming party and said, "Remember that you do not go unsupported. My brigade will follow close on your heels."

The regimental lineup had Holcomb's battery on the road, with the 21st Maine to its left, and looking left, the 116th New York, the 49th and 48th Massachusetts, followed by the 2nd Louisiana, and Dudley's brigade. Now they waited.

Shortly after 2:00 p.m., Augur commanded, "Now boys, charge, and reserve your fire until you get into the fort; give them cold steel, and as you charge cheer . . . ! Press on no matter who may fall." Lieutenant Colonel O'Brien then gave the order: "Come on boys . . ." The storming party emerged from the woods, turned to the right and up the road. "A small belt of timber to our left hid us from our foe."[608]

> The artillery had ceased firing; all was quiet till we passed that small belt and came in full view of the rebels. Then bullets, grape, and canister hurtled through the air, and men began to fall . . . For a few yards the field was smooth, but difficulties soon presented themselves. A deep ditch or ravine was passed, and we came to trees that had been felled in every direction. Over, under, around them we went. It was impossible to keep in line. The spaces between the trees were filled with twigs and branches, in many places knee-high. Foolishness to talk about cheering or the 'double-quick'. We had no strength for the former, aye, and no heart either. We had gone but a few rods ere our Yankee common sense assured us we must fail.

The fascine bearers could not keep up with the rest. "They looked more like loaded mules than men." The forlorn hope and the main forces now became mixed together. The state colors of the 48th Massachusetts "came passing by," wrote Henry Johns, "and I followed. Soon the standard-bearer was killed; an officer grasped the colors and waved them aloft. In less than a minute his blood had dyed the white silk of the banner."

---

606. Plummer, pp. 36–38; Bowen, *Massachusetts in the War*, pp. 648, 652, 653; Woodward, pp. 31, 32; Clark, pp. 84, 85; Bowen, p, 458; dman.ny.gov/ *Clippings, 161st Regiment*, p. 8; Irwin, pp. 179, 180.
607. Stevens, W. B., p. 136; Clark, pp. 88, 89. Clark's quote has been rearranged, a printer error suspected in the original text. The regimental lineup here does not agree with Irwin.
608. Johns, pp. 253-261; Plummer, p. 38.

Johns continues:

> We had been there about an hour, and as the fire slackened, Col. O'Brien came springing across the logs, waving his sword, shouting 'Charge! Boys, charge. In half a minute, just ahead of me, he fell dead.

He had turned to direct the charge and briefly faced his back to the enemy's fire and had taken a ball through the back. Examination of his body revealed that the ball had passed through and had flattened against a steel vest he was wearing. Next, it was Chapin, who was at first wounded in the knee, but remained on the field. Then, exactly one year after being wounded while serving in the Army of the Potomac in the Peninsular Campaign, he was killed by a bullet to the brain. Next it was Colonel Bartlett, commander of the 49th Massachusetts, "the only mounted man to be seen"[609] because while serving in the Army of the Potomac at the Battle of Ball's Bluff, he had lost a leg. Today he was twice wounded, once in the ankle of his remaining leg and in his wrist.

Adj. Joseph T. Woodward of the 21st Maine Regiment[610] wrote:

> To reach and scale them with the force remaining, exhausted as it was by the effort already made in the terrific heat of the day in the face of a foe admirably protected by fortifications and nearly equal in numbers, was impossible, though some of the stormers reached the ditch, there about six feet deep and ten feet wide, and placed their fascines in it. To retreat was extremely hazardous. On our left fires had started in the underbrush which added to the discomfort and danger. At this juncture, a line of supports came forward which met with the same resistance and secured no greater success. It was then evident that . . . the attack had failed.

The support he mentions was the 2nd Louisiana (black) of Dudley's brigade, "and a noble sight it was to see them bravely advance to our assistance; but it was of no avail."

Cease Fire

They knew not why, but "towards night . . . the bugle called out "cease firing" and immediately all strife stopped. Then commenced carrying off the dead and wounded."[611]

Major Burt of the 159th New York may have been the savior of the day. Unauthorized to do so, he called a truce, "which enabled him to bring off his dead

---

609. Stevens, W. B., p. 139.
610. Woodward, p. 34; Clark, p. 90.
611. Stevens, W. B., p. 140; Tiemann, p. 41; Hall, H. & J., p. 116; Stanyan, pp. 231, 232; *Defenders*, SHS, Vol. 14, pp. 323–325.

and wounded. The display of the white flag was mistaken for some distance along the line . . ." It is said that one Confederate regiment stacked arms on or near the works and fell back to await the surrender. However, when Gardner learned of it, he ordered that hostilities be resumed in half an hour. At least, this had allowed the men on both sides to emerge from their shelter for more than two hours, and no doubt, many lives were saved, particularly those wounded earlier in the day, who had suffered in the sun for hours without water.

Next day, an official truce[612] was finally effected. Considerable time was wasted by negotiations. It was initially agreed that it was to extend to 2:00 p.m. but had to be revised to 7:00 p.m. because Banks and Gardner had not finished their correspondence until after 3:00 p.m. A part of the negotiating involved Banks having to apologize for Burt's unauthorized truce.

Much delay was caused by Gardner's demand that the Union troops withdraw to no closer than eight hundred yards from the works. Banks refused, but in a subtle concession, he remonstrates: "The wounded men to whom my letter refers are on our left . . ." (where they were the most distant from Gardner). Gardner's reply did not press the issue, and he agreed to a truce. Banks then clarified the situation, saying, "I have been informed that there are also some dead and wounded on my right." All of this negotiation had consumed nine hours. It seems that if either man cared about the condition of the wounded still lying on the field, by now, he would have agreed to almost anything. The final words were from Banks: "I will agree to send all of the killed and wounded of your command that are within my lines or on my front to your lines, by unarmed parties, if you will consent to send the killed and wounded of my command within your lines or on your front to my exterior lines, by unarmed parties."

Such detail has been repeated here to allow the reader to try to grasp why the Native Guards dead and wounded were, in fact, ignored. A Confederate view of this is found in the history of Company K of the 1st Alabama Regiment.[613] "On the 28th there was a cessation of hostilities . . . for the purpose of burying the dead. Gen. Banks did not deem it worth while to bury the colored troops who 'fought nobly' and their bodies lay festering in the sun till the close of the siege . . ."

In the *History of the First Regiment, Alabama Volunteer Infantry,* the explanation given is:

> A brigade of negroes had charged the 39th Mississippi on our left; about half were killed outright on the field, and for the burial of these Gen. Banks never asked a flag of truce. They lay there in the hot sun and putrified and swelled until the stench became so unbearable to Col. Shelby . . . that he asked

---

612. O.R. Vol. 26/I, pp. 513–518; McMorries, p. 64.
613. Smith, D. P., p. 66; McMorries, p. 64.

Banks to allow him to bury them. Banks replied that he had no dead there.

Any reasonable person would be suspect of this explanation. Then again, with knowledge of Dwight's foibles, perhaps Banks really had been misinformed.

The truce worked to the particular advantage of the Confederate side. They were able to freely roam the field and collect the Enfields and Springfields left near the ramparts, a feat that they had attempted since the previous evening. Originally attracted to the scene by the cries of the Yankee wounded, they had ventured out of their fortifications to succor them with water. The discovery of the rifles was an important side benefit for the 1st Alabama Regiment, previously only armed with smoothbore flintlocks. "Our men quickly supplied themselves, and after this each man kept two loaded guns, his Enfield for 'long taw' and flint and steel for close quarters."

The day would claim 293 Union killed, 1,545 wounded, and 157 missing.[614] The heaviest total losses were in Augur's division, no doubt due to the bold charges of the forlorn hope. The heaviest loss, though, for any one regiment, was to Fearing's 8th New Hampshire, which during Weitzel's advance suffered 124 killed and wounded, or 42 percent.

The Native Guards 1st Regiment lost: 1st: 34 killed, 95 wounded
3rd: 6 killed, 38 wounded, 3 missing[615]

On the Confederate side, Beall, who faced Sherman's attack, allegedly reported the incomplete total of 68 killed, 194 wounded, and 96 missing, for a period up to June 1st. Steedman, who faced Weitzel's and Grover's attacks, reported that the totals of killed, wounded, and missing of his command as: 10th Arkansas (80), 15th Arkansas (70), and 1st Alabama (75), a total of 225, "or one man out of every four."

There were no casualties in the ranks of the 39th Mississippi and Wingfield's 9th Louisiana Partisan Rangers who had faced the Native Guards. Note that the reports partially overlap and do not agree. If the 1st Alabama losses, as reported by Steedman, are added to the Beall table, a total of 433 would result for the Confederate side. Clearly, Banks' attack on this fortified position, where the attacker is always at a disadvantage, and in such an inept, uncoordinated manner, caused nearly five times as many casualties as were taken by the defenders.

The unique story of black units in battle was set upon by the Northern press and was distorted by all sorts of exaggerations and inaccuracies. For example, the June 20th issue of *Harper's Weekly*: "On 29th an assault was made which was unsuccessful. The Second Louisiana (colored) regiment fought with extraordinary gallantry, losing in killed and wounded 600 out of 900 men."

---

614. Fox, p. 545; Stanyan, p. 245,
615. The numbers for the 1st Regiment are taken from Fox, chapter 6, p. 58. Numbers for the 3rd are not published in Fox. O.R. Vol. 26/I, p. 68, agrees with Fox and lists the 3rd. O.R. Vol. 26/I, p. 147; *Defenders*, SHS, Vol. 14, pp. 325, 326; McGregor, p. 418.

They grew worse. The June 27th issue of *Frank Leslie's Illustrated Newspaper* displayed a large engraving, spread across pages 216 and 217, which was captioned: "ASSAULT OF THE SECOND LOUISIANA (COLORED) REGIMENT ON THE CONFEDERATE WORKS AT PORT HUDSON, MAY 27TH 1863." The scene, figure 15, is entirely a figment of the artist's imagination, for the simple fact that the Second Louisiana *was never there*. They were stationed at Ship Island.[616] Also, no hand-to-hand combat, as is shown, took place in the real engagement. The 1st and 3rd Native Guards were repulsed entirely by rifle and artillery fire.

FIGURE 15

An honest appraisal of the performance of the Native Guards by an eyewitness, Capt. T. C. Prescott of the 8th New Hampshire Regiment, here quoted: "The two colored regiments . . . acquitted themselves like veterans, standing under heavy fire for hours, and this experiment with that class of soldiers proved their valor and worth . . ."[617]

---

616. Dyer, chapter 16, reg. 154.
617. Stanyan p. 230.

Chapter 10

The Siege Begins; June 10th; June 14th; Paine's Assault; Weitzel's Assault; Augur's Feint; Dwight's Assault; June 15th; Jackson; Taylor Again; The Last Days; Finally; "Record" 8/63; The Next Step

The Siege Begins

Seige is the term that has been used in the official records and in contemporary literature. However, there never was a siege, and it seems that Banks never intended one. When asked why assault when the place must inevitably be starved out in a few weeks, he is quoted as saying: "The people of the North demand blood, sir." Since the failure of the May 27th assault, every preparation was made for a June 14th assault, and after that failed, there were preparations for yet another.[618]

Having negotiated the terms of the truce, Banks' attention was briefly turned to Grant, who had written him on May 25th, explaining that he had decided to lay siege to Vicksburg after two bloody assaults had failed. This would have been a good example to Banks, had he paid attention.

Grant was worried about a buildup of Johnston's forces, and asked that Grierson be sent back immediately, along with "such force as you can spare." The note was delivered by a member of his staff, who was instructed to give "all the particulars of my present situation." He even sent two steamers along to carry the requested troops.

Banks' answer was troubling. He ignored the request for Grierson, countering with a request of his own for a brigade. The steamers were sent back empty.

Grant tried once more on the 29th, this time notifying Banks that he was sending Charles Dana to make the case. The missive from Grant passed down while another from Banks passed up. On the 29th, Banks wrote: "My force is far less than you imagine, and, with such detachments from it as would be necessary to protect New Orleans, while Port Hudson, Mobile, and Kirby Smith[619] are within a few days' movement…"

---

618. Hoffman, p. 70; O.R. Vol. 26/I, p. 520; O.R. Vol. 24/III, pp. 346, 359–360.
619. Lt. Gen. E. Kirby Smith, commander of the Confederate Trans-Mississippi Dept., at Shreveport; Richard Taylor's superior, and with whom Taylor often clashed.

Dana started on his mission on the 30th of May.[620] While passing down the river near Grand Gulf, he met the steamer bearing Banks' reply, and hearing of his latest negative response, Dana returned to Grant's headquarters. There the matter lay.

In the next few days, the regiments with Chickering and Morgan that had marched to Berwick City were brought up to Port Hudson, and most were returned to their divisions. Entrenching tools were distributed to those assigned to construct improved defenses, and as these were completed, the remainder of Keith's 1st Indiana Heavy Artillery was deployed.

If anything had been accomplished by all of the blood spilled on the 27th, it was that the Union lines had been advanced closer to the Confederate works. "Fort Babcock" was established opposite the 1st Alabama, in Steedman's line. The men of the 75th New York had sheltered there during the assault, and two fallen trees had provided enough protection to allow them to begin digging.[621] On June 10th, a battery of four mortars, commanded by Lt. Taylor of the 4th Massachusetts, was completed opposite Steedman's line near the Bennett house. It can be seen at position 3, at the very top of figure 1. According to Confederate reports, it was fired day and night, which "gave us a great annoyance."

Some reorganization was required in order to fill the impressive number of command positions vacated by those officers killed or wounded. Dwight[622] was assigned to the command of Sherman's division. Col. Thomas S. Clark, miraculously recovered from the cannonball wind that had knocked him down on the battlefield, replaced the wounded Brig. Gen. Neal Dow in command of the 1st Brigade; Lt. Col. Edward Bacon replaced Clark in command of the 6th Michigan; Lt. Col. James Smith replaced Col. D. S. Cowles (killed) in command of the 128th New York; Capt. Felix Agnus replaced Col. Abel Smith (killed) in command of the 165th New York Zouaves.

In Paine's division, Maj. James P. Richardson replaced Lt. Col. William M. Rodman (killed) in command of the 38th Massachusetts; Capt. William M. Barrett replaced Lt. Col. Oliver M. Lull (killed) in command of the 8th New Hampshire; and Lt. Col. Frederick A. Boardman replaced Col. Sidney A. Bean (mortally wounded) in command of the 4th Wisconsin.

In Augur's division, Col. Charles J. Paine, commander of the 2nd Louisiana, replaced Col. Edward P. Chapin (killed) in command of the 1st brigade. Lt. Col. Charles Everett replaced Paine as commander of the 2nd Louisiana. Maj. Charles T. Plunkett replaced Lt. Col. S. B. Sumner (wounded), after he had replaced Col. Wm. F. Bartlett (wounded) in command of the 49th Massachusetts.

---

620. Dana, p. 81; Irwin, pp. 186–187.
621. Hall, H. & J., p. 117; Stanyan, pp. 251–252, Defenders, SHS Vol. 14, p. 330. Figure 1 from Irwin, facing p. 192.
622. O.R. Vol. 26/I, pp. 17, 529–532, 632; Irwin, p. 182; Johns, p. 265.

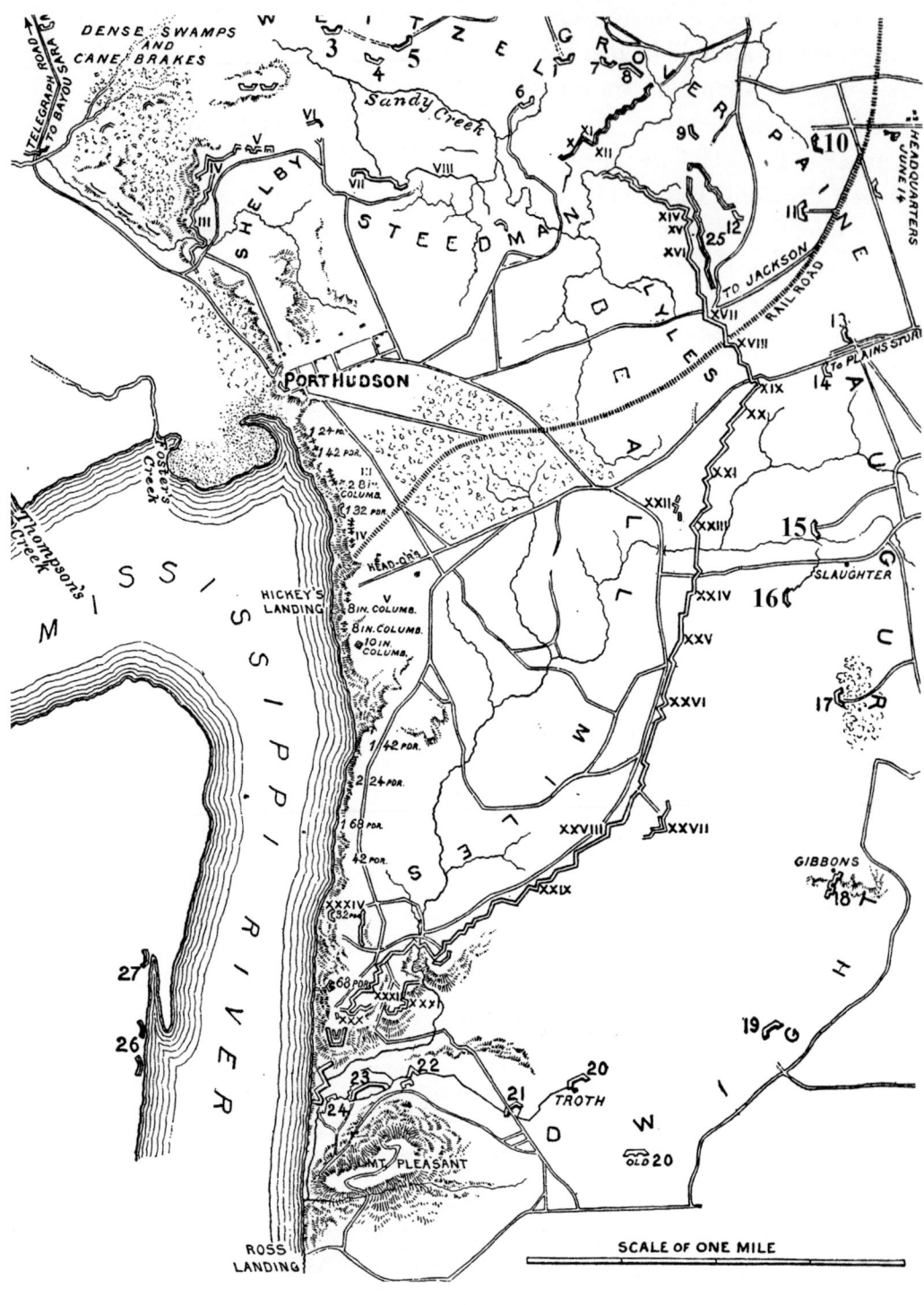

FIGURE 1

Other realignments: Grover got back four of the six regiments he had let go at the close of the Teche Campaign: the 22nd Maine, the 90th New York, the 52nd Massachusetts, and the 26th Maine, though he lost the 41st Massachusetts—now called the 3rd Massachusetts Cavalry—to Grierson. Of course, the 6th New York had gone home. Weitzel fell back into the command of his reserve brigade, to which the 114th New York had returned after the march to Berwick City. Paine lost the 4th Wisconsin (mounted) to Grierson, though he gained the 4th Massachusetts and the 16th New Hampshire. However, the 16th was found to be so disabled from its ordeal at Butte La Rose that it was replaced by the 28th Connecticut, brought up from Pensacola.

June 1st, Monday. From the *History of the 15th New Hampshire*, encamped on the Slaughter plantation:[623]

> Day mostly clear and pleasant; though at times it threatened rain but none fell…
>
> Regular siege operations are now on foot; heavy guns are brought up; the woods are full of shovels, picks, axes, wheelbarrows, and other tools. Our artillery and the fleet fire constantly on the enemy. There is no picket firing in the immediate front, the enemy for some reason remaining silent. Colonel Grierson is seen around our camp; he looks rough and soiled.

June 2nd, Tuesday.

> A most beautiful day. Our camp in the woods was shelled last night, and in the morning, just before roll-call five or six ten-inch shells were pitched over which fell right into our midst…

This was "Lady Davis" or the "Old Demoralizer" speaking.[624] It was a 10-inch Columbiad mounted on the locomotive turntable at the end of the railroad. It could thus fire in any direction. It did little damage, however, as its shells were either empty or failed to detonate.

> At 10 o'clock General Banks, with attendants, comes up to our front and views the situation with a glass.
>
> Thousands of negroes are picking and shoveling, as well as the soldiers, on the disputed open ground between us and the enemy, and gradually advancing the trenches and rifle pits toward their parapet…
>
> Captain Cordon was…on an eminence with some of the 6th Michigan boys, who were using "Henry" rifles; they were provided with telescopes. One of them said, "Look through this glass." The captain looked, and as one of our men fired he saw the bullet strike…and it seemed to go right through [the Confederate soldier foolish enough to have stood on the parapet]…
>
> We are nearly opposite their centre; [referring to figure 1, to the rear of Battery 16, near the Slaughter home] their works may be entirely silent

---

623. McGregor, pp. 419–422, 427, 434, 436–437, 440.
624. Sprague, p. 142; Tiemann, p. 43. The gun was mounted on a rail car, able to use the turntable or be moved along the railroad.

now, and none of their men are to be seen; but their works are manned; the Confederate soldiers lie thick behind them all armed and ready to fire on the instant.

While the 15th New Hampshire toiled, similar activity took place on Weitzel's front.[625] Navy Lt. Commander Terry and crews from the *Richmond* and *Essex* arrived with their guns at Battery 10, which was described as "748 yards from the enemy's works," see the upper right in figure 1, and figure 2 from Miller, F. T. Vol. 2 p. 219.

FIGURE 2. THE NAVY HELPS ON LAND

June 5th, Friday.

A battery of "rifled guns" was planted opposite the "slaughter pen" (the Bull Pen field referred to earlier) about 400 yards from the Confederate battery at the Bennett house. Battery 5 falls into this definition. It is found as having four rifled guns in a list of the batteries, see table 1, below.[626]

---

625. Stanyan, pp. 251–252; Hall, H. & J., p. 122; ORN 20, p. 257. These references do not enumerate the battery positions exactly, though the *15th New Hampshire History* (McGregor) does, providing a map nearly identical to Irwin's. Many battery positions were not constructed until after May 27th, and almost all of the Port Hudson maps available only show battery positions as of the close of the siege, in July.

626. Atlas, portion of plate 38. Note Duryea is listed at 6 and again at 12. It is assumed that his position at Battery 6 was on May 27th, and that he was subsequently moved to 12. Closson or Appleton will not move to No. 24 until June 25th, and it is noted that no exact position is shown for Battery L prior to that time, though they were "750 yards" from "right centre."

BATTERY		BATTERY	
1	(1) 30 pdr. Parrott	16	(4) 3 inch rifled, Bane
2	(2) 12 pdr. howitzers Phelps		(6) 12 pdr. Napoleon, Rawles
3	(4) siege mortars Taylor	17	(4) 8 inch siege howitzers, Rose
4	(1) 30pdr. Parrott, Harrower	18	(4) siege mortars, Hill
5	(4) 6 inch rifled, Healy	19	(6) 3 inch rifles, Hebard
6	(6) 12 pdr. Napoleons, Duryea	20	(4) 20 pdr. Parrotts, Roy
7	(2) 30 pdr. Parrotts, McLaflin	21	(4) 12 pdr. Napoleon, Bradley
8	(4) 12 pdr. Napoleon, Carruth	22	(8) inch siege howitzers, Baugh
	(4) 12 pdr. Napoleon, Bainbridge	23	(1) 10 inch siege mortar, Motte
	(2) 3 inch rifled, Norris	24	(2) 20 pdr. Parrotts, Hartley
9	(2) 12 pdr. rifled, Cox		(3) 24 pdr., Hinkle
10	(4) 9 inch Dalhgren's, Terry USN		(2) 9 inch Dahlgren Lt., Swann USN
11	(6) 20 pdr. Parrotts, Mack		(1) 8 inch howitzer, Lt. Glover
12	(6) 12 pdr. Napoleon, Duryea		(2) 12 pdr Napoleon, Closson
	(1) 20 pdr. Parrott, Duryea		(2) 10 pdr. Parrots, Appleton
13	(3) 24 pdr., Hadden		(3) 30 pdr. Parrots, Grimsley
14	(6) 6 pdr. Sawyer, Holcomb		(2) 10 inch siege mortars, Hamlen
15	(2) 20 pdr. Parrotts, Hamrick	25	Trench Cavalier (observation tower)
	(2) 24 pdr. Parrots, Harper		

TABLE 1

June 8th, Monday. Very hot:

One man wounded from the 128th New York. All men working night and day. Lt. Col. Blair, though wounded on the 27th returned, only to find that Dwight had arrested Col. Kingman, so Blair, with his right arm in a sling, and looking "weakened and emaciated," took command. The suspicion was that Kingman had been too forthright with Dwight in protesting that the May 27th assault was a futile, rash, and ill-considered affair. No charges were ever brought. Quoting Kingman: "It was a petty, spiteful, and cruel exercise of …authority, which I had no means of resisting or clearing up, as our term of enlistment had nearly expired."[627] – Dwight again.

June 9th, Tuesday.

Very hot, dry and dusty; good breeze that shakes the leaves; "Only coffee and hard bread for breakfast after shoveling and picking all night…Considerable bombarding but no one injured near our camp."

---

627. The 15th was one of Banks' 22 nine-months regiments. Though organized on Oct. 16th, 1862, they were not mustered into Federal service until Nov. 12th, 1862. Hence, their term of service was due to expire not in July, as they all thought, but on August 12th, 1863. From: *New Hampshire Adjutant General's Report,* Concord, 1865, Vol. I, p. xvi, and Vol. 2, Concord, 1866, p. 836.

June 10th, Wednesday.[628] Extracts from a letter home, written by Lt. Washington Perkins:

> We are building batteries and digging rifle pits all around them, and in a day or two I expect there will be one of the most terrific bombardments that has ever been known. Our rifle pits are within rifle shot of their parapet; we have been at work on them night and day. Their sharpshooters are firing on us all the time, and the balls are whizzing over our heads, but they don't hit many. They give us a shower of shells and grape, but we give them back ten fold. There is scarcely five minutes, day or night, but that we hear the roar of artillery, or of the bursting of shells.
>
> Our time is out July 16, and we expect to get home by that time.

This was a forecast of some trouble. The men calculated their time from when they were organized; the government calculated their time from when they were mustered into the Federal service; which was a month later.

June 10th

Not waiting for all of these gun positions, rifle pits and engineering work to be completed, a reconnaissance-in-force, of sorts, was ordered by Banks. It was prompted by the fact that Confederate fire had slackened by degrees. During the day, the Yankee sharpshooters were so effective that the Confederates had begun firing only at night. It was becoming difficult to "estimate their available ordnance."[629] The objects of the reconnaissance (Irwin describes it as a feigned attack) were stated to be: (1) harassment of the enemy, (2) inducing him to bring forward and expose his artillery, and (3) acquiring knowledge of the ground, which would allow pioneers to remove obstructions, if necessary.

The *History of the 75th New York Volunteers* describes the action, which began on the night of the 9th:

> [A] grand bombardment was begun…and kept up for thirty-six consecutive hours. Finding that this maneuver elicited little response, Banks resolved on a feint with infantry along the whole line…The ball opened on the left and the advance and skirmish fire soon became general, and the artillery, which for a short interval had been comparatively silent, now resumed their work…the hills and woods around Port Hudson resounded as they never had done before.

At midnight on the 10th, a continuous line of skirmishers was formed and ordered to sneak forward, in Banks' words, "[to] get within attacking distance of the works, in order to avoid the terrible losses incurred in moving over the ground

---

628. Stanyan, p. 252; McGregor, pp. 441, 445–446, 456, 461–462. Perkins does not understand that their time will be counted from their date of muster, which as noted, was nearly a month later.
629. Hall, H. & J., pp. 123–124; Irwin, p. 192; O.R. Vol. 26/I, pp. 14, 67, 131–133.

in front of the works."

Steedman describes it with a telling opening:[630]

> On the 10th of June a furious bombardment all day and night indicated to us an approaching attack, and at three o'clock in the morning of the 11th, a show of an assault was made near the centre of our line of fortifications, while, at the same time, the real attack was made on our left in the woods…During the fighting two regiments of the enemy, favored by the extreme darkness, crept up through a gorge among the abattis, penetrating within our lines of defence [*sic*]. Had they known the ground and been strongly reinforced, this movement might have proved disastrous to us.

As soon as the Confederates were seen to have manned their parapet, and were returning fire, the Union skirmishers were ordered to lie down, the sharpshooters in their rear conducting the return fire over their heads. From the 75th New York: "Toward morning a slow rain set in and the men were ordered to fall back."

The five companies of Weitzel's 12th Connecticut who were involved suffered casualties "greater in proportion to the number engaged than in any other single engagement during the entire siege." The 22nd Maine suffered some casualties. In all, incomplete reports totaled 2 killed, 53 wounded, and 65 missing.

The operation accomplished almost nothing, with the exception that on Grover's front the 131st New York was able to reposition itself closer to the works.[631]

June 12th.

> Very pleasant; not quite so hot. Work on the "great cotton battery built by Capt. Johnson…Battery 16," was becoming feverish. "Chief Engineer Bailey[632] became very nervous, and the men were urged to the utmost exertions; there were many more men on the work besides the Company D boys. It seems that Chief Engineer Bailey's orders were to have the work completed in such season that the battery [Hebard's] could drive into the works before light on the thirteenth."

June 13th.

> The day "rose in semi-tropical beauty, and during the day the sun shed down his fierce rays with a blinding glare and intolerable and pitiless heat." The work on the "12 gun… battery [16]" was not quite complete, but in the morning Hebard's 1st Vermont "with horses lashed to the keen gallop, went in in broad

---

630. Defenders, SHS used here and in the rest of the volume to abbreviate "Southern Historical Society Papers" Vol. 14, p. 326.
631. O.R. Vol. 26/I, p. 131.
632. Capt. Joseph Bailey, of the 4th Wisconsin, was eventually assigned to Banks' staff as acting military engineer of the 19th Army Corps. He was to become renowned for his dam on the Red River. O.R. Vol. 34/I, p. 406; O.R. Vol. 34/II, p. 544; Defenders, SHS Vol. 14, p. 326.

daylight..." They were subjected to a "terrible" fire with two men and one horse killed,[633] and three men wounded.

At 7:30 a.m., Banks wrote to Farragut, on the *Monongahela*:[634]

> I shall open a vigorous bombardment at exactly a quarter past 11 this morning, and continue it for exactly one hour. I respectfully request that you will aid us by throwing as many shells as you can into the place during that hour, commencing and ceasing fire with us. The bombardment will be immediately followed by a summons to surrender. If that is not listened to, I shall probably attack to-morrow morning, but of this I will give you notice.

Gardner refused the surrender demand, and at 9:00 p.m. Banks signaled Farragut to commence firing, with mortars only, at 11:00 p.m., and to cease at 2:00 a.m. on June 14th. The second assault was on and the defenders forewarned.

June 14th

A general council was held on the evening of June 13th at Banks' headquarters. Though preliminary orders for troop dispositions were sent out from headquarters at 8:45 p.m., and the signal to Farragut at 9:00, the final orders were not read and approved until 11:00, and the first copies were not sent out until 11:30. As a result, some regiments did not receive their orders until after they should have begun moving to their assigned positions.

The main attack[635] would be under the overall command of Grover, and would take place at the Priest Cap, indicated in figure 3 by the heavy arrow at the top. A second assault would be conducted by Dwight at the Union extreme left, the southern end of the Confederate defenses, as indicated by the heavy arrow at the bottom, and a feint would be conducted by Augur from the Plains Store Road, as shown by the light arrow.

Depending upon the course of events, Augur was to be prepared to reinforce either assault.

---

633. McGregor, p. 465. Table 1 lists Bane and Rawles, Battery G, 5th U.S. Artillery. They were later occupants, but apparently at this point it was Hebard.
634. ORN Ser. 1, Vol. 20, pp. 229–230; O.R. Vol. 26/I, pp. 552–553.
635. O.R. Vol. 26/I, pp. 554–555. Figure 3, Irwin, facing p. 192 (altered).

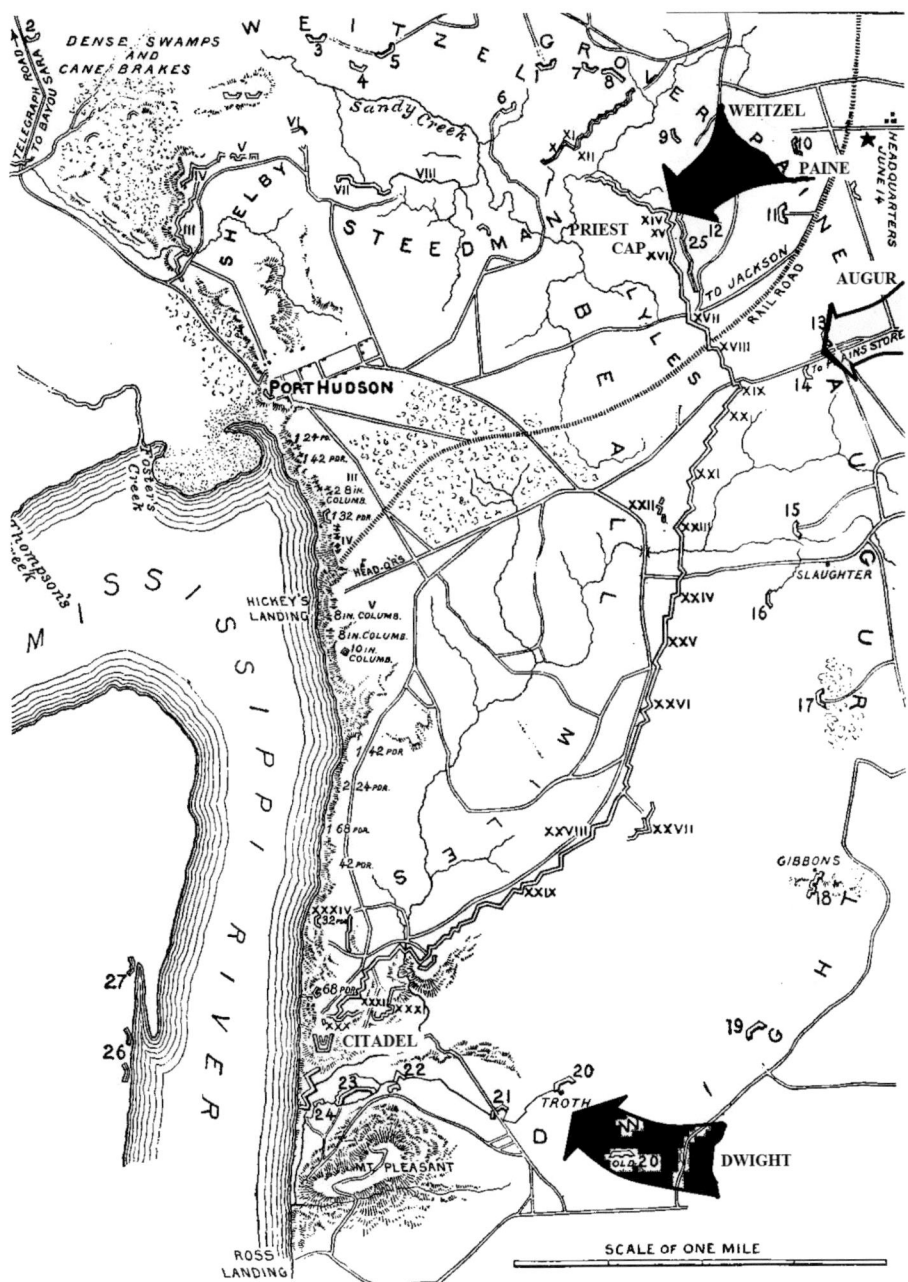

FIGURE 3

The attacks were to begin at 3:30 a.m. in a coordinated manner; with all watches set. The preliminary bombardment would begin at 3:00.[636] Nothing was vague and left to discretion as it had been on the 27th, *except* a caveat: The 3:30 a.m. advance by the skirmishers might be delayed until "as soon thereafter as General

---

636. Steedman's comment regarding the opening of the June 10th reconnaissance: "a bombardment indicated to us an approaching attack…"

Grover may find best." Another flaw related to time should have been evident, in that, there was but little of it allowed for the troops to get into position, particularly those transferred from other regiments. For example: the 48th Massachusetts from their location at Plains Store to Dwight, and the 50th Massachusetts from the Plains Store Road to Dwight. A lesser movement involved Dudley to Grover with the 161st and 174th New York regiments. The rough ground, and darkness, would make it difficult, even for those already in the vicinity, such at the 75th New York, to reach their final assault positions.[637]

Though Grover was named as in command of the main assault, the concept of the plan is credited to Weitzel, who believed that a concentrated attack on one point would penetrate the line "by a combination of force and stratagem[638]...A perfect surprise was intended." Ah, but it seems that the senior generals, and Grover in particular, were the ones who laid out the details, and worse, remained to direct. It was to be a deadly repeat of the 27th.

Since neither Weitzel nor Grover made any report after the battle, we have no clue as to whether anyone had argued against the use of a preliminary bombardment. Then again, it was to be general, and would not warn of an attack at a specific location. However, it was still a warning, and Gardner had the ready ability to transfer his troops as required.

Paine's whole division was to attack on the left of the Priest Cap, and be the "chief" assault, figure 4.[639] Weitzel's brigade, reinforced by Dudley's 3rd Brigade, with Grover's division in support, was to make an advance on the right through a gully, or "little ravine" which ran between Cox's Battery 9, and Duryea's Battery 12, into which a trench, or sap, about 200 yards in length, and dug to the depth of seven feet and a width of six feet, had been extended to within about 150 yards of the parapet.[640]

It is important to note that, in view of all this activity, Gardner had moved the 49th Alabama into place to reinforce Ben Johnson's 15th Arkansas. The 1st Mississippi was also in place near the 49th.

Griffiths' Field, through which Paine was to advance, was described as scrubby

---

637. Hall, H. & J., p. 125; Plummer, pp. 41–42. The 48th had marched from their bivouac at Plains Store; had worked all the day of the 13th entrenching a battery position on Augur's front, and then at dark, were ordered back to Plains Store. Upon arriving there, they received Banks' order to march to Dwight.
638. Hall, H. & J., pp. 124–125; dmna.ny.gov/75th New York, Clippings, p. 1, Letter to *N.Y. Herald*.
639. Irwin, p. 196.
640. Hall, H. & J., p. 124; McMorries, *Defense* map, portion. The Jackson Road is mislabeled on Irwin's map, and there are no terrain features whatsoever. The general arrangement of the McMorries map seems to better match the description of the attack. Labels have been added. The battery numbers have been changed to conform to Irwin's, and the Federal maps in general. (The Confederate map numbers the Federal batteries less by one, e.g., 12 is 11.)

undulating ground, a part of which was "formerly...cultivated,"[641] These words seem to forewarn of trouble, if not failure. The approach was nearly wide open and unobstructed. But of course, the advance was to be a stealthy surprise, at dark, and Weitzel's attack, not to mention those of Augur, or Dwight, were supposed to divert attention from Paine in a coordinated manner.

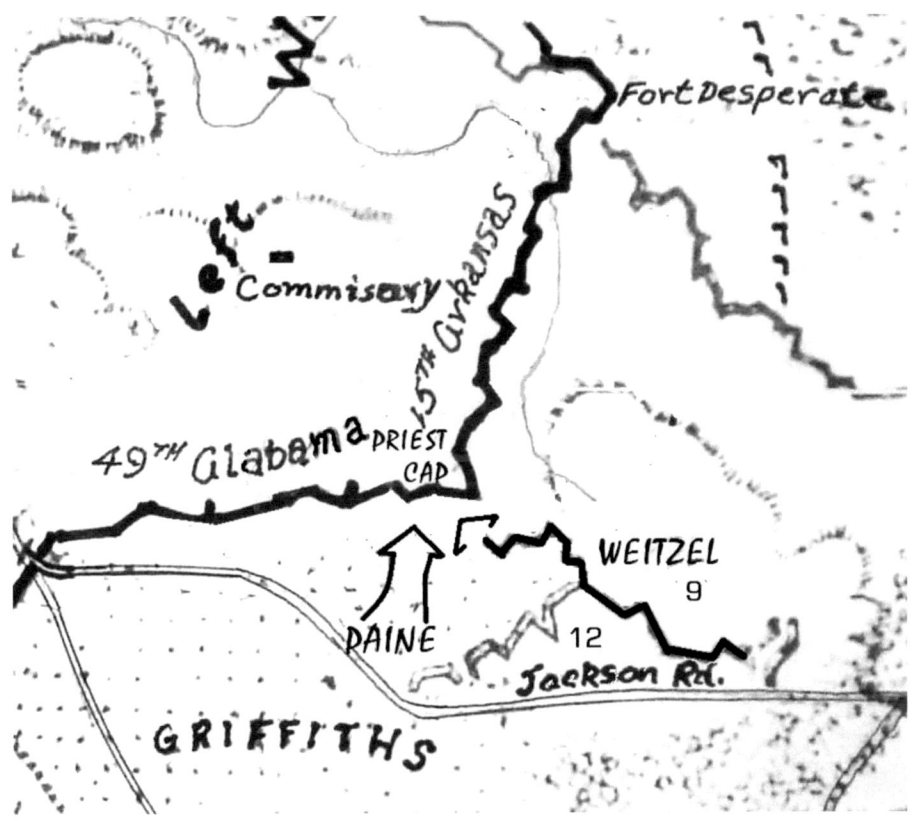

FIGURE 4

The sap through which Weitzel was to make his approach is shown in figure 4. It had been "pushed to within a few rods of the eminence" on which the powerful Confederate bastion stood. Weitzel's plan was to send a force into this passage by night, "which should begin the attack just before daylight." It is hard to imagine even portions of seven or eight regiments passing through a six-foot-wide sap, but that was the plan.

The general formation of the two attacking columns had been settled upon by the 11th of June,[642] and on June 12th Paine issued his General Order No. 64,[643]

---

641. Stanyan, pp. 254, 260; dmna.ny.gov/Clippings 75th New York, p. 1, letter to NY Herald, describes the distance from the end of the sap to the parapet as 150 yards.
642. Hanaburgh, p. 54; dmna.ny.gov/historic/reghist/civil/infantry/156thInf/156thInfCWN.htm.
643. Stanyan, pp. 254–256; Willis, p. 131. These two sources differ; Whitcomb, pp. 49–50; Ewer, p. 91; dmna.ny.gov/historic/reghist/civil/infantry/24thInf/24thInfScribner00Intro.htm; Powers, p. 105.

which set it straight as to where each of his regiments were to be positioned in his attacking column, and what their job was. The format of the order was as appears below, with the exception that changes to show the actual composition that existed on the morning of June 14th have been made. The hand grenades were percussion-fused, 6-pounder artillery shells, one per man, and the "cotton bags" were sand bags filled with 30 pounds of cotton, *two* per man.

Column of Attack

Front

8th New Hampshire, 4th Wisconsin as skirmishers, at intervals of two paces

Companies A&K, 4th Massachusetts and A, B, E,

& I, 110th New York, with hand grenades.

38th Massachusetts

53rd Massachusetts

31st Massachusetts and 60 men of the 173rd New York, with 400 cotton bags

156th New York

2nd Brigade, Fearing

133rd New York, 173rd New York

1st Brigade, Ferris

28th Connecticut,

50 pioneers, to level parapet

Nims' Battery

No such order of Weitzel's survives, but his column[644] was nearly identical to Paine's, with the 12th Connecticut and the 75th New York positioned at the head as skirmishers. Next, the 91st New York and 100 men of the 28th Connecticut, with hand grenades, the 24th Connecticut with cotton bags, followed by the 8th Vermont, the 114th New York, and the 160th New York. Grover's 1st brigade, reduced to the 1st Louisiana, 22nd Maine, and the 90th and 131st New York, were to follow. Grover's 2nd and 3rd brigades, with Dudley's 161st and 174th New York, were held in reserve. The 30th Massachusetts was to be initially deployed to support Terry's Battery (No.10).

Battery L and Duryea's Battery F were to be drawn up in the rear of the attacking column, and remain in reserve.

Paine's Assault

Roused a little before midnight, on this clear and cool evening, Paine's regiments began forming on the edge of the woods alongside the Jackson Road by 2:00 a.m., figure 4[645]. They were formed up by 3:00, with the 8th New Hampshire and

---

644. Irwin, p. 198; Crofutt and Morris, p. 412; Haskin, p. 193; *Massachusetts Adjutant General's Report*, for the year ending 1863, p. 800.

645. Powers, pp. 107, 108; Stanyan, pp. 259, 268.

4th Wisconsin in front, as planned. The cannonade had begun, the four IX-inch Dahlgrens of Terry's naval battery (10 in figure 3) throwing their 75-pound shells into the Confederate works over the heads of Paine's men.[646] Immediately to the south, Augur had promptly begun his feint. To the north, McLaflin's Indiana battery (7) had joined in the firing. Weitzel's troops had been on the march since 11:00 the night before, and the head of the column had arrived at its attack position. Today, unlike May 27th, a well coordinated attack was evidently poised to begin. Even a subsequent Confederate report would describe it as "simultaneous." However, that report notes: "The Federals at first pressed heavily on the right where the 49th Alabama was stationed…" This was Paine.

So we have the first intimation that the attack was not quite "simultaneous" after all, and that would prove to be fatal.

For a time, the front of Paine's column was shielded from the view of the Confederates by the thick hedges of osage orange on the approach, and somewhat by a fog that had developed toward dawn. There was little wind. At about 4:00, the order was given to go forward. They pressed through the wood, and found that the distance to the works was farther than expected, about 500 yards. It consisted of undulating ground, and a succession of ridges and ravines, in which there were fallen trees, and scrubby bushes and brambles. On the brow of every ridge they would be exposed to the fire of the Confederate riflemen, who by now would have been fully warned. Many men and officers were falling, even before reaching a clearer area. The advance lay checked about 90 yards from the face of the parapet. Paine, seeing that the advance had stalled, but that the first line of skirmishers was holding their positions, came forward from the head of the column of the 38th Massachusetts,[647] and "as loudly as I possibly could" gave the order to advance "at the first word of which the men sprang forward." Kimball, of the 53rd Massachusetts, repeated the order. Soon, General Paine fell, "… struck, *soon after daylight*, by a rifle ball… about fifty yards from the enemy's works…"

The statement is significant, since it gives us the time – 5:15 – soon after which Paine's assault faltered. It is also significant, in that it confirms the Confederate use of their "long taw," their Enfield rifles, picked from the field after the assault of May 27th. The significant numbers of these long range weapons now in the hands of the defenders no doubt contributed heavily to the high number of casualties in Paine's lead regiments, and was likely an unexpected surprise.

---

646. *Ordnance Instructions for the U.S. Navy*, U.S. Gov't Printing Office, Washington, 1866, p. 90; Bissell, p. 50. Stanyan, pp. 261, 269; Defenders, SHS, Vol. 14, p. 328; Irwin, p. 196; Willis, p. 137.

647. Stanyan, pp. 260–262; Powers, pp. 107–110; Willis, pp. 137–138. *Massachusetts Soldiers, Sailors and Marines in the Civil War*, Vol. 4, Norwood, MA, 1932, pp. 1, 620; *dmna.ny.gov, 133rd Regiment, Letters, Diary of a Williamsburgh Soldier–The Entrenchments*; Irwin, pp. 196–197; Defenders, SHS, Vol. 14, p. 328; McMorries, p. 66.

There was a pause in the action in the absence of Paine's commands, and the men simply sought cover. The hand grenades were a failure, many failing to explode when striking the soft earth of the Confederate entrenchments. They were then hurled back at the attackers. Company K of the 4th Massachusetts suffered three officers killed and nine enlisted men wounded, all on a date in time when many of them had reckoned their nine months service was up.[648]

A few of the 8th New Hampshire and the 4th Wisconsin men had crossed the ditch, which in places was only two feet deep, reached the parapet, went over, and were either immediately shot down or captured, but many lay dead in the ditch. From the "Defenders" article in the *Southern Historical Society Papers*: "The smoke was so thick that nothing could be seen more than twenty steps in advance, and before our troops were aware of it the enemy were pouring into the ditch and scaling our breastworks[649]…The ground in front of our works was blue with their uniforms…" The skirmish line was decimated; the 4th Wisconsin suffering 140 killed and wounded, or 63 percent, and as it turned out, the most of any regiment that day. The 8th New Hampshire was next, suffering 122 killed and wounded out of 217 present, or 56 percent; a dreadful repeat of May 27th when they lost 124. No regiment lost more at Port Hudson. The 38th Massachusetts had 90 men lying on the field either dead or dying. Next in line had come the 53rd Massachusetts, which now had 86 killed or wounded. The only advantage that accrued from their cotton bags was that the 31st Massachusetts had managed to make a breastwork of sorts at the site of their most advanced position, which they were able to hold, losing only 30 men.

The 133rd New York and the 173rd New York were mixed in with the advanced regiments when Paine fell. Maj. A. Power Galloway of the 173rd having been killed, the troops were rallied by L. D. H. Currie of the 133rd, who then had made a gallant charge. This last attempt to retrieve the situation failed, and left Currie and 58 others of the 133rd wounded, with 9 killed. Of the 173rd, there were 16 killed and an unknown number wounded.[650] Paine's assault had failed.

---

648. *Massachusetts Soldiers and Sailors*, Vol. 1, p. 228. Its companies had been mustered in at various dates in September, varying from the 1st to the 26th. Co. B, was the only one mustered in on the 1st, and both A and K, who were so cut up in the assault, were dated from the 23rd. Quite precisely then, only Company B had a complaint on the 14th of June. The government, i.e., Banks, now pushed their enlistment date as far forward as possible, reckoning that their time was from when their *field and staff officers* were mustered, and that was December 16th, 1862. Compiled records, M594, roll 77. They all were mustered out on August 28th, 1863, having been shipped out of Port Hudson on August 4th. Prior, a handful of the men had refused to do further service. They were arrested and threatened with doing time at Ship Island, but were instead dishonorably discharged.

649. Defenders, SHS, Vol. 14, p. 328; Powers, p. 108; Willis, p. 138, Stanyan, p. 260.

650. See *dmna.ny.gov./173rd Regiment*, Table. The table combines May 27th with June 14th for a total of 67 wounded; Croffut and Morris, p. 412. See also Peck, Lewis, p. 1.

Weitzel's Assault

The 75th New York, which was to head the column,[651] was quietly called to arms at 11:00 on the night of the 13th. After 60 rounds of ammunition had been issued to each man, and a breakfast of coffee and hardtack, they were on their way at midnight. They and the other of Weitzel's regiments in the attack would have to quietly withdraw from their positions along the ridge of the ravine north of the Sandy, skirt Fort Desperate, and then move south to be in position for the attack. They stumbled in darkness; the new moon would not emerge until June 16th.[652] On this rough, circuitous route through brush and woodland, it took the 75th three hours to get to their rendezvous, the entrance to the sap.

The 12th Connecticut was supposed to join them at the head of the skirmisher column, which was under the command of Col. Babcock of the 75th. Emerging from the sap, the 75th was to deploy to the right of the north of the angle of the Priest Cap, and the 12th to the left. The 91st New York, and 100 men of the 28th Connecticut were to follow with hand grenades, and the 24th Connecticut with cotton bags, to make a crossing of the ditch for the storming party. The rest of the brigade was to follow, the 8th Vermont, the 114th New York, and the 160th New York, all under the command of Colonel Smith of the 114th. The skirmishers and grenadiers, having driven the rebels back from the face of the parapet, would allow the cotton bags to be placed, and then the stormers would be expected to cross, scale the works without pausing to fire or reload, and charge into the works at the point of the bayonet.[653]

"The batteries of our friends had opened the ball." McLaflin's Indiana battery, (7 in figure 3) had joined in the firing, and it was now 4:00 a.m., exactly one hour before sunrise, the "transparent gray in the atmosphere which was the prelude of dawn, and which obscured objects without concealing them." There was enough light to just see the horizon through the early morning mist that had formed.[654] There would be only a few minutes before there was enough light for the rebels to aim and fire at the skirmishers who would emerge from the sap—150 yards or so from the ditch and the face of the parapet.

The 12th Connecticut had not arrived,[655] and after some delay, Colonel Babcock decided against waiting for them. Absent the 12th, there was another considerable delay as the result of having to rearrange the plans for deployment and inform the guides. Finally, the 75th advanced into the sap by double file. The skirmishers, the

---

651. Hall, H. & J., pp. 124–125. The 75th eventually headed the column on the right, the 12th Connecticut on the left. O.R. Vol. 26/I, pp. 132–133.
652. *http://aa.usno.navy.mil/moon phases & sunrise.*
653. Fitts, p. 125.
654. *aa.usno.navy.mil/*, beginning of nautical twilight for 1863 at Baton Rouge; Pellet, p. 115.
655. Sprague, p. 148; Fitts, pp. 127–128; Hall, H. & J., p. 125.

grenadiers, and the bag carriers were expected to crowd through this defile, and deploy at its end with sufficient stealth and speed to accomplish a surprise attack on the rebel parapet, at least sufficient to drive them away from the face, which would allow three more regiments sufficient time to filter through, mass, and carry the works.

The 12th finally arrived, having been misdirected by its guides, and entered, brushing past[656] the 91st New York and 28th Connecticut with their grenades, and deployed on the left as originally planned. Everyone then had to wring their hands while the bag carriers struggled through the sap. The 114th New York was just entering the sap when Weitzel's order "Fix Bayonets!" was heard. It was 5:00 a.m. – almost an hour after Paine's assault had begun. The delays had ruined the attempt at simultaneity, and the changing light, for 5:00 was sunrise, had ruined the chances for even somewhat of a surprise.

The 75th were in, and, the fighting having begun, those near the end of the sap became cautious, slowed, and it became clogged with humanity. "For God's sake, don't stop now; go on and let us get through…We can't, the fighting up front has choked up the road." Finally, the column moved slowly on, though now they had to make way for the stream of dripping wounded, aided or alone, which began to come out, their only safe path to the rear.[657]

Before the 114th made its way through it was something near 6:00 o'clock, and the ground was strewn with soldiers in blue; dead, dying, and too severely wounded to move. The noise would have prevented any distinction of the fact that on the left, Paine's assault had failed, but the fury of the Confederate fire in front forecast the fate of Weitzel's. The regiments were scattered in "hopeless confusion." Colonel Smith of the 114th was mortally wounded, and one-third of his regiment was disabled. The command now passed to Van Petten of the 160th New York, who ordered a second charge.

The aftermath is described by Capt. James F. Fitts, of Company F of the 114th New York:

> As the troops crowded up from the rear, they were sent forward to join in this bush-fighting; but there was no serious demonstration after the sun was an hour high. The battle was lost and the blood shed before sunrise…

There were some 49 killed or wounded (incomplete report) in the 75th New York, 96 in the 8th Vermont, 84 in the 12th Connecticut, 85 in the 114th New York, and 59 in the 91st New York.[658]

---

656. O.R. Vol. 26/I, p. 133; Croffut and Morris, p. 412; Hall, H.& J., p. 126.
657. Pellet, pp. 116–119; Fitts, pp. 120, 129–130.
658. Casualties derived from: Compiled Service Records, 75th NY and 91st NY, M594, roll 125; 12th CT, M597, roll 7; *dmna.ny.gov./History*; Carpenter, p. 124; Pellet, p. 119; Beecher, pp. 209–210.

There was palpable pause.[659] Sprague, in his history of the 13th Connecticut, illuminates: One of Banks' aides now appeared and informed Weitzel to "force an entrance at once into the rebel works at all hazards." "Yes," replied Weitzel; and then, to a staff officer, "Give my compliments to Colonel Holcomb, and tell him to go in immediately." Weitzel's "cool, yet unsatisfied and discouraged air astonished some of us, who looked for an impetuous charge by the favorite young general." Weitzel may have been discouraged to note that his own troops had failed and that he was now forced to turn to Grover's (Holcomb was temporarily commanding Grover's 1st Brigade.)[660] Another point could have been that Weitzel simply had concluded that the concept and timing of the assault had not been implemented as planned, and to continue it further would be futile, as was the last charge of the 114th New York.

Nevertheless, in went the 1st Brigade. By now it was 7:00 a.m.

Birge's 3rd Brigade, composed of the 13th and 25th Connecticut, the 26th Maine, and the 159th New York, "pressed forward to support Holcomb."[661] The 26th was either misdirected or uninformed of the sap, and charged straight into the ravine to its right and through the rough landscape. Noting the crowded sap, half of the 13th was split off, with five companies plunging forward on its left.

About half of the 3rd arrived in time to hear their old major (Holcomb was chosen from the 13th Connecticut by Butler to recruit the 1st Louisiana at New Orleans in July of 1862) "haranguing" his brigade at the front. The initial rush of the 13th had caused the Confederate firing to slacken, as "a few" of the Confederates were observed to turn and run back from the parapet, which gave a moment's time for Holcomb to roar: "All I ask of you is to follow me! Will you follow me?" "Yes! Yes!" was heard from a handful, the rest remaining silent. The interval also allowed somewhat of a line of battle to be formed by the mixture of the two brigades. Before this could be completed Holcomb, swinging his sword, gave the command, "Forward."

The line almost instantly lost any organization. Its center, with Holcomb leading, was exposed to the renewed fire from the parapet and he was shot down at once. The assault had failed.

None of the commanders still standing; Burt of the 159th New York, Morgan of the 90th New York, Jerrard of the 22nd Maine, Hubbard of the 26th Maine, or Day of the 131st New York; could agree on a plan, despite the urgings of Banks' staff for more assaults. The words of Col. Day sum up the situation: "It's too damned risky!" And no new assaults were made.

This was clearly "misbehavior before the enemy," the most serious of court martial offenses. However, no such charges are of record – additional assaults, as demanded by Banks' staffers, being regarded as futile and beyond the bounds of

---

659. Sprague, p. 314; O.R. Vol. 26/I, p. 130.
660. Compiled Service Record, M594, roll 67, 1st Louisiana; Beecher p. 208.
661. Sprague, pp. 151–157; Croffut and Morris, p. 413.

common sense. A Confederate view: "Again and again they reformed and charged, but…towards the last of the battle their officers could not get the Federals to leave their own breastworks. They were not cowards, but brave men. They saw no hope of storming our position successfully…"

Though they were in reserve, Battery L was close enough to the firing to have had two horses wounded. They remained in this position until June 20th, when one gun and its crew under Sergeant Becker was detached on an expedition to Jackson, one of two special assignments given to Battery L toward the close of the siege. Both are described later.

Augur's Feint

As noted, Augur's troops, reduced to only the 49th Massachusetts, the 21st Maine, the 2nd Louisiana, and the 116th New York, had begun their feigned attack promptly, at about two hours before daybreak.[662] The 116th crept out from its position in support of Holcomb's battery (14), moved south, and acting with the other regiments, sent half of their companies forward as skirmishers, creeping up through Slaughter's Field as close as possible to the Confederate parapet. They were not met with fire from the Confederate heavy guns, most of them having been disabled by the Union bombardments of the previous days. At the signal, the skirmishers opened up. The reserves were later called in, and they remained on the field for the rest of the day. They were withdrawn at nightfall.

There were light casualties, owing to the absence of Confederate artillery fire, and the fact that no charge upon the works had been made.

Dwight's Assault

Dwight issued his Special Orders No. 32 on June 11th.[663] Nothing was said about the location of the assault, except that: "The troops will be held in hand, and be prepared to move immediately to such point of assault as the Brigadier-General commanding shall designate." The usual details such as bag and fascine carriers were outlined, even ordering a detail of pontoniers, and another of engineers, to prepare a way for the entry of the artillery into the works. Two other sections of the orders were first, bizarre, and second, mysterious:

> 5. Colonel Clark will detail fifty picked men of the Sixth Michigan Volunteers… for a sudden attack upon the headquarters of General Gardiner [sic]…
> 6. Colonel Clark will also detail two hundred men of the same regiment, under command of the senior captain,[664] for an important and decisive movement.

---

662. Clark, p. 94.
663. Bacon, pp. 151–157; O.R. Vol. 26/I, p. 549; Hanaburgh, p. 55; Benedict, p. 16.
664. Capt. John Cordon, Michigan Adjutant General's Report, 1863, p. 33.

Revealing Dwight's contempt for the nine months units, section 7 ordered that they should lead the advance.

Dwight's troops consisted of his 1st and 3rd brigades, plus two regiments transferred to him on the night of June 13th. His 2nd Brigade had been assigned to the defenses of New Orleans. Col. Thomas S. Clark commanded the 1st Brigade, consisting of his own regiment, the 6th Michigan, plus the 15th New Hampshire, the 26th Connecticut, and the 128th and 165th New York. A temporary 2nd Brigade was commanded by Col. Lewis Benedict, just returned from his assignment with the 110th New York on the west side of the river. This consisted of the 162nd and 175th New York, 28th Maine, and 48th Massachusetts. The reserve consisted of four companies of the 28th Maine, the 177th New York, and the 50th Massachusetts.[665]

Summoned to Dwight's headquarters, the mysterious details were explained to Col. Edward Bacon, commanding the 6th Michigan. The assault would be two pronged; left, at the river, and right, along the Mt. Pleasant Road, figure 5. The left would consist of the 50 picked men mentioned, disguised as rebels, and guided by two deserters. Under the cover of darkness, they were to climb into the "citadel," the prominent Confederate fortification at the edge of the river, via steps that had been cut into the clay bank, and known to the deserters. The raiders were to then proceed immediately to General Gardner's headquarters and take him prisoner. The remaining 200 were to follow, and entering the Citadel, spike the guns. Bacon was to follow with the rest of his regiment, as Dwight had specified "five or six hundred infantry." All of this was to be formalized in later orders.

The right prong was to be the main assault, which would be carried by Clark and Benedict, with the balance of Dwight's regiments approaching along the Mt. Pleasant Road, an extension of which crosses a ravine at a bridge, and enters the fort through a sally port.

As a consequence of the tardy publication and promulgation of Banks' orders, noted earlier, many of the regiments all along the line were not informed of their role until very late. The remote location of Dwight's assault was even worse in this respect, though he was given the broad option of moving[666] "at such time after 3:30 a.m. to-morrow as he may deem most expedient." Thus, either by tardy orders or Dwight's own ineptitude, the assault would go forward in daylight.

For example, the 15th New Hampshire was not directed to begin to move into position until after midnight. The route to their assigned position with Dwight was almost four miles long, running from the woods in the rear of the Slaughter plantation, south to the Mount Pleasant Road. It was a wide circuit, ostensibly for stealth.

---

665. Compiled Service Records, 28th Maine, M594, roll 71. The remainder of the regiment was at either Donaldsonville or New Orleans; 177th New York, M594 roll 138, *Massachusetts Adjutant General's Report, 1863*, 50th Regiment, p. 440.

666. O.R. Vol. 26/I, p. 555

The latter portion of the route is shown in figure 5.[667] They made it to the rendezvous near the Mt. Pleasant Road at sunrise, too late, in the minds of the men, for an unseen approach. It didn't matter; Dwight had not yet given his order for the assault. There would be no surprise attack in darkness, as had been anticipated.[668]

The 6th Michigan, marching toward the rendezvous near the Mt. Pleasant Road, was met by Col. Clark with the late news that the raid on the Citadel was on, and the picked men required must be named. Final orders arrived with an aide who directed the 6th to follow the 14th Maine, which was to follow the picked men, all guided by two deserters. They headed toward the river, the road curving around Mt. Pleasant their likely path, shown as a dotted line. At the riverbank, they turned right and advanced along the mud and sand until coming within sight of the high bluff, atop which sat the Citadel/Confederate Battery XXXIII. It was now broad daylight, the column had been spotted, and the Confederate alarm had sounded.

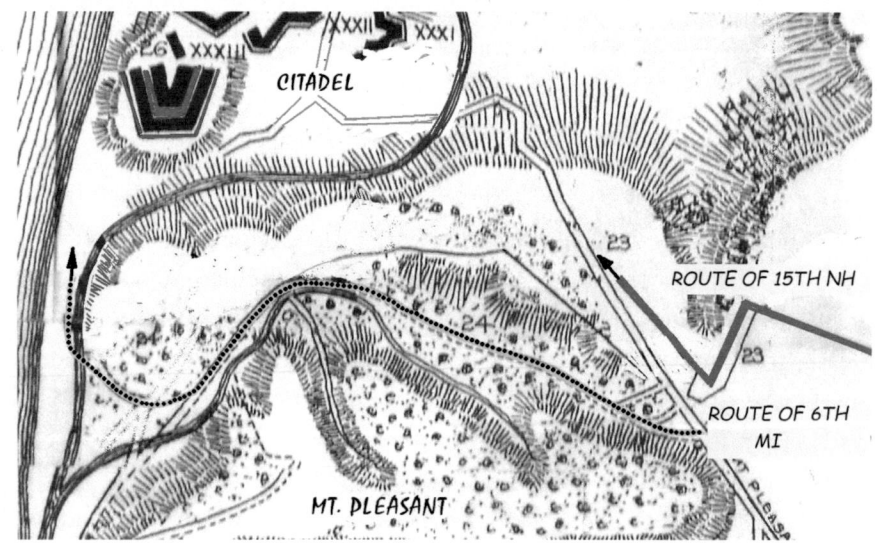

FIGURE 5

Quoting from Bacon:

> Dwight, in broad daylight, and at an hour when he ought to be sober, is about to sacrifice us, to carry out a scheme devised when he was drunk…
>
> Evidently the rebel officers have been ordering their men to withhold their fire, and are no longer able to enforce complete obedience. First one rifle shot, and the Minie ball comes singing slowly through the air over our heads. Another and another shot. Then a succession of scattering shots. . . . In a few moments more the rebel guns will be raking us, and there will be enough of us killed to make Dwight a great man. Captain Cordon's British common sense gets the better of every other feeling in his soul, and, taking whole responsibility, he says to his detachment, in unmistakable English, "Alt!" and the column never moves again.

---

667. Figure 5, a portion of a fold-out map from McGregor.
668. McGregor, p. 473, Bacon, pp. 176–177.

Colonel Bacon never countermanded the captain's order.

In the meantime, on the right, the advance of the 1st Brigade was announced by the scattered shots of the sharpshooters of the four companies of the 165th New York. Having learned the hard way on May 27th that wild charges are suicide, they were crawling in on their hands and knees, from stump to stump.[669] Skirmishers, a detail of 150 men from the 15th New Hampshire and the 26th Connecticut, followed in an attempt to drive in the rebel sharpshooters. The rebels had felt so safe atop their parapet twenty to forty feet above the ravine in front of it, that they had lingered outside their breastworks.[670]

The 128th New York followed in support, and the remainder of the 15th New Hampshire and the 26th Connecticut followed. They ran up to the edge of the ravine, which had been "entirely unseen and unsuspected until close in its proximity." Flanking left, figure 5, they found the way forward, along the Mt. Pleasant Road, "subjected to the most terrific fire" until "all fell into the ravine where the road makes its steep descent into its dense bottom." The bridge crossing the ravine had long since been destroyed. Now sheltered in the ravine, at least the Confederate fire passed over their heads, though it hit those approaching from behind.

The attack in "disorder,"[671] Benedict's troops were quickly sent in, led by the 162nd New York. The 175th New York followed, and then came the 48th Massachusetts. The few companies of the 28th Maine present never received the order to move, and never participated. The brigade reached the ravine where the 15th New Hampshire had earlier taken cover, and had been ordered by Col. Clark to "desist" from any attempt to storm "the ascent." With the commanding officer of the 175th, Col. Michael K. Bryan, killed, Benedict advanced no further, and ordering his troops to remain under cover, he walked all of the way back to Dwight's headquarters, where he reported "in person…the critical nature of his command." Dwight then ordered that the brigade was to lie where it was "until the darkness of night might favor its withdrawal."

The 177th New York, and the 50th Massachusetts, being in the rear, were never engaged.

One factor that stands out in Dwight's assault is that, having split his force, the numbers at each front were very small. Benedict's brigade only totaled 582 men. The 48th Massachusetts, one of Augur's regiments, had, on June 13th, been ordered to do pick-and-shovel work for General Arnold, the chief of artillery. After marching to Arnold's work site from their bivouac near Plains Store, and working

---

669. *Duryee Zouaves*, 1905, pp. 18–19. The place the Zouaves took is not clearly outlined. They are not even mentioned by Hanaburgh, or McGregor, and the only hint of their being there is in the *History of the 48th Massachusetts Regiment*, p. 43, who define them as the "red clothed fascine bearers." This, at least, puts them up front, as is recorded here.
670. McGregor, pp. 473–475; Hanaburgh, p. 55; Irwin p. 200.
671. *An Historical Sketch of the 162nd Regiment New York Volunteers*, pp. 16–18.

all day, they were ordered the three-and-a-half miles back to Plains Store. Before having the opportunity to cook supper, at 8:45 p.m., orders arrived to report to Dwight. Marching out immediately, it took until the "early morning" of the 14th to rendezvous with Dwight. After all of this running around, the 48th could muster only 175 men fit for the assault.[672]

Dwight's assault, too feeble or tardy to be of value as a diversion, nevertheless served one useful purpose; it introduced the generals to the value of Mt. Pleasant as a site for the placing of artillery. Its northern slope overlooked the southern flank of the Confederate defenses, and soon the largest Union battery complex at Port Hudson would be constructed there. Two sections of Battery L, among many others, would then occupy it on June 25th, 300 yards from the Citadel, and would remain there until the surrender.

June 15th

If ordered, those troops required to be withdrawn from the battlefield did so after dark on the night of the 14th and into the morning of the 15th. As well, at night, the recovery of the dead and wounded could proceed with renewed urgency. Many, many were not able to be removed during the day, as the Confederates had fired on any of the wounded that moved, and on the stretcher bearers; Halbert Paine's plight being a notable example. He would have to lie on the field for 14 hours, all the day in the blistering sun, witnessing two men killed as they attempted to retrieve him. From his diary:

> Two soldiers, whose names I have not been able to ascertain, attempted to reach me with a stretcher and fell near me. Private Patrick Cohen of the 133rd New York, lying wounded near me, tossed me a canteen cut from the dead body of a soldier. That doubtless saved my life.

It was nightfall before he was carried off. The Confederate sharpshooters continued their firing that night, and many wounded were still on the field by daylight.[673]

Banks sent a letter to Gardner[674] requesting him to allow medical supplies to be sent into Port Hudson "for the comfort of my wounded in your hands and of such of your own as you may desire to use them for." Gardner replied in the affirmative, and added details of what was already a disturbing situation: "I take the liberty to inform you (deeming that you are probably ignorant of the fact) that there are a few of your dead and wounded in the vicinity of my breastworks, and I have attempted to give succor to your wounded, but your sharpshooters have prevented it."

---

672. Plummer, pp. 41, 42; Stanyan, pp. 274, 358; *dmna,ny.gov*, 114th NY *Clippings*, 114th InfCWN4.pdf, "The Attack and Repulse at Port Hudson," *N.Y. Herald*, p. 40; O.R. Vol. 26/I, p. 554.
673. Irwin, pp. 200, 204; Fitts, pp. 130–131; Stanyan, p. 260.
674. O.R. Vol. 26/I, p. 557.

A stunned Banks seemed preoccupied with other matters, and this stinging defeat, added to that of May 27th, would aggravate any thinking person. Initially he did not react to Gardner's letter. Instead, General Order No. 49[675] called for more sacrifice, albeit voluntary. In part, it reads:

> We are at all points upon the threshold of his fortifications. One more advance and they are ours! For the last duty that victory imposes, the commanding general summons the bold men of the corps to the organization of a storming column of one thousand men …

Quoting Irwin: "It was not until the evening of the 16th that Banks could bring himself to ask for a suspension of hostilities for the relief of the suffering and the burial of the dead." Irwin is bending the truth, if evidence from the history of the 8th New Hampshire Volunteers can be counted on:[676]

> No flag of truce was had until the afternoon of the 17th, and then only in front of our own division. It was raised by the Confederates as they said that the stench was unbearable, and they proposed to deliver our dead and wounded to a certain point.

Beall reported to Gardner on June 17th, that 160 dead and only one wounded man had been delivered.[677] The one wounded man was Sgt. Charles E. Conant of Company F, 8th New Hampshire, "able to speak, though desperately wounded, who was parched with the dreadful pangs of thirst, and whose face, neck and hands had been completely fly-blown." He survived and was still living in Haverhill, Massachusetts, in 1892, when the *History of the 8th New Hampshire Regiment* was published.

The final tally of the Union casualties for June 14th was: 216 killed, 1,401 wounded, and 188 missing. Undoubtedly the missing were unidentified wounded that had died and were buried in mass graves, though some may have been deserters.[678]

The total, at 1,805, plus 1,995 for the May 27th assault, is a measure of the price that the attacker had, in that era, to pay for an assault on a fortified position, particularly if poorly executed.

Irwin comments:

The truth is …staff officers …forget that an assault upon an enemy behind entrenchments is not so much a battle as a battue, [the driving of game toward hunters] where one side stands to shoot and the other goes out to be shot…"

---

675. Stanyan, p. 270.
676. Stanyan, pp. 265, 283; Irwin, p. 204.
677. Defenders, SHS, Vol. 14, p. 333; Stanyan, pp. 265–267, 271. They had lain there for 60 hours or more. "Fly-blown" means attacked by blowflies and maggot filled; Irwin, p. 203.
678. Fox, p. 102, Irwin, pp. 207, 208.

Comparing Port Hudson with other assaults in history, such as those of Wellington, Irwin notes that: "... the losses of the assailants were in proportion less ... than at Port Hudson."

Recruiting for the 1,000-man forlorn hope began on June 16th, and went on until June 26th, when 893 officers and men had been accepted, including their leaders, Col. Henry W. Birge of the 13th Connecticut and two staff members.[679] The stormers were marched to a special encampment where training was to be conducted, and where special equipment, such as scaling ladders, were to be issued.[680]

For those who did not volunteer for the storming column, the drudgery of digging more and more trenches, for more and more batteries, was renewed, although aided by ever increasing numbers from the black regiments, both of infantry and engineers, who were being actively organized and equipped. As Henry Closson remembered, "...so far as mechanical ingenuity went, the Pilgrim, as usual, was unsurpassed. He might not be able to fight his way into Port Hudson, but he could dig there..." Saps were extended at key points, all intended to be loaded with explosives, and scheduled to be set off on July 9th.[681]

Jackson

On June 20th, the piece of Battery L under Sergeant Becker was detached to join the 2nd Rhode Island Cavalry and the 52nd Massachusetts Regiment in escorting a foraging train. The train consisted of some 140 or 150 covered wagons, which intended to gather corn and grain stored on two plantations some fourteen miles east, near Jackson Crossroads.

They were joined a few miles out by another party of 50 wagons escorted by a detachment of 200 cavalry from Grierson's command, also on the lookout for forage. Arriving at the crossroads at about 1:00 p.m. Becker went into battery on elevated ground on the southeast side of the crossing, which commanded a fine view of the plantations on either side of the road.

A Confederate private who had been taken prisoner informed of the presence of Generals Mouton and Hughes with a force of some 2,500 men encamped nearby, the same force that had attacked Grierson near Clinton two weeks before. They lay hidden in the woods and fields. Their attack on the train began within minutes. The panicked teamsters soon became disorganized in an attempt to return to a rear defensive position. As J. F. Moors, of the 52nd Massachusetts describes it:

---

679. O.R. Vol. 26/I, pp. 57–66, lists 976. Banks quotes 850 men (p. 46) several days later. See also Irwin, p. 213. The difference could reflect both sickness and the fact that 91 of the 1st and 3rd Native Guards had volunteered, but were not accepted.
680. Sprague, pp. 163–166; Haskin, p. 365
681. Irwin, pp. 219, 225; Stanyan, p. 287. Grant allowed such an operation a year later, at Petersburg. It failed miserably.

Among the many covered wagons in the train, and men and horses about our rendezvous, the Confederate generals had failed to notice the battery. We have only to swing around the muzzles of the guns to get the range, and open fire with shot and shell. This is done promptly, skillfully, and most effectively. The artillerymen do their whole duty, just in the nick of time.

The confederate attack was stunned, and repulsed, allowing a more orderly withdrawal. The train arrived back at Port Hudson late that night, minus some 60 wagons and 200 mules. No one in Battery L was injured, and Sergeant Becker was commended for his coolness in the action.[682]

Taylor Again

We recollect, from chapter 8, that Kirby Smith had been ordered "to do something" to relieve Pemberton at Vicksburg, and that he had reinforced General Taylor with Walker's Texas division. Taylor first responded by attacking Grant's supply points at Milliken's Bend, on June 7th.[683] He was driven off by the 29th Iowa and two Black regiments, with the aid of the navy gunboats *Choctaw* and *Lexington*.[684] Aware of a request from Johnston to come to the aid of Port Hudson, Taylor abandoned any further action against Grant, recognizing the "impossibility of approaching Vicksburg along Drury's point, *exposed to Federal gunboats for 7 miles*, from the west bank…" Taylor agreed with Johnston, the problem of Vicksburg was how to withdraw the garrison, not how to reinforce it.[685]

Taylor's reluctant Vicksburg move gave Banks relief from any real threat until mid-June. By then, Taylor had convinced Smith that an advance on the La Fourche would raise Banks' fear of losing New Orleans, and that he would then reduce or abandon his effort against Port Hudson, possibly allowing Gardner to break out and unite with Johnston.

Thus, Taylor planned a two-pronged attack on Berwick Bay, (A) attacking Brashear City, and (B) coming around to the rear of Brashear City at Bayou Boeuf, some six miles east on the New Orleans–Opelousas Railroad. Positioning the force destined for Bayou Boeuf would involve a cavalry raid. Col. James P. Major's three regiments of cavalry had just arrived at Alexandria from Texas, and to get them to Bayou Boeuf for a planned attack on the morning of June 23rd, would involve moving through a hundred miles of nominally Union controlled territory. Taylor met Major at Morgan's Ferry on the Atchafalaya, east of Washington. They crossed and passed down the Fordoche, to opposite Port Hudson, at Fausse Riviere, (near

---

682. Moors, pp. 193–199, Hosmer, pp. 200–204.
683. O.R. Vol. 22/II, pp. 854, 856–857, 904; Vol. 26/II, pp.41–43; Grant Vol. 1, pp. 544–545. Kirby Smith was influenced in his decision by his own need for ordnance, and it now must come from Richmond, his usual supply route from Texas being disrupted by the blockade.
684. ORN, Ser. 1, Vol. 25, p. 162.
685. Taylor, pp. 137–142. Italics added.

the Hermitage where Col. Benedict had been in command).[686]

Here, Taylor obtained guides. It would be impossible to sneak through the territory, so Major would conduct a raid on his way to Bayou Boeuf. Having moved on Waterloo, and made a demonstration on Hermitage, he proceeded down along the Gross Tete, figure 6.[687] On the 18th, he raided Plaquemine, burning three steamers, two flats, some cotton and taking 87 prisoners.[688] He then started down the west bank of the Mississippi, to Bayou Goula, destroying the Federal plantations and recapturing 1,000 of the former slaves working on them. Making a feint at Donaldsonville, but afraid to be detained by attacking Fort Butler, he moved toward Thibodeaux. A force sent forward captured Thibodeaux on the 20th. A stiff fight took place at La Fourche Crossing on June 21st. Remembering his date at Brashear City, Major disengaged on June 22nd and marched east, arriving at Boeuf Station at 4:20 on the morning of June 23rd.

The other half of the attack,[689] consisting of Gen. Thomas Green's cavalry brigade, and Maj. Sherod Hunter's Texas Cavalry, began their approach on the night of the 22nd. Green dismounted his men, entered an unoccupied Berwick City, and set up his gun batteries on the west side of the bay, facing the Union camp and Fort Buchanan at Brashear, about 800 yards across the bay. Forty-eight skiffs and flats had been collected for Hunter's 325 men to use to paddle from the Teche up into Grand Lake, and around behind Brashear City. Leaving on the evening of the 22nd, it took the "mosquito fleet" almost the entire night to quietly paddle the 12 miles around to their destination, reaching the shore at 5:30 a.m.

FIGURE 6

Green opened artillery fire into the Union encampment at dawn, and though the navy gunboat *Hollyhock* was stationed nearby, it was driven off by Green's guns. During Green's bombardment, Hunter's men approached the rear of Fort Buchanan in the open, upon a point where the low and meager earthworks were unfinished. Observers in the fort mistook the attackers as a part of a group of railroad hands reporting to assist in the defense, and hesitated to open fire. Hunter's men then advanced in a sudden bayonet charge,

---

686. O.R. Vol. 26/I, pp. 571, 608.
687. Figure 6 adapted from plate 156, atlas; O.R. Vol. 26/I, pp. 217–220.
688. O.R. Vol. 26/I, pp. 216–220. Irwin, p. 214, quotes 23 of the 28th Maine, with 14 of the Provost Guard escaping.
689. O.R. Vol. 26/I, pp. 215, 223–225; ORN, Vol. 20, pp. 310–311, 313, 316, 320; Duganne, pp. 112–113; Noel, *Campaign*, pp. 53–55.

and without firing a shot entered the campsite. It was all over by 7:30. Seventeen hundred officers and men, 11 heavy guns, and all of the Union supplies were captured. Three men from Battery L, privates George Chase, Joseph Kutschor, and Martin Stanners, who had been left there since the beginning of the Teche Campaign in April, were among the prisoners. Chase and Kutschor had been in hospital, and Stanners on detached service.

Fortunately, arrangements were made for parole of the enlisted men, and on the 27th, 1,360 were sent under guard toward Boutte Station, where they were received into the Union lines on July 3rd. The officers were held as prisoners, and were obliged to march from Brashear City to Alexandria, then to Shreveport, and finally to Camp Groce, in Texas.[690]

Kutschor and Stanners returned to Battery L, and went back into the lineup. There is no evidence that the terms of their parole were honored, in the sense that paroled men were prohibited from combat.

Chase did not return, and his fate remains a mystery. He is entered on every subsequent muster roll as "sick at Brashear City since April 22, 1863," until April, 1865, when he is dropped. No further record.

It is ironic that on June 21st Bank's AAG, Irwin, had sent an order to Emory, (now in command of the Defenses of New Orleans) to be forwarded to the Commander at Brashear, Lt. Col. Albert Stickney. It read, in part:

> The commanding general does not regard it as important that we should run any great risk to save Brashear. He desires that you will send orders to Brashear to get off everything of value there, and at Bayou Boeuf, including, especially, the guns, and, when pressed by the enemy, to retire on board the transports and proceed to New Orleans.

Emory's message forwarding the information was not telegraphed until the day of the attack, and a message was clicking in when one of Taylor's men destroyed the machine.[691]

Taylor estimated the value of what he had captured, and what should have been saved, at $2,000,000. As to Green, he followed those of the Union garrison who tried to escape east on the railroad. Blocked at Bayou Boeuf, 275[692] more surrendered, along with "four guns, ammunition, small arms, commissary and quartermaster's stores; and about 3,000 negroes."

Green promptly moved on to Donaldsonville and attacked Fort Butler, located on the Mississippi River at the mouth of the La Fourche, at 1:30 on the morning of June 28th.

---

690. O.R. Ser. II, Vol. 7, pp. 493–494; Duganne, pp. 182, 185–190, 203, 251.
691. Taylor, p. 142.
692. O.R. Vol. 26/I, pp. 216, 219. Depending on which report is quoted, either 275 or 435 surrendered.

Fort Butler was under the command of Maj. Joseph D. Bullen of the 28th Maine. Its defending garrison consisted of only companies F, G, and a part of H, of the 28th Maine, and 13 men of the 53rd Massachusetts. There were 150 convalescents from other regiments, "only about 130" of whom were fit enough to load and fire a musket.[693] The remainder of Bullen's men were either at stations in the New Orleans defenses, e.g., Plaquemine, or had been called to Port Hudson to be a part of Dwight's June 14th assault.

A telegram from Bullen received at Emory's headquarters at 8:00 on the morning of June 28th only hinted at the story of a rare and heroic defense, and this from a nine months' unit. "The enemy have attacked us and we have repulsed them. . . . I must have more men."

Banks responded quickly; the 1st Louisiana and "two sections of artillery" were promptly ordered to Donaldsonville, all under the command of a new arrival in the department, Gen. Charles P. Stone. The two sections of artillery were 4 guns and 24 men of Battery L, under the command of Lt. Franck Taylor. Though scheduled to arrive at midnight, they finally arrived at dawn on the 29th.

Immediately upon arrival, the reinforcements were deployed on a line just outside of the town, where they met a remnant of Green's men who had lingered, allegedly to recover their wounded, under a flag of truce. The truce was refused by Bullen, viewed as a ruse to save uninjured men who were still sheltered in the ditch near the fort. In an engagement of about an hour-and-a-half that followed, Green's men were driven off, with no losses to Battery L. That evening, Lieutenant Taylor and two of Battery L's guns returned to Port Hudson to rejoin the battery at its new position on Mount Pleasant. Sergeant Becker and the detachment of 23 men (listed in the muster roll in chapter 9) remained at Fort Butler until after the surrender of Port Hudson.

The Last Days

On June 25th, Battery L, and two of the IX-inch navy Dahlgrens that had occupied Battery 10, were moved into Battery 24 at Mount Pleasant. The position would quickly accommodate six other artillery companies, or portions thereof, with a total of 17 guns, see table 1. A representation of it, "The Great River Battery" that appeared in *Harper's Weekly*, is shown in figure 7.[694] The cotton bale construction is clearly visible, as is the Confederate Citadel, in the left background, separated by a deep ravine.

On June 26th, 1st Sgt. Lewis Keller's feat of shooting away the Confederate

---

693. Taylor, p. 147, destroys his credibility by saying "...two hundred and twenty-five negroes..."
694. Figure 7, *Harper's Weekly*, July 25th, 1863, p. 476; *Massachusetts Soldiers, Sailors and Marines, 13th Artillery*, Vol. 5, pp. 507, 519; *Record of the Massachusetts Volunteers* 1861–1865, Adjutant-General, Vol. 1, 1868, pp. 453–458; McGregor, p. 512.

flag three times was noted with satisfaction ("but it was raised no more") by the men of the 15th New Hampshire, who were in the rifle pits in front of Battery 23. On June 27th, while shelling the Confederate works across the ravine, Battery L's Pvt. John Casey was mortally wounded by a sharpshooter.

FIGURE 7

Banks wrote a summary of recent events to Halleck on June 29th. Though the enemy now had pushed Emory's forces all the way back to Algiers, Banks calmly related that: "The fall of Port Hudson will enable us to settle that affair very easily[695]…A few more days must decide the fate of this place."

On July 3rd Emory wrote to Banks that the situation was so dire at New Orleans that he must have reinforcements "immediately and at any cost." Banks advised Emory that he could not send him reinforcements, and in a calm rebuke, said: "I do not think…that the city is in peril." Next day, another message from Emory estimated that the forces threatening New Orleans were 13,000. Emory was correct, only his intelligence was the sum of all of Taylor's forces. Emory's source had failed to mention that, to Taylor's anger and chagrin, Kirby Smith had retained J. G. Walker's division outside of the New Orleans vicinity,[696] with the result that Taylor had only 4,000 to bring against the New Orleans defenses.

On July 6th Banks again wrote to Halleck: "The siege has been progressing rather slowly, indeed, but with all the rapidity attainable under the circumstances. Our approaches are pushed up to the ditch at the citadel on our left, and in front of the right priest-cap, where the assault of the 14th was made." That sap was within

---

695. O.R. Vol. 26/I, pp. 47–53.
696. O.R. Vol. 26/II, pp. 42, 110–111; Taylor, p. 139. Kirby Smith felt that even if Taylor captured New Orleans, he could not hold it. Taylor had the same doubt as well.

ten feet of the parapet on July 4th when the Confederates exploded a mine which collapsed it. Mines and counter-mines were being prepared in every approach.

Now, the storming column commanded by Birge was fully organized and ready, on 15 minutes' notice.[697]

Finally

On the morning of July 7th, the gunboat *General Price* reached Port Hudson with a dispatch from General Grant. It conveyed the news that Vicksburg had surrendered on July 4th, with 27,000 prisoners.

A note was prepared, wrapped around a "clod" of clay, and flung into the Confederate works. It said, "Vicksburg surrendered on the 4th of July." At first, it was regarded as "a damned Yankee lie," but when Gardner saw it, he sent a message asking for Banks' "official assurance" that it was true. By this time, the whole line along the Union side of the front had begun celebrating.

From the History of the 15th New Hampshire:[698]

> [A]t exactly high noon, by order, rousing cheers were given amid a general discharge of small arms and a grand salute fired by all our fleets and batteries, pouring a terrific iron hail upon the devoted foe within. These were the last shots fired upon Port Hudson.

Gardner, now convinced that the surrender at Vicksburg was true, agreed to an unconditional surrender, with Port Hudson to be occupied by the U.S. Army at 7:00 a.m. on the 9th of July.[699]

The original time specified for the occupation, 5:00 p.m. on July 8th, was postponed. This may have seemed innocent enough, but it was a ruse to allow some numbers of the Confederate officers to attempt escape, given the relaxed atmosphere after the cease-fire; a description of which is found in the account of Lt. James Feret, an engineer on Gardner's staff.[700] As soon as the agreement was signed,

> …the late combatants began to fraternize. Soldiers swarmed from their places of concealment on either side and met each other in the most cordial spirit. Groups of Federal soldiers were escorted round our works and shown the effects of their shots and entertained with such parts of the siege that they could not have learned before. In the same way our men went into the Federal lines and gazed with curiosity upon the work which had been giving them so much trouble, escorted by Federal soldiers who vied with each other in courtesy

---

697. Sprague, p. 168; Irwin, p. 226.
698. Stanyan, p. 554; Tiemann, p. 50, confirms it; McMorries' account, p. 68, says that the note was tied to a hand grenade. How could two sources vary to this degree?
699. O.R. Vol. 26/I, pp. 52, 54, 625; Hanaburgh, p. 71.
700. Stanyan, p. 290; Tiemann, p. 50.

and a display of magnanimous spirit.

There is ample evidence that a ruse was in operation by one device or another. Confederate Maj. S. L. Knox donned a private's uniform, and by obtaining the parole of a dead man, was able to pass unsuspected through the Federal lines. Simply sneaking out was another obvious option, and it happened to be aided by a heavy thunderstorm that came up that evening. Hence, other officers, as is stated in the *History of the 1st Alabama Regiment*, "after perilous adventure and much suffering from hunger and thirst, effected their escape through the enemy's pickets." There is record of at least two being captured and returned.[701] The 13th Connecticut intercepted Lt. Col. Lee of the 15th Arkansas, and Captain Hardee of Miles' Legion. They had been two days in the swamps without food.

Capt. C. M. Jackson, Gardner's acting assistant inspector-general, wrote a message addressed to Johnston, dated July 9th, giving the news of the surrender.[702] This was not a diary entry. It was a report that obviously he intended to deliver. Clearly, either he or a colleague was one of the number who had successfully taken advantage of the ruse. The contents of the first paragraph of his report is of interest:

> Port Hudson surrendered yesterday at 6 a.m. Our provisions were exhausted, and it was impossible for us to cut our way out, on account of the proximity of the enemy's works.

He goes on to note that during the siege there were 200 killed, between 300 and 400 wounded, and 200 died of disease, and at the time of surrender, there were only 2,500 men fit for duty.

Union records[703] list the number of Confederate troops surrendered as 5,935 enlisted men, who were paroled and released, and 406 officers, who were held prisoner.

Union casualties for the siege May 23–July 8, were:[704]

KILLED		WOUNDED		CAPTURED OR MISSING		TOTAL
OFFICERS	ENLISTED	OFFICERS	ENLISTED	OFFICERS	ENLISTED	
45	663	191	3,145	12	307	4,363

A chapter in the parole and release story echoes Irwin's earlier "useless fashion of the time" comment. McMorries, in the *History of the 1st Alabama Regiment*, reveals that all of the paroled enlisted men from Port Hudson were simply declared to be exchanged. This meant that all of the men, to their disgust, were immediately eligible for military service.[705]

The confederate commissioners of exchange used the reasoning that an

---

701. Sprague, p. 171; McMorries, E. Y., p. 69.
702. O.R. Vol. 26/I, p. 144; Moors, p. 190.
703. O.R. Vol. 26/I, p. 642.
704. O.R. Vol. 26/I, p. 70; Fox, p. 545.
705. McMorries, pp. 69–71, 94; Irwin, p. 233; CWSAC PA 002.

equivalent number of Federal prisoners had already been paroled and released by the Confederate Partisan Rangers - the likes of Mosby, Imboden, McNeil and others – a calculation that the Federal side refused to recognize.

Another sticking point was that, under a directive of Jefferson Davis, Federal officers who had commanded black soldiers, or otherwise were found with "abducted" slaves, would be held for trial. A later man-for-man proposal by Judge Ould, the Confederate commissioner, was rejected because it would not include those officers held for trial. Thus, the cartel ceased to function at this time, mired in bad blood.

It nearly collapsed when Benjamin Butler was appointed the Federal Commissioner of exchange in December. This War Department action couldn't have been more infuriating to the Confederacy, and they initially refused to deal with him. Butler was the man whom Jefferson Davis had proclaimed, on December 24th, 1862, [706] a "felon deserving capital punishment" for the hanging of William Mumford at New Orleans. In reality, it was the slave issue, for in the same declaration, Davis went on to announce a refusal to exchange former slaves serving as Union soldiers, saying they would: "be delivered over to the authorities of the respective States to which they belong, to be dealt with according to the laws of said States." These were code words for execution, because any escaped slave under arms was considered to be in "armed insurrection" punishable by death.

The extreme Confederate attitude is found in a June 13th, 1863 letter from E. Kirby Smith to Richard Taylor:

> I have been unofficially informed that some of your troops have captured negroes in arms. I hope that this may not be so, and that your subordinates . . . may have recognized the property of giving no quarter to armed negroes or their officers.

Though the cartel had sputtered on, Grant, on April 14th, 1864, only one month after he had become general-in-chief, ordered Butler "to decline all further negotiations."

The officer with the dead man's parole, Knox, would go on to command the reassembled 1st Alabama Regiment, which was to fight at Mobile, and with Johnston's army in the Carolinas, before finally surrendering on April 27th, 1865.

The surrender ceremonies began punctually at 7:00 on the morning of July 9th, figure 8.[707] They were simple and short. The Confederate troops, "looking ragged and rough," were drawn up in line, their left on the village, facing Banks' chief of staff, General Andrews, who accepted the surrender. The Stars and Bars was lowered from the flagstaff on the bluff, and Old Glory was raised. Duryea's battery

---

706. O.R. Ser. II, Vol. 5, pp. 795-797; O. R. Ser. II, Vol. 6, pp. 21,22; Northcott, pp. 185-187; O.R. Ser. II, Vol. 7, p. 50.
707. Figure 8, *Harper's Weekly*, cover, Aug. 8th, 1863; Irwin, p. 232; Sprague, p. 171; Stanyan, 293.

fired a salute, and the ceremonies were over.

As far as Banks was concerned, there was not a moment to lose before going after Taylor. As Irwin puts it: "The last echo of the salute to the colors had hardly died away when Weitzel, at the head of the First Division, marched off…and began embarking on board the transports…"[708]

FIGURE 8

This brief hour of relief was marred by confirmation of the rumor that the bodies of the 1st and 3rd Native Guards had lain on the field since the May 27th assault. Among them, the body of the captain of Company E of the 1st Native Guards, Andre Cailloux, was able to be identified, and it was brought to New Orleans for burial. The arrival of the body brought out an emotional reaction from the many Black civic societies of the city, Cailloux having been a member of the "Friends of the Order." A grand funeral and procession was organized in his honor. Figure 9 depicts the procession to the Bienville Street Cemetery.[709]

---

708. Irwin, p. 233; O.R. Vol. 26/I, pp. 626–627.
709. Wilson, pp. 214, 217; *Harper's Weekly*, August 22nd, 1863, pp. 549, 551; O.R. Vol. 26/I, p. 632.

FIGURE 9
Funeral procession for Andre Cailloux at New Orleans, July 29th, 1863

Banks' chief of staff, Andrews, was named to the command at Port Hudson, and the 6th, 7th, 8th, 9th, and 10th regiments of the Corps d'Afrique, under Ulmann, were to compose the garrison, along with some nine-months men whose terms were nearly up. After the arguments and threats of mutiny by the 4th and the 50th Massachusetts regiments over their expiration-of-service dates, they, and all of the other nine-months units had agreed to the later time of muster interpreted by Banks. This could have made a difference if Port Hudson had not surrendered as promptly, but now it mattered little. Nine were detached from their divisions and assigned to Andrews. There they would remain close to available transportation homeward.

The final act of significance to Port Hudson history was Banks' July 11th order to General Andrews: "The demolition of all the batteries and works of approach constructed by the United States forces for the recent reduction of Port Hudson will be commenced without delay..."[710]

Battery L saw the last of Port Hudson on July 11th, but rather than be transported directly to Donaldsonville with Grover, L was one of six other batteries under Chief of Artillery Arnold that marched. Escorted by the 3rd brigade, 3rd division, they endured the 25 miles of mud—it had rained since noon—to Baton Rouge. Here, Sergeant Becker's detachment from Donaldsonville joined them. It was July 15th before the combined company returned to Donaldsonville, thus

---

710. O.R. Vol. 26/I, p. 633.

missing the Battle of Kock's Plantation, which took place there on the 13th.

The three days at Baton Rouge were spent resting and leisurely walking about, in what had become one vast camp "with military garb displayed on every hand."[711] At Donaldsonville, they remained until the end of the month, then marched to Camp Kearney at Carrolton, and finally to the Apollo Stables in New Orleans. Here they remained until September 15th, after Taylor's forces had been chased out of the La Fourche, and Banks had been given orders from Washington as to what his priorities should be.

The Mississippi was now open, but not altogether due to the fall of either Vicksburg or Port Hudson. Included in the package was the defeat of the Confederate forces under Holmes in Arkansas. In an attempt to re-take Helena, Holmes attacked on July 4th, and was soundly defeated by the Union defenders under Prentiss, and only then was the Confederate threat permanently removed. The action consumed only one day.[712] As a symbol of the achievement, the steamboat *Imperial* arrived at New Orleans on July 16th. She had left St. Louis on the 8th and had passed over the whole course without having a shot fired at her. At last, "The Father of Waters" would again go un-vexed to the sea.

"Record" 8/63
1 JULY–31 AUGUST, 1863, PORT HUDSON & NEW ORLEANS

The Company remained at Port Hudson, La, until the 11th of July & then proceeded to Baton Rouge, La. arriving at 5 a.m. of the 12th. On the 15th left Baton Rouge and arrived at Donaldsonville, La. Left Donaldsonville. La. on the 29th and proceeded to Carrolton. Arriving on the 1st of August. Left Carrolton on the 9th and took up quarters at the Apollo Stables New Orleans, La. and remained until the present date.

Henry W. Closson	Capt. On furlough since Aug. 9th 1863
Franck E. Taylor	1st Lt. Commanding Battery and Asst. Comm'y of Musters S.O. no. 92 Hdqrts. Dept. of the Gulf 19th Army Corps April 9th 1863.
Edward L. Appleton	1st Lt.
J.A. Sanderson	2nd Lt. Appointed by promotion to Co. vice Gibbs promoted G.O. no. 73 W.D.A.G.O. Washington July 4th 1863 (never joined Co.)
Charles B. Slack	2nd Lt. Joined Co. by assignment S.O. no. 211 Hdqrts. Dept. of the Gulf 19th Army Corps
Detached:	
George Friedman	Pvt. On Det. Svc. at N.O. As artillerist. Left Co. May 24,'62.
Amelius Straub	Pvt. On det. svc. at Baton Rouge as cook in Univ. Hosp. since June 1,'63.
Absent with leave:	

---

711. Hanaburgh, pp. 81–82. Likely, all of the river transport available was taken by Weitzel and Grover. Imagine the space seven artillery batteries would occupy. Six guns on limbers, a forge, ammunition wagons, some 80 horses for each—over 90 wheeled vehicles—pulled by nearly 600 horses, if those ridden by the officers is included. O.R. Vol. 26/I p. 660; Whitcomb, p. 53.
712. Nicolay & Hay, Vol. 7, pp. 323, 327; O.R. Vol. 24/III, p. 492; Selby, p. 12.

Louis Lighna	Pvt. On furlough left Company August 21st for 14 days.	

Absent Without Leave:

William Fudge	Pvt. Since August 25, 1863	
George Harrison	Pvt. do.	
Dennis Moore	Pvt.	
William Mint	Pvt.	
Sholto O'Brien	Pvt.	

Absent in Confinement:

JohnMcKinney	Pvt. Since August 26, 1863	
Hiram Smith	Pvt. do.	
Peter Welsch	Pvt.	

In confinement

John Baker, James Campbell, Patrick Gibbons, John Kelly, James McCarthy, James Mahoney, Frank Morgan, Henry Williams

Deserted:

Michael Breen	Pvt.	At New Orleans	Aug. 27, 1863
William F. Brown	Pvt.	do.	Aug. 19, 1863
James Comfort	Pvt.		Aug. 19, 1863
Patrick Craffy	Pvt.		Aug. 19, 1863
Arthur Flynn	Pvt.		Aug. 19, 1863
Francis Jessop	Pvt.		Aug. 18 1863
George F. Leonard	Pvt.		Aug. 11, 1863
John Lewery	Pvt.		Aug. 10, 1863
John Lowry	Pvt.		Aug. 28, 1863
Christian Meese	Pvt.		Aug. 10, 1863
John H. Moran	Pvt.	Donaldsonville	do.
Michael O'Sullivan	Pvt.	New Orleans	
William Parketton	Pvt.	do.	Aug. 21, 1863
Michael Ranahan	Pvt.	do.	
John Roper	Pvt.	do.	
Henrick Schmidt	Pvt.		Aug. 19, 1863
Morgan L. Shapley	Pvt.		Aug. 21, 1863
Andrew Stoll	Pvt.		Aug. 13, 1863
Owen A. Wren	Pvt.		Aug. 21, 1863
William Wynne	Cpl.		Aug. 17, 1863

Discharged:

Edmond Cotterill	Pvt. By reason of reenlistment in the Battery, July 25, 1863
Philip H. Schneider	Pvt. By reason of reenlistment in the battery August 16, 1863

Died:

Bernard Farrell	Pvt. At Baton Rouge, La. July 1863 Cause and date not stated in record with Battery or at Regt'l Hdqrts.
Christopher Foley	Pvt. At Carrollton, La. drowned August 1, 1863
John Murphy	Pvt. do.
John Lanahan	Pvt. At Carrollton, cause and date not known, no record with Battery or at Regt'l Hdqrts.

John Casey	Pvt. From Monthly Return, on June 27th he was "Mortally wounded by Sharpshooters while at his gun." This record notes: Died on July 3, 1863 of wounds received in action at Port Hudson, La.

Joined: July 13, '63 From Detached Svc. at Donaldsonville:

Julius Becker	Sgt.				
William Demarest	Cpl.				
William Wynne	Cpl.				
James Beglan	Pvt.	Miles McDonough	Pvt.	Philip H. Schnieder	
James Comfort	do.	Daniel Moore	do.	Andrew Stoll	
Clark Dickson		Corneilus McEnearney		Michael Tieghe	
Daniel Howard		Christian Meese		Henry Ward	
Benjamin Hughs		John A. Nitschke		Joseph Wilkinson	
John Kelly		Sholto O'Brien		Thomas Wilcox	
John Kastenbader		Ephriam Orcutt			

Strength: 117, Sick: 14

Sick Present: 4 (not recorded) Sick Absent:
William C. Brunskill

Sick at Ft. Hamilton, NY, left Co. Sept. 17, 1861.

Wallace Wright	Sick at Pensacola since December 24, 1862
Charles Mansfield	Sick at Pensacola since December 24, 1862
George Chase	Sick at Brashear City since April 22, 1863
William Crowley	Sick at Baton Rouge, La. since July 13, 1863
Julius Becker	Sick in University Hospital since August 27, 1863
David J. Wicks	do.
William Brooks	Sick at Marine Hospital New Orleans, La. Since August 22, 1863
Henry Champion	Sick at University Hospital since August 27, 1863
John Deering	Sick at Marine Hospital New Orleans since August 28, 1863

The roll gives ample evidence of the battery having endured some rough service, with five deaths, and 14 sick. One man, John Moran, is missing from the list of those joined from service at Donaldsonville, and is listed as a deserter. He did not wait until the company got to New Orleans.

A sad ending for John Murphy. He had deserted at Fort Duncan in May of 1860 and been apprehended and returned to duty, and had remained in the company without further incident until his drowning. Details of either Murphy's or Foley's drownings are unknown.

Fees appear again, though none are noted for clothing or camp and garrison equipment. For ordnance, each of the following owe:

James Ahern $0.83, William Fudge $2.30, Benjamin O. Hall $0.83, John Miller $2.26, Hiram Smith $0.83.

The desertion of Corporal William Wynne is of note. It is hard to understand why one of the non-coms, higher paid, and in a respected and sought after position, would do so. He had replaced James Flynn on June 1st, 1863, after Flynn had been reduced to the ranks.

After a stay of two weeks at Donaldsonville, Battery L, Nim's Battery, and the 159th New York, were detailed as guard to a baggage wagon train under the command of Col. Molineaux that was headed to Carrollton. It departed Donaldsonville on July 29th, and proceeded down the east side of the La Fourche, seventy-three miles to Camp Kearny at Carrollton. It arrived on August 1st. The weather was hot and sultry, which exhausted the men and killed two or three horses.[713]

Capt. Closson was promoted to Brevet Major on July 8th, "For Gallant and Meritorious service at the Capture of Port Hudson, La." On October 4th, he was made chief of artillery of the 19th Army Corps. He would no longer be in command of Battery L on a day-to-day basis. He went on furlough on August 9th, and so did Pvt. Louis Lighna, the same day that the company arrived at New Orleans. Here, the troops would be given a chance to rest and recover, while Banks carried on correspondence with Grant and the authorities in Washington as to what to do next.

Note that upon L's arrival at New Orleans there were 20 desertions over the course of the month. Three were the very next day. The men had never had the chance to disembark from the *Che Kiang* and visit the storied Crescent City, even for a few hours, since arriving from Pensacola on Christmas Day of 1862. While at Donaldsonville, on duty with the 1st Louisiana, most all of whom hailed from New Orleans, the members of Battery L undoubtedly had the opportunity to learn about "the shrimp salads, the soft-shelled crabs, and the champagne of Moreau's,"[714] the gambling, the cock-fights, and more importantly, the best saloons and brothels in town. Having endured the hell of the siege, they simply went on an extended vacation, the army, the war, the government be damned. No doubt, many would return, and would appear on the next roll—wouldn't they?

The Next Step

There were congratulations all around. At the news of the fall of Vicksburg, it was first, Banks to Grant, on July 7th and then Halleck to Grant on July 11th, though Halleck was still worried by no news, up to then, from Port Hudson.[715] Then it was Banks to Grant, on July 8th, announcing the surrender at Port Hudson. After some further thought, Banks sent another message asking for 10,000 to 12,000 troops to assist in a move into Texas. This is his first mention of such a thing. Before receiving this, Grant had, on July 10th, sent word that he was assigning a division from his 13th Corps, Herron's, to Port Hudson, but once he received the news of the surrender, he reassigned Herron to occupy Yazoo City.

More correspondence followed, including news from Halleck to Banks that

---

713. Tiemann, p. 58; Whitcomb, p. 53.
714. Haskin, p. 367, Closson's reminiscences.
715. O.R. Vol. 24/III, pp. 490–492, 498–500, 519; O.R. Vol. 26/I, pp. 619, 624, 626, 644, 665, 709.

he could "get no more troops to send to you." But he did order Grant to transfer the 10,000 to 12,000 troops Banks desired.

After all, 21 of the nine-months regiments terms of service would expire by August, leaving Banks with 13,000 men, some 2,500 of whom would have to be returned to Pensacola, Key West, and forts Jackson and St. Philip. Thus, without any reinforcements, he would have only a net moveable force of 8,500.[716] Grant then sent Herron, and later, the whole reorganized 13th Corps, consisting of four divisions, temporarily under the command of Maj. Gen. C. C. Washburn.

It had been a memorable 4th of July. In addition to the surrender of Vicksburg, Meade had defeated Lee at Gettysburg. Celebrated in the newspapers, Meade's success was met in Washington by mixed joy and despair. Joy for the victory, and a promotion of Meade to a brigadier in the regular army, yet despair that he did not follow Lee, and crush him before he could retreat across the Potomac. Lee was not pursued, at least not vigorously, and on the night of July 14th, he successfully crossed the Potomac. Meade's general order after the victory spoke of "driving the enemy from our soil" and Lincoln, upon reading it, exclaimed: "This is a dreadful reminiscence of McClellan…the whole country is our soil."[717]

It appears that Lincoln never did learn about the details of Halleck's advance on Corinth, or of Banks' on the Teche, or he would have rightfully condemned them both. It was a similar story. A victory was defined as forcing the enemy to retreat. Now, Meade had simply carried on the Halleck tradition. It was all clear to Gideon Welles, who took an eyebrow-raising cynical view. In his diary entry for July 11th he writes:[718]

> I fear the rebel army will escape, and am compelled to believe that some of our generals are willing it should. They are contented to have the war continue. Never before have they been so served nor their importance so felt and magnified, and when the war is over but few of them will retain their present importance.

Regardless of all of this hand-wringing, the tipping point had been reached, and at least one Confederate officer in Lee's army recognized it. Capt. James Wood of the 37th Virginia Infantry Regiment, Stonewall Brigade, Army of Northern Virginia, later wrote:

> This battle and campaign was the crucial period of the Confederacy. It was an open secret, gained from rumor, that success would bring recognition of her independence by England, and later by France. Other nations would doubtless fall into line, and the blockade of her ports would soon have been raised. Credit and trade relations with other nations, the enlargement of her

---

716. Irwin, p. 258; O.R. Vol. 26/I, pp. 3, 603; Irwin, p. 259.
717. Nicolay & Hay, Vol. 7, pp. 278–279.
718. Welles, Vol. 1, p. 368.

armies and munitions of war, would follow; and permanent independence would be established. The high tide of Confederate hopes and prospects were now passed, and on the night of the 4th Lee retired to Hagerstown.[719]

On July 15th Augur was given a leave of absence from which he would never return.[720] It is revealed, in the *History of the 116th New York Volunteers*, that he, like Sherman, had "stoutly opposed the assault of May 27th… and afterwards was treated in so formal a manner by those at army headquarters that he deemed it best to withdraw as soon as possible." He was later assigned as the military administrator of Washington, where his activity there would soon prove to be, once again, a part of Battery L's fortunes. Weitzel succeeded him in command.

On July 18th, Grierson was finally ordered back to Grant.[721]

On July 24th, Halleck wrote to Banks discussing the overall strategy to be taken up next. "While your army is engaged in cleaning out Southwestern Louisiana, every preparation should be made for an expedition into Texas." Mobile was also mentioned. "The navy are very anxious for an attack on the latter place, but I think Texas much the most important. It is possible that Johnston may fall back toward Mobile, but I think he will unite with Bragg."

Grant was in favor of uniting as well—with Banks for an attack on Mobile, while the Confederate forces were in disorder. With Mobile as a base, the Union forces could then threaten Bragg, still opposite Rosecrans in Tennessee. Essentially, with the Anaconda Plan then completed, powerful Union forces could be brought to squeeze inland, to where the remaining Confederate armies were concentrated. Texas was outside of the Anaconda ring, and great strategists have to make tough decisions. Though Halleck correctly foresaw Bragg benefitting from the infusion of Confederate troops paroled from Vicksburg, thereby increasing the threat to the Army of the Cumberland, politics interfered. Texas was to be Banks' target. The President was interested in Texas, and on July 29th, he had asked Stanton to see Halleck about organizing an expedition. Great strategy was thus lost to political direction.

Quoting Grant: "The General-in-Chief having decided against me, the depletion of an army…commenced, as had been the case the year before after the fall of Corinth."

As has been chronicled, Texas was not only a source of raw materials and manpower, but it was the conduit for European arms landed in Mexico, which then were smuggled across the Rio Grande, brought to Louisiana, sent up the Red River and crossed into Mississippi. Now, with Union control of the river, this sort of traffic would be shut off, or nearly so. The blockade should have been able

---

719. Wood, p. 153.
720. O.R. Vol. 26/I. p. 642; Clark, p. 108.
721. O.R. Vol. 26/I, pp. 645, 652, 659; Grant, Vol. 1, pp. 578–579.

to take its toll on foreign ships attempting to drop off arms at the Texas ports. So, why should Lincoln be concerned about Texas?

The French, under Napoleon III, were still as full of the avarice that characterized the monarchies of Europe before, during, and after the time of our Civil War, and any opportunity would be cynically exploited, regardless of legal or moral considerations. The French army had landed at Vera Cruz in December of 1861, subsequent to threats about financial claims by French interests against the government of Mexico. They had advanced inland, but were checked at the Battle of Puebla, now celebrated as Cinco de Mayo. However, in September of 1862, the French were reinforced by 30,000 more troops under General Forey. They besieged and reduced Puebla in February of 1863, and entered Mexico City in June. The puppet government formed following the capitulation adopted monarchy, and offered the crown to an Austrian prince by the name of Maximilian.[722]

It was now Secretary of State Seward's job to make clear to the European courts, monarchies all, that the idea of a monarchy in Mexico, and the French presence there, was viewed adversely by the United States government. Seward foresaw Mexican interference in Texas. Napoleon III was to be warned about any coalition between "the Regency established in Mexico and the Insurgent Cabal in Richmond." A Union presence in Texas would be necessary to add credibility to Seward's message.

So it was that Lincoln wrote:

EXECUTIVE MANSION,
Washington, July 29, 1863.

HON. SECRETARY OF WAR:

SIR: Can we not renew the effort to organize a force to go to Western Texas?

Please consult with the General-in-Chief on the subject.

If the Governor of New Jersey shall furnish any new regiments, might not they be put into such an expedition? Please think of it.

I believe no local object is now more desirable.

Yours, truly,   A. LINCOLN.

---

722. *Encyclopedia Britannica*, Vol. 18, 1911, pp. 341–342; Nicolay & Hay, Vol. 7, pp. 400–402; Grant, Vol. 2, p. 545.

Chapter 11

The Sabine Pass Expedition; "Record" 10/63; A Land Route to Texas; The Rio Grande Expedition; "Record" 12/63; The Battle of Grand Coteau; "Record" 2/64; "Record" 4/64; Winter Quarters; The Red River Campaign

The Sabine Pass Expedition

Mobile was ruled out by Halleck in a letter to Banks on August 12th: "I fully appreciate the importance of the operation proposed by you…but there are reasons other than military why those heretofore directed should be undertaken first. On this matter we have no choice, but must carry out the views of the Government."[723]

Thus, Banks was forced into a move that had little or no military value. As Irwin[724] puts it: "To have overrun the whole state would not have shortened the war by a single day." However, Banks' job was political; it was to forestall potential interference in Texas by the French.

He chose to land on a part of the coast of Texas nearest to him, yet one of the "most important" seaports to the Confederacy – Sabine Pass.[725] Then Beaumont could be occupied, and from there Houston, which would control "all of the railway communications in Texas"[726]

General William B. Franklin, who had arrived in the department in late July, and had been assigned to the command of the 19th Army Corps, was to lead the expedition. It would be his first role since serving under Burnside at the Battle of Fredericksburg. There, he had been relieved, being accused of disobedience and negligence.[727]

Note that the remarks section of Battery L's August–October muster roll makes no mention of their participation. There is no doubt, however, that they were there, including batteries A and F, the 1st Indiana, 18th New York, and 2nd Massachusetts.[728]

---

723. O.R. Vol. 26/I, p. 675.
724. Irwin, p. 264.
725. O.R. Vol. 15, p. 143.
726. O.R. Vol. 26/I, pp. 18, 19.
727. *Encyclopedia Britannica*, Vol. 11, p. 33; O.R. Vol. 26/I, p. 658. Franklin was graduated at the head of his class from West Point in 1843. Though the charges were later disproved, a cloud hung over him at this time. O.R. Vol. 26/I, pp. 710, 711.
728. Haskin, pp. 214, 367, 554; O.R. Vol. 26/I, p. 721.

This was to be a formidable expedition, consisting of 14 regiments, 10 artillery batteries, and a detachment of the 21st Indiana, with eight heavy Parrotts. All of this was crammed aboard a combination of sailing ships, ocean steamers, and a half-dozen old river steamers, 23 in all, accompanied by three navy gunboats. The army troops were to be landed in combination with a navy attack on Fort Grigsby on the Sabine River, which defines the state line between Louisiana and Texas. It was supposed to be a surprise night approach and an early morning landing.

The troops were loaded into their transports at New Orleans and Algiers on September 4th and got underway on the 5th. Unfortunately, by a series of almost unbelievable, nay, incompetent, mishaps, the presence of the expedition off Sabine Pass was disclosed to the Confederate garrison on shore. As a result, the attacking navy gunboats were shot up, two surrendered, and there was no chance to land the troops. The mission was called off by General Franklin, and they were back home by the 11th.

It was not an auspicious beginning for General Franklin. He apparently had no thought whatever of Banks' *written* order that stated: "A landing, if found impracticable at the point now contemplated, should be attempted at any place in the vicinity where it may be found practicable . . ." In addition, a verbal order gave him the option to land 10 or 12 miles below (south of) Sabine Pass.[729]

It seems that Franklin's behavior was a case for removing him, but no hint of such an action has been found in the records.

"Record" 10/63
31 AUGUST–31 OCTOBER, 1863 BARRE'S LANDING, LOUISIANA

The Battery left New Orleans Sept. 15, '63 and proceeded by railroad to Brashear City and crossed to Berwick City, arriving on the 18th. Left Berwick City on the 26th and marched to Vermillion Bayou arriving on the 9th Oct. The Battery in action shelled the enemy. Crossed Vermillion Bayou on the 10th and marched to Barre's Landing, arriving on the 21st and remained there until 31 Oct. 1863.

Henry W. Closson	Capt. On det. Svc. Chief of Artillery 19th Army Corps S.O. no. 13 Hdqrts. 19th Army Corps Oct. 4, '63.
Franck E. Taylor	1st Lt. Commanding Battery and Ass't Comm'y of Musters S.O. no. 92 Hdqrts. Dept. of the Gulf 19th Army Corps Apr. 9, '63
Edward L. Appleton	1st Lt.
J.A. Sanderson	2nd Lt. On det. svc. with Gen. Lee Chief of Cavalry S.O. no. 250 Hdqrts. Dept. of the Gulf Oct.6, '63.
Charles B. Slack	2nd Lt. Joined Co. S.O. no. 211 Hdqrts. Dept of the Gulf Aug. 26, '63.
Attached:	
Charles Wheeler	Pvt. Asst. Com'y of musters clerk S.O. no. 20, Hdqrts. 19th Corps Oct. 12, '63.
Detached:	
William E. Scott	Cpl. On Det. Svc. Orderly Chief of Artillery

---

729. O.R. Vol. 26/I, pp. 287, 291, 303, 310; ORN Ser. I, Vol. 20, p. 546.

George Friedman	Pvt. On Det. Svc. at N.O. as Artillerist, Left Co. May 24,'63
Amelius Straub	Pvt. On Det. Svc. at Baton Rouge as cook in Univ. Hosp. since June 1,'63.
Thomas Newton	Pvt. On Det. Svc. in Ord. Dept. since Sept. 8, Per S.O. No. 4, Hdqrts. Dept. of Gulf Sept.'63. Due U.S. for Ord. $45.56. For 2nd reenlistment $3.00 per month

Absent with Leave:

Louis Lighna	Pvt. On furlough left Company August 21st for 14 days.

Absent in Confinement:

Michael O'Sullivan	In jail at New Orleans since Sept. 2,'63 (from desertion).

Absent Without Leave:

Joseph Kutschor	Pvt. Exchanged prisoner of war by virtue of G.O. no. 339, A.G.O. War Dept. Wash. Oct. 16,'63

Deserted:

William Fudge	Pvt. Deserted from N.O.	Aug. 25,'63	Due U.S. for ord.	$2.30
Hiram Smith	Pvt. do.	Sept. 10	do.	$7.55
George Kelly	Pvt. do.	Sept. 16	do.	$7.55
John M. Kastenbader	Pvt. from TerreBonne	Sept. 16	do.	$7.55
Clark Dickson	Pvt. from N.O.	Sept. 15	do.	$7.55
Charles E. Deal	Pvt. do.		do.	$7.55
Sholto O'Brien	Pvt.	Sept. 8	do.	$7.55

William Brooks	Pvt. Discharged from hospital and never joined Comp. Due U.S. for Ord. $47.10.
Henry Pelky	Pvt. from N.O. Sept. 15 Due U.S. for ord. $7.00.
William Mint	Pvt. from N.O. Sept. 8,'63 Due U.S. for ord. $7.55

Discharged:

Alexander J. Baby	Sgt. by Promotion to 2nd Lt. Gen'l Banks' bodyguard.
Charles Spangler	Pvt. Expiration of service
Denis Myers	Pvt. Expiration of service
Wallace D. Wright	Pvt. June 10,1863 at Ft. Pickens, Fla. for disability sick at Pensacola since December 24,'62.
Edward M. Laughlin	Pvt. Sept. 30,1863 from Berwick City by expiration of service. AWOL since Sept. 24 at N.O.

Died:

John Deering	Pvt. Died at Marine Hosp. Sep. 29,1863 of dysentery. No notice of decease until this month.
Hiram Hubbard	Pvt. Died in the field near Opelousas, La. Oct.26, 1863, of congestive fever.

Joined:

John Lowry	Pvt. From desertion at N.O. Sept. 3
Michael O'Sullivan	Pvt. In jail at N.O. since Sept. 2, 1863 Due U.S. for ordnance $43.36
Arthur Flynn	Pvt. From Berwick City Sept. 19, 1863 Due U. S. for ord. $1.63.
John H. Moran	Pvt. do. Sept. 19, 1863 Due U.S. for ordnance $41.63. To forfeit $8.00 of his monthly pay for 4 months pursuant to General Court Martial S.O. no. 3 Hdqrts 1st Div. 19th Army Corps, Sept. 28, 1863.
William F. Brown	Pvt. From desertion at Tarleton Plantation, Oct. 1,'63 Due U.S. for ord. $5.93. To forfeit $10.00 paid Constable for arrest.

Joseph H. Parslow	Pvt. do. Oct. 1, '63
Partick Craffy	Pvt. do. Oct. 2 Due U. S for ord. $1.63 To forfeit $10.00 paid Constable for arrest.
Owen A. Wren	Pvt. Oct. 2 Due U.S. for ord. $2.50. To forfeit $10.00 paid Constable for arrest
Martin Stanners	Pvt. Exchanged prisoner of war by virtue of G.O. 339 War Dept. Wash. Oct. 31, 1863

Strength: 108, Sick 5

Sick present: none

Sick absent:

William C. Brunskill	At Ft. Hamilton, Ny, left Co. Sept. 17, 1861.
Henry Champion	At University Hospital since August 27, 1863
William Crowley	At Baton Rouge, La. since July 13, 1863
George Chase	At Brashear City since April 22, 1863
Charles F. Mansfield	At Pensacola since December 24, 1862

Colored Cooks:

Henry Jefferson	31 Oct.'63	Barre's Landing, La. Cooking in Company since May 20, 1863.
Phillip Evens	31 Oct.'63	do.
Virgil Ayres	31 Oct.'63	do.

Why Joseph Kutschor was listed as AWOL and not in the "Joined" is an example of the rather haphazard way some of the muster roll records were kept, entries for the two-month period probably being made before subsequent information made them obsolete. Regardless, both Kutschor and Stanners were lucky men. A declaration of exchange had been issued by the Confederate agent for exchange, Col. Robert Ould, on September 12th, and the notice did not appear in U.S. Army General Orders until October 16th. The page from the O.R. is reproduced below.[730]

GENERAL ORDERS, } WAR DEPT., ADJT. GENERAL'S OFFICE
No. 339. } Washington, October 16, 1863.

1. A declaration of exchanges having been announced by R. Ould, esq., agent for exchange at Richmond, Va., dated Sepernber 12, 1863, it is hereby declared that all officers and men of the U.S. Army captured and paroled previous to the 1st of September, 1863, are duly exchanged. The officers and men herein declared exchanged will immediately be sent to join their respective regiments. By order of the Secretary of War:

E. D. TOWNSEND,
Assistant Adjutant-General

The information only makes the fate of George Chase more of a mystery. He had been sick at a place that fell to the enemy, much like Sgt. Charles Riley, so many months ago, and has not been heard from.

The New Orleans deserters who came back to the fold were not nearly as

---

730. O.R. Ser. II, Vol. 6, p. 383.

numerous as may have been hoped, and almost all of those eight who did, had been apprehended by the authorities. Only Arthur Flynn and Joseph Parslow seem to have escaped punishment. As had been known before, if you turned yourself in, you often escaped severe consequences, much as Francis Hagan, at Fort Duncan.

Fees for ordnance and clothing appear in great number, and only a few examples are given here. Out of 91 enlisted men, a handful owed money for clothing. George Howard owed the largest amount: $32.16. Thirteen owed the U.S. for ordnance, some three or four owed as much as three months' pay. Pvt. Thomas Newton owed the most; he is listed as "Detached," above. See also, John H. Moran, and others, in "Joined." Uncle Sam was still a cheap scrooge. Pretty soon, he'd have to pay dearly to keep these men happy enough to reenlist!

Battery L had arrived at New Orleans on August 10th. Since they remained until the beginning of September, it is likely that everyone had a chance to find time to visit the city. While there, there is evidence that they had met members of Farragut's squadron – those who were from ships in port for repairs - the *Gertrude, Sciota, Kennebec, Pinola, Virginia, Albatross,* and *Arizona* being among them.

Barroom heroes undoubtedly competed with stories of their recent adventures; the sailors in turn regaling their listeners with accounts of prize captures, and of the money shared by the crews. At least one member of Battery L became a believer.

At Pensacola, on December 20th, 1862, when he transferred from the 91st New York Infantry Regiment to Battery L, Enos Deal represented himself to Lieutenant Appleton as Charles. Now having experienced the dust and mud of the marches in the Teche Campaign, and the sweaty hell of the Siege of Port Hudson, it was time for a new Deal.

His plan was clear; at the first opportunity, he would desert and join the navy. That time came when Battery L was leaving New Orleans on September 15th. He walked away, and on September 22nd, under the alias of Charles E. Jones, joined to serve as a coal heaver on the USS *Kennebec*.[731]

The *Kennebec* was returned to service in November, with "Jones" aboard, and in December, she captured the blockade runners, *Marshall J. Smith* and *Grey Jacket*. "Jones" shared in the prize money. Other actions followed, including participation in Farragut's attack on Mobile in August of 1864, which is covered in chapter 15, because of the twist of fate that put Battery L's Commander, Henry Closson into the fray.

The departure of Baby, Newton, Spangler, and Wynne created vacancies which were filled by moving up David J. Wicks and William Demarest from corporal to sergeant, and the promotion of Michael White, Edward Cotterill, Charles E. Walton, and William E. Scott from private to corporal.

---

731. ORN Ser. I, Vol. 20, pp. 510, 627, 652; National Archives, Deal navy pension cert. no. 39985.

Even prior to the March, 1863 law alluded to in chapter 9 that authorized enlisting those of African descent as cooks, the company had informally accepted three escaped slaves into their midst over the course of the spring. They came into the company as "hangers on," as Thomas Newton, the commissary sergeant, described them, willing to work for their "keep," which consisted of their food, cast off clothing, and any favors from individuals.

Philip Ewens came into the company in "late January or early February" of 1863, while the company was at Baton Rouge. Called "Phil" or "the lawyer" for the fact that he questioned everything, he was properly named Phillipe Milton E. Ewens, a French Creole, born into slavery at New Orleans in about 1828.

Virgil Ayers came into the company while at Barre's landing, in April, and first worked as a teamster, "then later paid to cook." He was born into slavery at the Wicks plantation, near Franklin, on or about 1835, where he had worked as a field hand. He was described as a "quiet and well behaved fellow" by Sergeant Keller.

Henry Jefferson was the last, coming into the company "along the Atchafalaya River" on May 20th, while the company was moving to Port Hudson. He was born to a slave mother and a French father at New Orleans in about 1823, and was properly named Henri Louis Gerfie. He spoke French only, and given the soft pronunciation of his name, he was called "Jeff." When finally enlisted, Lieutenant Appleton assigned him as Jefferson. Described by Edmond Cotterill, the company clerk, as "different from the colored people in Louisiana," he claimed to have been in the original Native Guards enlisted by Governor Moore.

Word of the law evidently having filtered down, they are finally shown here as enlisted.[732]

The Battery L Record of Events is particularly barren of details for this period. As noted, the fact that the whole Battery, including all of its horses, went on the Sabine Pass Expedition is not even mentioned, and subsequent notes say little about what developed into only a feint on the Sabine by land, our description of which follows.

A Land Route to Texas

Immediately upon return from the Sabine Pass Expedition, portions of Franklin's troops, that is, Weitzel's, including Battery L, and Emory's divisions, were ordered to Brashear City to begin an expedition to Texas by one of the two inland routes available; (a) from Berwick City to Vermilionville, and then west along the plains to Niblett's Bluff, and the Sabine River, or, (b) via the Atchafalaya and the Red rivers to Alexandria and Shreveport, and then west into Marshall, Texas. The shorter southerly route is shown in figure 1.

Franklin's troops were to be followed by Grant's newly arrived 13th Corps. The

---

732. Pension files, National Archives; Ewens, 439866; Ayers, 433713; Jefferson, 43310.

13th, having been ordered south from Vicksburg on the 7th of August,[733] began arriving by steamer over the course of the next couple of weeks. On September 4th, Grant made his only visit to Banks' Department of the Gulf. In early August, Halleck had denied him a leave of absence to visit New Orleans and formulate a plan to "move against Mobile," but Grant now went on official business, to review the 11,000 men of his 13th Corps.

The occasion was before Grant was promoted to the command of the new "Military Division of the Mississippi,"[734] on October 17th, consisting of the combined departments of the Cumberland, the Ohio, and the Tennessee. For Grant, it would now be everything about Tennessee, and the trouble Rosecrans had gotten himself into at Chattanooga.

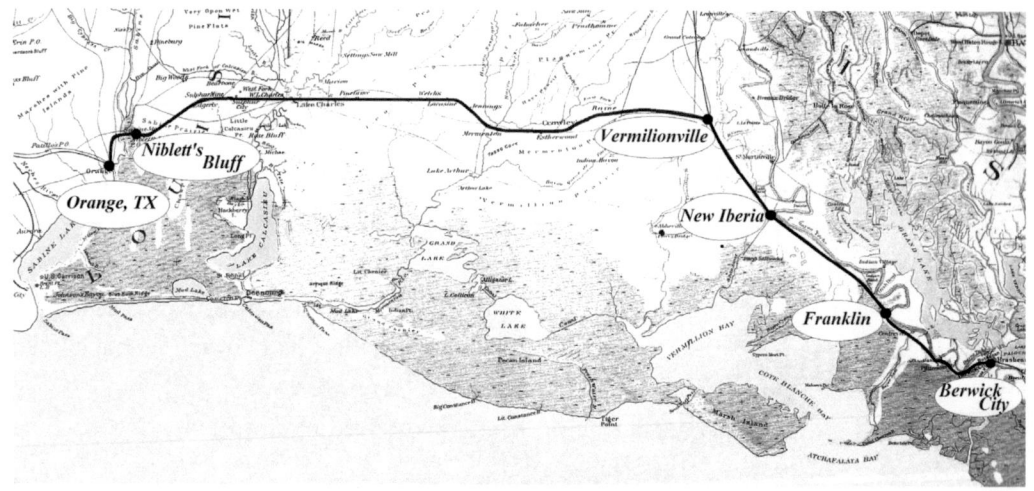

FIGURE 1

Grant's review of the 13th Corps was held on September 4th at Carrollton. A diary entry by a member of the 83rd Ohio captures the event, which serves to confirm the fact that Grant remained in the Department of the Gulf for an extended period of time, much longer than he would have planned.[735]

> The customary salute was fired and General Grant rode into the field, and was greeted with loud cheers. He rode slowly along the front of each line and galloped back at race track speed. His escort and visitors were strung out far behind, requiring some time for them to catch up and regain their places. There was one naval officer, dressed all in white, who went wherever his horse chose. He was a comical sight...
>
> The general had on his old brigadier coat and was in rather a marked contrast to the well formed and finely dressed General Banks...This was no

---

733. O.R. Vol. 24/III, p. 581; Grant, Vol. 1, pp. 579, 581; Figure 1, Atlas, plate 156, portion, altered.
734. Grant, Vol. 2, p. 17.
735. Marshall, pp. 107–108.

doubt intended to be his farewell to us, as he had been called east to be the main stay of the great Lincoln, which he most certainly became."

The wild and "vicious" horse Grant had been riding, which he had lost control of during the review, soon shied and fell over, pinning Grant on to the cobbled street. He was injured sufficiently to remain confined at New Orleans until the 13th.[736] The extended period undoubtedly was sufficient time to have raised the inevitable discussion of the Sabine Pass expedition, which had already begun when Grant arrived. Its failure only points up an expressed opinion of Grant's that Lincoln's wishes could have been fulfilled by not "wasting troops in western Louisiana and eastern Texas, by sending a garrison at once to Brownsville on the Rio Grande." Here he quietly suggests the occupation of only a single location in Texas.

Banks recommended a Rio Grande expedition to Halleck on the same day that Grant left,[737] and at the same time he gave orders to begin Halleck's preferred overland move. Grant's opinions were still ignored, as they had been since the beginning. To the ultimate benefit of the Union, this would finally change, but not for almost another year.

On July 16th, troops were loaded on cars destined for Brashear City. The ferrying of the troops and equipment across Berwick Bay took up nearly a week,[738] so that there were considerable numbers of troops in camp on both sides, doing what you do in the army, hurry up and wait. Here, the 19th Corps was joined by the 13th Corps, under the command of Gen. E. O. C. Ord. Now, the easterners had a chance to meet the westerners, most for the first time. The cultural gap was evident. From the 116th New York Regiment: "We soon found it impossible to live in peace with these western men, as they were constantly telling of their prowess, and what *they* would now do in this department if we paper collar soldiers would only let them alone. They were almost destitute of discipline, wore whatever dress they pleased, be it a uniform or not, and considered whatever they could "gobble" [plundering] as their inalienable right. Many hard words ended in harder blows…"

Banks left New Orleans and arrived in General Franklin's camp when it was at New Iberia, to be on hand to observe the march to Vermilion Bayou. It was reached on the morning of October 9th. The march was characterized by a lack of water, the army having passed away from the Teche. Their only supply of water was "buffalo holes," stagnant pools unfit for drinking. Here, just as in the spring, the enemy was found on the other side of the bayou, with the bridge destroyed. The artillery, including Battery L, was brought up. The enemy withdrew quickly upon being shelled. Franklin's force now camped, enjoying the fresh water, until

---

736. Grant, Vol. 1, pp. 582; O.R. Vol. 30/III, p. 594; Nicolay & Hay, Vol. 8, p. 326.
737. O.R. Vol. 26/I, pp. 19–20, 289.
738. Clark, pp. 129–131; Billings, pp. 105–106. "Paper collar" was an insult, such a soldier was a "beat" or a "Jonah": an effete shirker.

a new bridge could be built, which was crossed on the 10th. They then moved on to Carrion Crow Bayou. The 13th Corps, which had been following a day behind, pulled in and camped here also. The entire force under General Franklin was now about 19,500. It remained here, with Taylor's forces close by, and making daily reconnaissance's,[739] for nine days.

Banks had seen enough. He had tried Halleck's desired land scheme, and it was not going to work. The difficulties of a lack of water and supplies in the surrounding country, "which had been repeatedly overrun by the two armies, and which involved a march of…400 miles from Berwick Bay, with wagon transportation only…mostly upon a single road…" caused him to abandon the idea of either of the possible land approaches then and there. The enemy would be pushed for a while longer, until a seaborne expedition to Texas had shown some merit. Then Franklin would be ordered to withdraw.[740]

The Rio Grande Expedition

As early as October 2nd, Commodore Bell, commanding the West Gulf Blockading Squadron in Farragut's absence, had begun the preliminaries to the expedition, which reveals that Banks was seriously considering this plan, or even had already decided on it, regardless of the conditions of the land route. Bell ordered the USS *Tennessee* to prepare for special service off the Texas coast, and to bring along a captain of the Corps of Engineers to examine "[the] Brazos River, Pass Cavallo, Pass Aransas, Brazos Santiago, and Rio Grande bars…" and the defenses of Brazos Harbor.[741] The *Tennessee* returned with the information on the 12th, and the *Monongahela*, chosen to convoy the expedition, was ordered withdrawn from blockade duty on the 20th, to report to Bell. The *Owasco* and the *Virginia* would also accompany the expedition. This would be quite a contrast to the Sabine Pass Expedition; these were large heavily armed vessels, incapable of passing into shallow water. The troops would have to land through the surf.

His decision made, Banks sailed on October 26th, on the transport steamer *McCllellan*.[742] He was accompanied by the 2nd Division of the 13th Army Corps, commanded by Maj. Gen. N. J. T. Dana, with detachments of the 13th and 15th Maine regiments, and the 1st and 16th regiments of the Corps d'Afrique–a total of about 4,000 troops. Having encountered a violent "Norther" off Aransas pass, on the 13th, which scattered the flotilla, only some had reached Brazos Santiago on the 1st of November, but a landing was made at noon on the 2nd, despite the rough

---

739. Clark, p. 132; Irwin, p. 274; Hall, H. & J., pp. 150–152.
740. Irwin, pp. 274, 276–277; O.R. Vol. 26/I, pp. 19–20, 367, 771–772, 768. The route to Niblett's Bluff was 200 miles.
741. ORN Ser. 1, Vol. 20, pp. 606–607, 622–623, 636–637, 641, 643.
742. O.R. Vol. 26/I, pp. 292, 396–398, 428–429, 776, 783; ORN Ser. 1, Vol. 20, pp. 645–647.

conditions left from the gale. The small force of Confederate cavalry there offered no serious resistance, and a detachment from Company B of the 15th Maine, from the steamer *Clinton*, was the first to plant Old Glory on the shore.[743]

The next day, the *Monongahela* arrived, and went on to the mouth of the Rio Grande to assist with the landing of troops. Brownsville, figure 2, was occupied by the 94th Illinois Volunteers on the 6th.[744] The town was in turmoil, the Confederates having set fire to the U.S. barracks, which spread into private property. Banks arrived and made his headquarters there the same day. He was immediately immersed in Mexican politics, many of those opposed to the French having sought refuge in Brownsville. Juan Nepumuseno Cortina was one, the "marauder" familiar to Battery L in 1860. On November 9th Banks, with justifiable pride, was able to dispatch the following to Lincoln: "Sir: I am in occupation of Brazos Island, Point Isabel, and Brownsville…"

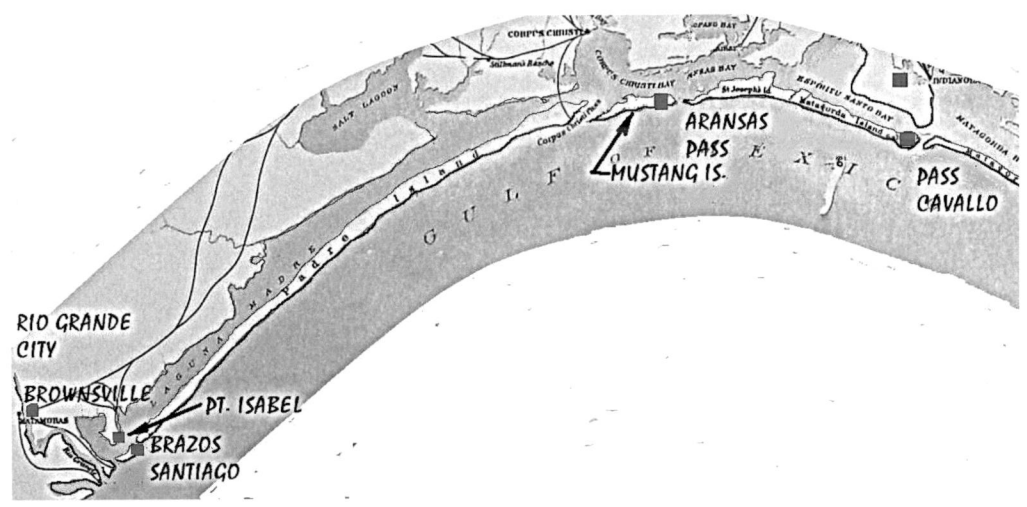

FIGURE 2

Banks left Brownsville for the next phase of the expedition on November 13th. It would be under the command of Brig. Gen. T. E. G. Ransom. After the embarkation of the 13th and 15th Maine, the 26th Iowa, the 8th Indiana, and an artillery battery, about 1,500 men, the expedition set off for Corpus Christi Pass, which was reached on the 16th. One transport that held some of the troops was the *Matamoras*, loaned to the expedition by Cortina, now a friend of the U.S.,[745] the thought being that its shallow draft would be able to pass over the bar at Corpus Christi. Finding the pass even shallower, the troops were unloaded through the surf and began the 22-mile march along Mustang Island, the target of the expedition being

---

743. *Harper's Weekly*, Nov. 28th, 1863, p. 754.
744. O.R. Vol. 26/I, pp. 399, 404, 407; Shorey, p. 56. Figure 2 derived from Atlas, plates 156 & 157.
745. O.R. Vol. 26/I, pp. 405, 409–410; Shorey, pp. 58–61.

Fort Semmes, which guarded the entrance to Aransas Pass. The fort surrendered without a fight on November 17th. Compare this determined action of Ransom to that of Franklin at Sabine Pass.

After a deserved rest, on November 23rd they were ferried across Aransas Pass to St. Joseph's Island, the objective being Fort Esperanza, which guarded the entrance to Pass Cavallo. Here, Ransom's troops were reorganized into a distinct brigade and were joined by a second, under Col. H. D. Washburn, all under the temporary command of Maj. Gen. C. C. Washburn, the brother of the governor of Maine. Delayed in their march by Cedar Bayou, which separates St. Joseph's Island from Matagorda Island, they were able to make a crossing on November 25th,[746] and the march was resumed on the 27th. Artillery was placed around the fort and shelling began on the 28th. By the night of the 29th, the Confederate garrison evacuated and blew up the fort. Coincident with the above operations, an expedition was sent up the Rio Grande to Ringgold Barracks (Rio Grande City) on the 23rd, and it was occupied by the 37th Illinois.[747] As of the end of December, there were troops at Rio Grande City, Brownsville, Pt. Isabel, St. Joseph's Island, Indianola, Matagorda Island, and Decrow's Point, at the tip of the Matagorda Peninsula. It was a brilliant expedition, Banks there in person, undergoing the hardships of the ship's passage in a gale, and in the landing and occupation of Brownsville. It was swift and decisive. Carried out almost entirely by Grant's officers and men, it pointed out their hardiness and their determined behavior, regardless of their sloppy uniforms. Grant had not stinted, he had sent some of those he held in highest regard; Ransom, an example.[748]

Grant later wrote:

> Suffice it to say, the close of the siege of Vicksburg found us with an army unsurpassed, in proportion to its numbers, taken as a whole of officers and men. A military education had been acquired which no other school could have given. Men who thought a company was quite enough for them to command properly at the beginning would have made good regimental or brigade commanders; most of the brigade commanders were equal to the command of a division, and one, Ransom, was equal to the command of a corps at least.

While all of this was taking place, Franklin was standing by with orders to send reinforcements if necessary, yet make no moves that would indicate a withdrawal. The Confederate forces in the Teche country were to be kept occupied and

---

746. O.R. Vol. 26/I, pp. 416–425, 446; Shorey, pp. 62–65.
747. National Archives, Compiled Records, 37th Illinois Infantry, Microcopy 594, roll 19, p. 1641; O.R. Vol. 26/I, pp. 847, 880, 898; Shorey, p. 68.
748. Grant, Vol. 1, p. 573; Shorey, p. 60: "During this brief campaign…General Ransom completely captured the affections of the Maine troops, and he and they were ever after close friends."

made to think that the overland attack toward Texas was still the objective, albeit changed to the lower route to Niblett's Bluff. As we have seen, the feint, as it had developed, had worked perfectly. At least up to the end of October, Taylor was not yet aware of the Rio Grande Expedition.[749]

"Record" 12/63
31 OCTOBER–31 DECEMBER, 1863 NEW IBERIA, LOUISIANA

Nov. 1 Marched from Barre's Landing to Carrion Crow Bayou 14 miles.
Nov. 2 " " Carrion Crow Bayou to Vermillion Bayou 16 miles.
Nov. 3 " " Vermillion Bayou to Carrion Crow Bayou 16 miles.
Nov. 4 " " Carrion Crow Bayou to Vermillion Bayou 16 miles.
Nov. 16 " " Vermillion Bayou to Camp Pratt 16 miles.
Nov. 17 " " Camp Pratt to New Iberia 5 ½–Remained until present date.

Henry W. Closson	Capt. Det. svc. Chief of Arty. 19th Army Corps.
Franck E. Taylor	1st Lt. Absent with leave S.O. no. 544 War Dept. Wash. Dec.28,'63.
Edward Appleton	1st Lt. Absent without leave since Dec. 28,'63.
James A. Sanderson	2nd Lt. Commanding Battery.

Transferred:
Charles B. Slack — 2nd Lt. Relieved from Duty with Battery Nov. 8, 1863 S.O. no. 4 Hdqrts. Dept. of the Gulf Oct. 20, 1863.

Attached:
Charles Wheeler — Pvt. Ass't. Com'y of Musters Clerk. Detached from the 8th Reg't Vermont Vols.

Detached:
Edmond Cotterill — Cpl. On Det. Svc. Clerk in A.G.O. Wash. S.O. no. 426, War Dept. A.G.O. Wash. Sept. 23,'63. Left Co. Nov. 20,'63.
Jeremiah Connell — Pvt. On Det. Svc. Orderly to Chief of Artillery 19th Army Corps. For 1st Re-Enlistment $2.00 per month
George Freidman — Pvt. On Det. Svc. at N.O. as Artillerist. Left Co. May 24,'62
Amelius Straub — Pvt. On Det. Svc. at Baton Rouge as cook in Univ. Hosp. since June 1,'63.

Absent with leave:
William E. Scott — Cpl. Absent with leave since Dec. 29,'63
Terence McGauly — Pvt. Absent with leave since Dec. 29,'63. Due U.S. for ord. $0.67.

Absent without leave:
Joseph Kutschor — Pvt. Deserted
William Brooks — Pvt. At New Orleans, from Marine Hospital, never joined company.

Discharged
Lewis Keller — 1st Sgt. Dec. 29,'63 At New Iberia, by order of Gen'l Franklin, S.O. no. 105 Hdqrts. 19th Army Corps Promoted 2nd Lt. 2nd Louisiana Cavalry.
Henry Champion — Pvt. Nov. 20,'63 At New Orleans, from University Hospital, for disability.
Thomas Newton — Pvt. Dec. 13,'63 At Port Hudson, by expiration of service.

---

749. O.R. Vol. 26/I, pp. 779.

Louis Lighna	Pvt. Oct. 15,'63 By expiration of service. Enlisted in 2nd Louisiana Vol. Cavalry.

Died:

James Hanney	Pvt. Nov. 2,'63 At Carrion Crow Bayou, of inflammation of the bowels.
John Baker	Pvt. Nov. 14,'63 At Vermillion Bayou from injuries incurred by a fall.
Hiram Hubbard	Pvt. Oct. 26,'63 At Barre's Landing, of congestive fever.

Joined from Desertion:

James Comfort	Pvt. Oct. 9,'63 At Vermillion Bayou. Due U.S. for ord. $41.63 to forfeit 1 mo. 19 days pay for time to make good to the U.S. time lost by desertion . S.O. # 83, Hdqrts. 19th Corps Dec. 5,'63.
Francis Jessop	Pvt. do.
Christian Meese	Pvt. do.
Michael Breen	Pvt. Oct. 10,'63 Arrested Aug. 26.
Sholto O'Brien	Pvt. Oct. 21,'63 At New Iberia.
Patrick Craffy	Pvt. Date and place not listed.
Joseph H. Parslow	Pvt. do.
Owen A. Wren	Pvt. do.

All of the above were court-martialed at the same time, December 5th, 1863, though most were returned in October, and are thus listed twice. All received forfeiture of pay for time lost, similar to Comfort, but no other penalty, because they returned voluntarily.

James Flynn	Pvt. Absent without leave until Dec. 30,'63. For 1st Re-enlistment $2.00 per mo.
John H. Moran	Pvt. In confinement. Due U.S. for ord. $41.63. To forfeit $8.00 per month from his pay for 4 months. S.O. no. 3 Hdqrts. 1st Div. 19th Corps Sept. 23,'63
John Lowry	Pvt. In confinement.

Colored Cooks: Henry Jefferson, Philip Evens, Virgil Ayres

Strength: 105 Sick: 5

Sick present: none

Sick Absent:

William C. Brunskill	Sick at Ft. Hamilton, NY left Co. Sept. 17, 1861
William Crowley	Sick at Baton Rouge since July 13,'63
George Chase	Sick at Brashear City since April 22,'63
Corneilus McEnearny	Sick in Gen. Hosp. New Iberia, since Nov. 17,'63
Charles F. Mansfield	Sick at Pensacola, Fla. Since Dec. 24,'62

For the first time, an officer, Appleton, is listed as absent without leave. With every other senior officer away for some reason, the command of the company has devolved to the most junior officer, Sanderson, who had arrived only a few weeks ago, an 1862 graduate[750] of West Point.

Lt. Slack, who was from Massachusetts, was transferred to Nims' Battery, and

---

750. Cullum, Vol. II, no. 84.

would be wounded on April 8th 1864, at the Battle of Mansfield.[751]

Cpl. Edmond Cotterill must have had exceptional skills to have been transferred to Washington.

Among those discharged, Keller had been promoted to the 2nd Louisiana Cavalry, and no one was immediately named to fill his place. Louis Lighna's record is finally updated. He followed Keller, the 2nd Louisiana Cavalry first being organized on November 25th at New Orleans. Though he never returned to Battery L, Lighna was still serving in the army, in the 1st Regiment of Artillery, Battery A, at Pensacola, in 1875,[752] where he was commended for his services and untiring zeal, which were "invaluable" during the yellow fever epidemic of that year. He had a son who died in the epidemic who is buried in Barrancas National Cemetery. Henry Champion had been discharged for disability, after being sick since August 27th. Compare this with William Brunskill, still sick at Fort Hamilton, New York, since September 1861.

There were no new deserters that had joined.

Joseph Kutschor, listed as an exchanged prisoner of war in the August–October roll, is now listed as deserted.

The Battle of Grand Coteau

In response to Banks' order to General Franklin to be prepared to send reinforcements to the Rio Grande Expedition, on the 27th of October he withdrew the 1st Division of the 13th Army Corps, under Lawler, from Opelousas to New Iberia.[753] That left the 3rd Division under McGinnis at Opelousas, and the 1st Brigade of the 4th Division under Burbridge at Barre's Landing. Both, including the 19th Corps, were ordered to pull out on the 31st of October, their march beginning the next day. The first move was to Carrion Crow Bayou, about 16 miles. McGinnis was to camp on the north side of the Bayou, and Burbridge, coming from the direction of the Teche, was to camp two miles away. The general location is shown in figure 3.[754] Weitzel, Grover, and Battery L, were to encamp on the south side.

Having been notified that the enemy was falling back from Opelousas, Taylor

---

751. O.R. Vol. 34/I, p. 462.
752. Haskin, p. 387. Requests for Lighna's service records at the National Archives have been returned with the comment "No Record." 2nd Louisiana Regiment history: *www.nps.gov/civilwar/search-rec*; Barrancas National Cemetery Records.
753. O.R. Vol. 26/I. pp. 779, 354, 357, 369. McGinnis was ill and did not participate. The corps commander, C. C. Washburn, was in his place. McGinnis's name has been used only to identify the 3rd Division. Note the hectic moves in the "Record of Events." The description given by Washburn as the position of Burbridge was north of "Muddy Bayou." Subsequently, in Burbridge's report, he describes his position as "3 miles from Carrion Crow Bayou, near the head of a small bayou that runs in the direction of Opelousas"–Bayou Bourbeau.
754. Figure 3, drawn by the author.

ordered Green to pursue and harass. It was Confederate General Thomas Green's cavalry again, the one and the same that had attacked Fort Butler in the La Fourche back in June. He was now reinforced by three regiments of Walker's Texas infantry.[755]

On the morning of November 2nd, the 19th Corps pulled out, its destination Vermilion Bayou, about 11 miles. The detachments of the 13th Corps under Washburn were left to hold their positions.

Burbridge's 1st Brigade, numbering about 1,625 men,[756] was camped on the north side of Bayou Bourbeau, with about three miles of open prairie separating them from Washburn, figure 3, prepared by the author.

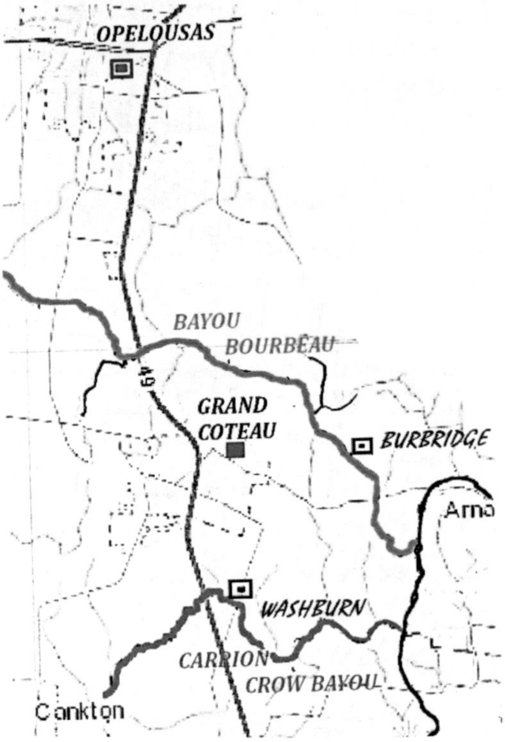

On November 2nd, Green advanced and tested Burbridge all day long. Washburn sent the 1st Brigade of the 3rd Division out to reinforce Burbridge, but finding the enemy had disappeared, they returned to their camp on Carrion Crow Bayou. However, enough was seen of Green's force for Burbridge to estimate it at 2,500. As it was later learned, Green's tentative moves on November 2nd were only because he was waiting for his infantry to come up, which was marching from Opelousas.

FIGURE 3

That night, six men from Burbridge's 1st Louisiana Cavalry deserted,[757] no doubt bringing news of Burbridge's strength and dispositions to Green. On the 3rd, skirmishing began at 10:00 a.m. and then tapered off, allowing Burbridge to think it safe enough to send away the 83rd Ohio on a foraging expedition. The troops were allowed to stack arms. Everyone's attention was then turned to the two paymasters who were in camp. In addition, the 23rd Wisconsin was to vote for state elections.

Their position was exposed, "in front, flank, or rear..."[758] and no mention is made of any earthworks having been dug, or any sort of defensive preparations made. Burbridge's concerns were put off by Washburn as nothing but "a scare." The following is taken from the history of the 96th Ohio:

---

755. Taylor, p. 150, O.R. Vol. 26/I, pp. 369, 393–394.
756. O.R. Vol. 26/I, pp. 360, 364. Note that each regiment averaged only 200 men.
757. O.R. Vol. 26/I, pp. 360, 365.
758. Woods, pp. 41–44.

At 12:00 our retreating cavalry gave notice that 'the Philistines were upon us.' The thrilling long-roll called every man to arms. In calm calculated haste, each donned his battle trappings, and with clockwork precision fell into line.

Washburn's report of the action having arrived at Weitzel's camp, the order was given, late on the evening of Nov. 3rd, for the 1st Division of the 19th Army Corps[759] to return to Carrion Crow Bayou. Battery L's record of events shows them having already arrived back there on the 3rd, but added detail is absent. After a forced march of five hours, the First Division arrived at 7:00 a.m., on the 4th, to find that all was quiet.[760] "Subsequent investigation showed that the attack was a shameful surprise, and was declared by General Weitzel "disgraceful to our arms." If the reader remembers the disputes between the easterners and their new western comrades, this now gave the "paper collar" easterners their opportunity to respond to who "knew a thing or two," given the daring heroism of Marland's section of Nims' battery, contrasted with the huge number of casualties in the ranks of the western regiments. A revised report had 125 killed, 129 wounded, and 562 captured and missing.

Green submitted a report listing[761] 22 killed, 103 wounded, and 55 missing.

Battery L and Weitzel's division quickly returned to Vermilionville, and remained there until they marched via Camp Pratt to New Iberia, where they remained until the 7th of January, 1864. Banks was still focused on Texas – he now had more than twice as many troops there as in Louisiana. The policies in Louisiana outlined in instructions given to General Franklin in October were still in effect:[762] "...hold your position in that quarter, and ascertain as much concerning the country...as possible, and...keep these headquarters well informed..."

By the end of December, with the exception of one brigade at Plaquemine, the entire 13th Corps had been shipped to Texas. A defensive stance was now a necessity; the 19th Corps, with a total of 8,500 men, including the cavalry, and its recently mounted infantry regiments, was in no position to hold all of this territory as well as to initiate a campaign against Taylor, whose forces were variously estimated at 6,000 or 10,000 depending on whether or not Green had returned to Texas.[763] Green, with about 3,000 men, indeed had. Another consideration was that the older three-year regiment's terms of enlistment were about to expire in January and February.

---

759. O.R. Vol. 26/I, p. 369; Pellet, p. 129.
760. Pellet, pp. 156–157. The 114th was a part of Weitzel's First Division, 19th Corps.
761. O.R. Vol. 26/I, pp. 359, 395.
762. O.R. Vol. 26/I, pp. 761, 891–892; Clark, pp. 134, 137.
763. O.R. Vol. 26/I, pp. 852, 863, 920; O.R. Vol. 26/II, pp. 260, 465; Irwin, pp. 272, 278.

"Record" 2/64
31 DECEMBER 1863-29 FEBRUARY 1864 FRANKLIN, LOUISIANA

Marched from New Iberia the 7th of January on route for Franklin, 53 miles. 8th marched 12 miles. 9th marched to Franklin 5 miles and remained there until present date.

Henry W. Closson	Capt. On det. svc. Chief of Arty. 19th Army Corps S.O. no. 13, Hdqrts. 19th Army Corps Oct. 4,'63.
Franck E. Taylor	1st Lt. Commdg. Battery and Asst. Comm. Of Musters, S.O. no. 92 Hdqrts. Dept. of the Gulf 19th Army Corps.
Edward L. Appleton	1st Lt. Commanding Battery
James A. Sanderson	2nd Lt.

Detached:

Edmond Cotterill	Cpl. On Det. Svc. Clerk in A.G.O. Wash. S.O. no. 426, War Dept. A.G.O. Wash. Sept. 23,'63. Left Co. Nov. 20,'63.
Jeremiah Connell	Pvt. On Det. Svc. Orderly Chief of Artillery
George Freidman	Pvt. On Det. Svc. at N.O. as Artillerist. Left Co. May 24,'62.
Amelius Straub	Pvt. On Det. Svc. at Baton Rouge as cook in Univ. Hosp. since June 1,'63.

Absent in Confinement:

Patrick Gibbons	Pvt. At Ship Island serving sentence of G.C.M. no. 18, Hdqrts. 1st Div. 19th Army Corps Dec. 31,'63. Left Co. Jan. 8,'64.
Michael O'Sullivan	Pvt. In jail at N.O. Left Co. Sept. 2,'63.
John Lewery	Pvt. Apprehended as a deserter and confined at New Orleans.

Absent without leave:

Joseph Kutschor	Pvt. Exchanged prisoner of war

Deserted: none

Discharged:

Charles F. Mansfield	Pvt. 10 Jan.'64 on Surgeon's certificate of disability, at Fort Pickens, Fla.
John Tomson	Pvt. 9 Jan.'64 by expiration of service at Franklin
Morris Galavan	Pvt. 14 Jan.'64 by expiration of service at Franklin
Martin Stanners	Pvt. 19 Jan.'64 by expiration of service at Franklin
Daniel McSweeny	Pvt. 11 Feb.'64 by expiration of service at Franklin.

Joined from Desertion:

John Lewery	Pvt. Feb.'64 Apprehended, in confinement at New Orleans.

Strength: 101 Sick: 8

Sick Present: Wm. Creed, Prosper Ferrari, Charles Jackel, Frank Morgan, Daniel Moore

Sick Absent:

William C. Brunskill	Sick at Ft. Hamilton, NY left Co. Sept. 17,'61
William Crowley	Sick at Baton Rouge since July 13,'63.
George Chase	Sick at Brashear City since April 22,'63.

The cooks are not listed.

Appleton had been back in command since January. No explanation for his absence appears in the muster roll record.

Why William Brunskill had not been discharged is of interest, considering that other chronic cases such as Mansfield's, who had been left sick at Pensacola since Dec. 24, '62, was finally dealt with here. Note that Chase is still missing.

As is noted in the "Record," Franklin's troops began their march out of New Iberia on January 7th, and it was abandoned on the night of the 8th, save a number of individuals with smallpox, who had to be left behind in the care of a surgeon.[764] By the 15th, the weather moderated, reminding them of early September in the north. One-month furloughs, and bounties were granted to those of the volunteer units that had re-enlisted for three years or the war, and before the end of the month, almost all of the 19th Corps had done so.[765]

It is noted that Battery L now had 101 men, down from its high of 149 a year before. As a consequence, their number of guns was temporarily reduced to four from six. In fact, all of the reorganized 1st Division and 2nd Division artillery units, which included Company A, 1st U.S., the 4th and 6th Massachusetts, and the 25th New York, had only four guns. The only artillery company that still had six guns was Nims', with their signature 6-pounder bronze rifles. They, however, had been assigned to Lee's cavalry, which would put them in extreme harm's way again.

"Record" 4/64
29 FEBRUARY–30 APRIL, 1864 FRANKLIN–ALEXANDRIA, LOUISIANA

March 16th struck camp at Franklin and marched 16 miles on Opelousas road. 17th Marched to Camp Pratt 18 miles. 18th Marched 3 miles beyond Carrion Crow Bayou, 18 miles. 19th Marched to plantation near Washington, 17 miles. 21st Marched 17 miles on Alexandria road. 22nd Marched to Holmsville, 14 miles. 23rd Marched to Cheneyville, 13 miles. 24th Marched to Welles plantation, 17 miles. 25th marched to Alexandria, 13 miles. 28th Marched from Alexandria on Nachitoches road, 18 miles. 29th Marched to Pine Woods, 9 miles. 30th Marched 3 miles beyond Cloutierville, 21 miles. 31st Marched 7 miles and camped on Cane River. 2nd Marched through Nachitoches, 6 miles. 6th Marched 14 miles on Texas road. 7th Marched to Pleasant Hill 22 miles. 8th Marched 10 miles camped for 2 hours, ordered to the front and lay in woods until morning. 9th Marched to Pleasant Hill, Battery in position, went into camp for 2 hours. Battery in action at P.M. 10th Marched 22 miles. 11th Marched to Grand Ecore. 21st Struck camp at 5 P.M. and remained harnessed until 3 A.M., then marched 24 miles and camped for about 4 hours, then marched to Cloutierville. 23rd Marched to pontoon on Cane River, crossed at midnight, camped until 6 A.M. the 24th then marched 16 miles. 25th Marched through Alexandria and remained until present date.

Henry W. Closson	Capt. On det. svc. Chief of Artillery 19th Army Corps S.O. no. 13, Hdqrts. 19th Army Corps OCT. 4,'63.
Franck E. Taylor	1st Lt. Commdg. Battery and Asst. Comm. Of Musters. S.O. no. 92 Hdqrts. Dept. of the Gulf 19th Army Corps.
Edward L. Appleton	1st Lt.

---

764. O.R. Vol. 34/II, p. 863.
765. O.R. Vol. 34/II, pp. 124, 132.

James A. Sanderson   2nd Lt.

Attached: [The 13th Massachusetts Battery was ordered attached on March 1st, by Special Orders no. 60. Seven new recruits joined later. Lincoln, Batson, Bragshaw, Connor, Kingsley, & Welsh joined on May 2nd at Alexandria, and Nichols on May 30th at Morganza. None of the names of the men in the 13th Massachusetts Battery were subsequently integrated into Battery L's Muster Roll. Only exceptions, such as sick, wounded, or killed, were penned in, and appear here in *italics*].

Sias, Chauncy R.	Sgt.	Oct. 13,'62 Boston	Carleton, William	Cpl.	Oct. 25,'62 Boston
Lincoln, James M.	Sgt.	Feb. 24,'64 Taunton	Fleming, Daniel H.	Cpl.	Oct. 21,'62 Boston
Merrill, Alfred K.	Sgt.	Nov. 26,'62 Charlestown	Hall, James F.	Cpl.	Oct. 1,'62 Charlestown
Betterton, George	Cpl.	Oct. 20,'62 Newton	Hesseltine, Charles	Cpl.	Oct. 30,'62
Dubois, Cesar	Musician	Dec. 26,'62 Boston	Hall, Ivory F.	Artificer	Jan. 24,'63 Boston
Berry, Thomas C.	Pvt.	Nov. 17.'62 Boston	Misener, James B.	Pvt.	Oct. 11,'62 Boston
Brown, Stephen E.	Pvt.	Oct. 28,'62 Boston	Mc Donald, John	Pvt.	Nov. 28,'62 Boston
Biaro, Italo	Pvt.	Jan. 21,63 Boston	McCostello, Michael	Pvt.	Mar. 30,'64 Foxboro
Batson, William	Pvt.	Mar. 29,'64 Chelsea	McCarrick, John O.	Pvt.	Mar. 31,'64 Mansfield
Bragshaw, William	Pvt.	Mar. 22,'64 Attelboro	Miller, George	Pvt.	Apr. 27,'64 Abington
Curtin, Patrick	Pvt.	Jan. 27,'63 Foxboro	Kingsley, Amos N.	Pvt.	Mar. 9,'64 Swansea
Connor, James	Pvt.	Mar. 11,'64 Bridgewater	Krall, Bartolomy	Pvt.	Apr. 8,'64 Boston
Connors, Patrick.	Pvt.	Mar. 13,'64 Scituate	Mee, Thomas	Pvt.	Mar. 31,'64 Natick
Dailey, Darby	Pvt.	Dec. 16,'62 Boston	O'Neil, Henry C.	Pvt.	Oct. 13,'62 Boston
Davis, George W.	Pvt.	Apr. 6,'64 Lancaster	Rivers, Harry	Pvt.	Oct. 17,'62 Boston
Edwards, James M.	Pvt.	Dec. 17,'62 Boston	Rivers, James H.	Pvt.	Oct. 22,'62 Boston
Falvey, Michael	Pvt.	Oct. 16,'62 Boston	Piper, Joseph	Pvt.	Apr. 1,'64 Swanzey
Ferguson, John	Pvt.	Oct. 21,'62 Boston	Simonds, John	Pvt.	Oct. 11,'62 Boston
Feeley, William	Pvt.	Nov. 28,'62 Stoughton	Sedgely, Robert	Pvt.	Nov. 28,'62 Boston
Fuchs, Henry	Pvt.	Nov. 25,'62 Boston	Sewall, James	Pvt.	Dec. 29,'62 N. Bedford
Gill, Robert	Pvt.	Oct. 22,'62 Boston	Smith, William	Pvt.	Apr. 8,'64 Lancaster
Graves, Ezekiel	Pvt.	Jan. 26,'63 Falmouth	Stevens, John	Pvt.	Apr. 21,'64 Dover
Hood, Charles	Pvt.	Oct. 28,'62 Boston	Timmins, John	Pvt.	Jan. 24,'63 E. Boston
Hartwell, James A.	Pvt.	Mar. 22,'64 Fitchburg	Thomas, Charles	Pvt.	Jan. 10,'63 Boston
Hesson, Michael	Pvt.	Apr. 25,'64 Arlington	Varner, Harry	Pvt.	Jan. 21,'63 Boston
Hiscock, George	Pvt.	Mar. 22,'64 Somerville	Wilkins, John C.	Pvt.	Oct. 8,'63 Brookline
Lynch, Thomas	Pvt.	Oct. 28,'62 Boston	Welsh, Benjamin C.	Pvt.	Mar. 9,'64 N'buryport
Larrabee, Thomas	Pvt.	Jan. 24,'63 Boston	Redding, George W.	Pvt.	Dec. 30,'62 Boston
Martin, Andrew	Pvt.	Oct. 21,'62 Boston	Roberts, Thomas	Pvt.	Jan. 8,'63 Boston
Mc Laughlin, John	Pvt.	Oct. 22,'62 Boston	Carney, David	Pvt.	May 19,'64 Newton
Murphy, Edward	Pvt.	Dec. 1,'62 Boston	Nichols, Edward A.	Pvt.	Feb. 16,'64 Boston

Detached:

Edmond Cotterill	Cpl. On Det. Svc. Clerk in A.G.O. Wash. S.O. no. 426, War Dept. A.G.O. Wash. Sept. 23,'63. Left Co. Nov. 20,'63.
Jeremiah Connell	Pvt. On Det. Svc. Orderly Chief of Artillery
Solomon Mongomery	Pvt. do.
George Freidman	Pvt. On Det. Svc. at N.O. as Artillerist, Left Co. May 24,'62.
Amelius Straub	Pvt. On Det. Svc. at Baton Rouge as cook in Univ. Hosp. since June 1,'63.

Absent in Confinement:

Patrick Gibbons	Pvt. At Ship Island serving sentence of G.C.M. no. 18, Hdqrts. 1st Div. 19th Army Corps Dec. 31,'63. Left Co. Jan. 8,'64
Michael O'Sullivan	Pvt. In jail at N.O. left Co. Sept. 2,'63.
John Lewery	Pvt. Apprehended as a deserter and confined at New Orleans.

Absent without leave:
George W. Redding   Pvt. Since April 9th 1864.

Casualties:[766] April 9th at Pleasant Hill:
Michael White	Sgt. wounded and missing.
Charles Hesseltine	*Cpl. wounded*
John Ferguson	Pvt. wounded
William Parks	Pvt. wounded
Thomas Clinton	Pvt. wounded [National Archives, M727-5, p. 340. *List of Killed Wounded and Missing in Batteries of 1st Div., 19th Army Corps at Battle of Pleasant Hill.*]

Died: April 9th at Pleasant Hill
James A. Sanderson	2nd Lt.
Sholto O'Brien	Pvt.
William H. Smith	Pvt.

Strength: 147 Sick:11

Sick Present: Joseph Parslow, David J. Wicks (total 4, remainder of record obscured)

Sick Absent: William C. Brunskill Sick at Ft. Hamilton, NY left Co. Sept. 17,'61
William Crowley	Sick at Baton Rouge since July 13,'63.
George Chase	Sick at Brashear City since April 22,'63.
Charles Jackel	Sick at Franklin, La, since March 12. 1864.
Henry Williams	do.
Churchill Moore	Sick at Franklin since March 12, 1864.
James Connor	Sick since May 8 '64

The cooks are not listed.

Winter Quarters

The two months at Franklin were pleasant, the most pleasant in recent memory. The people seemed friendly, and came out to see the drilling, the parades, and the elaborate decorations that the various regiments built. Emory's selection of their campsite, Camp Emory, was level, green, and "with just enough trees to shade it, but not mar its beauty or interfere…" with a proper layout.[767] Arbors and bowers of evergreens and Spanish moss enhanced the grounds, which "were laid out with exquisite taste."

The pause, which delighted the troops, was not for their rest and recuperation, or the weather, but for the fact that the government had no idea of what to do next. The coast of Texas had been occupied and the illicit trade out of Matamoras cut

---

766. William H. Smith is reported as killed in action on the Monthly Report. Names have been compared with the 13th Roster as published in: *Record of the Massachusetts Volunteers*, Adjutant-General, Boston, 1868, which indicates Smith as discharged by expiration of service on July 28th, 1865.
767. Clark, pp. 139–140.

off. Napoleon III had to have noted the Union victories at Gettysburg, Vicksburg, and Port Hudson, and that the Confederate cause was in difficulty. At this stage, he "had not the least notion of helping the unsuccessful," and the threat of his dabbling in Texas had faded. In fact, in late February, Secretary Seward, acting on the opinion that the "French rulers here…have no idea at this time of any interference in the war in our country," conveyed the President's order to Banks that he "avoid any collision between the forces under your command and either of the belligerents in Mexico…"[768]

While the troops were relaxing, and Halleck, prodded by Lincoln, was trying to promote the benefits of the occupation of "Western Texas," Banks was trying to complete yet another assignment, a political one, which had dated from August 24th, 1863.

Gen. George Shepley, of the 12th Maine Infantry, the Military Governor of Louisiana, had been notified by the War Department[769] that:

> Information has reached this Department that the loyal citizens of Louisiana desire to form a new State Constitution, and to re-establish Civil government in conformity with the Constitution and laws of the United States. To aid them in that purpose, the President directs…

Shepley was to inaugurate voter registration and order an election of delegates to a constitutional convention. *Much* discussion about how to accomplish this had ensued,[770] which was finally clarified by a Presidential Proclamation on December 8th. Lest that be misunderstood, on December 24th, direct orders from an exasperated Lincoln gave General Banks full authority over Shepley and the whole process.

Finally, an election for State officials was called for February 22nd, 1864. Appropriately, it was on George Washington's birthday, though at that time it was not a holiday. Hahn, the antislavery candidate, was elected governor. Now, with a state government "invested with the powers exercised hitherto by the military governor of Louisiana," Banks issued G.O. # 35, which announced elections for a Constitutional convention on March 28th.

Early in January, Halleck had sent Banks a long letter[771] which informed him that: "Generals Sherman and Steele (who had taken over the command of the Department of Arkansas in January) agree with me…that the Red River is the shortest and best line of defense for Louisiana and Arkansas and as a base of operations against Texas." He went on to opine against operations "mainly confined to the coast…" because Steele's line of defense would then have to be the Arkansas

---

768. Irwin, p. 266; O.R. Vol. 26/I, p. 659; O.R. 34/II, pp. 596, 596.
769. O.R. Ser. III, Vol. 3, pp. 232, 711, 712.
770. Nicolay & Hay Vol. 8, pp. 421-431, 434, 435.
771. O.R. Vol. 34/II, pp. 15, 16, 34, 41, 42.

River, "and most of Sherman's force may be required to keep open the Mississippi."

His proposed Red River Campaign was now waiting on the completion of Sherman's expedition to Meridian, Mississippi, which Sherman had suggested to Grant, to try to break the rail connections there to isolate Mississippi from the east, as well as to create the impression that Mobile was his objective. It worked, in that, Jefferson Davis peremptorily ordered Johnston to waste precious resources by sending reinforcements there.[772]

Sherman (who had replaced Grant in command of the Army of the Tennessee) then turned to planning for the campaign. He visited Banks at New Orleans from March 1st to the 3rd, and with the plan settled, he returned to Vicksburg to arrange for his loan of a full corps to Banks.

Sherman's March 4th summary memo of the meeting[773] stated that the troops from the Army of the Tennessee would rendezvous with Admiral Porter at the mouth of the Red River and that he would transport them up the river with the goal of meeting Banks at Alexandria on March 17th. Sherman's troops were "designed to operate by water," not having the encumbrances of wagons or horses. The next part of the memo was to prove all-governing, and indeed, fateful: "I calculate, and so report to General Grant, that this detachment of his forces in no event go beyond Shreveport, and that you spare them the moment you can, trying to get them back to the Mississippi in thirty days from the time that they actually enter Red River." Grant had planned on Sherman being available by April 15th to begin a move from the Tennessee River into the South, against Johnston and Atlanta.[774]

The same day, Sherman wrote a separate memo to Steele at Little Rock, which stated in part: "I understand that you have undertaken to act in concert, but the route and manner are not clear to me." Nor was it clear to Banks. Steele had been non-committal since revealing, in February,[775] that he also must protect the polls and the voters in state elections to be held on March 14th, "…and the President is very anxious it should be a success."

Upon reading this, Sherman responded that; "The civil election is nothing compared with the fruits of military success," and promised Banks: "I will …advise Steele to send you word at Alexandria by the 17th of his movement."

A letter to Banks from Steele, dated March 7th, added little. He was still committed to the elections, but "I am anxious to aid in your movement in every way possible." Another on the 10th conceded that he would send 7,000 troops via Washington, Arkansas, "and if necessary, from there to Shreveport."

Grant was confirmed a lieutenant-general on March 3rd, 1864.[776] Only two

---

772. Nicolay & Hay, Vol. 8, pp. 331–333.
773. O.R. Vol. 34/II, pp. 491, 494, 512.
774. Grant, Vol. 2, pp. 110, 120.
775. O.R. Vol. 34/II, pp. 448, 518, 546, 547, 576, 602, 603.
776. Nicolay & Hay, Vol. 8, pp. 334, 336, 346; Grant, Vol. 2, p. 114.

others had held the rank, George Washington and Winfield Scott, but Grant's victories at Chattanooga, which routed Bragg out of Tennessee for good and caused Jefferson Davis to remove him from command, had so impressed the Congress that Grant's mentor, Elihu Washburne introduced a bill to revive the rank. There could now be no doubt that Grant's desires would govern the strategy for the rest of the war, and in fact, the Red River Campaign could be said to be obsolete before it started, because Grant opposed it. It was a creature of Halleck, who was now out-ranked by Grant. If the campaign had not already been in effect before Grant learned about it, it might have been entirely cancelled. Grant's sole focus was on "Richmond, Lee and his army."[777] Unfortunately, the campaign was left to limp along, fatally fettered by the required return of Sherman's troops.

Finally apprised of the plan, a March 15th memo[778] written by Grant to Banks stated that under no circumstances could Banks hold Sherman's troops longer than the 30 days agreed to, and that they should be sent back "even if it should mean the abandonment of the main object of the expedition." The main object was then redefined:[779] "Should it prove successful, hold Shreveport and Red River with such force as you deem necessary and return the balance to the neighborhood of New Orleans."

Lest Banks had misunderstood the March 15th memo, another from Grant on the 31st clarified his orders: first, if the expedition against Shreveport was successful, the defense of the Red River should be turned over to General Steele and the navy, and, secondly: "That you abandon Texas entirely…" Grant was his own man now, conducting the war regardless of Lincoln's previous stance on Texas. Grant wanted to free Banks to go after Mobile.

The Red River Campaign

Banks' troops for the campaign[780] consisted of the 3rd and 4th Divisions of the 13th Army Corps, under the command of Ransom, which had been recalled from Texas, the 19th Army Corps under Franklin, consisting of Emory's 1st Division, Grover's 2nd Division, Lee's Cavalry Division, the Reserve Artillery under Closson, and a detachment of the Corps d'Afrique, under Dickey. In all, it totaled 23,000.

McPherson's[781] troops consisted of the 1st and 3rd Divisions of the 16th Army

---

777. Grant, Vol. 2, p. 146.
778. O.R. Vol. 34/I, p. 203; O.R. Vol. 34/II, p. 330; Irwin, p. 293.
779. Grant, with his practical knowledge, would have inherently known that forage stations would have to be set up along the route to keep the animals watered and fed if the expedition was to extend into the desert area of Texas. This would take weeks, and the caches would have to be guarded from Confederate destruction. Texas was an inconceivable goal for any 30-day expedition. Grant, Vol. 2, pp. 559–560.
780. O.R. Vol. 34/I, pp. 167–168; Irwin, pp. 285–286.
781. McPherson was named to the command of the Army of the Tennessee after Sherman was promoted to Grant's old command, The Military Division of the Mississippi, on March 18th.

Corps, under Mower, a Provisional Division of the 17th Army Corps, under T. Kilby Smith, and Ellet's Mississippi Marine Brigade, a total of 13,600, under the command of Brig. Gen. Andrew J. Smith.

It is noted that, at Grant's request, Ellet's Brigade had been transferred to the Army of the Tennessee in October of 1863, and Admiral Porter had been warned by Secretary Welles to "not interfere with its movements."[782] It was a remarkable demonstration of Grant's newly risen power and influence, given that Congress had acted to transfer the brigade to navy control the year before, in the face of Stanton's opposition.

Smith's troops were to be carried south on the Mississippi from Vicksburg to the Red, then to Alexandria. The transports would remain for the balance of the campaign.

Admiral Porter would meet Smith's troops at the mouth of the Red River, and by March 7th had assembled 19 ironclads there.[783]

A. J. Smith's transports pulled into view on March 11th. They started up the Red on the 12th. With obstructions in the river three miles below Taylor's Fort De Russy, which had been reoccupied, and the danger that the fort itself presented, a plan was made to attack it. A detachment from Porter was sent ahead to clear the obstructions, while the rest, including the army transports, turned down into the Atchafalaya, toward Simmesport, where the army landed and made a 30-mile march up behind the fort. Late on March 14th A. J. Smith's men reached it and attacked, just before nightfall, carrying it in 20 minutes, with only three killed and 35 wounded. Of about 350 occupants of the fort, 319 were captured, along with 10 guns.

Taylor's commentary on the attack doesn't agree, citing only 185 prisoners taken, but says: "Thus much for our Red River Gibraltar."

Porter's squadron then hurried on to Alexandria, the fastest boats arriving there on March 15th. It was occupied by the advance of Smith's command, General Mower, on the 16th, figure 4.[784]

Nicolay & Hay, Vol. 8, p. 346.
782. Society of Survivors, p. 332.
783. O.R. Vol. 34/I. pp. 486, 578; ORN Ser. 1, Vol. 26, pp. 24–27; O.R. Vol. 34/I. pp. 306–307; Taylor, pp. 136, 155.
784. Figure 4, U.S. Navy NH 59089 NR&L(M) 3684. A representation of the scene, but not specific to Mower; Irwin, p. 289.

FIGURE 4
UNION TRANSPORTS AT ALEXANDRIA

Vincent's Cavalry, familiar to Battery L from the Battle at Irish Bend, joined Taylor on the 19th. Taylor directed them to Henderson's Hill on Bayou Rapides, about 22 miles above Alexandria, and reinforced them with a battery of four guns. Learning of this, on March 21st, A. J. Smith sent Mower forward. It had been a cold and rainy day, and that night, Vincent's men sat by their camp fires. Their visible position and lack of security allowed them to be completely surprised, and 200 were captured, their four-gun battery included.[785] Chalk up another one for A. J. Smith.

Taylor, with a total force of 5,300 infantry, 300 artillery, and a force of cavalry now reduced to 300 from 500, was desperate for the return of Green from Texas, who had been detached during Banks' Rio Grande Expedition. Taylor would be able to do nothing but continue to fall back ahead of the advancing Yankee columns until Green appeared.[786]

All of this activity took place on the initiative of McPherson's men before the bulk of Banks' troops arrived. Though Lee and the cavalry had arrived on the 19th, Grover's troops did not begin to trickle in until the 26th. Battery L's "Record of Events" lists their arrival as on the 28th. What was the excuse? The controversial Franklin had been in charge; preparatory orders for the march were not issued until the 7th, and Grover was not told until the 10th. The march was leisurely, the 19th Army Corps rested the whole day of the 20th, and it rained heavily on the 25th making the march slower than ever. Banks broke away before the election, arriving on the 24th.

The late arrival of Banks and his troops caused criticism, but, as Irwin points

---

785. O.R. Vol. 34/I, pp. 177–178.
786. Taylor, pp. 156–157; O.R. Vol. 34/I, pp. 306–307, 511.

out, the water level of the river was so low that Porter's gunboats were not able to pass the rapids above Alexandria until April 3rd. Rising by only inches per day, it was not until then that 13 gunboats and 30 transports, were "in safety above the obstructions."

At Alexandria, a depletion of Banks' force took place. Gen. McPherson requested the recall of Ellet's Mississippi Marine Brigade for the defense of Vicksburg. Since Ellet's transports were not able to cross the rapids, and given that there had been an outbreak of smallpox among them, they were allowed to return. Thus, 3,000 were lost.[787] The slow rise of the river predicted that most of the deeper draft vessels would not be able to cross the rapids fully loaded, and some, never. Thus, supplies had to be unloaded and brought around the falls. To guard all of the stores and the line of transportation, Grover's division, another 3,000 troops, was detached.

Anticipating Porter's success, the remainder of the army departed on March 28th. They constantly skirmished with Taylor's forces, but no serious engagements were brought on.

At Natchitoches, figure 5, the march was halted for three days, waiting for signs that the river would rise sufficiently for Porter and the transports to push further on. The halt gave a chance for the troops to clean up after the dusty marching, and to visit the oldest settlement in the state. Many of the residences of the town were "almost regal in their splendor," the shift of the river to the east having left the town to survive as a resort for the wealthy. The mostly French inhabitants seemed to be agreeable to the Yankee presence, though "all this a few days later was explained as only their joy that we were being so easily led on to certain defeat."[788]

On April 6th, Porter finally determined that the river could be trusted only to float the lighter draft transports, including those of the 2,200 men of Kilby Smith's detachment, but not A. J. Smith's command. It was then that Banks' army resumed its march, now joined by Smith's 9,500 men of the 16th Army Corps. Porter, with half of his ironclads

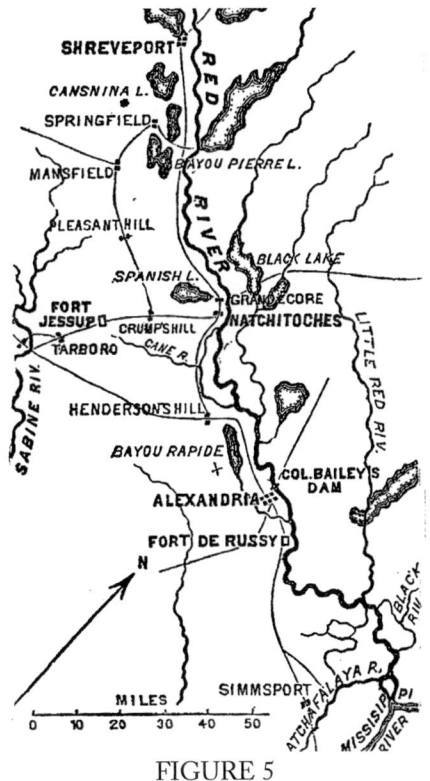

FIGURE 5

---

787. O.R. Vol. 34/I, pp. 197–198; Figure 5, Ewer p. 142; Irwin, p. 291; ORN Ser. I, Vol. 26, pp. 50–51.
788. Clark, p. 151.

unable to follow further upstream, left on the 7th. He expected to be at Springfield Landing, opposite Mansfield, in two days. There, the plan was for Kilby Smith to reconnoiter in the direction of Mansfield, and seize the bridge across Bayou Pierre.

It was from here at Natchitoches that the army diverted from the river, and took the stage road west toward Crump's Hill and Fort Jesup – the "Texas Road" as is written in the Battery L record. While Banks reports it as "the shortest and only practicable" route, he describes it as passing through "a barren, sandy country, with little water and less forage, the greater portion an unbroken pine forest."[789] Lee's cavalry had reconnoitered as far forward as thirty-six miles, to Pleasant Hill, on April 2nd, and reported (incorrectly) that Green's cavalry had arrived, (it had been ordered to report to Taylor on March 7th, but did not arrive at Pleasant Hill until April 6th, with about 3,100 men in total).[790] Prisoners taken at Crump's Hill were reported as from General Price's Arkansas command, which had been ordered to come to the support of Taylor on March 17th.

Thus, Banks knowingly stepped away from supplies on the river, the water in the river (no small item for the horses and mules), and the protection of Porter's guns; straight into a barren country which would soon have the greatest concentration of force E. Kirby Smith could supply to Taylor. Up to now, every advance by Banks had been met with a retreat by Taylor, and the conviction that pervaded the army was that Taylor was destined to retreat into Texas, or would, perhaps, make a final stand at Shreveport.[791] Little thought was given to the idea that once reinforced, Taylor would stand and fight.

Lee's Cavalry led the long column of the combined army, now under the command of General Franklin, confined as it was, to a narrow road thick with brush on its shoulders, in a dense pine forest. Lee's supply train, consisting of 250 six mule teams, followed immediately behind his 1st, 3rd, and 4th brigades. His 5th,[792] Gooding's, was detached as guards to the trains. Behind this, came Gen. Ransom's 3rd and 4th Divisions of the 13th Army Corps.[793] Next, the 19th Army Corps, which now consisted of only Emory's 1st Division, Battery L, and Dickey's brigade of the Corps d' Afrique–Grover's division having remained at Alexandria. A.J. Smith's detachment from the 16th Army Corps brought up the rear, and as a consequence, did not depart until April 7th. The overall column, including 900 wagon teams,

---

789. O.R. Vol. 34/I, pp. 198. This is his final report, written to Stanton, on April 6th, 1865. It is hence in hindsight, but the potential for danger in the conditions of the chosen route should have been obvious at the time.
790. O.R. Vol. 34/II, p. 1027; O.R. Vol. 34/I. pp. 445, 447–448, 479–480, 524, 552, 579.
791. Clark, p. 153; O.R. Vol. 34/III, p. 6.
792. Lee had 4 brigades: the 1st Lucas, 3rd Robinson, 4th Dudley, and 5th Gooding, O.R. Vol. 34/I p. 171.
793. O.R. Vol. 34/I, pp. 182, 236–238, 290.

being almost 20 miles long, was of such length that those preceding Smith had not passed out of Natchitoches until then.

The "unusual" placement of the wagon train of the cavalry, immediately behind it, was the subject of some discussion between generals Lee and Franklin, and to the detriment of the entire campaign, this issue was not resolved. Lee essentially wanted his supply train out of his way, considering the distances over which any cavalry often maneuvered. The proper action of the cavalry was, as soon as pressure on it was found to be too strong, not to attempt to hold the front on their own, but to fall back, and let the scattered command consolidate.[794] Though Lee wanted his train to be behind the infantry, he wanted it ahead of the infantry train. Franklin had disagreed, and if Lee was to move his train to the rear, it would have to be behind the infantry train.

One can describe this as so much petty nonsense, but what was worse, neither Banks nor Franklin seemed to appreciate the criticality of the issue. Banks had fallen into the trap that Lincoln had warned him about in late 1862. He had created too much "*impedimenta*"[795] viz., cavalry, which needed forage, a constant headache for the quartermaster to keep supplied in this barren country. Many of the 175 wagons and the 1,050 mules hitched to them which were lost in the subsequent battle at Sabine Crossroads were loaded with forage. In addition, their mere presence in the narrow road created the fatal danger that they would block an orderly withdrawal.

---

794. Haskin (Closson), p. 368.
795. O.R. Vol. 34/I, pp. 241, 454, 458. There are accounts of 250 or more wagons lost, but there were only 200 present. Lee, p. 452, reported 156 lost. The number quoted here is by the acting chief quartermaster. Eleven ambulances and 81 horses were also lost. Regarding "*impedimenta*," the 200 wagons contained 10 day's rations and three days forage, plus ammunition and camp equipage. Lee himself admits that they were loaded "mostly with forage."

Chapter 12

Wilson's Farm; Mansfield (Sabine Crossroads & Pleasant Grove); Pleasant Hill

Wilson's Farm

On April 7th, Lee's cavalry advanced, constantly skirmishing with Taylor's cavalry. They were halted when they encountered a strong force three miles beyond Pleasant Hill, at Wilson's Farm.[795]

A fight ensued, and Lee's leading 3rd Brigade, (Robinson) was initially pushed back until it was reinforced by the 1st Brigade (Lucas) which dismounted and charged the enemy. The enemy lines were broken and a pursuit followed, which ended at nightfall at St. Patrick's Bayou, near Carroll's Mill, some eight or ten miles beyond Pleasant Hill. Lee's casualties were 11 killed, 42 wounded and nine missing. From the 23 Confederate prisoners taken, it was determined that they were facing some 3,000 men and a portion of Green's cavalry. The question at this point was whether Taylor could, or would, bring up more.

During the early action, an aide to General Banks who had accompanied Lee was sent back to Pleasant Hill to inform General Franklin that, "General Lee was anxious to have a brigade of infantry sent out to his assistance." It was sent, but was inexplicably withdrawn before it arrived, when the firing was heard to cease. Lee was then instructed by Franklin to "proceed as far as possible, with your whole train, in order to give the infantry room to advance to-morrow."

The interrupted movement of the infantry; that is, its failure to come up to Lee's aid; was quickly settled by Banks upon his arrival at Franklin's headquarters at 11:00 that evening, and the 1st Brigade of the 4th Division of the 13th Army Corps, Emerson, was ordered to advance to Lee's support.[796] It seems that no one paid any attention to the information that Green's cavalry had arrived.

Franklin wanted to wait for all of Emory's division to arrive at Pleasant Hill. Emory's train had been slowed by a rainstorm which had made the road conditions bad. It did not arrive until late on the morning of the 8th. Then there was A. J. Smith, still further behind. Franklin wanted to try to close up the whole line,

---

795. O.R. Vol. 34/I, pp. 237, 257, 450, 454–455. Banks' report, p. 199, lists 14 killed, 39 wounded and nine missing; obviously, three wounded had subsequently died.
796. O.R. Vol. 34/I, pp. 264, 290.

at the expense of time. Franklin's original orders to Lee were to attack the enemy wherever he could find him, but not bring on a general engagement. If a general engagement was brought on, unless the troops were within supporting distance of those at the front, they would be beaten in detail. Yet Franklin ordered Lee to keep his train, which would give him no option for maneuver.

Undoubtedly, time was on Banks' mind, it was now only five days from the date he was told by Grant to wrap up the campaign. He wanted to push on and be in Shreveport by April 15th. Surely it was achievable, since they had thus far so easily pushed Taylor back.[797]

Mansfield (Sabine Crossroads)

Come the morning of April 8th, Col. Frank Emerson and his 1st Brigade, of the 4th Division, 13th Army Corps, accompanied by Col. W. J. Landram, the 4th Division commander, had reported to Lee at sunrise.[798] Lee quickly pushed forward the cavalry skirmish line, his 1st Brigade under Lucas, and formed the infantry in line of battle. They were successful in driving the enemy back, though with "severe" losses, and by noon, they had advanced some five or six miles, to a hilltop clearing called Honeycutt Hill, a part of the Moss plantations,[799] about three-quarters of a mile south of the junction of the road from the Sabine River and the road to Bayou Pierre, about three miles outside of Mansfield. The fighting had long since exhausted Emerson's Brigade, and they had run out of water, so a messenger had been sent to the rear to request a relieving force. The advance was halted at the "fine" position they now occupied. Though Emerson had done the complaining, the cavalry were also exhausted, not only from the day's efforts, but, quoting Henry Closson, by the "alternate pull and chase of the recent campaign, and had reached that pitch when the volunteer thinks he has done enough."

In the meantime, the rear of the column had moved ahead to St. Patrick's Bayou, about ten miles, and by 10:30 a.m., the bayou having been designated as the camping place for the day, had halted, an example of the slow pace ordered by Franklin. It was now about five-and-one-half miles behind Lee's position. The messenger having arrived, the relief force, the 2nd Brigade of the Fourth Division, 13th Army Corps, under Vance, was in motion by 11:00. Though delayed while passing the cavalry supply train, which was halted about a mile-and-a-half from the front, it arrived about noon and was deployed on the right, next to Emerson.

---

797. O.R. Vol. 34/III, p. 24; O.R. Vol. 34/I, p. 216. Banks received a reminder from Sherman on the 16th that the 30 days were up, by his calculation, on the 10th. It was written on April 3rd; Ewer, p. 144.
798. O.R. Vol. 34/I, pp. 265, 294, 456–457; Shorey, pp. 81–82; Honeycutt Hill, per display, Mansfield State Historic Site.
799. Benson, *The Battle...* p. 484; Haskin (Closson), p. 368.

There would be no relief for Emerson's 1st Brigade, which was ordered to remain in line.

It was apparent that there was a "heavy force" of the enemy in Lee's front. The 3rd Massachusetts Cavalry had run into Taylor's forces "massed in solid columns." Alerted to the situation, Banks and staff had arrived on the scene sometime after 1:00 p.m. and summoned Lee to report to him.[800] Lee explained the disposition of his forces vs. that of Taylor, and given the enemy's strength, in his opinion, the Federal army must either "fall back or be heavily re-enforced to advance." Light skirmishing continued on the flanks for some time, while Taylor's forces were observed to be massing on the right, along the road to Bayou St. Pierre, where the skirmishing became "sharp" after 2:30.

Banks took some time to react. The hilltop position from which he could observe the activity of the enemy, figure 1, shows that he had a good view of the dispositions Taylor was in the process of making. He could not have been deaf to the clink and thud, or the dust, of the thousands of Green's cavalry moving up, or the redeployment of Randal's brigade from the left of the Mansfield road to the right, to augment Mouton.[801] Did Banks conclude that it was already too late to make an orderly withdrawal?

After nearly an hour's consideration, Banks issued the fatal order. Lee was to dispose his troops "so as to advance to Mansfield." Banks makes no mention as to whether he had considered withdrawing, but Ransom put it this way: "It would have been impossible at that time to have retired from the position we occupied…"

He had noted the dangerous Union deployment, within the wings of a "V" where if one unit fell back, it would expose the flank of the defenders on the opposite side.[802] As later described by Taylor, Banks had put himself in a "vicious" position, and as "we invariably outnumbered the enemy at the fighting point" this was Taylor's cue to make the attack. Looking back, Kirby Smith had given Taylor a lukewarm approval to select "a position in rear where we can give him battle before he can march on and occupy Shreveport."[803]

In his official report, Lee could not have used the word "shocked" to describe his reaction to Banks' ineptitude, but says: "I immediately reported in person to General Banks, representing to him that the troops were already disposed for an advance, but that none could be made without bringing on an engagement. He then directed me to let them remain, and immediately sent an officer to the rear to hasten forward the infantry." This was to be Cameron with the 3rd Division of

---

800. Whitcomb, p. 68, gives 3:00 as the time Banks arrived, and goes on to give an account at odds with Lee's statements. Ransom's report, O.R. Vol. 34/I, p. 257, says 3:00, but Lee's report, p. 457, says "about 1:00 p.m."
801. Woods, p. 58.
802. O.R. Vol. 34/I, p. 292; Childers, pp. 515–518.
803. O.R. Vol. 34/I, pp. 199, 513, 526–528; Taylor, pp. 162–164; Ewer, p. 145; Marshall, p. 128.

the 13th Army Corps, and Emory's 1st Division, 19th Army Corps, which included Battery L.[804] This was the first intimation to Franklin, and those in the rear, of an impending battle.

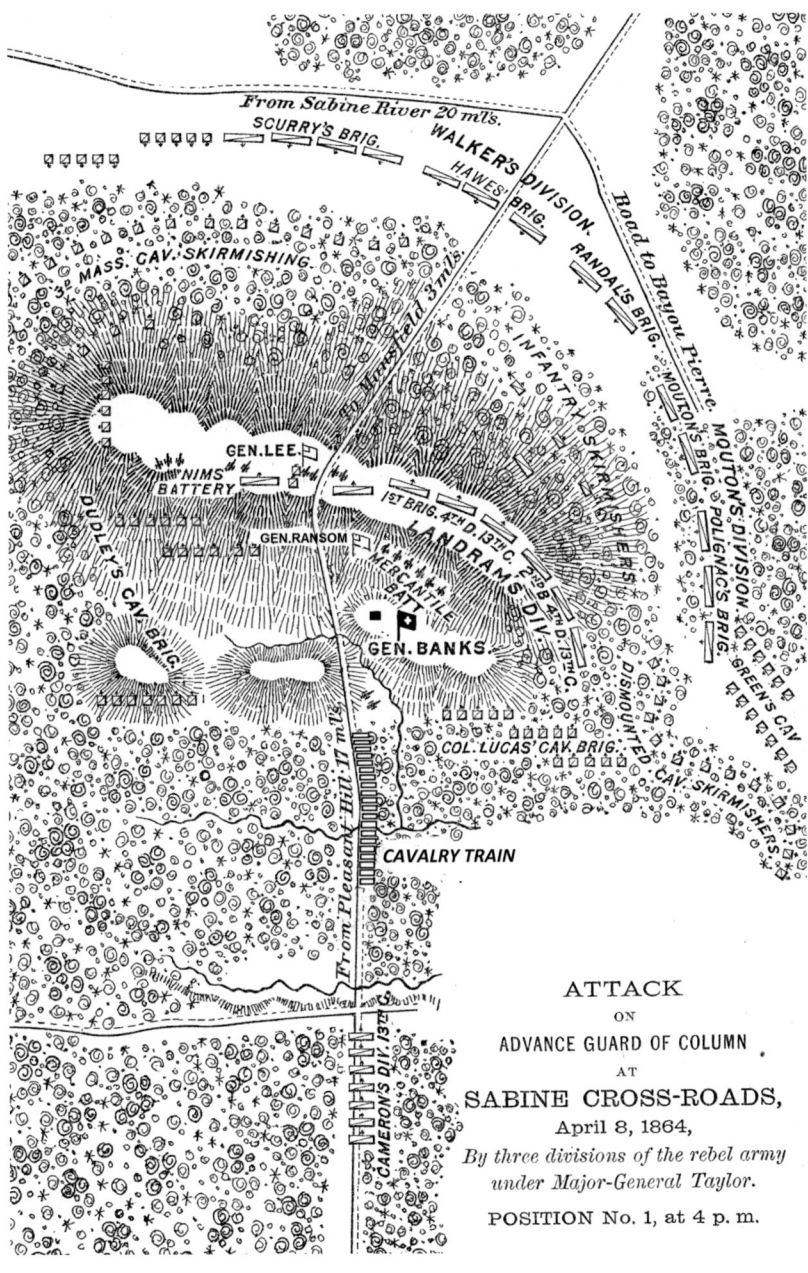

FIGURE 1

---

804. O.R. Vol. 34/I, pp. 273, 389–390; Irwin, p. 305.

The dispositions of the opposing forces are shown roughly in figure 1.[805] As to Banks' forces, the right; posted on a narrow, level, but broken plateau, one-fourth of a mile wide by a mile long, along which ran a fence which formed the boundary between two fields of the plantation; was Landram's 4th Division – Emerson's tired 1st Brigade, which had been joined by Vance's 2nd brigade. The "dismounted cavalry skirmishers," shown on the right, are the 16th Indiana. The 4th Division artillery, consisting of the Chicago Mercantile Battery and the 1st Indiana Battery, were ordered forward, and though initially sited in a plowed field to the left of the road, were quickly moved behind the 4th Division. The line on the right was extended by Lucas' cavalry acting as skirmishers.

Nims' six-gun battery, supplemented by two howitzers detached from the 6th Missouri Cavalry, were assigned to Lee. They were strung across the road at the top of the hill, supported by the 23rd Wisconsin from Emerson's Brigade, which constituted the extreme left of the defensive line. Dudley's cavalry, consisting of the 2nd New Hampshire, 3rd Massachusetts, 31st Massachusetts, and 2nd Illinois, all acting as skirmishers, covered the flank in the forest.[806]

At 4:00, Taylor ordered Mouton to open the attack on the Federal right.[807] At this time, Cameron's 3rd Division had not yet arrived. It had been five-plus miles back, at Bayou St. Patrick, and had received Banks' order forward at approximately 3:00. Though Cameron would bring his whole division, it was under-strength, totaling only 46 officers and 1,247 men, and they would prove insufficient to stem Taylor's tide.

Emory, with the 1st Division of the 19th Army Corps, was seven miles back, and would not receive the order forward until 3:40.

"The initial ardor" of Mouton's Louisianans, (though they may have been lukewarm to the Confederacy, they were "inflamed" by the recent Yankee outrages on their homes, and were concerned by camp rumors that if Banks succeeded, the Confederacy would abandon Louisiana) resulted in severe losses – 64 percent of the Consolidated Crescent Regiment. Mouton, carrying a regimental flag, was one of the first of the many Louisiana officers killed, having exposed themselves as examples to their men.[808]

The fierce resistance of Landram's Division, as directed by Gen. T. E. G. Ransom, for a time stopped Mouton's and Polignac's brigades, but, running short of

---

805. Figure 1, O.R. Vol. 34/I, pp. 227, 266, 279–280, 293, 298–302; Taylor, p. 163; Marshall, p. 132; Whitcomb, p. 69. Woods, pp. 55, 58, 66; Scott, R. B., p. 71.
806. Irwin, p. 302.
807. O.R. Vol. 34/I, p. 564.
808. Marshall, p. 132; O.R. Vol. 34/I, pp. 264, 302–303, 451, 462, 568. Mouton's brigade included the 28th Louisiana, Colonel Gray, and the Consolidated Crescent Regiment, including Clack's battalion–first seen at Irish Bend; Whitcomb, pp. 68–69; Scott, R. B., p. 69; Irwin, p. 304; Bentley, 1883, pp. 276–277; Benson, p. 486.

ammunition, the 83rd Ohio, the last of the line on the right, was in danger of being outflanked. Ransom's assistant adjutant-general, sent to order them to retire, was killed, and receiving no orders, they stood their ground. Many, including almost all of the 19th Kentucky, and 143 of the 77th Illinois, were taken prisoner. Adding in the 77th's 28 killed, wounded, and missing; it was left with only 130 on the roll after the battle, a loss of 57 percent.

On the other end of the line, the 23rd Wisconsin and Nims' battery were left nearly alone when the 67th Indiana (decimated at the Battle of Grand Coteau, and subsequently strengthened with non-veterans from the 60th Indiana, but still totaling only about 300 men), on their right, gave way. After having repulsed three successive charges of the enemy, with Nims' battery firing shell and canister, and opening wide gaps in the advancing enemy lines, within 20 minutes the left of Landram's line was ordered to retire. Three of Nims' guns were successfully removed from the hilltop, but three were left, their horses disabled. Lt. Snow was mortally wounded and captured; Lt. Slack was wounded slightly in the neck. The three guns that were removed were again prepared for action, but with the panic that overcame the retreat, and the enemy closely pressing, they were placed in the road, which soon became blocked.

Earlier, the wagon teams had been ordered not only to keep closed up to the front, but to remain facing forward. Subsequently, they had received conflicting orders to park on the sides of the road, or to face to the rear.[809] Now, overcome by the retreating mix of cavalry and a demoralized infantry, an order was received for the wagons to turn and retreat. The process of turning six-mule teams, in a rutted road little wider than the wagons themselves, was necessarily slow under normal circumstances, but now was utterly impossible. Nothing could pass the blockage, at the "slough" or creek, near Carroll's Mill, at which a bridge had been constructed, and over which much of the cavalry train had passed before it was halted. Thus, all of the train ahead of the creek was abandoned to the enemy, including the remaining three of Nims' guns, two guns of the 6th Missouri Battery, all six of those of the Chicago Mercantile Battery, the 1st Indiana Battery, and two of those of Rawles' Battery G of the 5th Artillery.

Quoting from *The Story of the Maine Fifteenth*:[810]

> A wild and utterly indescribable panic here ensued … Every man seemed to strike out for himself, eager to reach a safe place in the rear as rapidly as possible…the overturned wagons, the wagons faced to the rear, to the front, and but partially turned about, the fleeing cavalrymen, the frantic and riderless

---

809. O.R. Vol. 34/I, pp. 238, 267, 279–280, 293, 452, 458–459, 465; Irwin, p. 306.
810. Shorey, pp. 83–84; Shorey grants Banks the following: "It is but just to say…that at every point of this unfortunate and disastrous affair, Gen. Banks and all the subordinate commanders most gallantly acquitted themselves." Figure 2, O.R. Vol. 34/I, p. 228.

horses, the dead and wounded encumbering the way, the hatless officers, with drawn sabers, endeavoring to check the stampede, and the advancing and jubilant rebel hordes, pouring their hot-shot into the…crowd and rushing upon the "jam" with their glistening bayonets at the charge…

The "jam" is identified in figure 2 as *TRAIN*.

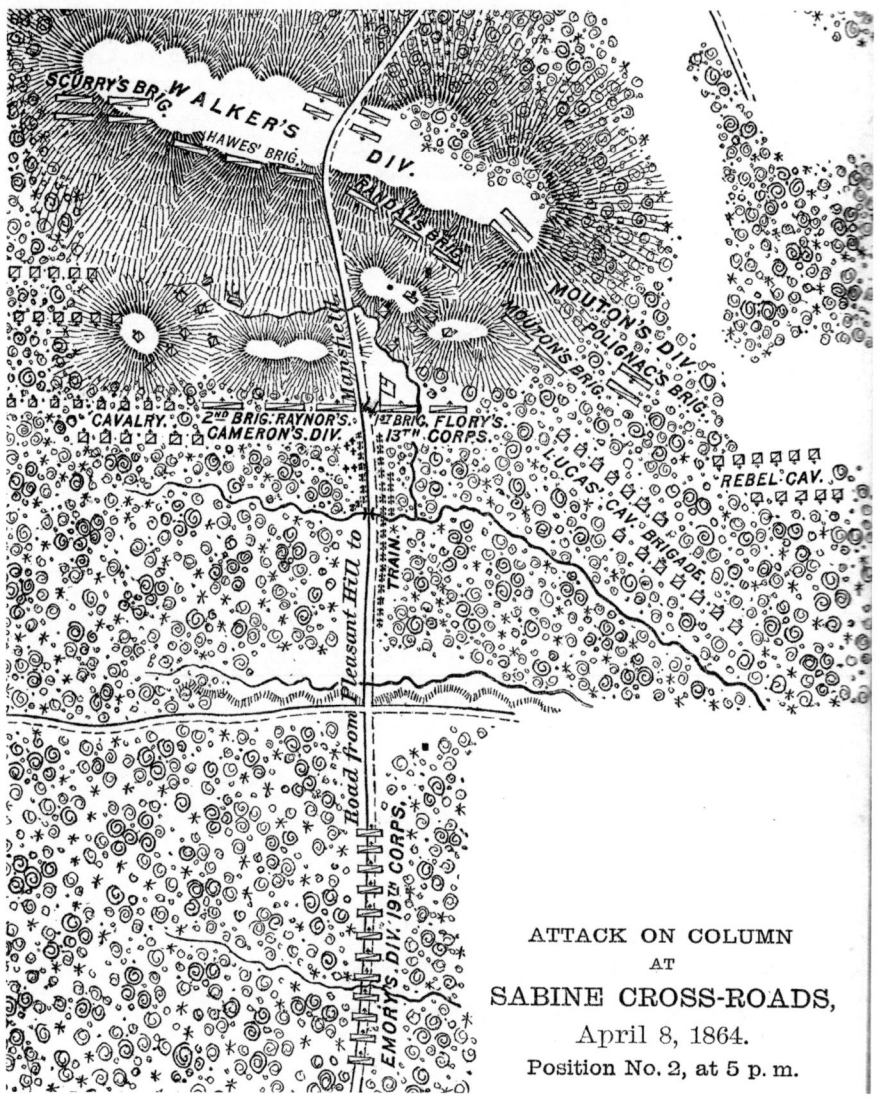

FIGURE 2

By now, generals Franklin and Cameron, with Cameron's 3rd Division, had arrived.[811] Cameron had advanced at the "double quick" for about a half-mile through the thick underbrush to the edge of the woods, the position shown in figure 2, where they saw the enemy advancing in "heavy force." Finding the "road

---

811. O.R. Vol. 34/I, pp. 228, 268, 273, 282–283; Irwin, p. 305; Cullum, Vol. 2, no. 1167.

so full of teams and stragglers on foot and on horseback as to make it impossible to move any farther" Cameron deployed in line of battle: Flory's 1st Brigade, composed of the 46th Indiana, and five companies of the 29th Wisconsin, on the right of the road, and Raynor's 2nd Brigade, composed of the 24th Iowa, 28th Iowa, and the 56th Ohio, on the left.[812] Here they met the "broken troops" of Landram's Division, with the news that both of its brigade commanders, Emerson and Vance, were wounded and taken prisoner, and a wounded General Ransom had been carried from the field.[813] Riding ahead to where Nims' battery still stood, Franklin described the situation as "discouraging." Hardly a minute later, a volley from Walker's advancing lines wounded Franklin and two of his staff officers, and Franklin's horse was killed. Remarkably, Franklin does not later mention the fact that he was wounded, though he wrote two separate reports of the battle. Franklin's leg wound would fester as time passed, much as Admiral Foote's had, after Fort Donelson, and ultimately require him to be replaced by Emory.

Ransom's line had repelled the advance of the enemy twice, but could not prevent an enemy flanking movement around its right rear, and with the 2nd Brigade having been pushed back, exposing the left of the 1st Brigade, and many of its men out of ammunition, it fell back, pursued by the rebel cavalry for a mile-and-a-half, to finally find refuge behind the line formed by Emory's Division,[814] thus saving the remnant which had not been killed or captured. Those who were prisoners would face captivity at Camp Ford, 140 miles away, near Tyler, Texas, and would have to endure a march of 16 days to get there, with little food, and no shelter.

As was earlier mentioned, Emory's 1st Division was seven miles back from Sabine Crossroads, and he did not receive his order forward until 3:40. The order contained the requirement to have his 6,000 troops carry two day's rations. The peculiar reference to the troops carrying rations gave no proper hint of the emergency, and, as Irwin says, if it was not for Emory's personal trait of sensing danger, and "had from the first hour of the campaign been apprehensive of some sudden attack that should find the army unprepared..." the issuance of the rations might have taken an hour, but it was rushed, and only took a "few minutes." Of course, each minute was critical beyond measure by now. Emory's numbers, going forward into line-of-battle at a point subsequently referred to as Peach Orchard or Pleasant Grove, which was about three miles behind the initial battle line, would finally stop Taylor.[815]

---

812. Bringhurst and Swigert, pp. 88–89; Blake, pp. 34–35.
813. Bentley, pp. 257–258; Shorey, pp. 84–86, Irwin, p. 305; O.R. Vol. 34/I, pp. 256–258, 261, 531; O.R. Vol 43/III, pp. 391, 472–474. Ransom later died of complications from his wound.
814. O.R. Vol. 34/I, p. 257; Bentley, p. 287; Bringhurst and Swigert, pp. 118–119.
815. Irwin, p, 307; O.R. Vol. 34/I, pp. 200, 421, 607; Taylor, pp. 162, 163, Woods, p. 66.

Mansfield (Pleasant Grove)

Leaving Battery L and the 153rd New York to remain with its wagon train, Emory's division, led by the 161st New York, was soon moving forward at the "double quick," and as they moved, the sound of battle became louder, "while demoralized camp followers, black by nature, and almost white from fear, skulking infantry soldiers by twos, and cavalry by squads passed…"[816] Then it was an ambulance carrying the wounded General Ransom. "It was plain that the most crushing disaster had occurred…" Emory, as angry and "savage as an infuriated bear" ordered the pace of the march increased to the utmost speed; the men of the 161st were ordered to fix bayonets without halting, and to use violence to open a path forward through the jam of fleeing humanity.

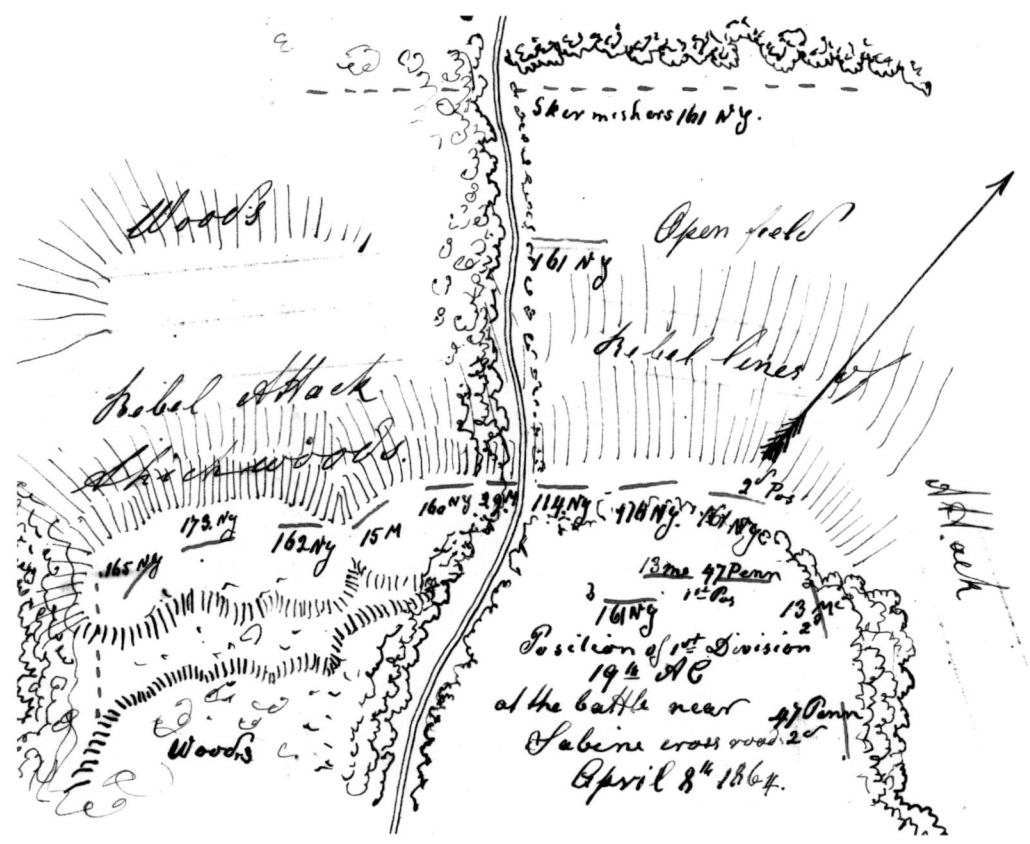

FIGURE 3

They soon came to a fenced farm and orchard, on high ground, which sloped away to a crossroad running parallel to a small stream, figure 3, an original sketch, from the NOAA Historical Maps and Charts Collection, no. 717-04-1864(2).

---

816. Clark, p. 155; Woods, p. 66; O.R. Vol. 34/I, p. 200; Irwin, p. 308; Shorey, p. 87; Pellet, p. 197; Taylor, p. 164.

Here, Emory halted to deploy his troops. The 161st New York was sent forward as skirmishers "to reconnoiter and check the enemy, until the stampeded cavalry and broken infantry could get to the rear…" and a line of battle could be formed. They had moved nearly across the clearing when Taylor's forces appeared at the far edge of the woods, which forced the 161st into an orderly retreat, loading, halting and firing, before finally reaching the 1st Division line, now formed, where it went into reserve. Emory's line was behind a fence. Most of Benedict's 3rd Brigade was deployed to the left, most of McMillan's 2nd was at the center, and to the right was a mix of Dwight's 1st and McMillan's.

From the history of the 96th Ohio, the last of Cameron's troops to pass within Emory's line:[817]

> Some of the defeated troops from the front, thrilled with admiration for their comrades who had with such gallantry and soldierly bearing come to their relief, rally and form a line in their immediate rear. The woods before them resound with shouts and musket volleys. The darkness [it was before sunset] is every where flecked with spectral rifle flashes…The full brass band of the Nineteenth Corps stands close by the road and pours into the ears of the advancing rebels the exhilarating notes of "Hail Columbia"…Time can be counted only by seconds, for the skirmishers, with heavy loss are hurled quickly back, followed instantly by the enemy in three compact columns, one in the road and one on either side.

The first rebel assault was made on the center, the position of the 29th Maine. From the history of the 114th New York:[818]

> Our whole Brigade line [which had been ordered to lie down, and fire from this posture] waited patiently until the advancing line was clearly visible and then delivered such a volley as is seldom heard. It seemed like the discharge of one piece. Its effect was instantaneous. The line was checked, and put in retreat…The reception was one they had not expected, and the sudden check disconcerted them.

An Elmira, New York, newspaper report for Wednesday morning, April 27th, 1864, under the title: "The Gallantry of the 161st " had it that the troops were ordered by General Dwight to not fire "…till they could see the enemy's eyes." From the 15th Maine: "The rebels, advancing up the slope of the hill, furnished an admirable target from our position, while their shots, for the most part, glanced over our heads and into the adjacent trees."

A second Confederate advance on the Federal right was made at about 6:15

---

817. Woods, p. 66.
818. Pellet, pp. 197, 198. Shorey, p. 89; Gould, p. 416; O.R. Vol. 34/I, pp. 416, 421–422; Letters, *http://dmna.gov/historic/reghist/civil/ infantry/161stInf/161stInfCWN.htm*

p.m. by Polignac's division, Polignac having taken over the command after Mouton was killed. The 13th Maine and the 47th Pennsylvania had to be moved from reserve to the right of the 116th New York, and finally, the 161st was ordered by Dwight returned into the line to stabilize it, on the left of the 13th Maine. A final determined Confederate attack on the left was then made, and was repulsed.[819]

No artillery was used by either side, so only the crack of the rifle and the boom of the musket had been heard on this evening. Taylor's artillery could not be brought up for the reason of the wagon jam through which he had advanced, and Emory had purposely left his artillery and baggage behind, though Henry Closson, the chief of artillery, had later ordered Battery L forward, but they did not arrive in time for the battle.

Emory closed this phase of his report on the battle with: "Nothing but the high discipline and morale of my division enabled me to form the line of battle under such discouraging circumstances." He reported 13 officers and 343 men as killed, wounded, or missing, no small number for this relatively small delaying action, which lasted less than 20 minutes. The 161st New York suffered seven killed and 36 wounded.

No single report contains the listing of casualties from the combined actions at Mansfield, that is, the battle at the Crossroads, and at Pleasant Grove, plus the casualties that continued to mount even after the battle, for the reason that Banks decided to precipitately retire from Pleasant Grove. After the close of the battle, Banks reports: "We were compelled, anticipating an attack next morning… either to await the advance of General Smith's corps, or to fall back to meet him. The want of water, the weakness of the position we held, and the uncertainty of General Smith being able to reach the position we occupied at day-break, led to the adoption of the second course."

It may have been well and good that Banks had made this decision, after he had held a "council of war," because Taylor's expected reinforcements, Gen. Thomas Churchill's division, of Price's command, had arrived. But, it was not well for the wounded left on the battlefield, unable to be moved due to a lack of ambulances to carry them. From the history of the 116th New York Volunteers:[820] "…it was decided to leave *all* our wounded on the field, and surgeons, with such medical supplies as were at hand, were left with them."

Thus, at about 10:00 p.m. orders were given to fall back upon Pleasant Hill, fifteen miles to the rear, and before midnight, with cartridge boxes replenished and under orders to maintain quiet, to the point where orders were whispered, the troops moved out toward Pleasant Hill. As a consequence of the stealth, many in the picket line "did not retire in season to rejoin their commands. Many of

---

819. Taylor, p. 163; O.R. Vol. 34/I, p. 392; Irwin, p. 310.
820. Clark, p. 159. Italics added by author.

these fell into the hands of the enemy, either on the picket line or while on the march… This in a measure accounts for the large number reported as missing."[821] For example, 27 men of Company C of the 165th New York were left on the picket line, and were taken prisoner.

From this confused picture, the accuracy of the reported Federal losses for the Battle of Mansfield are thrown into doubt. A summary report lists only the combined losses at both Mansfield and Pleasant Hill.[822] However, separate reports were made for Emory's 1st Division, Cameron's detachment of the 13th Army corps, and Lee. Lee did not differentiate between officers or men, so the table following, which summarizes the casualties, leaves his numbers separate. The total of all killed and wounded shown below is 694, with 1,541 missing.

Emory & 13th Corps	KILLED		WOUNDED		MISSING	
	OFFICERS	MEN	OFFICERS	MEN	OFFICERS	MEN
	6	68	27	304	72	1,325
Lee	39		250		144	

According to Irwin, Taylor claimed 1,000 as his total casualties, though this number has not been corroborated. More questions are raised by Taylor's claim of capturing 2,500 prisoners.[823] If only 1,541 Union men were missing, this number is hard to justify. Some further evidence of a lower number is found in the history of the 46th Indiana Regiment, where the number that were marched to Camp Ford, is stated as 1,250 officers and men, including its 70-year-old Chaplin, the Rev. Hamilton Robb.

In closing, it is fair to report that the only positive Yankee statement that can be found regarding the Battle of Mansfield was made by Lee:[824] "The ammunition train was saved." The more common statement would be on the order of: "It is impossible to measure the indignation of this army against Gen. Banks." Regardless of Banks' personal courage, and the brilliant repulse at Pleasant Grove, leaving the wounded behind was too much.

Pleasant Hill

Taylor makes no mention of whether he had planned to outflank the Federal army by heading them off at Pleasant Hill, but, at 3:30 a.m., when he discovered that

---

821. Shorey, p. 91; *Duryee Zouaves*, 1905, p. 26.
822. O.R. Vol. 34/I, pp. 258–261, 263–264, 452.
823. Taylor, p. 164; Bringhurst and Swigart, pp. 91, 118–119; The 46th Indiana had 86 missing; A. J. H. Duganne, himself a prisoner at Camp Ford, refers to, p. 363, "a first installment of 1186"; Irwin, p. 311.
824. O.R. Vol. 34/I, pp., 183, 392, 458; Bentley, p. 277.

Banks had retired from Pleasant Grove, he ordered Green's cavalry toward Pleasant Hill. They were to be followed by his new reinforcements, Churchill's two divisions of Parsons and Tappan, and then Walker and finally Polignac's weakened brigade.

Emory's troops began to pull in to Pleasant Hill at 7:00 a.m., and occupied the same campground they had left on April 7th.[825] The strain on the men and animals can only be measured by the fact that after marching ten miles the previous morning, they were ordered at the double quick seven miles forward to the rescue of the rest of the army, fought the Battle of Pleasant Grove, and now had marched all night, for fifteen miles, without stopping. Some were marching in their sleep, or asleep on their horses, and the road was strewn with stragglers and abandoned wagons and equipment.

Madison's 3rd Texas Cavalry, the leading element of Bee's cavalry under Green, after riding all night, caught up with the rear guard of Emory's column–which included Battery L—about three miles from Pleasant Hill, at about 9:00 a.m. The sound of whooping, yelling, rattling, and banging gave evidence that the rear was being harassed, and Gooding's cavalry, the only unit of Lee's not engaged at Mansfield, was then deployed to "prevent surprise." An engagement seemed to be a certainty.

Soon after Emory, Gen. A. J. Smith's 16th Corps arrived. A portion of Smith's command, Shaw's brigade, was then assigned to Emory, which relieved McMillan's brigade. McMillan had been positioned at the edge of a dense thicket, facing a field dotted with pine trees, through which passed the Mansfield Road. This was west of Dwight's position, near the old race track, on the northwest outskirts of town.[826] As Shaw's troops marched into position, Emory's troops stood quietly by, too tired to respond to a revival of some eastern–western insults, manifested by loud taunts, the likes of: "You fellers won't see us coming back…" in reference to their now having been posted at the most advanced position facing the enemy, as well as to the disastrous battle and retreat of the night before.

Bee, having picked off many stragglers, and while skirmishing with Gooding's cavalry, by 9:00 a.m. had driven to within about a mile of Pleasant Hill. Here, he first gained sight of what he described as "a line of battle across the fields in front of the village…The strong front by what was supposed to be a routed and retreating army rendered it prudent to reconnoiter the extent of this line before ordering a charge…" None was made, and soon after, when Green arrived and took command, another "close reconnaissance" resulted in Green doing something extraordinary for Green – he decided to make no move. As it happened, no action was taken until 4:00 p.m. when Taylor finally ordered an attack.

In the Federal ranks, it was learned with some surprise that the 13th Corps, with a portion of Dickey's brigade of the Corps d'Afrique, and most of the cavalry,

---

825. Taylor, p. 165; O.R. Vol. 34/I, pp. 565, 607, 617; Clark, p. 158; Shorey, pp. 91–92; Pellet, p. 207; *Duryee Zouaves*, 1905, p. 26; Haskin p. 195.
826. O.R. Vol. 34/I, p. 354.

(minus a detachment of 1,000 men ordered to report to General Franklin)[827] had been started to Grand Ecore. The decision to send this portion of the army on to Grand Ecore indicated that a decision to evacuate Pleasant Hill had been made even before the battle was fought. Heading back to Grand Ecore would leave Porter's fleet without support. A courier was sent to inform Porter of the decision. In fact, three separate messages were sent to Porter on April 9th; one from Banks, and two from Stone, Banks' chief of staff. [828]

Porter was now at Loggy Bayou near Springfield Landing ten miles south of Shreveport, as is shown in figure 4,[829] and 110 miles by the "narrow and snaggy" river north of Grand Ecore, with his flagship, the *Cricket*, and five others, the *Fort Hindman*, *Lexington*, *Osage*, *Neosho*, and *Chillicothe*, plus the 20 transports carrying Kilby Smith's command and supplies for Banks' army. He was in peril of losing this portion of his squadron, and he felt that Banks, this "political general" might abandon him: "I am not sure that Banks will not sacrifice my vessels now to expediency…"

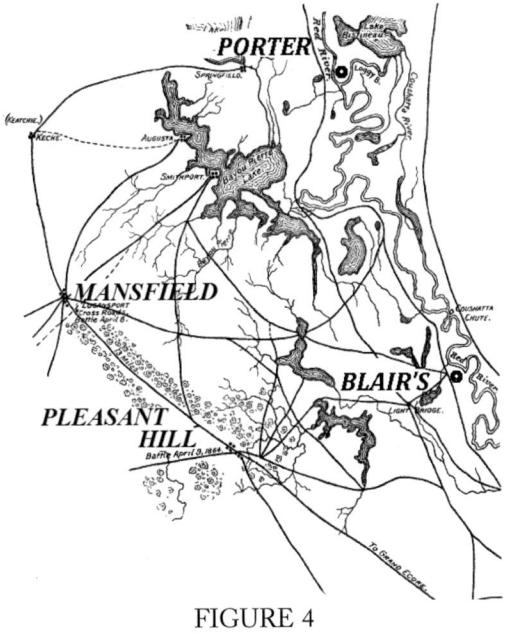

FIGURE 4

Since Kilby Smith's 2,237 troops were still with Porter, and Grover's division was still at Alexandria, plus the fact that the 13th Corps and others had been hurried toward Grand Ecore, only those of A. J. Smith/Mower, Lucas' 1st Cavalry, and Emory remained at Pleasant Hill, some 15,900.[830] Regardless, combined with the 13,200 of Taylor, the battle would rate as the largest of any fought west of the Mississippi.

Pleasant Hill, initially founded as a farming community in 1844, was situated on slightly elevated ground, or a plateau, about a mile wide[831] and two miles long, through which the road from Mansfield to Grand Ecore, about 35 miles distant, runs. It boasted about 200 inhabitants, many wealthy, such as the Childers,

---

827. O.R. Vol. 34/I, pp. 452, 459. The cavalry detachments were from the 1st Brigade, Lucas, and the 5th Brigade, Gooding; ORN Ser. 1, Vol. 26, p. 60.
828. O.R. Vol. 34/III, pp. 98, 152–153; O.R. Vol. 34/I, p. 383; ORN Ser. 1, Vol. 26, pp. 51, 60.
829. From ORN Ser. 1, Vol. 26, pp. 44c, 45.
830. O.R. Vol. 34/I, pp. 260, 292, 383; Irwin, pp. 292, 311. The 15,900 is derived by using 6,400 for Emory (his April return, O.R. Vol. 34, Pt. I, p. 168, which accounts for casualties) 8,500 for A. J. Smith, (less Kilby Smith's 2,200, using his March return, since they had not yet been in battle) and adding the 1,000 cavalry detachment. This agrees with Banks' estimate, p. 203. Taylor, p. 162.
831. Taylor, p. 217; John Scott, p. 202.

Jordans, Chapmans, Davises, Harrels, and Hamptons. The Childers mansion, figure 5, built in 1859 at a cost of $10,000, was a landmark on the southeast end of town. The village was the home of Pierce and Paine College. It had a hotel, and it was regarded as the "center of refinement and education for miles around."[832] Such was the luck of this elegant community to be the point where the Federal army would choose to make a stand. The Confederate and Union organizations for the battle are listed in tables I and II.[833]

CONFEDERATE
Maj. Gen. John G. Walker's Texas Division

1st Brigade Thomas Waul	12th 18th 22nd Infantry, 13th Cavalry
2nd Brigade Horace Randal	11th 14th Infantry, 28th & 6th Cavalry
3rd Brigade William R. Scurry	16th 17th 19th Infantry, & 16th Cavalry

Brig. Gen. Thomas J. Churchill–2 Divisions:
Brig. Gen. Mosby Parsons (Churchill's) Missouri

1st Brigade John H. Clark	8th 9th Infantry,
2nd Brigade S. P. Burns	10th 11th 12th 16th Infantry, 9th Bn. Sharpshooters
Brig. Gen. James C. Tappan	1st Div. Arkansas
Lucien C. Gause	26th 32nd 36th Infantry
H. L. Grinsted	19th 24th 27th 33rd 38th Infantry

Maj. Gen. Thomas Green, Cavalry Corps Commander
Brig. Gen. Hamilton Bee Cavalry Division Commander[834]
26th Texas Debray, 1st Texas Buchel, 37th Texas, Terrell
Brig. Gen. James Major Division Commander

Lane's Brigade 1st & 2nd Texas Partisan Rangers, 2nd & 3rd Arizona (Baylor Commanding)
Bagby's Brigade 4th 5th 7th 13th Texas

Brig. Gen. Camille J. Polignac's Division

Col. Henry Gray	28th Louisiana, Consolidated Crescent Regiment
Lt. Col. R. C. Stone	15th 17th 22nd 31st 34th Texas Cavalry (dismounted)
Artillery Maj. J. L. Brent	6th Louisiana, 4th 7th 9th 12th Texas 6th Arkansas Light Batteries

TABLE I

---

832. Childers, pp. 513–516; Blessington, p. 194.
833. From map, courtesy of the Mansfield State Historic Site, Scott Dearman; O.R. Vol. 34/I, pp. 169–172.
834. O.R. Vol. 34/I, pp. 601, 604, 606.

UNION
Maj. Gen. William H. Emory's 1st Division 19th Army corps

1st Brigade Brig. Gen. William Dwight	114th 116th 153rd 161st New York, 29th Maine
2nd Brigade Brig. Gen. James W. McMillan	13th, 15th Maine, 160th New York, 47th Pennsylvania
3rd Brigade Col. Lewis Benedict	30th Maine, 162nd 165th 173rd New York
Artillery Capt. Henry Closson	25th New York Light, 1st United States Battery L, 1st Vermont Light Cavalry (Detachment) 2nd Brigade

Col. Oliver P. Gooding

1st Brigade Col. Thomas I. Lucas	16th Indiana, 6th Missouri, 14th New York
5th Brigade Col. Oliver P. Gooding	2nd New York, 3rd Rhode Island

Brig. Gen. Andrew J. Smith (Detachment) Army of the Tennessee
Brig. Gen. Joseph H. Mower's 1st & 3rd Divisions, 16th Army Corps

1st Div. 2nd Brigade Col. Lucius F. Hubbard	5th Minnesota, 8th Wisconsin, 47th Illinois
1st Div. 3rd Brigade Col. Sylvester Hill	35th Iowa, 33rd Missouri
3rd Div. 1st Brigade Col. William F. Lynch	58th 119th Illinois, 89th Indiana
3rd Div. 2nd Brigade Col. William T. Shaw	14th 27th 32nd Iowa, 24th Missouri
3rd Div. 3rd Brigade Col. Risdon M. Moore	49th 117th Illinois, 178th New York

Artillery Capt. James F. Cockefair
Indiana Light 3rd & 9th Batteries

TABLE II

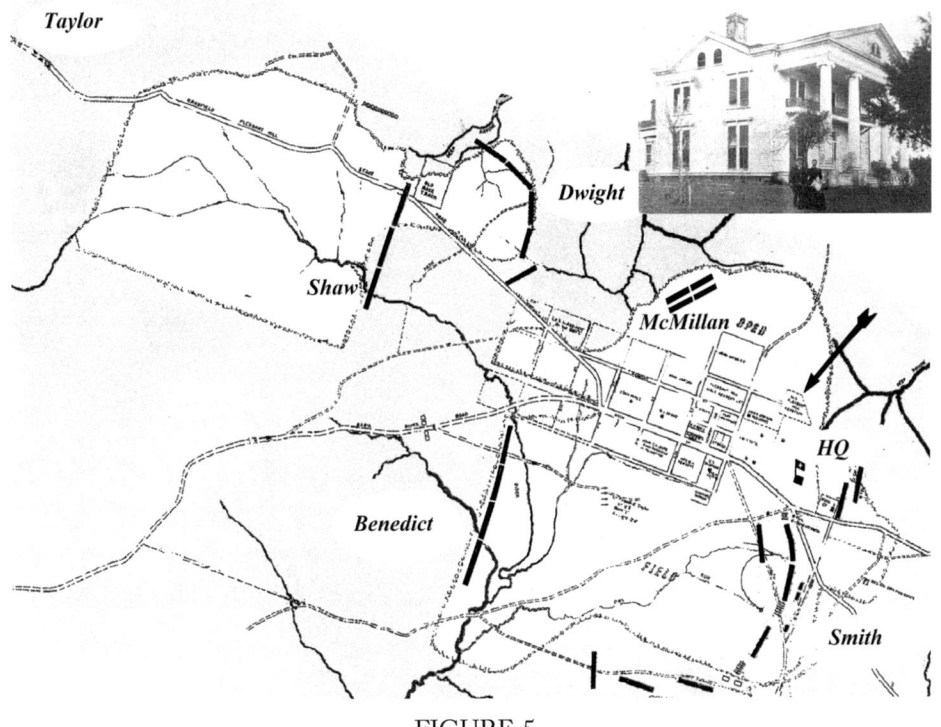

FIGURE 5

Figure 5[835] indicates what Taylor saw when he arrived; shown are the mid-morning positions of Shaw, Dwight, and Benedict, with McMillan and Smith having deployed in the rear.

---

835. Reproduced from a map obtained by kind permission from the Mansfield State Historic Site, altered by the author.

From his position on the Mansfield Road, Taylor described the Federal army as

> . . . extended across the open plateau, from College Hill on their left, to the right of the road to Mansfield. Winding along in front of this position was a gully cut by the winter rains but now dry, and bordered by a thick growth of young pines, with fallen timber interspersed. This was held by the enemy's advanced infantry, with his main line and guns on the plateau.

Taylor never flinched, despite recognizing A. J. Smith's powerful force as the one "not engaged on the previous day."

Shortly after "midday" the remainder of Taylor's troops arrived, Churchill in the lead. "At a glance," as Taylor says,[836] "it was clear that the men were exhausted and suffering from the heat," and his attack was postponed for two hours.

Like so many other Civil War sites, Port Hudson being an earlier example, the ravages of time have dramatically altered Pleasant Hill. Then, Taylor's view was clear, today the entire area is overgrown. Figure 5 shows the location of the village streets. Today, there is no village, and little evidence that there ever was one. One feature which can be found by today's visitor is the old cemetery, which is identified by the arrow. Unlike many other more popular battle sites, there is only one modest private monument, and there are no interpretive trails. Not an original village home remains; including the Childers mansion, which was used as Banks' headquarters, see the inset, and "HQ" near a black square with a flag. None of this was due to the war, but the fact that the residents tore down the village long after the war and moved it a few miles into Sabine Parish, so as to be closer to the new railroad.

Figure 6 is a more detailed version of figure 5, showing troop dispositions just prior to the battle. The Confederate regiments are shown in black, those of Emory's command using the more conventional representation of a black/white slashed rectangle, those of A. J. Smith's command, in a solid lighter shade, with the Union artillery indicated by slashed rectangles ◪. Shaw's placement of the 24th Missouri, at the upper left, on a hillock west of the old racetrack and next to Battery L, was by the approval of General Smith, but Battery L was soon called to the rear and put into park, to be replaced in that exposed position by the 25th New York Light Battery; all on the orders of General Emory.

---

836. Taylor, p. 166.

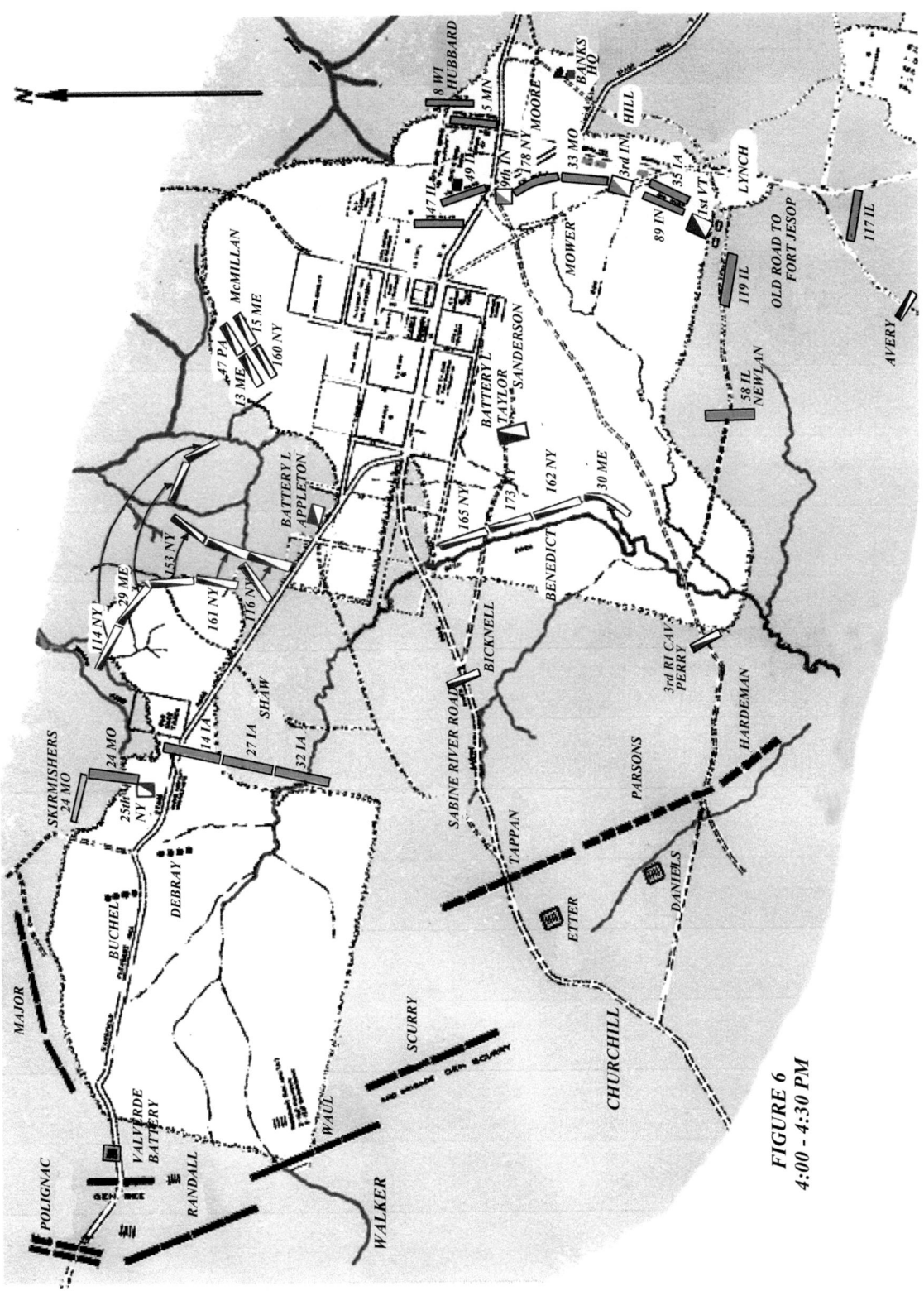

FIGURE 6
4:00 – 4:30 PM

FIGURE 6

General Smith then ordered Battery L into position in his line next to the 9th Indiana battery, and the 3rd Rhode Island Cavalry to their picket duty: Bicknell's troop on the Sabine River Road, and Perry and Avery on the other approaches.[837] This was remarkable cooperation for troops that were fighting together for the first time. However, the hurried arrangements left flaws in the dispositions. Three minutes later, *Emory* ordered Appleton's section to report to General Dwight, upon whom the battle had already opened, and the remaining two sections, under Taylor and Sanderson, to the ridge on the dangerous open slope behind Benedict's Brigade which had now deployed behind the ditch. It was a disastrous mistake. Not only would it take the lives of Lt. Sanderson and Privates Sholto O'Brien and William Smith, and wound five others, but it was directly in front of the field of fire of the 9th Indiana Battery and its supporting 178th New York Infantry, in Mower's line behind them. Neither dared to fire over, or through Battery L until Battery L's position was finally overrun at the point of the bayonet, and was abandoned.

As noted, Dwight's brigade was first camped along the edge of the clearing on the north side of the Mansfield Road, across from the old race track, the 114th New York being the right of his line. He did not even know who was in his front, until about 3:00 p.m., when Shaw, in command of the Missouri and Iowa troops deployed there, appeared at his headquarters. The enemy's skirmishers had already begun to flank the right of Shaw's line, and Shaw wanted Dwight to move forward to close the gap between them. He finally found Dwight, "near a house, *in the rear*," and Dwight agreed to make the changes. An hour later General Stone, of Banks' staff, appeared on the scene, agreed with Shaw, and rode off, apparently to hasten the changes. They were never made. Dwight instead *retired* by swinging around to the temporary positions shown (arrows and slashed partially white symbols) away from Shaw, much to Shaw's disgust.[838] Shaw's official report pulls no punches: "A few moments after five o'clock the enemy opened heavily on me with artillery… At the same time, General Dwight fell entirely out of my sight to the rear." A line from the history of the 29th Maine confirms the move: "Up to this time we had done nothing except to move a few steps down hill for shelter from the artillery fire and stray musket balls." The 29th was now facing the swamp to the north, with the 153rd New York on its left, and the 114th New York on its right. Dwight would soon be forced to face them about and advance to the road, after Major's dismounted cavalry had driven around and through Shaw—if the reader cares to peek ahead to the depiction of the Confederate advance, shown in figure 7.

Benedict's line is seen along the ditch. Originally, figure 5, positioned in the

---

837. O.R. Vol. 34/I, pp. 308–309, 318, 342, 354–355, 373, 472–473; Haskin, p. 196; Gould, p. 422.
838. O.R. Vol. 34/I, p. 355; Taylor, pp. 167–168. The time in various reports is often at odds, as has been seen earlier. Here, the time of the attack, which was "about 4 o'clock" according to the Confederate General X. B. DeBray, of the 26th Regiment of Cavalry, (see 'DeBray, SHS, Vol. 13, p. 158). Taylor says "soon thereafter" referring to 4:30; Gould, p. 422.

skirt of woods about three hundred yards to the west, Benedict's troops were ordered to this new position at 3:30, evidently too late to have an additional brigade sent to support their left, *as had twice been requested by Emory*. Having discovered that the ditch had at some previous time been deepened and straightened, and that its banks were covered by switch cane which was tall enough to hide a soldier lying down, the move out of the woods seemed desirable. The troops had a prepared rifle pit. Battery L, hastily placed in Benedict's rear, could then fire over the heads of those in the ditch, or lying near it. However, it would put Benedict far behind Shaw, with a gap similar to that between Shaw and Dwight, yet hung out four hundred yards ahead of A.J. Smith's line.

Smith's line, consisting of Mower's 1st and 3rd Divisions of the 16th Army Corps,[839] was basically a reserve position. Compared to the rather straightforward echelon line of General Taylor's dispositions, the Union defense looks peculiar. The right, Dwight, is heavily manned, though a swamp just to the north would have slowed or prevented any attack from that quarter.[840] It was therefore to be expected that the Confederate attack would be on the center and left of the Union line—if it came at all. There was, in fact, a brief delusion to the effect that it might not.[841] It is said that Banks, as he dismounted at the Childers Mansion, uttered the words: "There will be no fight today!" He was answered by the opening of Churchill's batteries.

Not Smith nor Emory, but General Stone had placed Lynch's 58th and 119th Illinois regiments hidden in the woods to Mower's left. Newlan's 58th was facing southwest, about one-quarter of a mile west of Kinney's 119th, their advanced positions and wide spacing would only slow, but could warn of, any Confederate approach from that quarter (which Taylor, in fact, had intended).[842] Little could either of these two commanders have imagined their almost accidental key role in the upcoming battle. They would be one jaw of a trap which would turn the tide. On the right of Mower, after it was relieved by Shaw, Emory had placed McMillan's brigade in reserve. Redeployed after the battle was initiated, they would be the other jaw of the trap.

At 4:30, it was reported to Taylor that Churchill's troops, ordered to the extreme right to gain the Fort Jesup Road, hence outflanking both Benedict and Mower, were nearly in position to begin their attack. To divert the attention of the Federals, Major Brent was ordered to advance the guns of the Valverde Battery, Moseley's 7th Texas, and West's 6th Louisiana, to open on the 25th New York Battery.[843] The sound of Churchill's guns signaled that he had begun his attack, and

---

839. O.R. Vol. 34/I. p. 308.
840. Swamp shown on map of *32nd Iowa History*, Scott, J., p. 288.
841. Scott, J., pp. 203–204.
842. O.R. Vol. 34/I, pp. 338–342, 344–351, 566.
843. Taylor, p. 167; Benson, p. 493; O.R. Vol. 34/I. p. 368.

Walker then began his advance.

A projectile screamed over gun no. 2 of the 25th New York, and it returned fire immediately. The 25th was doing "most excellent service" until the two batteries further to the Confederate right opened on them. At this juncture, about 5:00, the 25th fell back.[844] Seeing this, General Green assumed Shaw's line had broken, and ordered Bee's cavalry to charge. As Bee's 350 troopers issued from the woods, Buchel's 1st Texas Cavalry on the north side of the Mansfield Road and De Bray's 26th Texas Cavalry on the south—all west Texas men, superbly mounted and thoroughly disciplined—they made a magnificent spectacle. As they charged across the clearing toward the hillock where the 25th New York had stood, they could not see the skirmishers of the 24th Missouri, posted behind a fence at a right angle to the main line of the 24th Missouri and the 14th Iowa, who were lying on the ground, screened by the woods behind them. Holding their fire until the enemy had approached to within "50 paces," Shaw's troops then let loose a deadly barrage of musket fire which stopped the cavalry in its tracks, the horses and riders almost falling within Shaw's lines. In Bee's words: "The command was literally swept away…" Buchel was mortally wounded, and fell, a prisoner, into the hands of the 14th Iowa.[845] Bee and many other officers were wounded. DeBray's horse was killed, and fell, injuring DeBray. The cavalry drew back, never to charge again. It was then dismounted, and assigned to Polignac, who was still in reserve.

Walker was now pressing his attack on the 32nd and 27th Iowa regiments, on Shaw's left, south of the Mansfield Road. The fire of the 32nd was so destructive that Walker bypassed it in the gap to its left (south). Col. John Scott of the regiment later reported that the Confederates were "working across the rear of my position, so that in a short time the battle was in full force far in my rear."[846]

In the meantime, the 14th Iowa and the 24th Missouri felt the repeated attacks of Major's and Bagby's dismounted cavalry, but their resistance was so stubborn that Taylor was forced to call up Polignac's Texas troops. Mouton's old brigade, the 28th Louisiana and the Crescent Regiment, so devastated at Mansfield, and now under the command of Louisiana's Governor Allen, remained in reserve, and were never used.[847]

After repeated charges, Major's Texans gained a position on the right flank of the 24th Missouri.[848] The Missourian's ammunition was running short, and they

---

844. O.R. Vol. 34/I, pp. 355, 360, 369, 567, 608. Shaw thought that the battery fell back without orders, prompting Shaw's condemnation in his report.
845. Benson, p. 496. The position of the 24th Missouri skirmishers is shown in figure 6 by a dark rectangle at a right angle to the lighter rectangle representing the 24th Missouri.
846. O.R. Vol. 34/I, p. 366.
847. Sliger, p. 458; Blessington, p. 194. Having taken so many casualties at Mansfield, they were exempted.
848. O.R. Vol. 34/I, pp. 355–356, 363, 369; Gould, pp. 422–423; Pellet, p. 210.

fell back into the open field behind them, taking heavy casualties. The cross-fire extended to the 14th Iowa, whose commanding officer was killed and its adjutant wounded. At this point, orders from A. J. Smith came to Shaw to retire. So many of the officers of his regiments were either killed or disabled that Shaw could find no one to communicate with, and was compelled to seek out the scattered men individually amid the smoke and heavy brush, and order them out. Shaw was not able to reach Col. Scott of the 32nd Iowa, and soon the enemy occupied Shaw's recently vacated positions. Now, the 32nd found itself taking fire on its right, as well as its front and left. Though almost out of ammunition, they were able to hold their position until, at around sunset, an hour later, the tide turned, and the final general advance of Smith's line drove the retreating enemy past them.[849]

Figure 7 shows these and other movements, but cannot clearly depict the withdrawal of Shaw's troops. They simply melted away, through Dwight's lines, leaving the 32nd an island in a sea of the attackers, and almost ignored. That anyone could survive, and for much time go unseen, is testimony to the condition of the "clear" areas of this portion of the battlefield. They were fallow farm fields, overgrown with brush, and were now clouded with smoke. When the fighting had swept behind the 32nd Iowa, it threatened Dwight, who then repositioned the 153rd New York, 29th Maine, 114th New York, and 161st New York to the south, across the Mansfield Road. The 29th Maine men, as they marched south, were crossed by some of Shaw's retreating troops. Their narrative:

> Just before Shaw's brigade broke, our brigade was ordered toward the center of the field, and how the other regiments got there we cannot tell, but ours came by the right face and filed to the right—our late rear—and in that order (four abreast) went jumping over logs and brushing through the bushes, but whether it was north, south, east or west, who could tell?

From the 114th New York: "The smoke of battle hung over us so densely that the sun was entirely obscured."[850]

Again, from the 29th Maine:

> We had not gone far before we saw the men of Shaw's brigade coming on the run from our right hand…After Shaw's troops had gone through us we continued to march as before with our flank to the unseen enemy.

The reader will note that a few faint lines in figure 7 attempt to depict the route of the withdrawal of Shaw's Missouri and Iowa regiments. Though they were scattered, Shaw insists that they were reformed behind Dwight, as is depicted by the symbols 27 IA, 14 IA, and 24 MO, within the old village. Quoting from Shaw's

---

849. 6:41 p.m., per the calculations of U.S. naval Observatory for April 9th, 1864. *http://aa.unso.navy.mil/cgi-bin/aa_rstablew.pl.*
850. Pellet, p. 210; Gould, pp. 422–423.

writing years later:[851]

> About one-fourth of a mile from where my line was first formed I struck General Dwight's brigade just forming across the Mansfield road... I passed through his lines with the 14th Iowa and 24th Mo., and formed in line of battle within ten paces of Dwight's line.[852]

This is interpreted in figure 7 as the position on the left flank of the 161st New York.

As the 29th Maine boys came to the Mansfield Road, they spotted an abandoned Napoleon gun. This was clearly one of the two of Appleton's section of Battery L, since the 25th New York, the only other artillery unit in the immediate area, was equipped with 3-inch Southworth rifles.[853]

Having been assigned to Dwight at the onset of the battle, and since Dwight had found, quoting Appleton, "...no place where we could be of use and told me to take the section back, but directly ordered that it should be halted." Accordingly, it was halted in the road some distance in rear of the infantry, in figure 7, in the road near the 161st New York. From Appleton:

> Soon after, the enemy charged across the open field in our rear and through the woods on the flank of us, and in reversing the pieces in order to get into position, both poles were broken and it became impossible to unlimber them so that they could be fired. The road at that point was very narrow, and the carriage wheels in a ditch. In the mean time, the enemy were advancing on our flank and rear, keeping a hot fire of musketry, and I found it impossible to get the pieces into a position to fire, though the most strenuous efforts were made...I therefore ordered, when the enemy were within a few yards of the guns, the horses to be unhitched and the cannoneers to fall back with them... Shortly after,[854] the enemy were repulsed and the pieces drawn to the rear... The caisson horses were mostly killed or wounded in the charge; otherwise repairs were easily effected, and the section was ready for action again early in the evening.

Moving south, the next sight that the 29th encountered only hinted at the fate of Benedict's brigade. A spot of bright red was seen beyond Appleton's abandoned gun. It was a dead Zouave, "lying at full length in the sand," clad in the brilliant red pantaloons of the 165th New York.[855] Benedict's troops had evidently broken and withdrawn just as Shaw's had. Dead Zouaves "like sacred roses, dotted all

---

851. Scott, J., p. 183.
852. Emory's map of the battle: O.R. Vol. 34/I, p. 391.
853. O.R. Vol. 34/I, p. 406.
854. O.R. Vol. 34/I, pp. 411–412.
855. Gould, p. 423; The 165th was the only known Zouave unit in the 19th Corps still dressed as such, at this time, Benson, p. 502.

along the slope from the great ditch where Benedict fell, up to the crest of the hill on which stood the village…" Churchill had attacked Benedict's line, consisting of the 165th, 173rd, 162nd New York and the 30th Maine, at about 5:00, though this was counter to Taylor's plan. Churchill was supposed to have hit the left flank of Mower's 16th Army detachment, by swinging further south and east as far as the Fort Jesup Road; but he was either misdirected by his guide, or upon crossing the Sabine River Road, and the road south of it, he assumed he was already on "the enemy's left."

Bursting from the woods into the clear, Churchill's Arkansas and Missouri troops immediately swept forward, charging down the slope to the ditch, where they fell upon Benedict.[856] The advance is represented in figure 7 by the dashed black arrows. Under orders from Taylor to rely on the bayonet "as we had neither ammunition nor time to waste," a pitched battle briefly took place, the closely mingled troops preventing Battery L from further firing. With Benedict killed and other officers wounded or missing, the lines of the 3rd Brigade were driven back in disorder. The 165th New York broke first, then the 162nd, and then the 173rd, which suffered severely for its stand, losing 200 of 400 men. Seeing the others go, and nearly surrounded, the 30th Maine retired, "a portion" landing south in Mower's line (Lynch's Brigade of the 58th and 119th Illinois), the lower portion of Smith's "crescent", as it has been referred to.[857] The others were re-formed, though short more than 400 casualties, behind the upper portion of Mower's line in Smith's crescent. The arrows approximate their retreat, and the grey/white slashed symbols indicate where they reformed. The upper portion of the crescent now consisted of McMillan's brigade, which, as noted, had been re-deployed by Emory.[858]

Walker, on Churchill's left, had moved through the woods, and was "entertained" long enough by Shaw to fail to support Churchill.[859] In addition, Walker's troops faltered for a time when Walker was wounded and was personally guided from the field by General Taylor. Eventually, his troops slid past the 32nd Iowa, as was earlier described, and then appeared behind Dwight, Appleton, and upon a portion of Benedict's line.

---

856. Benson, p. 494; O.R. Vol. 34/I, pp. 431–433, 602, O.R. Vol. 53, pp. 477–478; Peck, p. 3 New York Infantry, *Sketch, 162nd Regiment*, p. 26.
857. O.R. Vol. 34/I, p. 346. Another specific reference as to where Benedict's troops retreated is found in Benson, p. 494. "They were driven through the 9th Indiana Battery and through the 178th New York in Smith's crescent…"
858. O.R. Vol. 34/I, pp. 566–567; Taylor, pp. 167–168.
859. O.R. Vol. 34/I, p. 553.

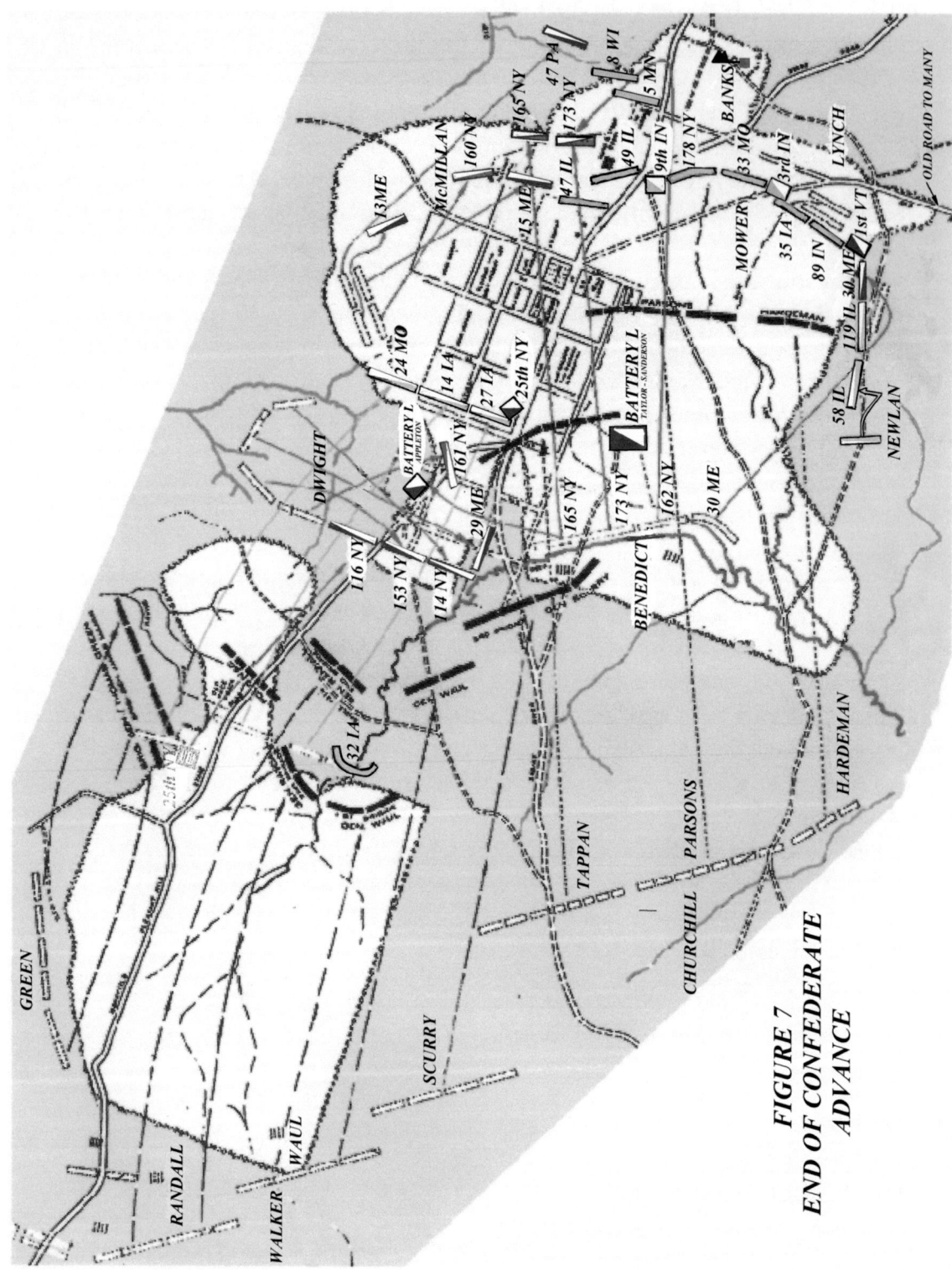

FIGURE 7
END OF CONFEDERATE ADVANCE

FIGURE 7

Benedict's retreating men can be seen sweeping through Taylor's section of Battery L, leaving it to its fate. Hung out in open space, the two sections, one commanded by 1st Lt. Taylor and the other by 2nd Lt. Sanderson, in spite of the canister which they poured into the advancing foe, the enemy reached 40 yards away, and the cannoneers were forced to retreat.[860] Most of the limber horses had been either killed or wounded, so three of the four Napoleons had to be abandoned. Only then were the guns of the 9th Indiana Battery able to open on Churchill's advancing horde.

The Confederate positions (black) indicate their maximum forward advance. It is instructive to refer to the report of Maj. Thomas Newlan, commanding the 58th Illinois Infantry Volunteers,[861] which was located at the lower "horn" of Smith's crescent shaped line. They had been placed there, facing southwest, expecting to be attacked from that direction. This revelation is at least a salutation to the sagacity with which this portion of the Union defense was planned by General Stone. It was an exact forecast of Taylor's intent. As Maj. Newlan of the 58th Illinois reports:

> ...in this position we were not attacked as expected. About 4:00 p.m. heavy skirmishing commenced on our right, and a few minutes afterward the rebels charged Benedict's brigade... [which had been posted about] one hundred and fifty paces to our right...in the ravine, and on the ridge in their rear four guns of a regular battery. [Battery L's sections of Taylor and Sanderson]
>
> [Benedict's brigade] delivered one or two volleys and fled in disorder. Being hid by an undergrowth of pine, the enemy did not observe us, but passed by our right flank, two hundred yards distant, like an irresistible avalanche, pursuing the retreating brigade toward the center of the crescent. For the briefest moment, the whole field of battle was filled, from where Shaw's line had been, up to the edge of the village, with a great southern army[862] ...
>
> Observing this, seeing the battery referred to captured, and fearing that we would be cut off and captured, I fell back about 100 paces, changed front so as to face the enemy's flank, and immediately opened fire with deadly effect. In a few minutes the enemy began to stagger under our fire, and finally broke in disorder.

The 58th charged, and joined by the 89th Indiana, drove the Confederates back to the ditch, capturing many prisoners.

At this point, as A.J. Smith reports:[863] "Seizing the opportunity, I ordered a charge by the whole line..." Smith's charge, according to a member of the 4th Texas Cavalry, admits that this turned the tide of the battle:[864]

---

860. Haskin (Closson) p. 196; O.R. Vol. 34/I, p. 410
861. O.R. Vol. 34/I, pp. 340, 350.
862. Benson, pp. 496–497.
863. O.R. Vol. 34/I, p. 309
864. Noel, T., *A Campaign...* p. 80.

Parsons, with the Missouri troops in our centre, were driving everything before them. Just at this juncture, A.J. Smith arrived. They were thrown against our right and centre, under Churchill. The shock was too great. Ten to one they could not stand; and back they came to be rallied no more that evening.

Walker's men, seeing themselves flanked, joined in the retreat, and in their confusion, with the oncoming darkness, a quote from General Taylor defines the end of the battle:[865] "[A]n idea prevailed that we were firing on each other…At nightfall I withdrew the troops."

A. J. Smith finishes: "We drove them back, desperately fighting, step by step across the field, through the wood and into the open field beyond." From Benson, of the 32nd Iowa: "Rolling like a sea of brown, along the roads west and southwest."

General Taylor's losses are only partially reported. E. Kirby Smith, in an April 11th letter[866] to General Price, who had been facing Steele in Arkansas, offers: "Our loss has been heavy." Irwin says that Taylor reported 1,500, which cannot be confirmed. Only Parsons and Tappan of Churchill's division wrote reports which include loss figures. Churchill reported seven officers and 58 men killed, with 393 wounded, which is less than the total reported by his two brigade commanders.[867] His force, at 4,300, represents only about a third of Taylor's total. If it was assumed that Walker's, Major's, Green's, and Polignac's troops took similar percentages of casualties, Taylor's total would calculate to be 1,374.

The losses of the Union army were 152 killed, 859 wounded, and 495 missing, for a total of 1,506. The casualties were about equal between those of Emory, at 725, and Smith at 753. The highest casualties were taken, as may have been expected, by the two most exposed positions on the battlefield, and those which were overrun; namely the brigades of Shaw and Benedict, at 447 and 483 respectively.[868]

There was some recrimination in the Union ranks, the westerners blaming the easterners. We have already mentioned Col. Shaw of the 14th Iowa, and his disgust with General Dwight. The perceived lack of guts of Benedict's brigade is hinted at in the report of Newlan, of the 58th Illinois, when he reports: "Rebels charged Benedict's brigade, which delivered one or two volleys and fled in disorder."

The position that Benedict's line had been assigned was an advanced one, much as was Shaw's. Neither position was tenable. Normal procedure would dictate a fall back, once seeing that they would be overwhelmed. Shaw was ordered to withdraw, Benedict was killed. No doubt Benedict's command would have, or should have been so ordered. Having seen their commander shot in the head[869] in the first few moments of the action, was a considerable factor leading to confusion.

---

865. O.R. Vol. 34/I, p. 568.
866. O.R. Vol. 34/III, p. 759.
867. O.R. Vol. 34/I, pp. 604–605. Parsons reported 321, and Tappan, 201; Irwin, p. 322.
868. Irwin, p. 321; O.R. Vol. 34/I, pp. 260, 313, 413, 432; Phisterer, pp. 3890, 3921, 3966.
869. Pellet, p. 215; Clark, p. 163; Benson, Van Dyke's Narrative, p. 524.

The nearly identical losses taken by the opponents, at 1,500, leads one to agree with a final assessment of the battle, which was made by Confederate Gen. X. B. DeBray:[870]

> This was, at best, a drawn battle. Both armies held the ground they occupied in the morning, but General Taylor, apprehending a renewal of the contest on the next day, knowing that water was not accessible where his troops stood, determined to fall back to a creek five miles distant, there to select a position. DeBray's and Buchel's regiments were left on the battle-field, with instructions to observe the enemy, and, if necessary, to retire slowly before his advance.

Nightfall found the Federals in control of the battlefield, and thus there was no exact repeat of the situation at Mansfield, where the Union wounded had been outrightly abandoned to the Confederates. Some of the dead were buried, and, at least, some walking wounded did not fall prisoner into Confederate hands. Nevertheless, darkness and the precipitate departure of Banks' army from the scene, at about 2:00 a.m. on April 10th, left an overwhelming number of the wounded, estimated to be 400,[871] lying in agony hoping to be recovered to hospital.

And then there were the animals. Rider-less horses, some wounded, were still wandering or dragging themselves around. Remember that most of the limber horses of Franck Taylor's sections of Battery L were either killed or wounded, as were the caisson horses of Appleton. The three limbers and two caissons that were lost were drawn by 30 horses, 27 of whom were wounded or killed. The wounded, left still in harness, some still standing, faithfully waited hours for the slap of the reins and the command to move, though the dead and maimed hitched with them would make it impossible.[872]

A party of General Taylor's men appeared at the battlefield that morning with a flag of truce, prepared to bury their dead. They were astonished to find only the Union surgeons and the rear guard of Banks' army, Colonel Lucas' cavalry.

Banks's order to abandon the wounded and retire that morning was protested by A. J. Smith, who is said to have favored remaining at Pleasant Hill long enough to care for the wounded, then continuing the advance. He considered arresting Banks, and proposed to Franklin that he take command. Franklin later acknowledged that Smith had approached him, but he avoided the matter of the alleged mutiny, only writing that he opposed the idea of an advance.[873] He apparently was opposed to the idea of serving any longer under Banks, which was confirmed in his response to a letter of inquiry from Col. John Scott. Franklin wrote: "The idea of

---

870. DeBray, X. B., p. 159.
871. Benson, *The Battle...* p. 503.
872. Blessington, p. 201.
873. Scott, J., pp. 231–237; Irwin, p. 322; O.R. Vol. 34/I, p. 184.

an advance after what I had just experienced under Banks' generalship was odious to me..."

Banks' report of the retreat, written on April 13th, says, in part:

> There was not water for man nor beast, except as the now exhausted wells had afforded during the day, for miles around... These considerations... the exhaustion of rations, and the failure to effect a connection with the fleet... made it necessary for the army ...to retreat to a point where it would be certain in communicating with the fleet and where it would have an opportunity of reorganization." [He was apparently still in contemplation of continuing the campaign, but at this time could offer only the vague], "upon a line differing somewhat from that adopted first and rendering the column less dependent upon a river proverbially as treacherous as the enemies we fight."

As to Battery L, Franck Taylor's official report gives few details of the battle, and does not name his casualties. He had filed a separate casualty report that was incomplete, but another, figure 8, was prepared by members of the 1st Division staff,[874] which contains the casualties taken by Battery L, including members of the 13th Massachusetts Battery, the 25th New York, and the 1st Vermont.

The data for Battery L reads: Killed, Pvt. Sholto O'Brien, and from the 13th Massachusetts Battery, Pvt. William H. Smith. Wounded and missing: 2nd Lt. James A. Sanderson, Sgt. Michael White, Pvt. Thomas Clinton, Pvt. William Parks, and from the 13th Massachusetts Battery: wounded, Cpl. William Hesseltine, Pvt. John Ferguson, and missing, Pvt. J. Redding.[875]

Lt. Sanderson, known as "Sep" at West Point, had stood by his guns and was horribly mangled by a shell of case shot which exploded on the cantel of his saddle. Though mortally wounded, he remained "between his pieces...still giving his commands." He was later taken to the hospital, died that night, and was buried out under the pines.

---

874. Figure 8, National Archives *Regimental Returns*, 1st Artillery, January 1861–December 1870, M727-5, p. 318.
875. There is a George Reading listed in the 13th Massachusetts Battery, (Record of Massachusetts Volunteers, Vol. 1, p. 457) who served from December 30th, 1862, until his discharge on July 28th, 1865, by expiration of service. The service dates identify him as the man with Battery L. Absent any other record, he clearly was returned from missing; Haskin, p. 364; Schaff, Morris, pp. 89, 240.

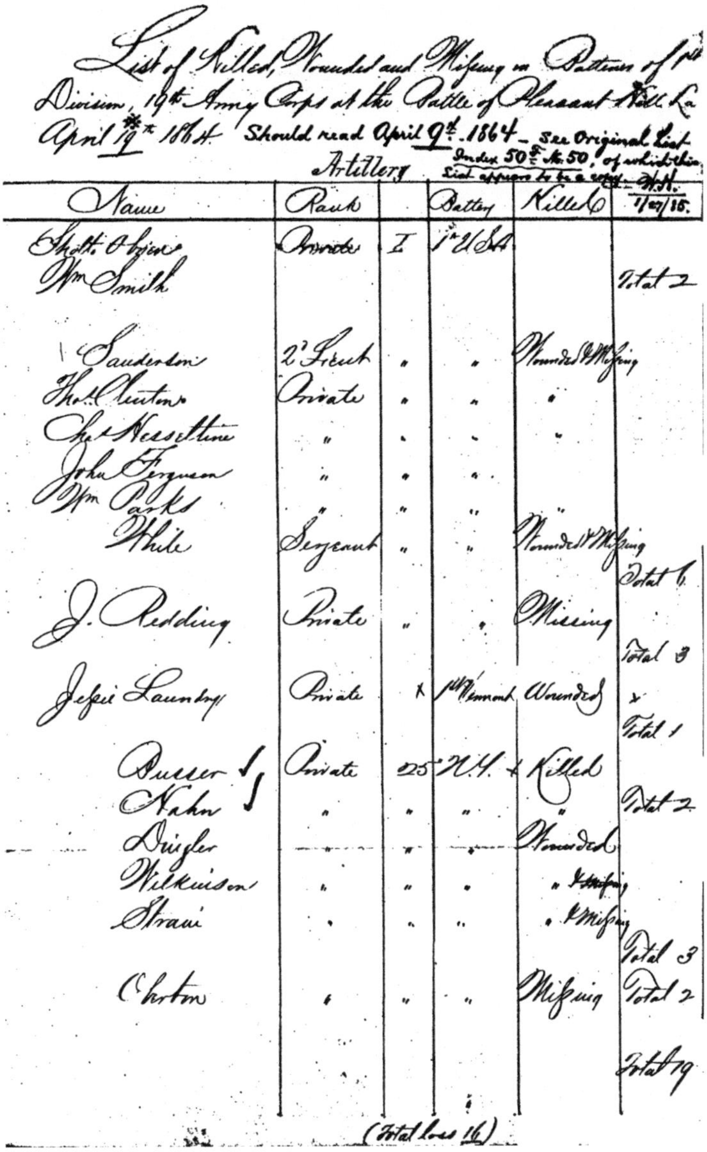

List of Killed, Wounded and Missing in Batteries of 1st Division, 19th Army Corps at the Battle of Pleasant Hill La April 19th 1864. Should read April 9th 1864 — See Original List Index 50 ℔ № 50, of which this list appears to be a copy. M.N.

Artillery

Name	Rank	Battery	Killed	7/27/15
Scott O'Bryan	Private	I	1st US A	
Wm Smith				Total 2
Sanderson	2d Lieut	"	"	Wounded & Missing
Thos Clenton	Private	"	"	"
Chas Hesseltine	"	"	"	"
John Ferguson	"	"	"	"
Wm Parks	"	"	"	"
White	Sergeant	"	"	Wounded & Missing
				Total 6
J. Redding	Private	"	"	Missing
				Total 3
Jesse Laundry	Private	x	1st Vermont	Wounded
				Total 1
Busser	Private	25	N.H.	Killed
Nahn	"	"	"	
				Total 2
Düsler		"	"	Wounded
Wilkinson	"	"	"	Missing
Strain	"	"	"	Missing
				Total 3
Chrtm	"	"	"	Missing
				Total 2
				Total 19

(Total loss 16)

FIGURE 8

Chapter 13

April 10th; Grand Ecore; Cane River;
Porter's Passage to Alexandria; The Dam at Alexandria; "Record" 6/64; Marksville
& Mansura; Canby; Last Days in the Gulf

April 10th

The portion of the "Record" for February–April, which, to date, bears upon the bloodiest battle Battery L had ever been caught up in, gives no hint of the difficulties during the battle, or subsequently. Having had to bury Lt. Sanderson, and privates O'Brien and Smith, someone had to see that a wounded Thomas Clinton was cared for, all the while knowing that White, Hesseltine, Parks, Ferguson, and Redding were missing, perhaps prisoners, or perhaps lying unseen in the brush, dying. There was frantic activity, all the while working in the dark to unhitch wounded and dead horses from their limbers and caissons,[876] and replace broken poles and wheels before their abandoned guns could be removed from the battlefield; all of this, after the stress of a battle, and without proper rest or food since the 8th. Then the order came to march again, after midnight, for 22 miles. They were reduced to towing only four guns, after 27 of their horses had been killed and 11 others were unserviceable. It is surmised that lightly wounded horses were coaxed into duty, without any treatment whatever. The severely wounded animals had to be ignored.

The march continued for several miles in the almost absolute darkness of the chill night. The road was bad; worst of all,[877] seeing how the campaign had been mismanaged, the men were in a dark mood. The thoughts were for the wounded that had been left on the battlefield, now for a second time, yet for a new reason. The medical supply train had been sent toward Grand Ecore with the rest of the wagon train, leaving no facilities available for removal or treatment. Reflecting the opinion of A. J. Smith, a most decided victory had been gained; the disgrace of Mansfield had been wiped away, and the troops had looked forward to chasing the defeated enemy to Shreveport, but that was to be denied.

April 10th was a day of some other significant events. Only two hours after

---

876. O.R. Vol. 34/I, pp. 411–412.
877. Lufkin, pp. 86–87; Blessington, p. 201; Benson, *The Battle...* p. 500; Pellet, p. 215; Clark, p. 163; Scott, J., pp. 215–218, 233–236; Heath, pp. 520, 522.

their arrival, that afternoon, near Loggy Bayou, where they were blocked from further ascent of the river by the sunken Confederate steamer *New Falls City*, Porter's gunboats and Kilby Smith's transports received Bank's message.[878]

E. Kirby Smith appeared at Taylor's camp after the battle. He had ridden the 65 miles from Shreveport on the 9th, but did not reach there in time to witness the battle. Satisfied that Banks was in full retreat, he was concerned that Taylor would not be able to supply his troops in a pursuit. Quoting Smith: "The country below[879] Natchitoches had been completely desolated and stripped of supplies. The navigation of the river was obstructed, and even had our whole force been available for pursuit it could not have been subsisted below Natchitoches." Incredibly, the *New Falls City*, sunk to block the advance of Porter's gunboats and Kilby Smith's transports, was now operating in the reverse. It would block the shipment of any supplies to Taylor from Shreveport. While discussing their options, Smith convinced Taylor that the better alternative was to break off the "main body of our infantry" and use it to pursue Steele, then at Prairie D'Ane, Arkansas.

In accordance with Halleck's original plan, Steele's 7th Army Corps,[880] about 7,000 troops, had marched out of Little Rock on March 23rd in the direction of Shreveport, more than 200 miles distant. At about the same time, another column of about 5,000 under General Thayer left Fort Smith, planning to unite with Steele at Arkadelphia, also a march of more than 200 miles. Swollen spring streams and bad roads delayed Thayer, and initially Steele marched on alone. As has been mentioned, Steele knew almost nothing of Banks' fortunes and vice versa.

After repeated skirmishes along the road forward, April 2nd at Terre Noir Bayou, April 4th at Elkin's Ford, on the Little Missouri, and on April 10th, an engagement at Prairie d'Ane, about 100 miles southwest of Little Rock, Steele was still more than 100 miles from Shreveport, yet Kirby Smith viewed him as "in position to march upon our base and destroy our depots and shops…" To this argument Taylor agreed, and requested that he might "accompany" the troops.

The argument was, however, only a portion of E. Kirby Smith's thoughts. His private opinion, as expressed to Jefferson Davis, confirmed the general sentiment of the Union troops that had fought at Pleasant Hill, as well as that of A. J. Smith, viz., that Taylor's force was completely worn out, and should have been relentlessly pursued. Kirby Smith:

> To my great relief I found in the morning that the enemy had fallen back during the night. He continued his retreat to Grand Ecore, where he entrenched himself and remained until the return of his fleet and its passage over the bars,

---

878. O.R. Vol. 34/I, pp. 204, 380, 381, 385, 388; ORN Ser. I, Vol. 26, p. 51.
879. O.R. Vol. 34/I, pp. 480, 485; E. Kirby Smith to Jefferson Davis.
880. Byers, p. 284; O.R. Vol. 34/I, p. 653; Vol, 34/III, p. 39; *www.nps.gov/abpp/battles/* AR012, AR013.

made especially difficult this season by the unusual fall of the river.

The question may be asked why the enemy was not pursued at once. I answer, because our troops were completely paralyzed by the repulse at Pleasant Hill.

Thus, Steele in Arkansas being his boss's preferred target, Taylor detached Walker's Texas division and Churchill's two divisions, Parson's Missourians, and Tappan's Arkansas infantry, who departed Mansfield on the morning of the 14th to march to Shreveport. Taylor would retain his cavalry, already sent in pursuit of Porter's fleet, and Polignac's infantry, which had been sent on the road to Natchitoches, in support of the cavalry.[881] Taylor, figuring to accompany his troops on the expedition to reinforce Sterling Price in Arkansas, arrived at Smith's headquarters on the 15th, where he learned that the fight at Prairie d' Ane had forced Steele to divert his march away from Shreveport, toward Camden. Regardless of the fact that Price seemed capable of handling Steele, Smith still wanted to press on with his expedition, and had decided to command it himself. Taylor would remain at Shreveport in nominal command.

In the meantime, the remnant of Taylor's command had pursued the fleet. Bagby to Grand Bayou Landing 15 miles below Loggy Bayou, and Green to Blair's Landing. On the east side of the river, Gen. Liddell, commander of the sub-district of North Louisiana, with Harrison's brigade of cavalry, and two sections of artillery, harassed the fleet almost constantly, following it downriver from the 11th to the 15th. However, Porter labeled Harrison's force as "careful about coming within range" and their effect only slowed the fleet at one point on the 13th. On April 14th, they again endured musketry fire from the east bank, below Campti, 20 miles above Grand Ecore, but were finally met by a detachment from A. J. Smith, which assured their safe passage to Grand Ecore the next day.

April 10th was the day that Gen. W. T. Sherman was to have his troops back. So determined was he that he sent Brig. Gen. J. M. Corse on a special mission to visit Banks, with orders for A. J. and T. K. Smith to prepare for the spring campaigns in the east.[882] To Banks he wrote:

> I beg you will expedite their return to Vicksburg, if they have not already started, and I want them, if possible, in the same boats they used up Red River, as it will save the time otherwise consumed in the transfer to others boats. All is well in this quarter and I hope by the time you turn against Mobile our forces will again act to the same end…General Grant, now having lawful control, will doubtless see that all minor objects are disregarded, and all the armies acting on a common plan.

The earlier initiatives in Texas and in the Red River, were now "minor objects,"

---

881. O.R. Vol. 34/I, pp. 388–389, 570, 572, 634; Taylor, p. 177.
882. O.R. Vol. 34/III, pp. 24–25, 244.

and were to be canceled.

These were serious tidings to a man in a deep quandary. The fleet had had trouble grounding in the shallows arriving at Grand Ecore, and the river was continuing to fall. The low level of the river below Grand Ecore and the falls at Alexandria threatened to trap it. There could be no question of an advance unless the river rose, and to that, even the ever-aggressive Porter agreed.[883] However, as to whether Steele was successful in advancing and ultimately joining with Banks, at this time, it was impossible to say. It had come down to the old problem of communications, always behind by 15 to 20 days. It was easier for Steele to communicate with Sherman and Grant than Banks, and vice versa. Any messages between Steele and Banks would have to pass through enemy territory or else go by the circuitous route of courier to Vicksburg and Memphis, to finally be telegraphed from there. Though Banks had heard nothing from Steele, one point was very clear. On March 27th, Banks had received from Grant the fateful March 15th order, and as we recall, Grant had followed up on March 31st, with an order to turn over the defense of the Red River to Steele and the navy, abandon all of Texas except the Rio Grande, and then prepare to move against Mobile. Curiously, Grant left Sherman to inform Steele of this order, which was not sent until April 7th. In it, Sherman informed Steele of his recall of A. J. Smith, and then said: "Your forces and General Banks' conjoined…would be able to accomplish all that should be attempted this spring. But if General Grant has also recalled Banks' command to be directed on Mobile, as I suppose he has done or will, you will not have enough men to accomplish all that should be done." He went on to say: "I have recommended to General Grant to give you all available forces in Kansas and Missouri…"[884]

Grand Ecore

It was clear that Banks' campaign was "ruined," as Irwin puts it,[885] and the only thing to do was to try to wrap it up before May 1st as Grant had directed. The fleet must be extricated from the falling river, and for the moment, at least, Grand Ecore would be fortified to form a defensive line while the fleet passed down. To do this, however, Banks would have to have the security of A. J. Smith's troops, at least until he could recall troops from Texas, and accept the chance of thinning the number of troops defending the rest of the department. McClernand was ordered to leave Texas with nearly all the troops at Pass Cavallo, and four regiments of the 13th Corps at Baton Rouge were ordered to Grand Ecore. Grover's 2nd and 3rd brigades were sent to Grand Ecore. The 133rd New York, serving in the defenses

---

883. ORN Ser. I, Vol. 26, p. 64; O.R. Vol. 34/I, pp. 11, 203, 216.
884. O.R. Vol. 34/III, pp. 76, 77.
885. Irwin, p. 326; O.R. Vol. 34/I, p. 174; O.R. Vol. 34/III, pp. 125, 126-128, 194, 195, 254, 269. The 20th U.S. Infantry (colored), from Port Hudson, was sent to Pass Cavallo to replace those troops of the 13th Corps who were ordered to New Orleans.

of New Orleans, was sent to report to Grover at Alexandria. Grover's 1st Brigade, under Nickerson, left Carrollton, also for Alexandria.[886]

A hint of Grant's suspicion and distrust of what was going on in Banks' department was his April 15th order to Halleck to have Major General David Hunter sent to look over Banks' shoulder, to "impress upon the general particularly two points": the importance of an advance on Mobile at the earliest possible time, and that Banks detach enough troops for the expedition. "You will remain with general Banks until his move from New Orleans is commenced, and a landing is effected at Pascagoula…"[887]

On the 16th when Smith reported that his transports had all arrived at Grand Ecore, and that he was ready to respond to Sherman's order to leave, Banks refused.[888] Smith obeyed, and with the others sent as reinforcements, continued building a defensive line around Grand Ecore, which consisted of two miles of felled trees covered with earth.

The struggle of bringing the transports and Porter's gunboats down to Grand Ecore and then to Alexandria, had resulted in the sinking of the massive *Eastport*, which was alleged to have struck a torpedo eight miles below Grand Ecore, on April 15th.[889] Attempts to re-float her by pumping had to await the arrival of pump boats. On the 21st she was finally pumped out, bulkheaded, and floated. After 57 more miles and four more days of dragging her over increasing numbers of sand bars and log jams, her five support boats all the while taking considerable Confederate fire, the *Eastport* became impossibly stuck on a bed of sunken logs, and rather than have her fall into Confederate hands, she was blown up at 2:10 p.m. on April 26th.

The name of Lt. Col. Joseph Bailey, the acting military engineer in the department, is revealed in a later report by Banks, commenting on the *Eastport* affair. Bailey, familiar to almost all of the troops at Port Hudson, where he supervised the construction of many of the gun battery positions, and who was credited with the salvage of the Confederate steamers *Starlight* and *Red Chief*, which were captured in Thompson's Creek, had suggested the construction of wing-dams to raise the level of the river, and thus float the *Eastport* free. Army aid was ignored.[890] A sour Banks writes; "No counsel from army officers was regarded in nautical affairs."

---

886. Hanaburgh, p. 101.
887. O.R. Vol. 34/III, pp. 160, 169, 190–192. It is noted that Grant sent this order to Halleck from his headquarters at Culpeper, Virginia. Such details were left to Chief of Staff Halleck. Grant was all business, and notably remained out of Washington, and away from Lincoln and Stanton, where there was a chance that they might interfere with his plans. From p. 123 of Vol. 2 of Grant's *Memoirs*: "I did not communicate my plans to the President, nor did I to the Secretary of War or to General Halleck."
888. O.R. Vol. 34/III, p. 175; Beecher, p. 327; Pellet, p. 221; Sprague, p. 192;
889. ORN Ser. 1, Vol. 26, pp. 68, 72, 76, 81–82, 84–87, 110, 167, 781.
890. O.R. Vol. 34/I, pp. 206, 403.

On April 21st, the minute that Banks had heard that the *Eastport* was afloat, he prepared to march out of Grand Ecore, already having waited ten days for Porter. Marching down the island created by the Cane River and the Red River at the furious pace set by Birge, who was leading a temporary division,[891] forged ahead of Porter, and left him unsupported—an accidental scenario of what Porter had earlier feared regarding his whole fleet while at Loggy Bayou.[892] Kilby Smith, in a letter to Porter dated April 25th, remarked: "General Smith and I both protest at being hurried away. I feel as if we were shamefully deserting you."[893]

Cane River

Banks' haste was the result of the fact that he had been warned that Taylor had sent troops around from his front at Grand Ecore to his rear. This was either to intercept the passage of the fleet at the mouth of the Cane River, attack Alexandria, or obstruct the march toward Alexandria at Monett's Bluff.[894] To Richard Taylor, it was all of the above, depending on how events unfolded. On the 22nd, Taylor sent Bee to the crossing at Monett's bluff.

The Battery L "Record" says:

> 21st struck camp at 5 P.M. and remained harnessed until 3 A.M. 22nd, then marched 24 miles and camped for about 4 hours, then marched to Cloutierville. 23rd Marched to pontoon on Cane River ...

It says nothing about the fact that its commanding officer, as chief of artillery, was a key in the battle at the crossing, or that it was held in reserve, standing the whole day at the ready.

About three miles out of Cloutierville, Gooding's cavalry, at the head of the column, met and engaged Bee's pickets. At about 4:00 a.m.[895] as they approached the crossing, Bee's Texans – Debray, Bagby, and Major – opened on them from batteries hastily placed on the high bluffs on the south side of the river. Closson then deployed the reserve artillery to shell the enemy guns. It was part of a plan which would cover the move of a column of four regiments led by Birge which crossed the river at a ford about three miles above, and out of sight of the enemy.

Debray's flank was then assaulted at the point of the bayonet. He fell back, and followed Major and Bagby who had fled. It was a long and costly engagement.

---

891. Sprague, p. 193: "Twenty five miles" from 5:00 p.m. on the April 21st to 3:00 a.m. on April 22nd; Hanaburgh, p. 103, Taylor, p, 182. Cloutierville was thirty-two miles, where the army finally halted.
892. ORN Ser. 1, Vol. 26, p. 56.
893. O.R. Vol. 34/III, p. 279.
894. O.R. Vol. 34/I, p. 190; Taylor, pp. 180–182; O.R.. Vol. 34/III, pp. 235–236.
895. O.R. Vol. 34/I, pp. 190, 207–208, 394–397, 406–407, 418–419, 610–612; Pellet, pp. 226, 228; Clark, pp. 169–170; DeBray, SHS, Vol. 13, p. 161.

It had taken Closson until after 2:00 p.m. to deploy and fire from long range, then move in over swampy ground to fire at closer range, before he knocked out all of the enemy guns, and the flank assault had taken 153 casualties.[896]

Battery L's "Record" notes that they finally crossed at midnight.

Porter's Passage to Alexandria

Leaving the *Eastport* on April 26th, the five support boats were attacked by guerrillas as they headed downstream, but repelled them.[897] However, at a point five miles above the entrance to Cane River, to which point Taylor had ordered the four guns of Capt. F. O. Cornay's St. Mary's Cannoneers, Porter was really cut up. Cornay's battery, remembered from the battle at Bisland and Grover's passage to Irish Bend, was impressive—attested to by the fact that Porter claimed that Cornay had 18 guns, and that the flagship, the *Cricket*, had been struck 38 times. The incident was a memorial to the skill and daring of Cornay, who was killed, and a reminder of Gideon Welles' comment that exaggeration was a Porter family "infirmity."

There was another incident on the 27th. Between the two days, Porter's flagship had lost 25 killed and wounded, the *Juliet* lost 15, the *Fort Hindman* 2, the *Champion No. 5*, one killed and all of the crew taken prisoner, and the *Champion No. 3*, whose boiler was shot through and exploded, had four of the crew scalded to death, the cook wounded; and of about 200 contraband passengers, men, women, and children—taken from the plantations along the river—there were 100 dead and 87 badly scalded, all of whom died within 24 hours. The *No. 5* was abandoned while on fire. The *No. 3* was captured, repaired and put into service by the Confederates.

The Dam at Alexandria

Grant's man, Hunter, arrived on April 27th, eight-and-one-half days out from Washington. It was at the worst possible time—the crisis over the entrapment of the fleet at the rapids at Alexandria.[898] Hunter took little time to reach a judgment. The next day, he reported back to Grant that the situation was "complicated, precarious, and perplexing…" He precipitously recommended destroying the boats, and went on to say: "Why this expedition was ordered I cannot imagine."

April 28th was the day that Grant had received Banks' latest report, dated April 17th. In it, Banks tried to promote the idea of enlarging the campaign, to prevent the enemy from threatening Arkansas and Missouri. He would again advance and capture Shreveport. Grant was unmoved. He wrote to Halleck: "I do not see that better orders can be given than those sent a few days ago…(*retrace your steps, etc.*)… General Banks, by his failure, has absorbed 10,000 veteran troops that should now

---

896. *162nd New York Infantry*, p. 28; Peck, p. 3, Haskin, p. 370; O.R. Vol. 34/I, pp. 434–435, 438–441; Clark, pp. 226–227.
897. ORN Ser. 1, Vol. 26, pp. 74–87, 176–177; Taylor, pp. 183–184; Welles, Vol. 1, p. 157.
898. O.R Vol. 34/I, p. 191; Vol. 34/III, p. 316.

be with General Sherman and 30,000 of his own that would have been moving toward Mobile, and this without accomplishing any good result." He had already made up his mind to sack Banks,[899] and asked for the President's concurrence in a telegram dated April 22nd. Grant reveals that he had had a low opinion of Banks for nine months, which would correspond with the time he had spent with Banks at New Orleans.

As to Porter's gunboats, his troubles while dropping down from Grand Ecore were only the end of the beginning. The rapids at Alexandria began at an upper falls, about a mile above the town, and followed an S-shaped channel of solid rock to a lower falls, figure 1. Through these the water was running at a depth of 3 feet 4 inches, and Porter's gunboats required 7 feet. There were 10 in all, the ironclads *Mound City*, *Louisville*, *Pittsburg*, *Carondelet*, *Chillicothe*, *Osage*, *Neosho*, and *Ozark*, plus the tinclads *Lexington* and *Fort Hindman*, all "blockaded" as he put it, above the upper falls.[900]

Porter was angry, and in a letter to Welles, he blamed Banks' incompetence in managing the campaign. He offers nothing of his own as a plan of action, except to suggest a massive relief effort. Perhaps Porter could see his legacy of being a naval hero slipping away, as well as his feast at the trough of prize money. Here, he transparently asserts: "I have sacrificed all private interests, all desires of a personal nature…"[901] Rescuing Porter was essential, the loss of this portion of the Mississippi squadron would be an embarrassment, and vessels worth nearly $2,000,000 could not be left to fall into Confederate hands. Even if they were destroyed, the salvage of their iron would be valuable to the Confederacy.

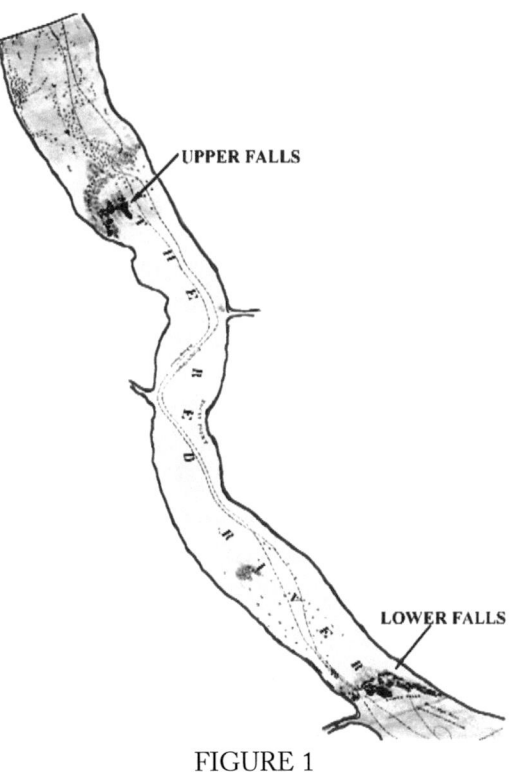

FIGURE 1

---

899. O.R. Vol. 34/III, p. 252
900. ORN Ser. 1, Vol. 26, pp. 92, 94–95. Figure 1 adapted from NOAA Historical Collection no. TO 1921-05-1864.
901. ORN Ser. 1, Vol. 26, pp. 35, 318, 342, 363, 394, 412, 460, 556. Prize claims were made for captured cotton, molasses, iron, vessels, and horses, among others; Value of gunboats, ibid, pp. 132, 159.

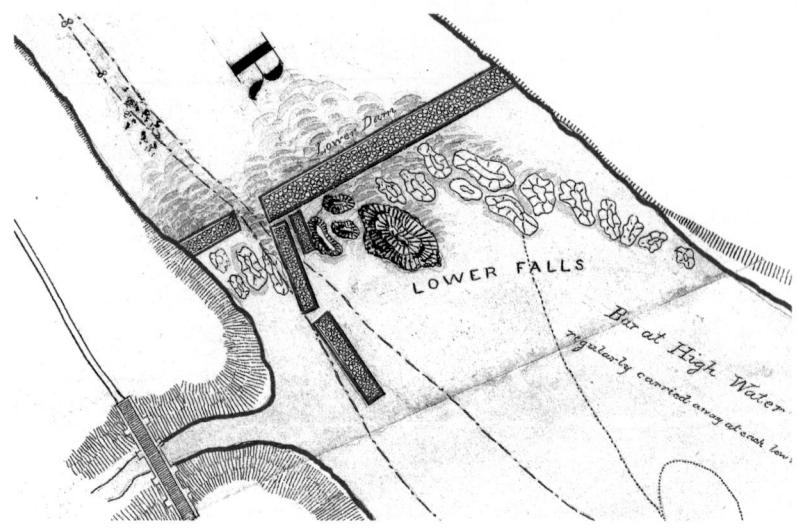

FIGURE 2

Aside from constructing defensive deployments around the perimeter of Alexandria, which occupied the next few days,[902] a plan to dam the river to raise the water level sufficiently to allow the passage of the gunboats took shape. Blasting the solid rock bed of the channel to deepen it was ruled out, given that it was estimated to take 30 days. A dam was judged feasible by the engineering staff, and the supervision of the project was given to Lt. Col. Joseph Bailey.

The work began on May 1st, and the initial plan, accomplished by some 3,000 men working day and night, was completed on May 8th. It consisted of building a dam across the river at the lower falls, figure 2,[903] to back up the water level behind it to create a pool which would extend all the way to the upper falls. At the point chosen, the river was 758 feet wide. The longest part of the span extended from the northeast side. It consisted of large trees laid with their tops facing the current, their rooted ends tied with cross logs, and the whole covered with brush, and weighted down with stone and brick, which was obtained by tearing down neighboring buildings. There were fewer trees on the south side, so that portion was constructed of a crib filled with stone, brick, and pieces of machinery collected from sugar houses and cotton gins in the area. To narrow the gap between the two wings, which was about 150 feet, four heavy navy coal barges were sunk in it, which narrowed it, yet left a sufficient gap through which the boats could pass. At completion, the water was found to be pooled to a depth of almost eight feet at the upper falls, the water having risen almost five-and-a-half feet. The three light draft gunboats, the *Osage*, *Neosho*, and *Fort Hindman*, having gotten steam up, soon passed the upper falls and entered the pool created by the dam.

---

902. ORN Ser. 1, Vol. 26, pp. 95, 97–98; O.R. Vol. 34/I, p. 209; ORN Ser. 1, Vol. 26, p. 95; Pellet, p. 231.
903. Figure 2, portion of figure 1; Irwin, pp. 338–340.

The water level on the dam now was at its maximum, and the velocity of the flow through the narrow opening was tremendous, but the dam held; yet there was no further reaction from the rest of Porter's fleet. This tardy response, which Porter conveniently omits in his reports, soon rendered the hope for the successful passage of the rest of the fleet to be dashed. Early in the morning of May 9th, the surging water swung aside two of the coal barges, leaving a gap 66 feet wide. Warned of the failure by Banks, who personally observed it, Porter is credited with riding north to the upper falls and ordering the *Lexington* to run immediately into the rapids, and through the chute at the dam. After a heart-stopping run through the foam, it was safe below. Seeing the example, the three light-drafts followed; figure 3, from the cover of *Harper's Weekly* of June 18th, 1864.

FIGURE 3

It was too late to move the remaining six gunboats and two tugs still above the upper falls. The rush of water out of the dam lowered the pool behind it and the level at the upper falls soon dropped below six feet. Judging that a repair to the lower dam would only wash out as it already had, Bailey rose to the occasion, and proposed a second dam at the upper falls, sufficient to raise the level there to allow the gunboats to pass into the channel of the lower pool and finally through the lower dam. Construction of an upper dam took three more days. This confined enough water to the channel to raise it to slightly more than six feet, but still insufficient to pass the gunboats without a herculean effort to lighten them. This the navy heartily entered into, having seen the results of the army's earlier effort. Lightened by the removal of guns, cargo, ammunition and plating, the draft of the remaining gunboats was reduced sufficiently to have them hauled through the crooked and narrow channel between the two dams. By the 13th of May all had passed safely south of the lower falls.[904]

This was the signal for Banks to leave.

---

904. ORN Ser. 1, Vol. 26, pp. 130–133; Clark, p. 177.

"Record" 6/64
30 APRIL–30 JUNE 1864 ALEXANDRIA–NEW ORLEANS, LOUISIANA

May 13th Marched from Alexandria 12 miles. 14th Marched 12 miles. 15th Marched 10 miles. 16th Marched to Marksville 15 miles. 17th Marched to Simmesport 10 miles. 19th Crossed Atchafalaya & marched 3 miles. 20th Marched 10 miles. 21st Marched 18 miles. 22nd Marched to Morganza 3½ miles. Remained until June 28th; proceeded on board S.B. *Universe* to New Orleans, arrived on the 29th.

Henry W. Closson	Capt. Det. svc. Chief of Arty, 19th Army Corps
Franck E. Taylor	1st Lt. Commanding Battery and Asst. Comm. of Musters
Edward L. Appleton	1st Lt. On det. svc. Commanding Battery "C" 2nd US Arty. S.O. no 36 Hdqrts. 19th Army Corps June 11, 1864

Joined:
George Kelly	Pvt.	From desertion, May 19, 1864, at Simmesport. Apprehended at Pensacola and sent under guard to Company.
Andrew Stoll	Pvt.	do.

Detached:
Edmond Cotterill	Cpl. On Det. Svc. Clerk in A.G.O. Wash. S.O. no. 426, War Dept. A.G.O. Wash. Sept. 23,'63. Left Co. Nov. 20,'63.
Solomon Mongomery	Pvt. On Det. Svc. Orderly Chief of Artillery
George Freidman	Pvt. On Det. Svc. at N.O. as Artillerist, Left Co. May 24,'62.
Amelius Straub	Pvt. On Det. Svc. at Baton Rouge as cook in Univ. Hosp. since June 1,'63.

Absent in Confinement:
Patrick Gibbons	Pvt. At Ship Island serving sentence of G.C.M. no. 18, Hdqrts. 1st Div. 19th Army Corps Dec. 31,'63. Left Co. Jan. 8,'64.
Michael O'Sullivan	Pvt. In jail at N.O. since Sept. 2,'63.
John Lewery	Pvt. Apprehended as a deserter and confined at New Orleans.
John H. Moran	Pvt. In confinement, awaiting sentence.
Peter Welsh	Pvt. In confinement.

Absent without Leave:
Michael White	Sgt. Wounded and missing since April 9, 1864.
George W. Redding	Pvt. 13th Mass. Since April 9th 1864.
William F. Brown	Pvt. Since June 29, 1864
Joseph Kutschor	Pvt. Since June 29, 1864
John Kelly	Pvt. Since June 29, 1864

Died:
James McCarthy	Pvt. Of typhoid fever at Morganza, June 8, 1864.

Strength: 99 Sick: 9

Present Sick:
Patrick Craffy, Joseph Eisle, Joseph H. Parslow.

Absent Sick:
William Brunskill	Sick at Ft. Hamilton, NY left Co. Sept. 17, 1861.
William Crowley	Sick at Baton Rouge since July 13, 1863.
George Chase	Sick at Brashear City since April 22, 1863.
Charles Jackel	Sick at Franklin since March 12, 1864

Churchill Moore	Sick at New Orleans Left Co. April 20, 1864.
Henry Williams	Sick at New Orleans Left Co. March 12, 1864.

Sgt. Demarest is listed as: "under arrest," no explanation given.

Cooks: Virgil Ayers, Phillip Evens, Henry Jefferson.

The cooks are listed without rank, and in fact they were never given an official rank. They were now being paid $6 each per month. Here, there appears the note: "Error on last payroll." No explanation is given.

The "Joined" column shows deserters George Kelly and Andrew Stoll as returned on May 19th at Simmesport, but they had been apprehended at Pensacola. They were part of the large number of men who had deserted, see chapter 10, at New Orleans in 1863; Kelly on August 13th, and Stoll on September 16th. How they managed to remain in New Orleans and Pensacola for eight months is material for a detective story in itself. One factor might have been that they were dropped from the muster rolls for the intervening period—reason unknown.

The sick list is remarkably small, as was the ten sick on the previous muster roll, though the men were worked, fought, and marched to exhaustion. The extended periods of sickness for some of those absent, Brunskill for almost three years, Chase for a year, and Crowley nearly a year, reminds us, again, of the medieval state of medicine in that era. It did not discriminate—remember Admiral Foote, who died of what today would be called a minor leg wound, and now we witness the illness of General Franklin, also with a leg wound.

McClernand's command had arrived at Alexandria from Texas on April 29th, and gone into the defensive line beside A. J. Smith. While there, with all of the attention focused on the dam, the Federal army lay confined behind its works, leaving Taylor free to roam the countryside.[905] Though there were local skirmishes, no determined effort was made by Banks to destroy Taylor.

Unchallenged, Taylor divided his reduced force to try to block traffic in the Red River below Alexandria and cut off Banks' communication with the Mississippi. He sent Major with about 1,000 men 25 miles south to David's Ferry, which Major reached on April 30th. The next day, he captured and sank the transport *Emma*, and on May 3rd, the *City Belle*, carrying 425 officers and men of the 120th Ohio, many of whom were either killed or wounded. The steamer was then sunk across the channel and the river blocked. On May 5th, the *John Warner*, carrying the 56th Ohio, was captured. In the same action the *Covington* was disabled, then abandoned and set on fire, and the *Signal* was captured and sunk across the channel. The river remained closed for 15 days. With the army pulling out of Alexandria on May 13th, and following the river course until the 15th, it passed the points where Major's batteries had been placed and the river blocked. Major was then forced to flee, though he remained, along with Polignac and Bagby, hovering at the front

---

905. Irwin, pp. 342–343; Taylor, pp. 185–186; ORN Ser. 1, Vol. 26, pp. 112, 116–123, 134.

and flanks of the Federal column.[906]

As at Grand Ecore, columns of smoke were seen as the army left Alexandria. Banks, in his 1865 report of the campaign, makes note of the fact that a fire was started in a building on the levee, which had been occupied by refugees or soldiers, and that the wind on that day made it impossible to extinguish. The fact that "a considerable portion" of the town was consumed was not intentional. This is corroborated by Lt. Edward Cunningham, Kirby Smith's aide-de-camp, who wrote: "My opinion is that they did not intend total destruction."[907] In addition, more widespread fires, specifically on 19 plantations, were noted by a member of the 4th Texas cavalry, Theophilus Noel, as he stood on McNutt's Hill 12 miles northwest of Alexandria.[908] He makes the startling point that these were lit by native Louisiana residents "who embraced this opportunity of revenge on the rich planters and their cruel overseers, who had fenced them off from water and had taken their cattle, as did the lords in the feudal days of the dark ages."

Marksville & Mansura

On May 15th, the army left the course of the river, and entering Avoyelles Prairie, turned toward Marksville, an easier march to Simmesport, per the Battery L record. Here, the advance of Banks' column, Col. Thomas Lucas' 1st and 3rd cavalry brigades, with Battery F of the 1st U.S. Artillery, moved forward at a trot, "pieces and caissons jumping and pounding against the cypress knees until it seemed that much more of such traveling would knock everything to pieces, but finally clear daylight ahead and an open prairie with the village of Marksville in the distance."[909] Here stood Polignac, with a brigade, who slowly fell back through the village, and for two miles beyond, on the rolling prairie. At sunset, he turned and formed across the prairie, from the swamps and bordering woods on the left to the edge of woods on the right—a line of mounted riflemen extending a mile-and-a-half, all in plain view.

A battery of four of Polignac's rifled guns, posted in his center, played upon the approaching Federals. Lucas's brigade then formed up and charged. The Confederate line broke, and doubled back into the woods on either side. From Haskin, in command of Battery F: "The whole performance was in plain sight and was one of the most exciting it was ever my fortune to witness." Darkness ended the engagement.[910]

The main body of the army had encamped two miles back, but now it was hurried up to encamp near the cavalry, and retain the ground gained. On the

---

906. Clark, p. 179; Irwin, p. 344; Taylor, p. 191, Pellet, p. 233; O.R. Vol. 34/I, p. 212; O.R. Vol. 34/III, p. 568.
907. O.R. Vol. 34/I, pp. 212, 558.
908. Noel, *Autobiography*, pp. 143–144.
909. Haskin, pp. 555–556.
910. O.R. Vol. 34/I, p. 447; O.R. Vol. 34/III, p. 517; Haskin, pp. 198–199, 371.

morning of the 16th, the march resumed, and after several hours of skirmishing on the rolling prairie, the Federal column was confronted by a torrent of projectiles from ten of Polignac's guns[911] placed in battery near the village of Mansura. It was a grand and symbolic last stand, taken before the Federal army could pass through on the road to Simmesport. Vastly outnumbered, Polignac traded artillery shots with Lucas' cavalry, and Battery F, at the front, while the army drew up in line of battle, Mower on the right, Emory in the center, and Lawler, now in command of McClernand's troops, on the left. They were all in full view, from flank to flank. As they advanced, Polignac began a steady withdrawal, away to the Federal right, toward Cheneyville. A spectacular scene, on a beautiful clear day, with neither tree nor fence to obstruct the view, as thousands of men in blue moved forward with mathematical precision. Major Henry Closson,[912] present, and seeing the Confederate artillery fire taking its toll on Battery F as it entered Mansura, sent in the 1st Vermont Battery to its relief, and the engagement ended.

Simmesport and the Atchafalaya were reached the next day; finding the transports and gunboats waiting. Crossing the swollen river, now some six- or seven-hundred yards wide, would have to await the completion of another of Col. Bailey's inventions, a pontoon bridge consisting of steamers lashed together.[913]

Ever the aggressor, Taylor attacked the rear of the Federal army on May 18th, near Yellow Bayou. A. J. Smith's command, under Mower, had been stationed there to cover the crossing of the Atchafalaya. Mower advanced, and fought one of the sharpest engagements of the campaign. Mower lost 267 in killed, wounded, or missing, and Confederate returns report 452 of theirs as killed or wounded.[914]

The crossing was made on the 19th, two days short of their crossing a year before, when at the end of the Teche Campaign, they had turned to Port Hudson.

It had been a long and exhausting year; one with satisfaction at the completion of the Siege, and the consequent opening of the Mississippi, but indelibly marred by memories of the Red River Campaign's failure, and of the fruitless waste, "this sink of shame," the invasion of Texas. The quote is from Irwin, who, like most in the department, knew that the Texas invasion was Halleck's creation, as had been urged by Seward and Lincoln. Though it had been faithfully carried out by Banks, he was now blamed for it by Grant; and Halleck, now bending in Grant's wind, had quickly turned on Banks. Lincoln was more faithful. He was very reluctant to relieve Banks, which, no doubt, resulted in the awkward creation of a new department—the Military Division of West Mississippi,[915] to include

---

911. Irwin, pp. 344–345; Clark, pp. 178, 180; Pellet, p. 234; Ewer, pp. 182–183.
912. Brevet Major Closson had been chief of artillery of the 19th Corps since October of 1863. He still is listed in Battery L records as captain, his permanent rank. O.R. Vol. 34/I, p. 408.
913. Irwin, pp. 346–348; Clark, p. 181.
914. Irwin, p. 346; Taylor, p. 191.
915. O.R. Vol. 34/III, pp. 490, 543; Pellet, p. 235; Irwin, p. 348.

Steele's Department of Arkansas, and Banks' Department of the Gulf. However, both Steele and Banks had a new boss, Maj. Gen. E. R. S. Canby, who assumed command on May 11th. In fact, Banks was left no command at all, but would remain squeezed between a subordinate, Franklin, and a junior officer, Canby, as his superior.[916]

From the 114th New York regiment:

> There were many opinions expressed regarding the step, but a majority of the enlisted strength of our army sympathized with the former. But it was only to be expected, and is decidedly "American." A single success is proof of Napoleonic genius; a single defeat, of total incompetency. General Banks was not an exception to the rule.

In Taylor's words: "[On] May 19, 1864, the enemy crossed the Atchafalaya and was beyond our reach. Here, at the place where it had opened more than two months before, the campaign closed."[917] Not quite, at the rear of the column were A. J. Smith's troops, who only completed the crossing on May 20th, after which the steamboats were unlashed and the "bridge" was no more. It was then that Smith's troops were ordered back to Vicksburg.[918] The remainder of the army continued the march, arriving at Morganza, on the Mississippi, 51 miles above Baton Rouge, on May 22nd.

Last Days in the Gulf

On June 7th Battery L was ordered to New York, and on June 11th, Lt. Edward Appleton was ordered on detached service, to serve as commander of Battery C, 2nd U.S. Artillery.[919] Battery L was to be accompanied by three of the other regular batteries in the department: A & F, 1st U.S., and C, 2nd U.S. The order for Battery L to leave was perhaps "out of the blue" as it were, because L, with the 13th Massachusetts men attached, was nearly up to strength. Of the other companies, none were in full war organization, something Emory and Canby felt that was needed if they were to remain. They were to be transferred out, for reorganization, on the recommendation of Maj. Henry Closson, the chief of artillery, with the concurrence of Emory.[920] Closson would remain behind, the 19th Corps still

---

916. Canby was promoted to major-general just before taking over the control of the new department, making him equal in rank to Banks, *but*, junior by date. Background on the replacement of Banks can be found at: O.R. Vol. 34/III, pp. 253, 278, 293–294, 331–333, 409, 580, 600, 615, 631.
917. Taylor, p. 191.
918. O.R. Vol. 34/I, pp. 212, 645–646, 680, 695; Stanyan, p. 482; Tiemann, p. 78. Canby took over command on May 19th, at Simmesport.
919. O.R. Vol. 34/IV, p. 256, Battery L June monthly return, see "Record," this chapter.
920. O.R. Vol. 34/IV, pp. 306–307, 333, 358–360.

requiring his services as chief of artillery.

Batteries A and F had only 78 and 53 men, respectively, and had been reduced to four guns. Until Appleton was assigned, Battery C had no officers, their commander, Rodgers, being sick. Battery L was weak, on paper at least, *if* the men of the 13th Massachusetts Battery were discounted. There were 99 regulars and 61 of the Massachusetts men (listed separately) on Battery L's June return. The brief integration of the state volunteers with a regular unit was evidently something out of the ordinary, and perhaps something that the larger system could not accommodate. The Massachusetts men were never listed on any of Battery L's muster rolls while attached to L, except as casualties. In fact, if anyone outside of the Department of the Gulf read the returns of the 1st Regiment, they would have had no clue as to the association of Battery L and the 13th Massachusetts Battery. The combination was apparently a creature of the Department of the Gulf, and may never have been known or officially approved at Washington. That said, on July 1st, the 13th Massachusetts was again made an independent command, when its commanding officer, Capt. Charles Hamlin, returned to Louisiana.[921]

Though the three batteries in Emory's command had been ordered out, considerations regarding turning over their equipment would delay their departure and events in the east, specifically, Confederate Gen. Jubal Early's advance on Washington, would cause a surprising turn. On June 30th, a substantial detachment of the 19th Army Corps would be boarding steamers for movement to New Orleans. They also had been ordered east, and the urgency of the matter caused them to leave Louisiana before Battery L.

---

921. *Massachusetts Soldiers, Sailors and Marines*, Vol. 5, pp. 507–508; Bowen, p. 854; O.R. Vol. 34/IV, pp. 440, 464.

Chapter 14

"Record" 8/64; Camp Barry; Washington Threatened; The 19th Corps Arrives;
After Early; Sheridan Appointed;
The Army of the Shenandoah; Weapons; First Moves; Early Reinforced

"Record" 8/64
30 JUNE–31 AUGUST, 1864, NEW ORLEANS, LA–NEAR BERRYVILLE, VIRGINIA

Battery ordered to Fort Schuyler, New York for refit and the mustering of new recruits. The Company left New Orleans on board the steamer Yazoo on July 27th and arrived at New York Harbor on August 4th. Thence ordered to Camp Barry, near Washington, D.C. Left Washington on the 15th and marched to Harper's Ferry, Va. Engaged the enemy on the 28th & 29th near Smithfield, Va. Camped near Berryville, Va. on the 31st.[922]

Henry W. Closson	Capt. Det. svc. at New Orleans on General Granger's staff.
Franck E. Taylor	1st lt. Commanding Battery
Edward L. Appleton	1st Lt. On recruiting service at New York S.O. 185 Hdqtrs. Dept of the East, New York City Aug. 4, 1864.
Detached:	
Edmond Cotterill	Cpl. On Det. Svc. Clerk in A.G.O. Wash. S.O. no. 426, War Dept.A.G.O., Wash. Sept. 23,'63. Left Co. Nov. 20,'63.
William Demarest	Sgt. On recruiting service, S.O. 185 Hdqtrs. Dept of the East, New York City
Benjamin O. Hall	Pvt. do. do.
Michael Teighe	Pvt. do. do.
Rueben Townsend	Pvt. do. do.
Edmund Anglin	Pvt. In Battery G 5th US Artillery at New Orleans.
Arthur Flynn	Pvt. do. do.
Michael Kenny	Pvt. do. do.
John Meyer	Pvt. do. do.
Joseph Smith	Pvt. do. do.
Solomon Montgomery	Pvt. [Previously orderly to the Chief of Artillery, he does not appear on this roll, he reappears on the next roll as a private.]
Absent in confinement:	
Patrick Gibbons	Pvt. At Ship Island, serving sentence of Gen. Court Martial S.O. no. 18,

---

922. The "Record of Events" portion of the muster roll is blank. The narrative above is a combination taken from Haskin, p. 199, and the July monthly return.

Hdqtrs. 1st Div. 19th Army Corps Dec. 31, 1863.

John Lewery	Pvt. In confinement at New Orleans, La., apprehended as a deserter.

Absent Missing:

Michael White	Sgt. Wounded and missing since April 9, 1864. Deserted:
Henry Wilkson	Cpl. Deserted July 27, 1864 Location not specified.
William F. Brown	Pvt. Deserted July 2, 1864 do.
Benjamin Hughes*	Pvt. do. July 2, 1864 do.
George Harrison	Pvt. do. July 23, 1864 do.
Francis Jessop	Pvt. do. Aug. 11, 1864 Philadelphia
Joseph Kutschor	Pvt. do. July 2, 1864 Location not specified.
John Lowry	Pvt. do. July 23, 1864 do.
Daniel Moore	Pvt. do. Aug. 11, 1864 Philadelphia.
John H. Moran	Pvt. do. July 4, 1864 Location not specified.
Michael Olvany	Pvt. do. Aug. 11, 1864 Philadelphia.

* This man deserted from this organization and enlisted Aug. 5, 1864 under the name of John Wilson (as substitute) in Company M, 1st Missouri Light Artillery Vols. in violation of the 23rd (now 50th) Article of War.

Discharged:

George Friedman	Pvt. By promotion to 2nd Lt. Corps d' Afrique S.O. no. 200 Hdqtrs. A.G.O. Washington June 7, 1864.
William E. Scott	Cpl. By reason of reenlistment in the Battery, July 11, 1864.
Henry Wilkson	Cpl. By reason of reenlistment in the Battery, July 18, 1864.
Owen A. Wren	Cpl. By reason of reenlistment in the Battery July 11, 1864.
Ludwig Rupprecht	Musician do.
James Ahern	Pvt. By reason of reenlistment in the Battery July 18, 1864.
James Beglan	Pvt. do.
John Burke	Pvt. do.
Patrick Craffy	Pvt. do.
Patrick Donnely	Pvt. do.
William Creed	Pvt. do.
Patrick Cummings	Pvt. do.
George Howard	Pvt. do.
Miles McDonough	Pvt. do.
Andrew Stoll	Pvt. do.
Reuben Townsend	Pvt. do.
Henry Williams	Pvt. do.
Prosper Ferrari	Pvt. By reason of enlistment in the Battery, July 19, 1864.
George Hadley	Pvt. do.
Michael Olvany	Pvt. do.
Wm. V. Thompson	Pvt. do.
Henry A. Ward	Pvt. do.

Died:

Michael O'Sullivan	Pvt. In parish prison New Orleans, La., Oct. 2, 1863, of chronic diarrhea.
Joseph H. Parslow	Pvt. Killed in action near Smithfield, West Virginia.

Strength: 86 Sick: 7

Present Sick: none

Absent Sick:

William Brunskill	Pvt. At Ft. Hamilton, NY left Co. Sept. 17, 1861.
George Chase	Pvt. At Brashear City, La., since April 22, 1863.
William Crowley	Pvt. At Baton Rouge since July 13, 1863.
Daniel Howard	Pvt. At Harper's Ferry, Va., since Aug. 25, 1864.
Charles Jackel	Pvt. At Franklin, La., since March 12, 1864.
John McKenny	Pvt. At Harper's Ferry, Va., since Aug. 26, 1864.
Churchill Moore	Pvt. At New Orleans, since April 20, 1864.

Colored Cooks: William Jefferson, Phillip Evens, Virgil Ayers. Paid $12.00 each, with the note: "Error on last payroll."

Note that there are only two illnesses that are recent. Others are the typical long-lasting debilities of the time, William Brunskill being the extraordinary example. What a way to go: Michael O'Sullivan, dead in prison; doubtlessly from unsanitary conditions, poor food – or maltreatment?

It could have been predicted, having not had a roof over their heads for the past year, much less having enjoyed even the most rudimentary comforts of civilization, that what had happened upon their arrival at New Orleans back in 1863 was repeated on August 4th at New York. It meant going out on a "vacation." The number of those who are listed as deserted, at ten, however, is somewhat lower in proportion to the number of those who deserted at New Orleans. Then there were 141 listed in the battery, and 20 deserted. Older and wiser? Wiser seems reasonable. Sixty-four had reenlisted since, and were due installment bounty payments. Earlier reenlistments were for a bounty of $100, and only $25 had been paid on installment. However, the government having become near desperate to retain veterans, the bounty now was $400, and no installment had been paid to date. Another minor point is that four of the men were assigned to recruiting service at New York! Who would desert if they had gotten wind of the potential for that plum?

Yet, Cpl. Henry Wilkson deserted, as did Michael Olvany, just after reenlisting and getting the government's $400 promise. Wilkson may have been influenced by those damned assessments for camp and garrison equipment, and ordinance! He owed $40; though Olvany only owed $3.04. Note that Olvany, Moore, Moran, and Jessop deserted at Philadelphia, while the battery was on its way to Washington. They had not returned, voluntarily or otherwise, by the end of the year.

Another way to get ready money in hand, though, was to become a substitute. The $300 offered a substitute attracted Benjamin Hughes, who illegally, but for a time successfully, deserted and reenlisted, under the assumed name of John Wilson, in the 1st Missouri Light Artillery, as was over-written into the muster roll record—*in 1890*. This was likely when he applied for a pension.

Friedman, discussed earlier, here left the Battery L books forever, promoted to 2nd lieutenant.

Camp Barry–Consolidation with Battery K

The steamer *Yazoo* pulled into New York Harbor on August 3rd with the four artillery batteries.[923] Reporting at headquarters, Department of the East, they were immediately ordered to the Light Artillery Depot and Camp of Instruction,[924] also known as Camp Barry, on the Bladensburg Turnpike, northeast of Washington. Here Battery L was consolidated with Battery K of the 1st U.S. Artillery. In the absence of new recruits from the failure of the draft, consolidating the men of the two batteries was the device chosen to obtain something close to full war organization, though they still would have only four guns.

K had been there since the 14th of July, arriving from duty with the Army of the Potomac, where it had served in Grant's spring campaign, the "big licks" designed to end the war. From their new association, Battery L was introduced to the intensive campaign that Grant had begun six weeks earlier, and the fact that Battery K had been severely cut up in Wilson's raid.

On June 21st, aggressively using the cavalry, Grant ordered Brig. Gen. James H. Wilson's 3rd Division of Sheridan's[925] cavalry, of Meade's army, and Kautz's cavalry division of Butler's army, on a raid to destroy the Weldon and Southside Railroads, the major supply routes to Petersburg and Richmond, figure 1.[926] Battery K was assigned to Wilson's cavalry.

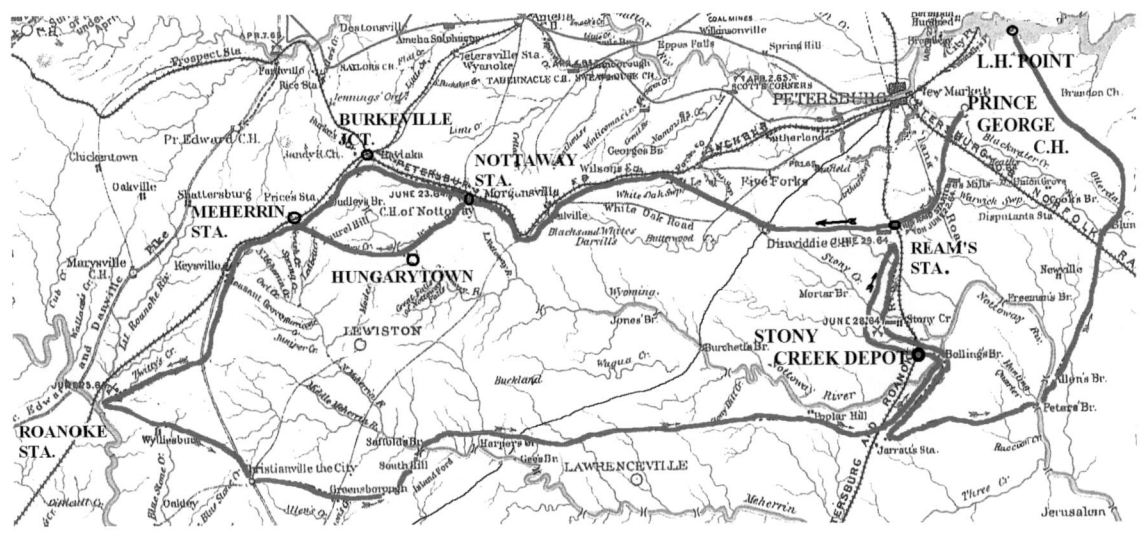

FIGURE 1

Leaving their camp in the vicinity of Prince George Court House early on

---

923. Simpson, p. 322; Haskin, p. 199.
924. O.R. Vol. 43/I, p. 975.
925. O.R. Vol. 36/I, pp. 26, 119, 208–209; O.R. Vol. 40/I, pp. 625–633; Sheridan, Vol. 1, pp. 438–445.
926. Figure 1, adapted from Atlas, plate 74; Wilson's Report, O.R. Vol. 40/I, pp. 625–632.

the morning of June 22nd, they reached Ream's Station, on the Weldon Railroad, destroying it. They then proceeded west, passing though Dinwiddie Court House, to a point on the Southside Railroad about 14 miles west of Petersburg, where they moved deliberately, tearing up the track toward Nottaway Station. Moving south, through Hungarytown, they struck the Richmond and Danville Railroad at Meherrin Station. On the morning of the 25th, they proceeded toward the Staunton River, and the bridge near Roanoke Station. That evening, the advance ran up against Confederate batteries and entrenchments on the south bank of the river.

Beaten, the column turned to the eastward and took almost a direct line back toward Ream's Station, where it expected to find the left wing of Meade's army. The column reached Stony Creek on the 28th without meeting significant resistance, though Confederate forces were now gathering all around it. Here, a small force of the enemy was found at Stony Creek Depot, which was met and overcome, allowing the wearied column, which had barely subsisted off the land for the past eight days, to move on to Ream's Station. Crossing Rowanty Creek, the enemy was met again, and fierce fighting ensued. Facing them was Gen. Wade Hampton, with his and Fitzhugh Lee's cavalry. Wilson then discovered that two brigades of infantry under Confederate Gen. William Mahone had moved in behind them. By 10 o'clock that night they found themselves surrounded.

Wilson ordered a retreat, back down the road, and across the Nottaway. The movement began at midnight. Re-crossing the Rowanty, the column was attacked at Stony Creek, and the whole rear of the column was thrown into confusion. The only route out was through a wood which obstructed the passage of the guns and wagons, and all 12 guns of the three batteries accompanying the raid were abandoned. All of Wilson's wagons were lost, and out of 5,500 men, there were 1,800 casualties. The battered and worn column finally reached Light House Point on July 2nd.

As to Battery K, Capt. Maynadier and Lt. Egan went missing; Sgt. McNamara[927] was wounded and missing, and 35 others were captured or missing, though revised figures show 16 men and officers as prisoners of war. The battery lost all of its 125 horses. On July 11th it was sent to Camp Barry to refit. It had no officers. Capt. William Graham was on detached service at Regimental Headquarters, Concord, NH; 1st Lt. Tulley McCrea was absent on leave, and 2nd Lt. Jacob Counselman had been on detached service since May 14th as a Brevet Lt. Col. of the 1st Regiment of Maryland Cavalry.

Regarding the enlisted men, there were 64, but 16 were on detached service, sick, or absent with or without leave, and 15 were prisoners of war; this left but 33 enlisted men for duty.

---

927. Regimental casualty list, M727-5, pp. 332–333. Haskin; pp. 207, 562; has 15 men, plus the two officers as prisoners. The General Assembly of Maryland, p. 704.

The huge attrition described above was growing, and it had taken place everywhere. The July strength for Battery L was 90, as listed on the monthly report, but only 74 officers and men were present for duty. A relatively new and urgent reason for men to be absent had now appeared. One officer, Appleton, and five enlisted men were on detached service, on recruiting duty in New York.

The result of the recruiting efforts for Battery L were worse than dismal. Only two men were recruited. One was the former first sergeant of the battery, Lewis Keller, who had been discharged on December 29th, 1863, by reason of promotion to Second Lieutenant in the 2nd Louisiana Volunteer Cavalry. The other was a member of the detachment itself, Sgt. William Demarest.

Equally dismal recruiting results were reported elsewhere. The 116th New York Regiment, soon to arrive in Virginia with the 19th Army Corps, had detailed a lieutenant colonel, a captain, a lieutenant, and six enlisted men to return home[928] "for the purpose of bringing to the regiment such conscripts as should be assigned to it." They left on August 12th, 1863, fully expecting the operation of the new draft to fill their needs. They did not return to the regiment until it was encamped at Morganza, in June of 1864. In ten months, they had not recruited a single soul. We see now the utter failure of the draft, and the reason for the $400 bonus.

The $400 for reenlistment had worked, and most of the five-year regulars, plus the first group of the 300,000 three year volunteers, the veterans, had reenlisted.[929] However, the Union army was still running short of men. The draft was not producing the men that were expected because its provisions for exemption were lenient enough to allow thousands to avoid service. In the end, the draft only produced 52,068 men and 75,429 substitutes. Upon the payment of a $300 commutation, 86,724 were excused.

Washington Threatened

Now came the last, and greatest, threat to Washington. It was precipitated by Lee's reaction to Grant's spring campaign, part of a series of seesaw movements between Grant and Lee that would cause the four batteries and the detachment of the 19th Corps to be transferred to Virginia.

With Butler's Army of the James having failed in an attempt to enter Petersburg[930] on June 9th, the Army of the Potomac crossed the James River and began moving to support Butler and renew the assault, which began on June 15th. After some initial successes, Lee rushed reinforcements to the defending general, Beauregard, and when the Union 2nd, 5th, and 11th corps attacked on June 18th, they were

---

928. Clark, pp. 116, 186.
929. Fox, pp. 526, 532–533.
930. CWSAC: Antietam, MD003, Valley Campaign VA101-106, Petersburg 1, VA098, Petersburg 2, VA068.

repulsed with heavy casualties. The siege of Petersburg had begun. Thus, as Grant puts it, "comparative quiet reigned about Petersburg until late July."[931]

Another part of Grant's spring campaign was his order to Franz Sigel, head of the Department of West Virginia, at Winchester, Virginia, to destroy the Confederate railroad complex at Lynchburg. Advancing down the Valley Pike, Sigel met Maj. Gen. John C. Breckinridge at New Market, and was soundly defeated. Sigel was replaced[932] by Maj. Gen. David Hunter, who took over the direction of the initiative. He ultimately failed and retreated into West Virginia. This move left the Valley once again open to the Confederacy. Before Lee even knew of the direction of Hunter's retreat, he had ordered Lt. Gen. Jubal A. Early to move north, "and if opportunity offered, to follow him into Maryland" and threaten Washington.[933] Perhaps the threat would cause Grant to detach troops from the siege of Petersburg.

FIGURE 2

---

931. Grant, Vol. 2, p. 303.
932. O.R. Vol. 37/I, pp. 1, 5–6: CWSAC VA110.
933. Sheridan, Vol. I, p. 457; O.R. Vol. 37/I, pp. 180–182, 191–196 769; CWSAC #MD007, DC00; Grant, Vol. 2, pp. 304–306; Early, *Memoir*, pp. 41–45.

Hunter "disposed of," Early moved north on June 23rd. Often traveling by two roads; his cavalry on one, and his infantry on a parallel one; Early reached Staunton on June 26th and 27th. Unopposed, he marched rapidly down the valley, the track shown in figure 2,[934] reaching Winchester on July 2nd. He captured Martinsburg on July 3rd, causing Sigel's[935] force there to retreat across the Potomac to Shepherdstown, and the garrison at Harper's Ferry to retreat across to Maryland heights. The Confederate cavalry occupied Boonsboro, and a detachment out of McCausland's cavalry drove a portion of the 6th U.S. Cavalry from Hagerstown on July 6th, and $20,000 was levied against the inhabitants. Breckinridge occupied Frederick on the 9th, where the levy demanded was $200,000.[936]

Early met his first real resistance on July 9th. Gen. Lewis Wallace, commanding the Middle Department, headquarters at Baltimore, who had been reinforced by Rickett's Division of Wright's 6th Corps, sent from Grant, had put up a defensive line on the Monocacy River, just east of Frederick. Wallace's hastily formed force of about 3,350 were outflanked and defeated by Early's 12,000.[937]

Early then moved toward Washington via Rockville, and making his headquarters at Silver Spring, his troops pushed along the 7th Street Road, up to the defenses near Fort Stevens, late on the 11th, figure 3.[938] His troops were exhausted, the weather being "excessively hot…and the dust so dense…" that he had had to slacken his pace, after making 30 miles on the 10th. They were not in a condition to make an attack, and they rested until the next day.

Wallace's action on the Monocacy had delayed Early long enough to allow Wright, with the two remaining divisions of the 6th Corps, to arrive near Fort Stevens, and on July 12th, Wright pushed out from the defenses in the face of destructive Confederate fire, and drove Early's pickets out from the house and orchard grove which sheltered them, back about a mile. The determined action resulted in Wright's troops suffering 280 casualties.[939]

---

934. Long, pp. 122–123. Figure 2 from Plate 81, map 11; Atlas.
935. O.R. Vol. 37/I, pp. 7, 199, 347, 349.
936. O.R. Vol. 37/I, pp. 170, 336–337, 349
937. Nicolay and Hay Vol. 9, pp. 161, 169; Early, *Sketch*, p. 381, reports 2,000 cavalry, 10,000 of the 2nd Corps, and 2,250 of Breckinridge's command. O.R. Vol. 37/I, p. 191; O.R. Vol. 37/II, pp. 158–159.
938. Figure 3 from Miller, Vol. III, p. 155.
939. O.R. Vol. 36/I, p. 28.

FIGURE 3

The President, as usual, was keenly interested in the action, and had ridden out to Fort Stevens on the afternoon of the 11th. When the first of Early's troops arrived, Lincoln was standing near the parapet, his tall figure making him a conspicuous target, until he was advised to withdraw. He returned on the 12th and stood in similar danger, watching as Wright's men moved out. Whizzing bullets mortally wounded an officer standing within three feet of him, and he was again advised by Wright to take cover.[940]

An excerpt from Early's official report:

> I determined at first to make an assault, but before it could be made it became apparent that the enemy had been strongly re-enforced, and we knew that the Sixth Corps had arrived from Grant's army, and after consultation with my division commanders I became satisfied that the assault, even if successful, would be attended with such great sacrifice as would insure the destruction of my whole force.

Having steadily marched from near Lynchburg, Virginia since the 23rd of June; battled at Monocacy; then pressed on to Washington, in summer heat, over roads choking with dust, the decision to recognize that effort as a failure was undoubtedly a painful one. Nevertheless, Early satisfied himself that he had "given the Federal authorities a terrible fright."

Lee's comments on the original object of the expedition, (written, of course, for his superiors) refuses to admit of failure, and he retained Early in command:

---

940. Nicolay and Hay, Vol. 9, pp. 172–173.

"threatening Washington and Baltimore General Grant would be compelled either to weaken himself so much for their protection as to afford us an opportunity to attack him…." It did not happen.

The Nineteenth Corps Arrives

As it happened, Emory, with the advance of the 19th Corps, consisting of four companies of the 114th New York, and the 153rd New York, had arrived on the steamer *Crescent* at Fortress Monroe on the afternoon of July 12th.[941] He was immediately ordered to Washington and arrived at Fort Saratoga that night. He then encamped at Camp Barry.

After Early

At the height of the tension, on July 10th, as Early was approaching Fort Stevens, Lincoln had appealed to Grant to come to the scene in person. This time, however, the Federal authorities were not frightened and Lincoln was not urging more defense, but offense. He was proposing that Grant come and supervise a "vigorous effort to destroy the enemy's force in this vicinity. This is what I think, upon your suggestion, and is not an order." The suggestion referred to was Grant's. It was: "Forces enough to defeat all that Early has with him should get in his rear south of him, and follow him up sharply…"

At Grant's suggestion, rather than going into the defenses of Washington, Halleck ordered Wright to chase after Early. Typical of Halleck, he interjected his own interpretation of the situation, warning Wright to be *cautious*.[942]

Cautious or not, it was too late to prevent Early from successfully crossing back into the Shenandoah; reaching Berryville on July 17th. It was left to Crook's infantry and Averell's cavalry, referred to as the Army of West Virginia, under Hunter, to try to deal with Early.

Sheridan Appointed

Dana, now the assistant secretary of war, and as we know, always free to express his opinion to Grant, on July 24th telegraphed Grant's chief of staff regarding Early's raid, and the defense of Washington: "Wright and Crook accomplished nothing, and Wright started back as soon as he got where he might have done something worth while."[943] It all sounded like business as usual.

---

941. O.R. Vol. 37/I, pp. 347-349; Grant, Vol. 2, pp. 304–306. Pellet, pp. 243–245; Early, *A Memoir…*, p. 302.
942. O.R. Vol. 37/I, p. 348; O.R. Vol. 37/II, pp. 207, 210, 258–261, 264, 284, 285, 287, 291, 345, 547.
943. O.R. Vol. 37/II, pp. 374, 408, 427, 433.

Grant must have been impatient with the progress of things as early as July 18th, when he had suggested to Halleck that the departments of the Susquehanna (Pennsylvania), West Virginia, and Washington be merged. Though he did not name the Middle Department (Maryland), it was clear that one vigorous commander would cut through all of the agonizing communication and ego problems that had hampered the response to Early. Grant had suggested Franklin as the man. The response from Halleck was immediately negative: "General Franklin would not give satisfaction. The President ordered him to be tried for negligence and disobedience of orders when here before, but General McClellan assumed the responsibility of his repeated delays in obeying orders."

The reader is left to judge Halleck's response from Franklin's performance in the Department of the Gulf.

On the 25th, Grant tried again. He wrote directly to Lincoln, outlining the same proposal, and explained that he didn't care who was to be placed in command, and mentioned Meade. The "Middle Division" as it would be called, if run by Meade, would be "used to the very best advantage from a personal examination of the ground, and would adopt means of getting the earliest information of any advance of the enemy, and would prepare to meet it." The letter was delivered to the President by Grant's chief of staff, Rawlins, who could convey "more information… than I could give you in a letter." Part of that information was of such a nature that he would "not care to commit to paper…"

Clearly, the comment about "personal examination of the ground" referred to Halleck. We have seen continued evidence of the fact that Halleck had never visited the field, save for the disastrous advance to Corinth. Since, he had sat in the War Department relying on telegrams and dispatches, which tended to be several days late for the reason of cut telegraph wires, or the inability of a fighter in the field to promptly put pen to paper.

The rest of Grant's note, the part that he cared to not commit to paper, is hinted at in his *Memoirs*. It evidently referred to long previous and continuing interference from Washington, i.e., both Halleck and Stanton.[944]

The same day, Lincoln responded that he would like to meet with Grant at Fortress Monroe, "after Thursday," which would have been July 28th. This was too quick for Grant, and he begged off, saying: "I am commencing movements for which I hope favorable results." He was referring to the preparations for the explosion of a mine under the Confederate breastworks of Petersburg, and a diversionary raid on Richmond, designed to draw away some of Lee's forces from the front at Petersburg, which has now come to be known as Deep Bottom.[945]

Elements of Early's command had skirmished with Averell's cavalry at

---

944. Grant, Vol. 2, p. 317.
945. CWSAC VA069.

Stephenson's Depot, about six miles northeast of Winchester, on July 20th, figure 4,[946] and Early had struck a part of Hunter's command, under Crook, at Kernstown, in force, on the 24th,[947] handing him a severe defeat. Crook had been flanked, and shaken. Crook doubted, for a while, that he would be able to retreat to Harper's Ferry and Halleck had taken the initiative to send Wright back out, with orders to unite with Hunter's forces, wherever they could be found.[948] As it happened, Crook was able to get to safety at Maryland Heights and Harper's Ferry.

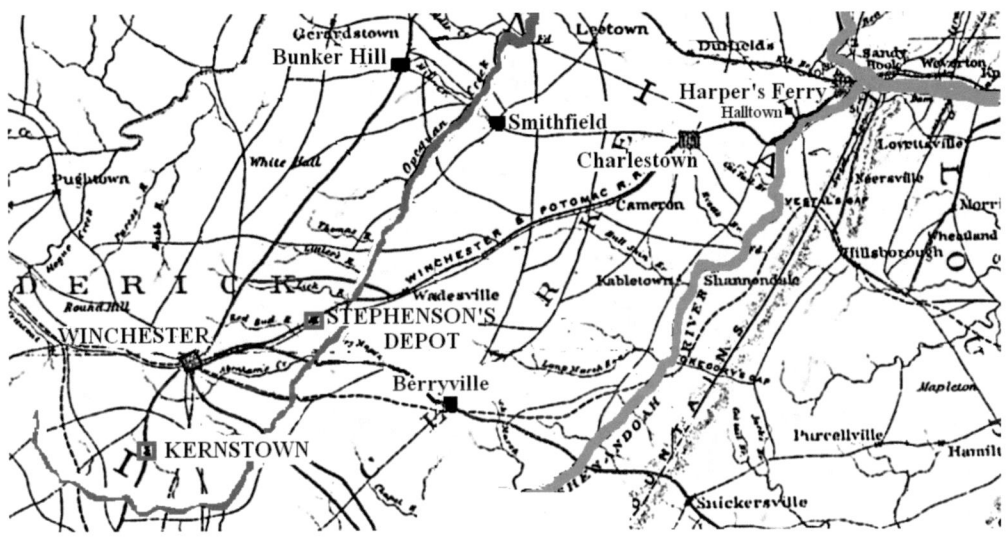

FIGURE 4

Circumstances were such that at least two of Grant's earlier suggestions were adopted immediately: (1) a single commander, and (2) a combined command. Incredibly, on the 27th, Stanton, as directed by Lincoln, named Halleck to the command of the combined Middle, Susquehanna, Washington, and West Virginia Departments.[949] Grant reveals in his *Memoirs* that he had suggested Sheridan, but the suggestion was rejected by Stanton on the grounds that Sheridan was too young for so important a command, though he was now 34 and had been in the regular army for 11 years. After Corinth, he had distinguished himself at Booneville, under Halleck, and under Rosecrans at Murfreesboro, Chickamauga, and Chattanooga. On Rosecrans' recommendation, he had risen to brevet major general. He was appointed commander of the cavalry corps of the Army of the Potomac after Grant had taken command, only after Grant had asked Halleck for "an active and energetic man, full of spirit and vigor and life."

---

946. Figure 4, Atlas, plate 27, portion of map 1, altered.
947. O.R. Vol. 37/II, pp. 408, 436, 445–446, 459; Early, *Sketch*, pp. 398–400; CWSAC VA116.
948. O.R. Vol. 37/II, p. 456.
949. O.R. Vol. 37/II, pp. 463, 470, 486, 509, 511, 515, 525, 558; Grant, Vol. 2, p. 317; *Encyclopedia Britannica*, 1911, Vol. 24, p. 847.

In the meantime, Early was left free to harvest grain and hay in the Shenandoah Valley and run it to Culpeper, Virginia, by train. At this point, Early had knowledge of Hunter's actions in the burning of the houses[950] of several prominent Confederate officials in the Valley, as well as the Virginia Military Institute. He determined on a retaliatory attack. Having control of the territory as far north as Martinsburg, he decided to strike at Chambersburg, Pennsylvania, and beyond, if possible. On the 29th two of his cavalry brigades, one under McCausland and the other under Bradley T. Johnson, crossed the Potomac. Entering Chambersburg on the 30th, McCausland ordered it burned when his ransom demand[951] for $100,000 in gold or $500,000 in greenbacks, could not be met. He moved on to Hancock, Maryland, where he demanded $30,000. Not having obtained it, he ordered the town burned. However, Col. Harry Gilmor, a Confederate Marylander, commanding the 2nd Maryland Cavalry (formerly Partisan Rangers), objected, and with the permission of Johnson, stationed his own troops so as to prevent it. Averell's arrival shortly thereafter chased them out of the place, and it remained unharmed. McCausland was reported as drunk during most of these proceedings.

Johnson's official report, which includes acts before and after the Chambersburg and Hancock raids, terms the conduct of McCausland's troops as "outrageous."

Chased by Averell, McCausland fell back into West Virginia. Near Moorefield, Johnson was offended that the barbarism had continued. He relates: "A lieutenant knocked down and kicked an aged woman who has two sons in the Confederate army, and after choking the sister locked her in the stable and set fire to it. This was because the two women would not give up horses he and his fellow thieves wished to steal." Here were Confederate soldiers stealing from Confederate sympathizers. What was going on? There was little doubt that the South was suffering and the Confederate soldier was becoming desperate, frustrated, and often barefoot.[952] Most significantly though, many in McCausland's cavalry were former guerrillas, with lawless habits inherited from what had been authorized in the Confederate Partisan Ranger Law.

By the 1st of August, with Early finally driven away by Averell's cavalry, Grant had had enough. His "movements," that he had earlier described to Lincoln for which he hoped "favorable results" had proved to be a "stupendous"[953] failure. The mine explosion in front of Petersburg was ineptly followed up, the troops entering

---

950. Early, *Sketch*, pp. 401–402.
951. Nicolay and Hay, Vol. 9, pp. 176, 177; O.R. Vol. 37/II, pp. 525, 534, 542; Gilmor, p. 209, says $200,000 in gold, or its equivalent in greenbacks. Reference to Hancock, p. 213; O.R. Vol. 37/I, pp. 354–356; O.R. Vol. 43/I, pp. 7–8.
952. O.R. Vol. 33, p. 1275, O.R. Vol. 36/II, p. 821; O.R. Vol. 37/I, pp. 13, 120; O.R. Vol. 43/I p. 558; Pollard, pp. 332–334; Foote, SHS, Vol. 31, pp. 237–239; Butler, p. 610.
953. Grant, Vol. 2, p. 315; CWSAC no. VA069, VA070; Grant, Vol. 2, pp. 313–314; O.R. Vol. 40/I, p. 563.

the massive 175-by-50-foot crater, rather than skirting it, leaving them to be picked off as fish in a barrel, which resulted in almost 4,000 men as casualties or prisoners. In addition, the diversionary raid to Deep Bottom had failed to draw away enough of Lee's force to make a difference.

Perhaps the necessary action to solve the problem of Early in the Shenandoah Valley was really an opportunity in disguise. After all, it was Lee's breadbasket, and had been his avenue into the north ever since 1862. The Shenandoah would not be given up easily by Lee, and Early was soon reinforced. If Early could be decisively defeated in the open country and good roads of the Valley, it would be a major setback to the Army of Northern Virginia, something that now seemed impossible to accomplish, given the stalemate at Petersburg.

Grant now instructed Halleck that he "wanted" Gen. Philip Sheridan put in temporary command of all of the troops in the field, "with instructions to put himself south of the enemy and follow him to the death." This was the first time that Grant had not made a "suggestion." Up to this point, Halleck and Grant had showed great deference to each other's authority,[954] but this telegram offered no options.

During Early's invasion, Lincoln had resolutely kept from interfering in any of the decisions made by either Halleck or Grant, and he once again said nothing for or against Sheridan. However, upon reading Grant's telegram, he replied (and here, it is quoted in full, to picture the President's level of despair at the lack of Halleck's decisiveness, and the lack of aggressiveness of those in the field):

> I see your dispatch in which you say "I want Sheridan put in command of all the troops in the field, with instructions to put himself south of the enemy and follow him to the death. Wherever the enemy goes let our troops go also." This, I think, is exactly right as to how our forces should move, but please look over the dispatches you have received from here ever since you made that order, and discover, if you can, that there is any idea in the head of anyone here of "putting our army south of the enemy," or of "following him to the death" in any direction. I repeat to you it will neither be done nor attempted, unless you watch it every day and force it.

The one aspect of Grant's appointment of Sheridan that was not settled was the reaction of Hunter. Prompted by Lincoln's remarks, Grant replied that he "would start in two hours for Washington." Instead, he went directly to Monocacy Station, near Frederick, Maryland, where he found Hunter.[955]

As had happened earlier, when Wright had been assigned to the overall

---

954. Nicolay and Hay, Vol. 9, pp. 172, 180; O.R. Vol. 37/II, pp. 582, 591.
955. Grant, Vol. 2, pp. 319–320; Nicolay and Hay, Vol. 9, pp. 181–182. The idea that Hunter was gallant in stepping aside is put forward by Nicolay and Hay, and also in Sheridan's *Memoirs*, Vol. 1, pp. 465–466.

command of the troops chasing after Early, Hunter offered to resign. Previously, he had been assuaged by his being retained as department commander, having his subordinates, Crook and Averell, in the field. A similar desk job was now offered Hunter, with Sheridan to be in the field, but Hunter "gallantly" declined; he allegedly stated that the greater good was for him to step aside and let Sheridan run his own show. Grant only too quickly said "very well then" and summoned Sheridan from Washington to meet him.[956] Having met Lincoln and Stanton just before he left, Sheridan was provided with a special train.

The three generals met the next day, August 6th. Though he was to be relieved, Hunter had not hesitated to order the first phase of Grant's written instructions—a concentration of all of his forces at Halltown, southwest of Harper's Ferry. As a result, Monocacy Station was virtually deserted, save Grant, Sheridan, Hunter, and their staffs. Grant, with a few words, handed Sheridan his letter of instructions. It was, in fact, still addressed to Hunter.

By Presidential order, Sheridan was officially made commander of the Middle Military Division on August 7th. The troops in his command, which came to be called the Army of the Shenandoah, would consist of Wright's 6th Corps, Emory's 19th Corps detachment (remember that only two of its divisions, under Dwight and Grover had come north), the Army of West Virginia, under Crook, and a cavalry corps, eventually organized into three divisions, Merritt's, Averell's, and Wilson's, under the overall command of Alfred Torbert. Torbert, Merritt, and Wilson were sent by Grant from the Army of the Potomac, and Averell was transferred from Crook.

Looking back at these events, after the President's prompting, it can be said that this was the point in time when Grant gained full control of the war as general-in-chief.

The Army of the Shenandoah

Sheridan was detached from the Army of the Potomac, where he had commanded the cavalry corps. Almost his whole experience in the war had been with the cavalry, having been promoted to the command of the 2nd Michigan on May 25th, 1862.[957]

In earlier discussions with Halleck, prior to his appointment to head the whole Army of the Shenandoah, then assuming he would serve under Hunter, Sheridan had expressed his preference for the cavalry alone. He would not need the 6th Corps. His reasoning was not his inherent love of the cavalry, but his opinion that in the open country of the Maryland and northern Virginia area, with its fine

---

956. Sheridan, Vol. 1, pp. 464–466, 472–474; O.R. Vol. 37/II, pp. 572, 582–583.; O.R. Vol. 43/I, pp. 110–111.
957. Sheridan, Vol. 1, p. 141.

macadamized roads, the cavalry would be most effective. Fortunately, he was given control of the infantry as well, and events would prove that he would badly need it.

Grover's division, the 2nd, had not yet joined Emory's 1st, which was now in the vicinity southwest of Harper's Ferry, at Charles Town. Grover, who had been stationed at Washington, was ordered to join Sheridan on the 12th.[958] On August 14th, Augur, in command at Washington, was notified that Batteries K & L, having been consolidated at Camp Barry, and wondering what would become of them, were to form a reserve battery for the cavalry of General Sheridan's command, and were to temporarily join Grover's division, "and proceed with it." This was exciting news; official notification that their place in the scheme of things for the rest of the war was to be with the cavalry. It was the first time ever that Battery L had specifically been assigned to support cavalry.

The march of the column of Grover's Division with Battery K–L arrived at Halltown, four miles outside of Harper's Ferry, on the 22nd, where the two 19th Corps divisions were united under Emory. Here, Battery K–L was assigned to the command of Brig. Gen. Alfred Torbert's 1st Cavalry Division, under Brig. Gen. Wesley Merritt. It would be the last of Battery L's association with the 19th Army Corps.[959]

Sheridan would now have a much larger force than that of Hunter and Wright when they made their previously aborted attempts to deal with Early. Sheridan mentions 26,000; Grant mentions 30,000, but of that total, "8,000" were cavalry. However, these declarations are misleading. Both Grant and Sheridan were only thinking of the new detachments from the Army of the Potomac which would follow Sheridan into the Valley. There already was Crook's Department of West Virginia, with 12,436 infantry and Averell's 6,472 cavalry, which are listed in figure 5. There were thousands of others in his new Middle Military Division, but as Sheridan notes: "Baltimore, Washington, Harper's Ferry, Hagerstown, Frederick, Cumberland, and a score of other points; besides the strong detachments that it took to keep the Baltimore and Ohio Railroad open through the mountains of West Virginia, and escorts for my trains, absorbed so many men that the column which could be made available for field operations was small when compared with the showing on paper."

This reminds us of the price to be paid to hold territory. There is no question that holding this particular home ground was a sound and necessary strategy, but think back to the wisdom of Grant wanting Banks to pull out of Texas and most of Louisiana. The war could never be won by trying to occupy those vast stretches beyond the Mississippi.

With Crook and Averell added in, Sheridan had more like 38,500 infantry

---

958. O.R. Vol. 43/I, pp. 40, 79, 724, 727, 760–761, 776, 793; Pellet, p. 249.
959. O.R. Vol. 43/I, pp. 421, 516, 987.

and 15,000 cavalry. The artillery attached to the cavalry would raise the total force available for use in the field to 57,000; out of a total of 90,000 in the department listed as "present for duty."[960]

Abstract from return of the Middle Military Division, Maj. Gen. Philip H. Sheridan, U. S. Army, commanding, for the month of August, 1864.

Command.	Present for duty.		Aggregate present.	Aggregate present and absent.	Pieces of artillery.		Headquarters.
	Officers.	Men.			Heavy.	Field.	
General headquarters	24	120	216	389			In the field.
Department of Washington (Augur):							Washington, D. C.
Staff and infantry	483	18,597	23,076	26,572			
Cavalry	120	3,402	5,752	7,180			
Artillery	218	5,895	7,753	9,854	736	279	
Detachment of Signal Corps	7	68	75	81			
Total	828	27,962	36,656	43,687	736	279	
Department of the Susquehanna (Couch):							Chambersburg, Pa.
Staff and infantry	138	1,872	2,228	3,670			
Cavalry	13	401	427	464			
Artillery	5	181	194	296		12	
Detachment of Signal Corps	2	56	59	66			
Total	158	2,510	2,908	4,496		12	
Middle Department (Wallace):							Baltimore, Md.
Staff and infantry	260	5,067	5,963	8,362			
Cavalry	19	297	347	660			
Artillery	8	219	294	329		6	
Total	287	5,583	6,604	9,351		6	
Department of West Virginia (Crook):							In the field.
Staff and infantry	538	11,898	14,032	23,443			
Cavalry	241	6,231	8,457	14,722			
Artillery	83	2,877	3,521	4,478			
Total	862	21,006	26,010	42,643			
Sixth Army Corps (Wright):							In the field.
Staff and infantry	635	11,333	15,717	29,599			
Artillery	24	623	697	812		24	
Total	659	11,956	16,414	30,411		24	
Detachment Nineteenth Army Corps (Emory):							In the field.
Staff and infantry	657	12,068	14,187	21,081			
Artillery	15	436	458	559		20	
Total	672	12,504	14,645	21,640		20	
Cavalry forces (Torbert):							In the field.
Cavalry and staff	371	7,891	10,347	20,028		32	
Artillery	22	611	701	979		24	
Total	393	8,502	11,048	21,007		56	
Grand total	3,883	90,143	114,501	173,624	736	397	

FIGURE 5

---

960. Figure 5, O.R. Vol. 43/I, p. 974; Sheridan, Vol. 1, p. 475; Grant, Vol. 2, p. 321.

Imagine having 18,500 horses to feed (15,000 cavalry and 3,500 artillery), requiring an absolutely stupendous amount of grain and forage. The 1862 caution of Lincoln to Banks about the "thousand wagons to feed the animals that draw them…" was still true, but here in the Shenandoah, supply lines were shorter by far than in Louisiana, and the expense of all of the "impedimenta" for such a magnificent cavalry force seems to have never entered into anyone's calculation. Economy had gone out the window. The national purpose was now to win at any cost. We remember the enormous expense of the $400 reenlistment bonuses offered to thousands of veterans this past month. But, it had successfully retained the first volunteers, and what was money when the very substance of the Nation was at stake?

The July 23rd issue of *Harper's Weekly,* p. 467[961] ends a long discussion of the military situation with:

> The duration of the war proves quite as much the national purpose as the rebel pluck. If it would have been foolish to relinquish our cause before we had proved our quality, it would be the height of folly to do so when that quality has been established. If it were wrong to cry for quarter before we had taken Paducah, and were not sure of St. Louis, it can hardly be right when we have one hand on Atlanta and the other on Richmond.

Weapons

The subject of the tools with which the armies or the navies of either side had at their disposal has been only briefly touched upon, because the soldier seldom had the inclination or the time to make any technical observations. Remember that Bragg at Pensacola had been supplied with a formidable array of guns in April of 1861? It was the London Times reporter, Russell, who observed that only a handful were useful. When Harvey Brown arrived, to his credit, he recognized that he must be supplied with rifled cannons to be able to do any damage to Bragg. Also, during the bombardments, Admiral McKean noted his need for heavy rifled guns.

At Irish Bend, it was Colonel Day of the 131st New York who noted that many of his "pieces" were defective and of short range. Also, we learned that the 26th Maine carried old smoothbore muskets, the same short range muzzleloaders as Taylor's men.

It was a picture of the contending sides emerging from opposite states of unpreparedness - the North with shortages and the South with obsolescence.

When Secretary Floyd looted the Northern arsenals and sent their contents to the South, he personally may have recognized that most of those small arms were obsolescent, but this was lost upon the "fire eaters." The same could be said of the

---

961. Welles, Vol. 2, pp. 70–71.

heavy weapons found in those coastal forts which the Confederacy had occupied.

Thus, though Josiah Gorgas, the Confederate chief of ordnance, had sent hundreds of guns to Bragg. The manufacture of useful modern weapons of any kind, either small or large, would confound the Confederacy for the rest of the war.

The Union army briefly carried obsolescent small arms because it had to scour Europe for anything available, before sufficient numbers could be manufactured at home. Floyd had done some damage. The Union recognized its deficiencies, and quickly took steps to remedy them.

There were two centers for the supervision of the procurement and development of weapons – the navy Bureau of Ordnance and the army Ordnance Department.

The command of the Washington Navy Yard was given to the respected John Dahlgren in August of 1861,[962] and in July of 1862 he was given the command of the navy Bureau of Ordnance. The development of the line of heavy Dahlgren shipboard guns and the purchase of Sharps and Hankins breechloader carbines for boat service is attributable to him.[963]

The Army Ordnance Department went on to develop guns for mobile use, one of the more notable being the 3" Ordnance Rifle, two of whom were issued to Battery L at Pensacola in December of 1862.[964]

Regarding small arms, though, the Union was overwhelmed with the need to procure a reliable standard weapon, which was intended to answer the likes of Colonel Day, stuck with a mix of foreign weapons purchased out of necessity at the outset of the war. The Chief of Ordnance, James Ripley, no doubt influenced by the problems of the complex but innovative Maynard Patent rifled musket, adopted in 1855, refused to countenance anything new or complex. Thus, throughout the war, most of the Union army carried the Model 1861 Springfield, or the British Enfield, a very similar design. These were, however, semi-obsolescent.

Ripley had refused to consider one of the most revolutionary weapons of all time – the breechloader. Spencer breech loading carbines were not ordered in any numbers until Ripley retired in 1863. If anyone had carried them prior to 1864 it was either because his state had purchased them or the soldier had done so himself. Col. James Kidd recalls that two divisions of the Michigan cavalry carried Spencer rifles at Gettysburg, purchased by Governor Blair.[965]

However, by mid-1864, those late 1863 orders were being filled, and the question became: To whom should they be issued? This was answered in a heartbeat by Gen. James H. Wilson.

In January of 1864, Wilson, a respected member of Grant's staff, was

---

962. Dahlgren, p. 46.
963. *Congressional Globe*, 38th Congress, 1st Session, p. 851.
964. Haskin, p. 188.
965. Kidd, pp. 128, 270, 281; The Michiganders traded their rifles for carbines in July 1864.

temporarily assigned to head the Cavalry Bureau in Washington.[966] While the intent of the assignment had been to root out corruption related to the purchases of horses, Wilson also saw to it that the standard cavalry arm would henceforth be the Spencer carbine. On April 17, 1864, his reforms complete, he was assigned to the command of Sheridan's 3rd Cavalry Division. The question remained as to when the cavalry actually received the new weapons, but Wilson assures us that it was by the time Sheridan entered the Shenandoah.

Selected infantry units were issued the Spencer rifle, including the 37th Massachusetts, a part of the 6th Corps, because they had requested them.[967]

First Moves

Early, concerned that McCausland would not be able to withdraw from the Chambersburg raid unscathed, and to keep Hunter guessing, moved out on the 4th of August from Bunker Hill, figure 6,[968] where he had been encamped since the first of the month. He sent Breckinridge across the time-worn fords of the Potomac at Shepherdstown, and occupied Sharpsburg, just as Lee had done in 1862. To Williamsport, through which Lee had retreated after Gettysburg, Early sent Rodes' and Ramseur's divisions.

These moves did nothing to protect Johnson's and McCausland's brigades from Averell's attack, which routed them, and hearing of the news that Sheridan's cavalry had entered the Valley, Early retreated back to Bunker Hill on August 7th.

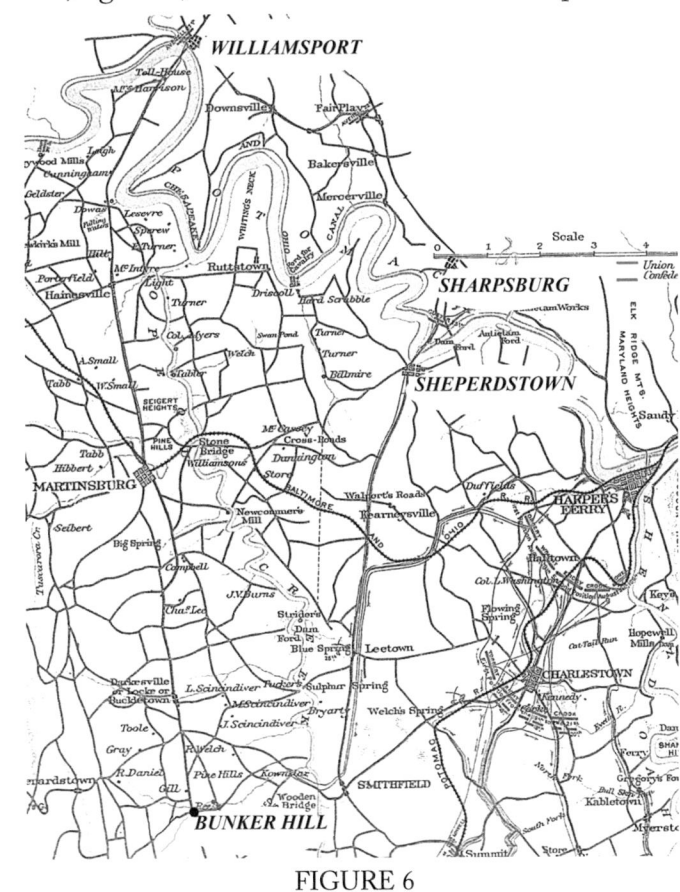

FIGURE 6

---

966. Wilson, pp. 326, 331, 332, 367, 374.
967. O.R. Vol. 43/I, pp. 516, 770, 789, 797, 817, 860, 923; Kidd, pp. 77, 78; Bowen, J. P., pp. 354, 355; House Documents, Purchases of Arms, 1861, pp. 168-170.
968. Figure 6, portion of Atlas, plate 83, map 1, altered; Sheridan's *Memoirs*, Vol. 1, pp. 476–478.

Taking only three days to study the situation, Sheridan moved out from Halltown on the 10th of August; even before all of his force had gathered. He subsequently found that he could not safely hold any ground, and in little more than a week was back where he started. Sheridan may not then have understood that he had, in fact, gained something even more valuable. His careful initial moves drew Early to the fatally flawed conclusion that Sheridan was merely another one of the timid generals from the Army of the Potomac.[969]

Figure 7 shows the area of the Valley that Sheridan initially occupied and from which Early retired. At the center, the area around Winchester would become the focus of almost every subsequent action, and Fisher's Hill (Strasburg) at the bottom, the defensive line that Early would rely on.

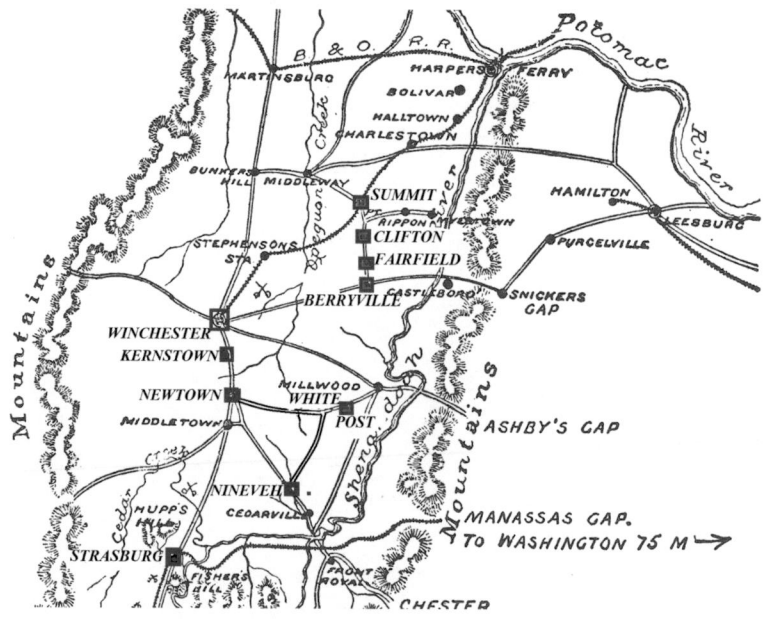

FIGURE 7

On the 11th, Merritt's cavalry discovered that Early was retreating up (south) the Valley Pike, toward Kernstown. Torbert then moved toward Newtown, to strike Early's flank and harass him. Merritt met Gordon's division of Early's infantry, which held the ground east of Newtown until nightfall. Lowell's brigade, of Merritt's division, followed through Winchester on the Valley Pike, while Crook moved toward Nineveh south of White Post, and the rest of the infantry remained within supporting distance of Crook.[970]

The next morning, Early retired further south from Newtown, through Middletown, to Hupp's Hill, and seeing Sheridan's forces still advancing in much heavier concentration than he had yet encountered, he decided to withdraw to Fisher's Hill, two miles south of Strasburg.

---

969. Early, *Sketch*, p. 414.
970. Sheridan, Vol. 2, pp. 479–480; Early, *Sketch*, p. 406, Nicolay and Hay, Vol. 9, p. 293.

Early Reinforced

Early was now made aware that Lt. Gen. Richard H. Anderson, with infantry cavalry, and artillery units from Lee's Army, had been sent to reinforce him. Anderson having arrived at Culpeper on the 12th, Early requested him to move to Front Royal, just west of Manassas Gap, to cover the approaches, and prevent Sheridan from advancing from the east, potentially getting behind him through the Luray Valley.

It was at Fisher's Hill that McCausland and Johnson finally joined Early with the remnants of their brigades, having been routed, as previously noted, by Averell on the 7th. Escaping from Hancock, Maryland, they were pursued all the way to Moorfield, West Virginia. They then moved on to Mount Airy, west of Middletown, and finally reported to Early. They had lost all of their artillery, a setback which Early later wrote had materially weakened his cavalry[971] for the rest of the campaign. This was a significant accomplishment for Averell, though he and Sheridan would soon clash and Averell's career would come to an end.

While at Fisher's Hill, Early set up a signal and observation station on Signal Knob, on Three Top Mountain, where the Massanutten range ends. It overlooked the Valley for miles. Early would use it to his advantage in the future.

On August 14th, Sheridan received, through Halleck, a message from Grant, considered of such importance that it was delivered by a full colonel, who was escorted by a regiment of cavalry. It warned of Anderson's corps having been sent by Lee. Grant decided that this increase in Early's force would be "too much for General Sheridan to attack."[972] It now became Sheridan's time to be concerned. He might become trapped and overwhelmed by a superior force. Wilson's cavalry, and Grover's division of the 19th Corps with Battery K–L, had not yet joined him, and he had had to leave numbers of troops to garrison points taken in his advance up the Valley, including Winchester.

Deciding that there was no defensible ground in the Valley except at Halltown, Sheridan began a withdrawal, which would also allow time for the remainder of his army to join him. On the way north, he had been instructed by Grant to "destroy and carry off the crops, animals, negroes, and all men under fifty years of age capable of bearing arms. In this way you will get many of Mosby's men." All male citizens under fifty can fairly be held as prisoners of war, and not as citizen prisoners. If not already soldiers, they will be made so the moment the rebel army gets hold of them." The Confederate conscription age at this time had been extended to all males from the age of eighteen to fifty-five.[973]

---

971. Gilmor, pp. 225–226; Early, *Sketch*, pp. 405, 407.
972. Nicolay and Hay, Vol. 9, p. 293; Sheridan, Vol. 2, pp. 482–483, 489; O.R. Vol. 43/I, pp. 792, 811.
973. Pollard, p. 331.

After Anderson had left, Lee decided to send even more troops to Early. On the 11th, Lee promoted Maj. Gen. Wade Hampton to the command of the cavalry of the Army of Northern Virginia, and ordered him to report to Anderson. Advised of Lee's movements of reinforcements, Grant played the diversionary raid card once again, and it was once again at Deep Bottom. Deep Bottom was fearfully close to Richmond, only some 20 miles, and bound to get Lee's attention.

On August 12th, Grant ordered Gen. W. S. Hancock, commander of the 2nd Corps of the Army of the Potomac, to prepare, in the utmost secrecy, for a move that would appear as the transfer of his troops to Washington. On the night of the 13th, the 2nd Corps, along with 9,000 men of the 10th Corps, and Gregg's Cavalry, were embarked on steamers and landed on the north side of the river, at Deep Bottom.[974] Hancock's instructions were to threaten, but not to bring on a battle. In addition, a coordinating attack on the Weldon Railroad, near Petersburg, was made on the 17th by Warren's 5th Corps. After six days, and some sharp fighting, it looked like Hancock's advance had run up against lines of the enemy that were, even if successfully attacked, deemed difficult to hold, and of no decisive advantage.[975] News of Warren's success, known as the Battle of Globe Tavern, allowed Hancock to be recalled. Hancock's casualties were 95 killed, 553 wounded, and 267 missing, a total of 915. This time, the effect of a Deep Bottom raid was nearly instantaneous. To Hampton from Lee, dated August 14th: "Halt your command and return toward Richmond."[976]

---

974. O.R. Vol. 42/II, pp. 131–132, 135–137, 140–141, 148, 153, 160, 162, 167, 172–173, 210–211; Grant, Vol. 2, pp. 321–322.
975. O.R. Vol. 42/II, pp. 226, 250, 301, 326–327; O.R. Vol. 42/I, pp. 216–221; CWSAC VA071.
976. O.R. Vol. 42/II, p. 1177.

Chapter 15

Cedarville/Winchester/Summit Point; The Public Mind; Berryville; Smithfield Crossing; "Record" 10/64; Anderson; Winchester

Cedarville/Winchester/Summit Point

The retreat north continued, driving livestock and burning grain, as ordered by Grant. It was a complicated affair, and was costly, in that numerous bloody skirmishes took place, some adding up to be named battles, with substantial casualties. Devin's brigade, to whom Battery K–L was eventually attached, was attacked near Cedarville, outside of Front Royal, on the north side of the Shenandoah River. Fitz Lee's cavalry and a brigade of Kershaw's division, part of the reinforcements arriving for Early, were repulsed, but yet another brigade of infantry was seen wading the river. Custer's 3rd brigade was called in and their repeating carbines were put to use with deadly effect. Though this resulted in a loss of some 500 to the Confederate side, as compared to 71 on the Federal, other of these small seesaw battles were Confederate victories. For example, the Jersey brigade of the 6th Corps, which, on the 17th had reported to Torbert at Winchester to act as a rear guard as the army retired, was attacked in the afternoon and lost 250 men as prisoners.[977]

It was a difficult time, as Sheridan observes: "Early's whole army had followed us from Fisher's Hill, in concert with Anderson and Fitzhugh Lee from Front Royal, and the two columns joined near Winchester on the morning of the 18th."

Sheridan continued to fall back, moving the 6th Corps to Flowing Spring, west of Charles Town, with Emory at Welch's Spring, just south. Emory now had Grover, who had arrived at midnight. Merritt fell back to Berryville, and Wilson to Summit Point.[978]

On August 19th, Early had moved to occupy Bunker Hill, figure 1, and now, on the 21st, thinking that Sheridan had taken a position at Summit Point, moved from Bunker Hill across the Opequan, through Smithfield (Middleway) aiming to hit Sheridan's rear. This was coordinated with Anderson, who moved out from Winchester toward Berryville. At Cameron's depot, southwest of Charles Town, Early ran up against the first two brigades of Wilson's 3rd Division of cavalry, who,

---

977. O.R. Vol. 43/I, pp. 155, 165, 172, 423, 472, 516; Merritt, pp. 502,503; Sheridan Vol. 1, p.490.
978. Tiemann, p. 92; O.R. Vol. 43/I, p. 44; Sheridan, Vol. 1, pp. 490–491; Irwin, p. 375.

after a sharp fight, were ordered to retire. Pressing on, about a mile north, Early then met Getty's Division of Wright's 6th Corps. Heavy skirmishing ensued, Getty with a loss of 260 killed and wounded.[979] That night, the infantry retired to Halltown, Sheridan's strong point, about five miles further north, which was covered by the heavy guns on Maryland Heights behind Harper's Ferry.

On the morning of August 22nd, the cavalry followed through Charles Town, toward Halltown. However, Merritt's Division, with Battery K-L now joined, was ordered north to near Shepherdstown.

Early then pushed up to Sheridan's position, and for the next two days probed its defenses. Finding no suitable place to attack, he resolved on a plan to place Anderson, with Kershaw's Division, McCausland's cavalry, and a regiment of Fitzhugh Lee's cavalry, in position before Sheridan, to mask a move north toward Maryland. On the 25th, Early moved his infantry through Leetown and Kearneysville to Shepherdstown, while sending the rest of Fitzhugh Lee's cavalry to Williamsport via Martinsburg.[980] Early gives the reason that he did so was "to keep up the fear of an invasion of Maryland and Pennsylvania."

Sheridan recalls that he had anticipated Early attempting such a move, and had planned to fall on his rear if he did.[981]

In retrospect, Sheridan spoiled his opportunity to "fall" on Early by being curious as to what had happened to the bulk of Fitzhugh Lee's cavalry, missing from Anderson's position near Charles Town. His orders to Torbert, with Merritt's and Wilson's cavalry, to go out to Kearneysville and Leetown, resulted in Torbert's clashing with not only the Confederate cavalry, but Breckinridge's infantry. The engagement that resulted overpowered Torbert's forces, which had to scatter, delaying any report to Sheridan. By the time Sheridan could react, Early withdrew back to Bunker Hill, Anderson returning to Stephenson's Depot.

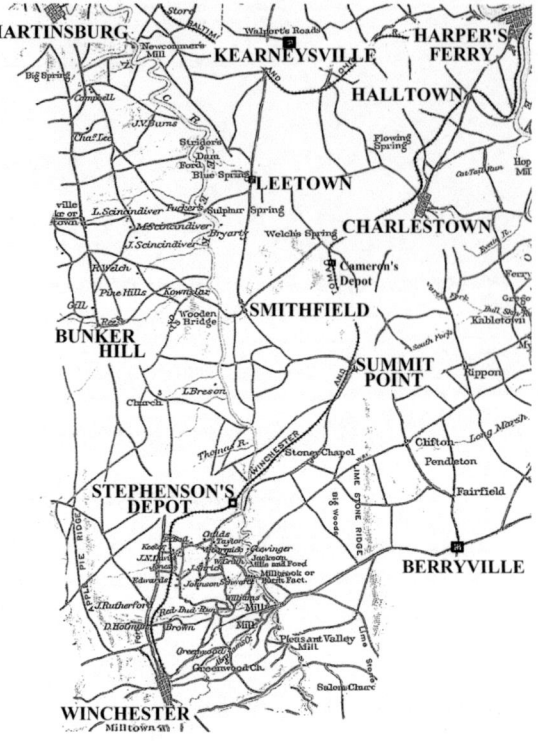

FIGURE 1

---

979. O.R. Vol. 43/I, pp. 156, 416, 473; Merritt, p. 504; Sheridan, Vol. 1, p. 491, Figure 1, portion of plate 69, map 1, altered.
980. Early, *Sketch*, pp. 409, 410; Sheridan, Vol. 1, p. 493; O.R. Vol. 43/I, p. 425; Irwin, p, 375.
981. Sheridan, Vol. 1. pp. 493-495.

The Public Mind

Though Sheridan's retreat back to Halltown caused concern, good news soon came in from Alabama and Georgia. From Mobile, the public[982] was soon treated to the news that Farragut, in a combined operation with an army force provided by Canby, had on August 5th, made one of his classic daring runs, this time past the guns of Forts Morgan and Gaines, into Mobile Bay. First reports of the feat were known in Washington as early as August 10th from Confederate newspapers, but Secretary Welles did not receive Farragut's report until the 23rd. He had captured the formidable Confederate ram, the *Tennessee*, and the gunboat *Selma*.

The cost was great in terms of damage to Farragut's ships, but these details would remain to be disclosed later. The monitor USS *Tecumseh*, leading the fleet into the bay, struck a torpedo and sank, carrying almost all of her crew to the bottom. A revised count of the casualties was 52 killed and 170 wounded.

During the attack, observing the *Tecumseh* sink, and not understanding the reason for the lead ship, the *Brooklyn*, having slowed, Farragut says: "I determined, at once, as I had originally intended, to take the lead…I steamed through between the buoys where the torpedoes were supposed to have been sunk…but believing that their having been some time in the water, they were probably innocuous, I determined to take the chance of their explosion." Thus, we have the Admiral's modest explanation for what he was famously, and only somewhat erroneously, quoted as ordering: "Damn the torpedoes, full speed ahead!"[983]

The *Tennessee*, almost impervious to Farragut's gunfire, survived the passage of the fleet, and when it was observed to be undamaged in their rear, the fleet turned, with orders to run her down at full speed. The *Monongahela*, freed from the *Kennebec* (with Enos/Charles Deal, alias Jones, aboard), which had been lashed to her side during the passage of Fort Morgan, struck her first, and though the *Tennessee* may have been injured badly, she remained afloat. The *Hartford*, and the three remaining monitors in the fleet again bearing down on her, at 10:00 a.m. she finally surrendered.

Of particular interest to the members of Battery L was what had become of Henry Closson. As we know, he had remained in the Department of the Gulf, and had been assigned to the staff of Maj. Gen Gordon Granger, as planning for a raid on Mobile had already begun.[984]

The troops under Granger were to be landed on Dauphin Island, to threaten Fort Gaines, at the harbor entrance, before Farragut was to attempt his run.

By August 5th, a landing site had been constructed close by the besiegers

---

982. *Harper's Weekly*, August 27th, 1864, p. 547, Welles, Vol. 2, pp. 101, 115; ORN Ser. 1, Vol. 21, pp. 405–407, 414–418, 520–521.
983. Farragut, pp. 416–417.
984. O.R. Vol. 41/II, pp. 65, 104, 105, 326, 566, 759, 760.

before Fort Gaines, and the heavier guns were begun to be landed, all under the direction of Henry Closson.[985] On the morning of the 6th, the *Chickasaw* moved into position and shelled Fort Gaines, and Granger ordered the siege guns to open on it.[986] On the morning of the 7th, Colonel Anderson, Fort Gaines' commanding officer, requested terms from Farragut. The fort was occupied by Union troops the next day, after an unconditional surrender.

Granger then planned to immediately move to the attack of Fort Morgan, on the peninsula opposite. He landed there on August 9th though Fort Morgan held out until the 23rd. On August 24th, Closson, his job done, was ordered to return to "his battery wherever it may be."

Reports from Sherman were soon to be heard, and combined with Farragut's exploit, it is clear that any news of Sheridan's retreat was sandwiched into a narrow context, yet it was close at hand, and Early's army was still regarded as a great menace. The price of gold reached an all time high.[987]

Smithfield Crossing[988]

On August 27th, all of Early's infantry was in position in his old camps at Bunker Hill, with his cavalry holding Leetown and Smithfield. Anderson was south, at Stephenson's Depot.[989] Next day, Sunday, the bugles in Merritt's Division near Shepherdstown blew reveille at 3:00 a.m., to be ready to go out on a reconnaissance. At daylight, they headed south toward Kearneysville, marching in typical Sheridan fashion, the three brigades abreast; Custer's 1st on the right, Devin's 2nd on the left, and the Reserve Brigade, then commanded by Gibbs,[990] on the Pike. Outside of Leetown, the advance of the Reserve Brigade, the 2nd U.S. Cavalry, ran into the camp of Harry Gilmor's and portions of Bradley Johnson's Rangers on the Smithfield Pike. Supported by the 6th Pennsylvania, Merritt attacked, and pressed the Rangers back to a wood outside of Smithfield, where Early's artillery opened on them. Williston's 2nd U.S. Battery D,[991] one of the two artillery batteries assigned to Merritt's command at this time – the other being Taylor's Battery K–L – was then brought up and placed upon "an eminence" to the left, soon silenced the Confederates. Threatened by a move on their left flank by Custer, the entire enemy force then retired down the road and across the turnpike bridge on Opequan

---

985. O.R. Vol 52/I, p. 582.
986. ORN Ser. 1, Vol. 21, pp. 414, 514, 524; O.R. Vol. 41/II, pp. 592–593, 631, 832.
987. Wilson, J. H., p. 540.
988. So named in popular references, CWSAC WV015; Smithfield is now known as Middleway; Cheney, N., pp. 214–215; Lee, Wm. O., pp. 98, 99.
989. Sheridan, Vol. 1, pp. 495–496; O.R. Vol. 43/I, p. 1025.
990. O.R. Vol. 43/I, pp. 94–95, Gibbs assumed command on August 10th, and Lowell on September 8th.
991. O.R. Vol. 43/I, pp. 469, 470, 486, 488, 571, 946, 953, 966, 987.

Creek. Darkness coming on, the Union cavalry then encamped for the night on the heights on the east side of the creek, within sight of Early's campfires.

At dawn on August 29th, Custer boldly crossed the creek, and advanced about a mile before running into Early's cavalry, which Custer drove back. But he then ran smack into Early's infantry, Gordon's and Ramseur's, which had already been marching from Bunker Hill. Early had become uneasy that Merritt had possession of the bridge and fords,[992] and had ordered his troops to the attack. Custer's Brigade came swiftly back, some across the covered bridge, and others spurring their horses into making the four-foot jump from the bank into the creek. They then formed on the right of the Reserve Brigade, which was in position to the right of Devin's Brigade, (under the temporary command of Col. Louis P. Di Cesnola), which was lined up to face the enemy, as is indicated by the small black positions captioned "Merritt" on the right in figure 2.[993]

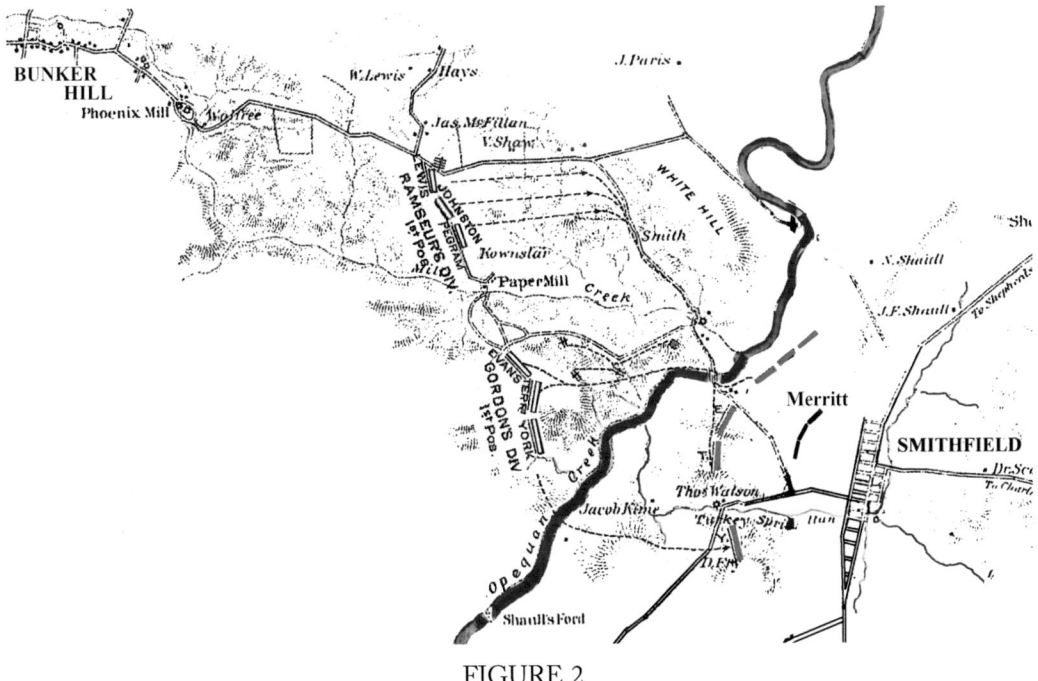

FIGURE 2

At about 10:00 a.m., Early's force, indicated by the positions in white, and his moves, the dotted lines, appeared on the far side of the creek, and by 11:30, he had artillery in place on either side of the road, which began a furious and well-directed fire at Merritt's line.

This was not the type of fight that any cavalry liked to be forced into. They

---

992. Gracey, pp. 292–293; Bowen, J. R., pp. 220–221; Cheney, p. 215; Haskin, p. 213; Early, *Sketch*, p. 410.

993. Figure 2, Atlas, portion of plate 82, map 7, altered. Original drawn by Jedediah Hotchkiss, Early's mapmaker. Devin was wounded at Cedarville.

should perhaps have withdrawn immediately, but were directly ordered to make a stand by Merritt, regardless of the fact that the 6th Pennsylvania was out of carbine ammunition. "Use your pistols," he said. Even those regiments still with ammunition were at a disadvantage, due to the fact that the long-range rifle fire of Early's infantry was "far more effective than the cavalry carbines…" Their Spencer repeaters, designed for close fast action, were of lesser range, compared to the Enfields and Springfields which the Confederates now had appropriated in increased numbers. The cavalry revolvers, handy for multiple shots in furious action while in the saddle, were even more limited and near useless. Of course, their sabers, in this case, were nothing but an encumbrance.

The important item that differed here, though, was that Merritt's cavalry now had its own artillery, albeit only two batteries, and they had been available to promptly respond. Early describes this initial phase of the engagement, with the two sides separated by the creek, as a sharp artillery duel.

At 12:30, Gordon's Division executed a flank movement, by fording the Opequan south of the bridge. Thus threatened, and "notwithstanding the…shells, grape, and canister poured into their ranks by our artillery, on they came like demons." Their regiments then formed in front of the defenders. Sometime after 2:00 p.m. so many of Merritt's defenders had run out of their carbine ammunition that the three brigades were finally ordered to withdraw. Thus began a slow, deliberate withdrawal, without confusion, the men protecting themselves with their pistols, and, of course, covered by their artillery. Over the space of some three hours they were drawn back through the town, to an open ridge about two miles east of Smithfield. There, at 5:00 p.m. they were greeted by Sheridan. Sheridan and his staff had arrived just before, and as was his custom, he was right there at the front, ahead of the reinforcements – Rickett's 3rd Division of the 6th Corps.[994] On horseback, in the open, Sheridan's little group was fired upon, and Sheridan's acting medical director was hit by a Minie ball and killed. The retreat was halted, and breastworks of fence rails were hastily constructed, while the enemy occupied the village in heavy force. Now, with something like equal numbers between the combatants, the tide turned, and Early's entire force was quickly pushed back across the Opequan.

Specifics of the participation of Battery K–L in this, its first action in the Shenandoah, are not reported. Any report Franck Taylor may have written has not been found, and Col. Di Cesnola, in temporary command of the 2nd Brigade, only refers to placing his "pieces" and consulting with his battery officers, without naming them. In contrast, the action of Williston's battery in support of the Reserve Brigade is reported in some detail. We only know from the Battery L monthly report that Pvt. Joseph H. Parslow was killed, and from the regimental casualty

---

994. O.R. Vol. 43/I, pp. 469, 489; Parslow death: "*Record*" 8/64, ch. 12.

report that privates Brown and Jesse W. Smith of Battery K were wounded.

"Record" 10/64
31 AUGUST–31 OCTOBER 1864 THE SHENANDOAH VALLEY CAMPAIGN

Battery engaged at the Battle of Winchester, Va. Sept. 19, 1864. In action at Fisher's Hill, Va. Sept. 23. At Mt. Jackson, Va. Sept. 24. At New Market, Va. Sept. 25th. At Port Republic, Va. Sept. 27th & 28th. Engaged in the Cavalry Fight near Strasburg, Va. Oct. 8th & 9th. Pursued the enemy to Woostock, Va. Participated in the battle of Cedar Creek, Va. Oct, 19, 1864. Camped near Middletown on the 20th. Relieved from duty in the field and ordered to Reserve camp in Pleasant Valley, Md. Oct 26th and arrived there on the 29th October, 1864.

Henry W. Closson	Capt. Joined Co. Oct. 25,'64 from absence with leave. Sp. Orders 347 War Dept. A.G.O. Washington Oct. 14,'64.
Franck E. Taylor	1st Lt. Relieved of duty with Battery and assigned to duty as Chf. Arty. & Ord. (in the field) of Cavalry, War Dept. A.G.O. Washington Oct. 14,'64.
Edward L. Appleton	1st Lt. Relieved from duty Gen. recruiting svc. & ordered to join Co. S.O. no. 315, War Dept. A.G.O. Washington, Sept. 22,'64. Absent without leave.

Detached:
Charles Cooke	Pvt. Abs. on det. svc. As orderly for Chf. Arty. of the Cav. Mid. Mil. Div. since Oct. 28, 1864.
Michael Kenny	Pvt. In Battery G 5th US Artillery at New Orleans.
John Meyer	Pvt. In Battery G 5th US Artillery at New Orleans.

Absent in confinement:
Patrick Gibbons	Pvt. At Ship Island, serving sentence of General Court martial S.O. no. 18, Hdqtrs. 1st Div. 19th Army Corps Dec. 31, 1863.
John Lewery	Pvt. Abs. confined in New Orleans, La. Apprehended as a deserter.

Deserted: none.

Discharged:
Julius Becker	1st Sgt. Oct. 12,'64 Strasburg, Va. By expiration of service
Michael White	Sgt. Oct. 7,'64 New Orleans, La. do.
David J. Wicks	Sgt. Oct. 25,'64 Middletown, Va. do.
Edmund Anglin	Pvt. Oct. 19,'64 New Orleans, La. do.
William Brunskill	Pvt. Oct. 19,'64 New York Harbor do.
Issac T. Cain	Artificer Oct. 4,'64 Harrisonburg, Va. do.
Jeremiah Connell	Pvt. Sept. 13,'64 Smithfield, Va. do.
Edmond Cotterill	Pvt. Sept. 18,'64 Washington, DC By order of War Dept., A.G.O.
James Flynn	Pvt. Sept. 13,'64 Smithfield, Va. By expiration of service.
Joseph Smith	Pvt. Oct. 11,'64 New Orleans, La. do.

Died:
Rowland Card	Pvt. Oct. 19,'64 Of wounds received in action at Cedar Creek, Va.

Joined:
William Demarest	Sgt. Oct. 14,'64 Reenlisted in the company while on recruiting duty at New York.
Lewis Keller	Cpl. Oct. 28,'64 From enlistment in New York.

Benjamin O. Hall	Pvt. Oct. 14,'64	From recruiting duty in New York.
Michael Teighe	Pvt. Oct. 14,'64	do.
Rueben Townsend	Pvt. Oct. 14,'64	do.
John McKenny	Pvt. Sept. 18,'64	From sick at new Orleans.

Strength: 76 Sick: 11

Sick Present: none

Sick Absent:

Ludwig Rupprecht	Bugler Oct. 9,'64 Wounded in action at Strasburg, Va.
James Campbell	Pvt. Sept. 19,'64 Wounded in action at Winchester, Va.
George Chase	Pvt. April 22,'63 At Brashear City, La.
Patrick Craffy	Pvt. Sept. 26,'64 At Philadelphia, Pa.
William Crowley	Pvt. July 13,'63. At Baton Rouge, La.
Charles Jackel	Pvt. April 20,'64 At New Orleans, La.
John Kelly	Pvt. Sept. 19,'64 Wounded in action at Winchester, Va.
Churchill Moore	Pvt. April 20,'64 At New Orleans, La.
Warren P. Shaw	Pvt. Oct. 12,'64 (no entry)
Andrew Stoll	Pvt. Sept 25,'64 Wounded in action near New Market, Va.
John C. Wood	Pvt. Sept. 19,'64 Wounded in action at Winchester, Va.

The cooks are listed on the muster roll, though separately from the numbered roster, and are not counted as part of the 76-man strength. Their pay due is listed with the comment, "$13.00, error on last payroll"; their pay finally had achieved parity with that of a private.

Items of note are: 1. Henry Closson finally arrived back from his duty at Mobile, and some additional time on leave. He did not, however, get back in time to participate in any of the battles in the Shenandoah Valley. He was brevetted a lieutenant colonel on August 23rd, 1864: "For Gallant and Meritorious Services at the Battle of Ft. Morgan, Ala."[995] 2. It was now Franck Taylor's turn to become chief of artillery, but only of the cavalry. 3. Edward Appleton, who had gone to New York on recruiting duty, was declared AWOL. He had been ordered to rejoin Battery L on September 22nd, more than five weeks earlier. 4. William Brunskill was finally discharged, not on a disability, but by expiration of service. After being taken sick at Fort Pickens on September 10th, 1861, and sent to the Hospital at Fort Hamilton, New York, on September 17th, 1861, he had been in the hospital for more than three years out of his five-year enlistment. Though Michael White's term of service may have expired, his listing here as discharged is in error. He is still missing – since the battle at Pleasant Hill.

Anderson

Early had been receiving "loud" calls from Lee, according to Sheridan's memoirs, to

---

995. Cullum, Vol. II, no. 1638, Vol. 2, p. 580.

return Anderson and Kershaw's division.[996] No records have been found to contradict Sheridan's statement about the force of Lee's remarks, but Early confirms that he had received a letter *requesting* Kershaw and Anderson, and that, after consulting Early, Anderson had decided to return, taking Kershaw and Fitzhugh Lee's cavalry with him. He moved on September 3rd, towards Berryville, intending to cross the Blue Ridge at Ashby's Gap. To cover the move, Early had agreed to move toward Charles Town. The fruit of the Grant/Sheridan policy was about to ripen. With Early stripped of his reinforcements, Sheridan could begin an aggressive campaign.

In the meantime, Sheridan had begun a general advance south, moving Wright to Clifton, Crook to Berryville, and Torbert to White Post, with orders to reconnoiter as far as the Front Royal Pike. Crook had just gotten into position about an hour before sunset, when firing was heard along his front west of the town. It was Anderson, who had blundered into Crook. The next day, Early came to Anderson's rescue, only to discover that he now faced a whole new line of Sheridan's, which extended all the way to Summit Point. Appreciating Sheridan's strength and position, Early and Anderson then both decided to withdraw back to Winchester and Stephenson's Depot. More reconnaissances were sent out by Sheridan on the several days succeeding, but it was not definitely learned that this incident involved Anderson's and Kershaw's attempt to withdraw from the Valley until they tried again, and succeeded, on September 14th.

On September 5th, news was received of the President's message of September 3rd, giving the national thanks to Admiral Farragut and generals Canby and Granger for the success of the Mobile expedition, and in the surrender of Fort Morgan on August 23rd.[997] A 100-gun salute was to be fired at the navy yard in Washington on the 5th, and at "each arsenal and navy yard in the United States…" another was to be fired on the 6th. On the 7th, yet another was to be fired at the arsenals in honor of Sherman's occupation of Atlanta on September 2nd. Lincoln also requested that: "In all places of worship in the United States thanksgiving be offered to Him for His mercy in preserving [the Union]."

The news of the victories came at a point in time during and just after the Chicago Democratic Convention, which had opened on August 29th, and had closed on September 3rd. It had nominated George B. McClellan.[998] The effect of the news, as commented upon in *Harper's Weekly*, was: "There is not a man who did not feel that McClellan's chances were diminished by the glad tidings from Atlanta…" The presidential campaign season had now opened, and Grant understood that the administration was afraid of any setback, hence all of the caution Sheridan had been warned to exercise up to this time was still in effect. Caution involved knowing more about Anderson's plans.

---

996. O.R. Vol. 43/II, p. 862; Sheridan, Vol. 1, p. 498; Early, *Sketch*, pp. 410–411.
997. Cheney, p. 217; ORN Ser. 1, Vol. 21, pp. 538, 543–544; O.R. Vol. 38/I, p. 127.
998. *Harper's Weekly*, Sept. 10th, 1864, p. 579; ibid., Sept. 17th, 1864, p. 594; O.R. Vol. 43/I, p. 811.

The positions of both Early's forces and Sheridan's remained essentially unchanged from September 3rd until September 17th, though Sheridan's cavalry was employed every day in harassing the enemy. Sheridan purposely remained some six miles back from Early's positions behind (west of) Opequan Creek, and if this expanse could be dominated by Union scouting parties, no enemy pickets would be able to warn of any future general move Sheridan might make.[999] On one occasion, September 13th, McIntosh's brigade, of Wilson's division, advanced on the Berryville Pike, crossed the Opequan, and when within two miles of Winchester, at Abraham's Creek, captured the entire 8th South Carolina Infantry, a part of Kershaw's division, which decisively proved that Anderson had not yet left. No move on Early would be prudent yet.

Even this reconnaissance did not get the real information necessary: *when* Kershaw and Anderson were planning to leave, if at all. After weeks of inaccurate information, much of it passed on from Washington,[1000] Sheridan felt that he needed a better approach. Up to now, the gathering of information, particularly that of enemy positions or troop movements, had been by reconnaissances, which only incidentally captured prisoners for interrogation. Other methods involved interviewing "doubtful citizens and Confederate deserters." The more detailed information on enemy plans was elusive. Sheridan concluded that his own soldiers who volunteered for specific intelligence gathering duty would be more accurate and trustworthy, and that they should be organized.[1001] They were consequently formed into a battalion, dressed in Confederate uniforms as the occasion required, and were to be paid in an incentive scheme in proportion to the value of the intelligence gathered.

In only a few days, the new scouts had learned of a possible way of getting information out of Winchester. An "old colored man" living near Millwood, southeast of Winchester, had been given a permit by the Confederate commander to pass in and out three times a week for the purpose of selling vegetables. The scouts had "sounded this man, and finding him both loyal and shrewd, suggested that he might be made useful…" Fortunately, General Crook was acquainted with many of the Union people residing in Winchester, and on the 15th of September, Sheridan turned to him to recommend someone to contact. He mentioned Miss Rebecca Wright, a Quaker teacher who ran a small private school, with the caveat that her pro-Union sentiment was well known to the Confederate authorities, and that she was under constant surveillance. Sheridan wasted no time, that same night directing his scouts to bring the old man to his headquarters. The interview convinced Sheridan "of the negro's fidelity." He said that he knew Miss Wright well.

---

999. O.R. Vol. 43/I, pp. 24, 46, 87, 427, 517; Early, p. 419. Early began a raid on Martinsburg on the 17th. Abraham's Creek is today referred to as Abram's Creek.
1000. O.R. Vol. 43/II, pp. 18, 21–22; Sheridan, Vol. 2, pp. 2, 8.
1001. Sheridan, Vol. 2, pp. 2–6; O.R. Vol. 43/II, p. 90.

After some persuasion, he agreed to carry a letter, wait for any answer, and return. The letter was written on tissue paper, and compressed into a small pellet, which was wrapped in tin-foil, the object being to protect it while it was carried in the old man's mouth. In the event he was stopped, he could swallow it. The message was delivered to her home the next day. It said:

> I learn from General Crook that you are a loyal lady, and still love the old flag. Can you inform me of the position of Early's forces, the number and divisions in his army, the number and strength of any or all of them, and his probable or reported intentions? Have any more troops arrived from Richmond, or are any more coming, or reported to be coming?
>
> I am, very respectfully, your most obedient servant,
>
> P. H. Sheridan, Major-General Commanding
>
> You can trust the bearer.

Upon opening the tin-foil, the message was found to be readable, and the old man departed, telling Miss Wright that he would return later for her reply. She was startled by the perils involved, but after consulting her mother, decided to become involved, though it might put her life in jeopardy. Incredibly, only the night before, a convalescent Confederate officer had visited her mother's house, and in conversation, revealed the fact that Kershaw's division and Cutshaw's batallion of artillery had already left to rejoin Lee. She answered:

> September 16, 1864.
>
> I have no communication whatever with the rebels, but will tell you what I know. The division of General Kershaw, and Cutshaw's artillery, twelve guns and men, General Anderson commanding, have been sent away, and no more are expected, as they cannot be spared from Richmond. I do not know how the troops are situated but the force is much smaller than represented. I will take pleasure hereafter in learning all I can of their strength and position, and the bearer can call again.
>
> Very respectfully yours,

* * * * *

This finally resolved the conflicting reports and rumors that had gone before. Sheridan decided to bring Early to battle the next day. He had a plan to attack him south of Winchester, at Newtown, and thus seal off his withdrawal up the Valley.

On the 15th, Grant decided to visit Sheridan, with the purpose to have him "attack Early, or drive him out of the Valley and destroy that source of supplies for Lee's army."[1002]

He sent a courier (no telegram, which would be read in Washington) to Sheridan that he was coming, and then left, directly for Charles Town, ten miles south of Harper's Ferry. He went in person, for the reason that: "I knew it was impossible for me to get orders through Washington to Sheridan to make a move, because

---

1002. Grant, Vol. 2, p. 327

they would be stopped there and such orders as Halleck's caution (and that of the Secretary of War) would be given instead, no doubt contradictory to mine." Grant's telegram reached Sheridan in time for him to call off his planned attack, and "defer action" until he had met with Grant.

In their meeting, Sheridan laid out his plan. It consisted of approaching Early from the south via Newtown, hitting his right along the Valley Pike, his only outlet from the Valley, and his communication with Lee.[1003] Grant was sufficiently impressed with the plan to not question it, and did not disclose his own plan, which he had drawn up on a piece of paper, still in his vest pocket.[1004] The only question that Grant asked, was *when* Sheridan could launch the attack. Knowing that Sheridan's wagon teams and most of his supplies were at Harper's Ferry, Grant asked if he could be ready by the following Tuesday. This was on Friday. Sheridan answered that he could be off whenever the general should say "Go in,"—before daylight on Monday, the 19th, if necessary. So delighted was Grant with this answer that he simply said "Go in!"

Note the careful planning regarding the teams. The horses and wagons were held at their point of supply, so that forage would not have to be delivered to them by yet more teams.

Winchester

Sheridan had reported that once the two armies had drawn up facing each other, Early at Winchester, and Sheridan at Berryville, there were no further substantial movements. One that escaped all of his reconnaissances and scouting was on the 17th. It was then that Early had detached Rodes' and Gordon's divisions, Lomax's cavalry, and Braxton's artillery, all under the command of Breckinridge, to take yet another crack at the B&O Railroad at Martinsburg. Next day, while Rodes remained at Bunker Hill, Gordon, with a part of Lomax's cavalry (Jackson's), reached Martinsburg, driving Averell from the town, and across the Opequan. Fortunately, Averell reported the fact to Sheridan at noon.[1005] While there, Early learned from the Union telegraph office that Grant had visited Sheridan, and, as a result, suspected an "early move." He therefore rushed Rodes back to Stephenson's Depot that evening, and ordered Gordon back to Bunker Hill, with orders to return to Stephenson's Depot on the 19th.

Sheridan, now aware that Early had sent off the two divisions, but *not aware* that they had been ordered back, then decided to alter his plan to attack via Newtown. His original orders had been issued on Sunday afternoon. They were

---

1003. Nicolay and Hay, Vol. 9, p. 299.
1004. Grant, Vol. 2, p. 583; Irwin, p. 377.
1005. Early, *Sketch*, p. 419; Sheridan, Vol. 2, pp. 9–10; O.R. Vol. 43/I, pp. 25, 46, 518, 554, 1027; O.R. Vol. 43/II, pp. 106–107; Park, p. 90; Pellet, pp. 252, –253; Early, *Battles and Leaders* Vol. 4, p. 522.

canceled and new ones issued. Instead of heading for Newtown, Sheridan now planned to take advantage of the "disjointed state of the enemy giving me an opportunity to take him in detail…" and planned to go directly west, along the Berryville Pike. Not knowing that Early had planned the return of his divisions would *almost* prove fatal to Sheridan. It certainly increased the intensity of the subsequent battle and Sheridan's casualties.[1006]

The plan was for Torbert, on the flank, to advance with Merritt's cavalry from its location at Summit Point, to carry the crossings of the Opequan at Seiver's and Locke's fords, and form a junction with Averell, who was to advance south from near Darkesville, to where he had returned on the night of the 18th. Marching south via the Pike, through Bunker Hill, he was to join Torbert near Stephenson's Depot, figure 3.[1007] Sheridan makes no mention of the fact that Merritt was, apparently, instructed to delay those Confederate forces found at the fords, so as to prevent them from freely retiring back to reinforce Breckinridge at Stephenson's Depot.[1008]

Wilson's cavalry was to move west on the Berryville Pike, cross the Opequan, drive in the enemy's pickets, and attack the enemy positions to cover the advance of the infantry. The 6th Corps was to follow, the 19th Corps behind it, with Crook, who was originally to go to Newtown, to be held in reserve.

Regardless of the fact that Early had ordered back his two divisions, Rodes' division had only reached Stephenson's Depot, five or six miles north of Winchester, and Gordon's division, the only one of the two that had entered Martinsburg, was allowed to stop overnight at Bunker Hill.[1009] Hence, neither were within supporting distance of Ramseur, posted east of Winchester. Gordon, having to march about eight miles to Stephenson's Depot from Bunker Hill, would not even reach there until about 8:00 a.m. Thus, the situation still looked satisfactory enough to Sheridan, at daylight on the 19th, to continue with his "straight in" plan. As we shall see, regardless of the great victory claimed for the outcome of the battle, its

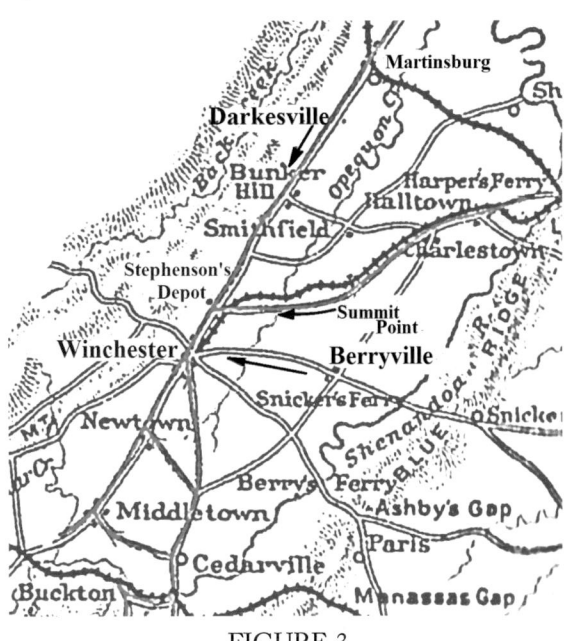

FIGURE 3

1006. Sheridan, Vol. 2, pp. 11–14.
1007. Atlas, plate 85, map 1, portion, altered.
1008. O.R. Vol. 43/I, pp. 444, 455.
1009. O.R. Vol. 43/I, p. 554; Sheridan, pp. 83 89; Naval Observatory sunrise for Winchester, VA, Sept. 19th, 1864, was 5:58.

commencement was delayed, and it was long, difficult, and bloody. Worse, note figure 3, Sheridan had left the door open for Early to escape south, with Crook held in reserve, and subsequently needed at Winchester, rather than driving to Newtown.

Since the cavalry was the arm to which Battery K–L was assigned, its organization for the battle is shown in some detail in figure 4.[1010] Note that the artillery is listed separately, unassigned to any particular brigade. They were intended to be flexibly assigned as required, and in the coming battle, elements of Battery K–L served under both Devin and Custer.

CAVALRY: Br.-Gen. Alfred T. A. Torbert. Escort: 1st Rhode Island Maj. Wm. H. Turner, Jr.

FIRST DIVISION: Br.-Gen. Wesley Merritt

FIRST BRIGADE: Br.-Gen. George A. Custer
    1st Michigan, Col. Peter Stagg
    Michigan, Maj. Smith H. Hastings
    6th Michigan, Col. James H. Kidd
    7th Michigan, Maj. Melvin Brewer
    29th New York, Maj. Charles J. Seymour

SECOND BRIGADE: Col. Thomas C. Devin
    4th New York, Maj. August Hourand
    Maj. Edward Schwartz
    6th New York, Maj. Wm. E. Beardsley
    9th New York, Lt.-Col. George S. Nichols
    1st New York Dragoons, Col. Alfred Gibbs
    17th Pennsylvania, Maj. Coe Durland

RESERVE BRIGADE: Col. Charles R. Lowell
    2nd Massachusetts, Lt.-Col. Casper Crowninshield
    1st United States, Capt. Eugene M. Baker
    2nd United States, Capt. Theophilus M. Rodenbough
    Capt. Robert S. Smith
    5th United States, Lt. Gustavus Urban

SECOND DIVISION: Br.-Gen. Wm. W. Averell

FIRST BRIGADE: Col. James M. Schoonmaker
    8th Ohio (det.), Col. Alpheus S. Moore
    14th Pennsylvania, Capt. Ashbell. F. Duncan
    22nd Pennsylvania, Lt.-Col. Andrew J. Greenfield

SECOND BRIGADE: Col. Henry Capehart
    1st New York, Maj. Timothy Quinn
    1st West Virginia, Maj. Harvey Farabee
    2nd West Virginia, Lt.-Col. John J. Hoffman
    3rd West Virginia, Maj. John S. Witcher
    Division Artillery: 5th United States, Battery L, Lt. Julian V. Weir

---

1010. Sheridan, Vol. 2, pp.11–21; O.R. Vol. 43/I, pp. 107–112. *http://dmna.ny.gov/historic/regthist/civi/cavalry/4thCav/4thCavHistSketch.htm*

THIRD DIVISION: Br.-Ben. James H. Wilson

FIRST BRIGADE: Br.-Gen. John B. McIntosh
   Lt.-Col. George A. Purington
   1st Connecticut, Maj. George O. Marcy
   3rd New Jersey, Maj. Wm. P. Robeson, Jr.
   2nd New York, Capt. Walter C. Hull
   5th New York, Maj. Abram H. Krom
   2nd Ohio, Lt.-Col. George A. Purington
   Maj. A. Bayard Nettleton
   18th Pennsylvania, Lt.-Col. Wm. P. Brinton
   Maj. John W. Phillips

SECOND BRIGADE: Br.-Gen. George H. Chapman
   3rd Indiana, (2 companies) Lt. Benjamin F. Gilbert
   1st New Hampshire, (batallion), Col. John L. Thompson
   8th New York, Lt.-Col. Wm. H. Benjamin
   22nd New York, Maj. Caleb Moore
   1st Vermont, Col. Wm. Wells

HORSE ARTILLERY: Capt. La Rhett Livingston
   1st US, Batteries K&L, Lt. Franck E. Taylor
   2nd US, Batteries B&L, Capt. Charles H. Peirce
   2nd US, Battery D, Lt. Edward B. Williston
   3rd US, Batteries C&F, Capt. Dunbar R. Ransom
   4th US, Battery C, Lt. Terence Reilly

FIGURE 4

Figure 5 gives an abbreviated organization of the Union infantry. Units on duty elsewhere are omitted.

6th ARMY CORPS, Maj.-Gen. Horatio G. Wright
1st DIVISION, Br.-Gen. David A. Russell
   Br.-Gen. Emory Upton
   Col. Oliver Edwards
2nd DIVISION, Br.-Gen. George W. Getty
3rd DIVISION, Br.-Gen. James B. Ricketts
Artillery Brigade, Col. Charles H. Tompkins
19th ARMY CORPS, Bvt. Maj.-Gen. William H. Emory
1st DIVISION, Br.-Gen. Wm. Dwight
Artillery, Lt. John V. Grant
2nd DIVISION, Br.-Gen. Cuvier Grover
Artillery, Albert W. Bradbury
Reserve Artillery, Capt. Elijah D. Taft
8th ARMY CORPS (THE ARMY OF WEST VIRGINIA)
   Bvt. Maj.-Gen George Crook
1st DIVISION, Col. Joseph Thoburn
2nd DIVISION, Col. Isaac Duval
   Col. Rutherford B. Hayes
Artillery Brigade, Capt. Henry A. Du Pont

FIGURE 5

Figure 6[1011] gives an abbreviated organization of Early's army.

1011. O.R. Vol. 42/II, p. 1213, Vol. 43/I, pp. 1002–1003, 1011; remainder corroborated by: Garnett, SHS, Vol. 31, *Battle of Winchester*, p. 63; Park, p. 25; Early, *Battles and Leaders*, Vol. 4, pp. 522, 523.

ARMY OF THE VALLEY, Lt.-Gen. Jubal A. Early
    Rodes'   Division II Corps, Army of Northern Virginia (Lee)
    Ramseur's   "   "
    Gordon's   "   (Breckinridge)[1012]
    Wharton's   "   (Breckinridge)
CAVALRY: Maj.-Gen. Fitzhugh Lee
    Lomax's   Division
    Lee's   "
ARTILLERY Col. T.H. Carter
    Braxton's   Battalion
    Nelson's   "
    King's   "

FIGURE 6

The battle that is known as 3rd Winchester, or Opequan, began on September 19th, at 2:00 a.m., when Wilson's 3rd Division of cavalry, with McIntosh's 1st brigade in advance, moved to the Berryville Pike, crossed the Opequan, passed through the two-mile-plus long defile, or canyon, through which the road ran, and before dawn, drove in the Confederate cavalry pickets in front of Stephen D. Ramseur's infantry division.[1013] Ramseur's division had been posted to the east of Winchester, near Abraham's Creek, while Early's three others that had been on the raid to Martinsburg, were at, or approaching, Stephenson's Depot, not in position near Ramseur. Reaching the high ground beyond the canyon at about 3:00 a.m., McIntosh's brigade drove Ramseur from his position. It was the same place where McIntosh had captured the 8th South Carolina regiment on September 13th. Wilson then disposed his force so as to hold the ground and the creek crossing until the 6th Army Corps and the 19th Army Corps could come into position.

Also at 2:00 a.m., Merritt's 1st Division,[1014] Custer's 1st brigade in advance, had started out from its encampments at Summit Point, intending to reach the crossing of the Opequan at Seiver's Ford, on the Charles Town Road, south of where the railroad crosses, before daylight and unobserved. Custer's march went by the most direct route, cross-country, regardless of roads. Custer arrived first. The other two brigades, Lowell's and Devin's, followed, in the rear of the division train, and did not arrive until soon after daybreak. Custer was then ordered to take Locke's ford, about one-and-a-half miles to the north. A skirmish there on the 13th and another on the 15th had brought Breckinridge's attention to the fords and he had dispatched Wharton's division, consisting of Forsberg's, Smith's, and Patton's

---

1012. O.R. Vol. 37/I, p. 768; O.R Vol. 39/II, p. 877. Breckinridge, assigned to the command of Gordon's and Wharton's Division on June 27th, was assigned to the command of the Department of Southwestern Virginia and East Tennessee, on September 27th.
1013. O.R. Vol. 43/I, pp. 518, 574, 1025; Early, *Memoir*, pp. 89–90, 415; Haskin, p. 379.
1014. O.R. Vol. 43/I, pp. 443, 454–455, 481–482, 490; Early, *Battles and Leaders,* Vol. 4, p. 523.

brigades, figure 7,[1015] supported by King's battalion of artillery, to the location. Their skirmishers were found to be entrenched on the west side, alert to the Yankee arrival. Thus, all hopes of securing the crossings unopposed were lost, and the attempt at stealth was a waste.

At Custer's front, a mounted charge by the 25th New York and the 7th Michigan was organized, which failed.

Another charge was organized, and by order of General Merritt, one section of Battery K–L, under the command of Lt. John McGilvray,[1016] accompanied by the 4th New York Cavalry as its support, was detached from Devin and reported to Custer. It was directed to a position on a hill opposite the ford, where they "rendered valuable service," according to Devin.[1017]

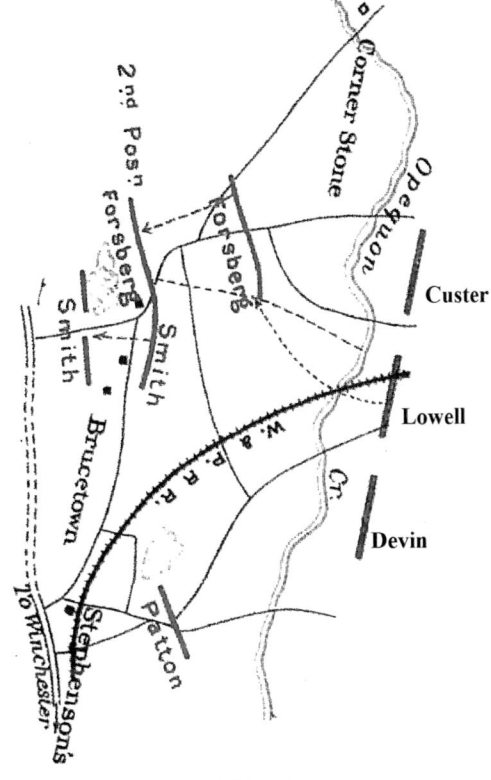

FIGURE 7

Actually, the enemy suffered little from their fire, being protected as they were by breastworks.

Preparations for this final attack were just about completed when it was discovered that the Confederate skirmishers were withdrawing. With this news, the second charge succeeded, the rifle pits were captured, and a "considerable" number of prisoners taken.[1018] As Custer reports: "The enemy retired about one mile from the ford in the direction of Winchester and took position behind a heavy line of earth-works, protected by a formidable cheval-de-frise."

In the meantime, below, at Seiver's ford, Lowell had crossed and established himself on the opposite bank. It was now after sunrise, and as Merritt writes: "The

---

1015. Atlas, plate 85, map 12. Altered and reoriented to approx. correct north-south. O.R. Vol. 43/I pp. 466, 467, 490.
1016. Haskin, p. 213; Hall, Besley, and Wood, p. 223.
1017. O.R. Vol. 43/I, p. 482; Denison, F., p. 389. No report by anyone in Battery K or L has been found. We know nothing of whether casualties were taken in this action.
1018. Kidd, J. H., pp. 387–389; O.R. Vol. 43/I, p. 455. We have quoted Custer as to the position to which Forsberg withdrew. Atlas, plate 99, map 1, allegedly to scale, shows less distance. Drawn in 1873, it is of help only in corroborating the time at which the Confederates completely withdrew, which was at 9:00 a.m., per Atlas plate 85, map. 12. O.R. Vol. 43/II, p. 113: The Confederate withdrawal was observed by Custer almost an hour later, and reported to Torbert by a member of his escort, Capt. John Rogers, as occurring at 9:50.

rich crimson of that fine autumnal morning was fading away into the broad light of day when the booming of guns on the left gave sign that the attack was being made by our infantry."[1019]

Custer's brigade now began to consolidate along the ridge in the positions across the creek which had been vacated by the enemy. Custer reports: "Prisoners captured at the ford represented themselves as belonging to Breckinridge's corps… which was posted behind the works confronting us. Deeming this information reliable…I contented myself with annoying the enemy with artillery and skirmishers until the other brigades of the division, having effected a crossing…established a connection with my left."

The connection between Lowell's and Custer's brigades finally made, a combined charge into the face of the enemy defenses was organized. Lowell leading the 2nd US Cavalry and Custer the 1st Michigan, 7th Michigan, and the 25th New York. It failed, as could have been predicted, but the threat of the attack, the artillery firing, and the skirmishing, to this point, had prevented Breckinridge's force from extricating itself and returning to reinforce Early, as he had already ordered.[1020]

Sheridan's personal memoirs mention nothing about using Merritt's cavalry so as to detain Breckinridge, but both Custer and Merritt make a point of it in their reports, and the strategy is recorded in Humphreys' *Field Camp, Hospital and Prison*. Sheridan brought on the battle with the knowledge that Early had divided his forces, and holding Breckinridge at the Opequan kept him away from the main field of battle. The trouble was, at some point, who was detaining whom? While tied up with Breckinridge's/Wharton's forces, Merritt was delayed in uniting with Averell. We shall see that Averell's advance from Darkesville was a somewhat similar story, though he was confronted by a much smaller force, as Gordon's division had already left. Averell mentions nothing about enjoining any Confederate force in order to detain it.

At the lines of Custer and Lowell, the artillery firing and skirmishing continued until Custer observed that, unseen, Forsberg had massed his force in the rear and withdrawn, to a "2nd pos'n." as shown in figure 7.[1021]

At about 11:00, a charge upon this new line was made by the combined forces of portions of the 1st and Reserve brigades, and like the others, was unsuccessful. But time had passed, and Early had finally peremptorily ordered Breckinridge[1022] to: "Move your whole force back toward Winchester and put it on the Martinsburg road about a mile from town." The order had its time of dispatch noted as 11:40 a.m.

1019. O.R. Vol. 43/I, pp. 443, 462, 482.
1020. Early, *Sketch*, p. 424; O.R. Vol. 43/I, pp. 444, 490; Humphreys, pp. 158–159; Sheridan, Vol. 2, p. 10.
1021. O.R. Vol. 43/I, pp. 427, 444, 456; Vol. 43/II, p. 113.
1022. O.R. Vol. 43/II, p. 876.

At this point, Custer, without orders, advanced, hoping to move beyond Wharton's infantry, and strike him in reverse. At about 1:30 p.m., Torbert gave the order for a general advance, and Devin's Brigade crossed the Opequan without opposition. Joining with Lowell, they moved together toward Winchester. McCausland's brigade of cavalry, and Patton's infantry, which was left with it in support, repeatedly clashed with Lowell and Devin as they advanced.[1023]

As to Averell, at 5:00 a.m. he had crossed the Opequan and headed toward Darkesville, driving Imboden's cavalry pickets, the 23rd Virginia Cavalry, under the command of Col. Charles T. O'Ferrall, steadily before him until he reached Bunker Hill, where the Confederate defenders made a determined stand.[1024] Bunker Hill, twelve miles from Winchester, was where Gordon had encamped after leaving from the Martinsburg raid, and here had remained another of Imboden's units, the 62nd Virginia, under the command of George S. Smith. Smith and O'Ferrall couldn't hope to stave off Averell's overwhelming force, and despite their stubborn resistance, Averell pushed through, reaching the area just above Stephenson's Depot into which Torbert/Merritt were pushing Wharton.[1025]

Thus are outlined the actions of Sheridan's 1st, 2nd and Reserve divisions of cavalry on Early's left, and Battery K–L's involvement. These actions took all the morning. Frequent references as to the heavy cannonading[1026] heard from further south, was thought to be proof that a larger battle had been taking place for more than four hours. Actually, the battle was still only developing. Wilson's 3rd division of cavalry had reached the high ground on the west side of the Opequan on which stood the earthworks occupied by Ramseur's division at dawn, and taken possession. Here, Wilson awaited the advance of Wright's 6th Corps. As stated previously, Emory's 19th Corps detachment followed behind the 6th, and Crook's Army of West Virginia, held in reserve, now brought up the rear.

The 6th had been awakened with orders to move since 1:00 a.m., but, preceded by Wilson, it had to wait until 4:30 before marching from camp.[1027] Proceeding directly across the farm fields, led by the 2nd Division, and followed by Ricketts' 3rd Division, with Russell's 1st Division taking up the rear, the 6th Corps arrived at the Berryville Pike about two miles east of the Opequan. Here they met Emory and the 19th Corps, which had already arrived. Emory was told to halt until the 6th

---

1023. O.R. Vol. 43/I, pp.482, 498; Bowen, J. R., p. 229, Cheney, pp. 219–220; Altas, plate 99, map 1; Hawkins, pp. 11–12; O.R Vol. 43/I, p. 597, *www.vmi.edu>Archives>*. The Patton referred to here was Col. George S. Patton, of the 22nd Virginia Regiment, who was mortally wounded later in the day. He was the grandfather of the illustrious World War II general, George S., Jr. Grandfather, father, and son all graduated from the Virginia Military Institute.
1024. O'Ferrall, pp. 89, 94, 96, 114–115; O.R. Vol. 43/II, p. 1247.
1025. Hawkins, p. 153; Early, *Sketch*, p. 424.
1026. Nichols, pp. 182–183.
1027. Paine, p. 256; O.R. Vol. 43/I, pp. 149, 279; O.R. Vol. 43/II, pp. 146–147.

Corps had passed, including its ordnance and ambulance trains.

The narrow, winding canyon through which the Berryville Pike passed, slowed the passage of the 6th Corps, the troops being pressed between the steep sides of the road. It was not until 8:00,[1028] that they finally came forward into the battle area held by Wilson. As they emerged, they were targeted by Ramseur's line, supported by Nelson's artillery. The 6th was not fully deployed until 9:00, and Battery M of the 5th US Artillery, reports that it did not begin to return fire on the enemy until 10:00.[1029]

In the meantime, Emory, as we know, always prompt and ready, had to sit, and watch, and wait. He was "swearing mad." He was angry enough to finally order his lead division, Grover's 2nd, to push forward past and through the 6th Corps train. Nevertheless, it was two hours before the 19th Corps arrived, though it had managed to push up to the rear of the 6th Corps column, and even managed to get in position on the field a little earlier than the 1st Division of the 6th Corps, which had become "mixed up with the artillery and wagon trains." Operations were suspended until all could be fully deployed.

Thus, all that noise which was heard by Merritt, Custer, and Battery K–L, was not a battle, but an artillery bombardment by Ramseur, and the rifles of his skirmishers, while the Federal army struggled to deploy.[1030] The 19th Corps was deployed in two lines to the right of the 6th, all the while closely supervised by Sheridan, who, insisting upon every order being executed to the letter, contributed to even more delay.[1031] It was after 11:00 before the lines were formed as they are depicted in figure 8.[1032]

Note the scope of the battlefield. Ignoring the distant positions of Merritt's cavalry, and Averell who is to the north, the distance from Wilson's position south of Abraham's Creek to the left of Gordon's line on Red Bud Run is almost four miles. Note also that the Berryville Pike, the choke point through which the Federal army took so long to negotiate, is in its rear. It is a point where, if a retreat became necessary, Sheridan could become bottled up, with worse disarray than had occurred during the advance. The potential for Early to strike for the entrance to the canyon and cut Sheridan off was a real possibility. Then Crook could not reinforce, nor could Sheridan retreat. The potential for a situation similar to the Battle of Mansfield was evident. It almost happened.[1033]

1028. O.R. Vol. 43/I, p. 518; Haines, p. 257.
1029. O.R. Vol. 43/II, p. 278; Flinn, p. 177; O.R. Vol. 43/I, p. 318; Early, *Sketch*, p. 421.
1030. Bennett, p. 175.
1031. Clark, p. 219.
1032. Figure 8 Nicolay and Hay, Vol. 9, p. 302, considerably altered, to conform to other texts and maps. See: Atlas, plate 99, map 1; also: F. M. Buffum, facing p. 212.
1033. Sprague, p. 227.

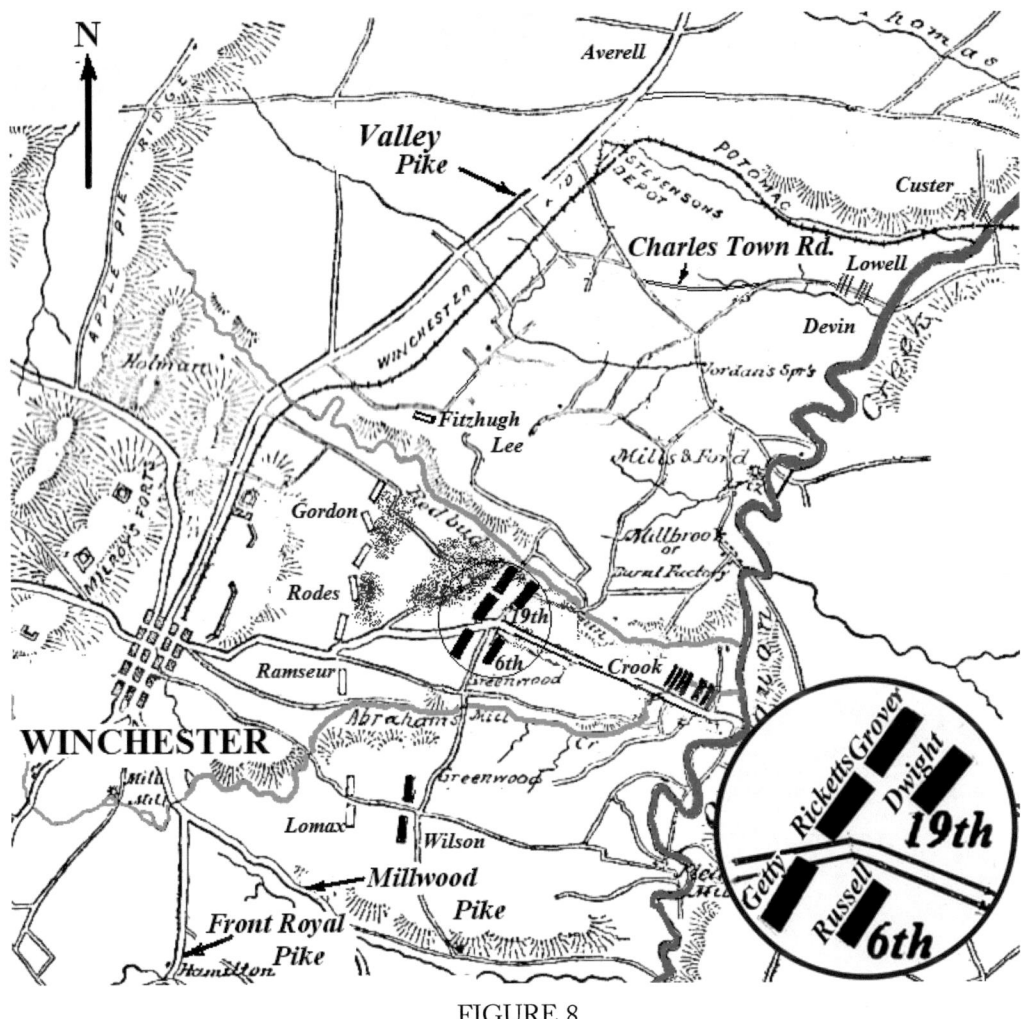

FIGURE 8

Sheridan's choice of advancing along the Berryville Pike had allowed adequate time for Gordon's division to fall into line, Rodes' division arriving last, save Breckinridge's, rendering Sheridan's strategy of meeting and beating the enemy in detail a failure.

In figure 8, the positions of the Union troops are in black, the Confederate positions are uncolored. Wilson is on the south side of Abraham's Creek, a tributary of the Opequan, near the Millwood Pike, to where he moved after the 6th Corps occupied his initial position on the ridge.[1034] The 19th Corps is to the right of the 6th, in an open field behind a heavy piece of woods, Grover in front, Dwight in reserve. Their right rests next to the swampy lowland of Red Bud Run. Crook's 8th Corps is seen in reserve, lined up on the Pike, at the entrance to the canyon, in a position to act as a turning column, by advancing towards the Valley Pike south of Winchester. Sheridan was, at this time at least, still hoping to carry out this part of his original plan.

1034. Clark, pp. 218–219; Sheridan, Vol. 2, pp. 14, 20, 24; O.R. Vol. 43/I, p. 279.

Opposed to Wilson was a part of Lomax's cavalry, consisting of Jackson's and a part of B. T. Johnson's brigades. Opposite the 6th Corps was Ramseur, partially covered by heavy woods, with Nelson's artillery posted on its line. Gordon, who, despite the difficulties of initially missing Early's message to report, had finally arrived from his position at Stephenson's Depot a little after 10:00, and was posted in heavy timber, Red Bud Run on his left, opposite the 19th Corps. To the north of the Red Bud, was Fitzhugh Lee's cavalry, and a battery of horse artillery.[1035] Rodes, arriving after Gordon, formed on Gordon's right, covered by heavy brush and cornfields.

Strangely, all firing had tapered off, and not a warlike sound could be heard. Orton Clark of the 116th New York, which was in Dwight's division, remembers: "All was still as death, to which every man felt he was possibly advancing." Déjà vu, Port Hudson – they were about to make a frontal assault on a fully prepared enemy. It wasn't supposed to have been this way.

Finally, at 11:40, Sheridan's bugle call for the advance was heard.

As the two Federal corps lines moved forward, the Confederates opened fire from their hidden positions all along their front. Considerable ground was gained at first. The 6th Corps, the divisions of Getty and Ricketts, guided on the Pike, pressed back Ramseur's infantry, while Wilson, cooperating, faced Lomax's cavalry. The Pike swung gently to the left, and as the lead division of the 19th, Grover's, drove forward into the woods in its front, a separation between the two corps lines began. Grover's division broke into a clearing beyond a wood, and the pace of Birge's brigade was stepped up to reach a second wood in his front, which anchored Early's left, Evans' brigade of Gordon's division. Evans was quickly broken up, and in pursuing him to within range of seven of Braxton's guns, the continuity of Grover's line was lost. This created a gap at the vital point in the Union line where it covered the Pike's entrance to the gorge.[1036]

The success of Grover's advance had created an opening for Early. Early ordered Battles' Brigade of Rodes' Division into Grover's exposed left flank. Under the weight of the fire from Braxton's artillery, Battle's advance, and a terrible flanking fire from Fitzhugh Lee's battery, stationed on an eminence across the Red Bud not over 600 yards away, Grover's divisions were compelled to retreat.[1037]

It was here that Lt. Col. Willoughby Babcock of the 75th New York, which was a part of Birge's 1st Brigade, was killed. When the news finally was available to those in Battery L, their thoughts were undoubtedly not only ones of respect and sadness, but also of how long ago it was when he was with them as the provost-marshall at Pensacola, and of the fruitless assaults at Port Hudson.

1035. Haines, p. 258; Early, *Sketch*, pp. 420–422; Sheridan, Vol. 2, pp. 21–22; O.R. Vol. 43/I, pp. 150, 222, 266, 279–280, 318–319; Early, p. 422.
1036. Clark, p. 220, O.R. Vol. 43/I, p. 222; Pellet, p. 253; Buffum, p. 212.
1037. Park, pp. 90–91; Hall, H. and J., p. 211.

Though *most* of Grover's left was separated from the 6th Corps line, that is, from Ricketts' division, by about 500 yards, the 156th New York, a part of Sharpe's 3rd Brigade, never lost contact, as Grover points out in his official report.[1038] Ricketts fell back, leaving the 156th stranded, taking all of its casualties there in a few minutes.[1039] Ricketts had taken action to close the gap by ordering three regiments of his 2nd Brigade into it, but to no avail. "The bloody but victorious advance was changed into a bloody and ominus [*sic*] retreat."

"Grover's and Ricketts' commands reached the base from which they had advanced in a state of confusion which threatened wide-spread disaster." Emory, Grover, and others, made tremendous efforts to stop the flight,[1040] but 6th and 19th Corps men were crowding together up the Berryville Pike, while to the right and left of it, the fields were dotted with wounded.

Seeing that the enemy had gotten to their right and rear, Col. Edwards ordered the 3rd Brigade of Russell's Division to advance, with bayonets drawn, to the right side of the road, through the mass of the fleeing fugitives. Opening fire at 150 yards, they "drove the enemy back handsomely," the 37th Massachusetts hitting the enemy from the right and rear, their Spencer repeaters enabling them to "… defeat more than five times their number". It was here that General Russell was killed by a shell fragment to the head, testimony to the "… hot and continuous fire from the enemy's artillery…" which they had to endure.[1041]

Regardless of their potent firepower, the 37th paid heavily for their success, with 92 casualties, though not quite as heavily as the 156th New York, with a revised figure of 111.

On Emory's front, the enemy progress was gradually checked by two small efforts, and a final bloody one that completely recovered all of the lost ground. First, Grover, finally appreciating the value of artillery, ordered up the 1st Maine Battery, which slowed the Confederates, then the 131st New York, posted by Emory in a wooded ravine, made a flank movement on the advancing Confederate column, and poured such a volley into its backs that it recoiled.[1042] Seizing the opportunity, Grover's 2nd and 3rd Brigades made a second charge, which recovered a large portion of the lost ground. When their ammunition was expended, they were relieved by Dwight's 1st Division, which had been drawn up in column behind as a reserve. Leading it was the 114th New York, which checked the advance, taking 188 casualties out of a total of 315 men present, the highest percentage of any regiment that

---

1038. O.R. Vol. 43/I, p. 319.
1039. De Forest, p. 196.
1040. Sprague, p. 231.
1041. O.R. Vol. 43/I pp. 112, 115,184, 185; O.R. Vol. 43/II, p. 927; Bowen, J. L., pp. 354, 355, 380. On July 14th, while the 37th was in Washington, it was issued the Spencer repeater. The 2nd Rhode Island Battalion also received them, presumably at the same time.
1042. Hanaburgh, p. 147; DeForest, p. 197; Flinn, p. 182.

day, and ranking with the highest in any of the other notorious battles of the war, such as Antietam, Gettysburg, or Chickamauga.[1043]

As may have been expected, there was controversy over Dwight's behavior during the battle. Grover brought charges against him, among them misbehavior before the enemy. This because he was not seen on the battlefield between noon and 3:30, and was found in the rear "beyond the falling shot of the enemy." This was in addition to suspicious behavior discovered days earlier, in which movements of the enemy, reported to him, were not forwarded to Sheridan's headquarters. In the end, Dwight survived the controversy.

At noon, Sheridan had given up the idea of using Crook as a turning force, and had ordered him to the front.[1044] The time required to make the two-mile advance through the canyon "so blockaded by ammunition wagons, battery wagons, forges, ambulances and stragglers going to the rear…" was such that he was not in place until about 3:00.

Captain James Garnett, an ordnance officer in Rodes' Division, observed in his diary: "Up to 3 o'clock[1045] we had whipped the enemy well, and but for cavalry we might have held our own against succeeding attacks." Three p.m. would prove to be the turning point.

In the interval that this great infantry battle had taken place, Merritt's cavalry had broken out from its delaying action in front of Breckinridge, as much due to its own actions, or as we have seen, to Breckinridge being ordered to withdraw and come to Early's aid. He did not reach Early until 2:00.[1046] However, essentially freed of entanglement (Breckinridge left only Patton's infantry and Payne's cavalry behind) Merritt moved quickly. Custer arrived just outside of Stephenson's Depot at 1:00. Typical of Custer, he reported:

> In the absence of instructions I ordered a general advance, intending, if not opposed, to move beyond the enemy's left flank and strike him in reverse. I directed my advance toward Stephenson's Depot and met with no enemy until two miles of that point, where I encountered Lomax's division of cavalry,[1047] which at that time was engaged with Averell's division, advancing on my right on the Martinsburg Pike. Our appearance was unexpected and caused such confusion…that though charged by inferior numbers, they at no time waited for us to approach within pistol range, but broke and fled.

By 2:00 Merritt (Custer, Lowell, and Devin) had linked up with Averell

---

1043. Pellett, p. 256; Fox, p. 36; Sheridan, Vol. 2, pp. 23, 24; O.R. Vol. 43/I, pp. 150, 300–307; O.R. Vol. 43/II, p. 30.
1044. O.R. Vol. 43/I, pp. 361, 280–281; Sheridan, Vol. 2, p. 30.
1045. SHS, Vol. 31, Richmond, VA, 1903, Garnett, pp. 62–63.
1046. Early, *Sketch*, p. 424.
1047. Lomax's division was a part of Fitzhugh Lee's cavalry. At this time, Lomax had four brigades: McCausland's, Imboden's, Johnson's, and Jackson's. O.R. Vol. 43/I, p. 566.

(Schoonmaker and Powell) near the junction of the Charles Town road and the Valley Pike, just south of Stephenson's Depot,[1048] figure 9, where the Federal cavalry is shown arrayed at the top. Here were some 8,000 superbly equipped mounted troops, and counting the artillery that accompanied them, their total made nearly 9,000.[1049] One can only imagine the spectacle that this force presented.

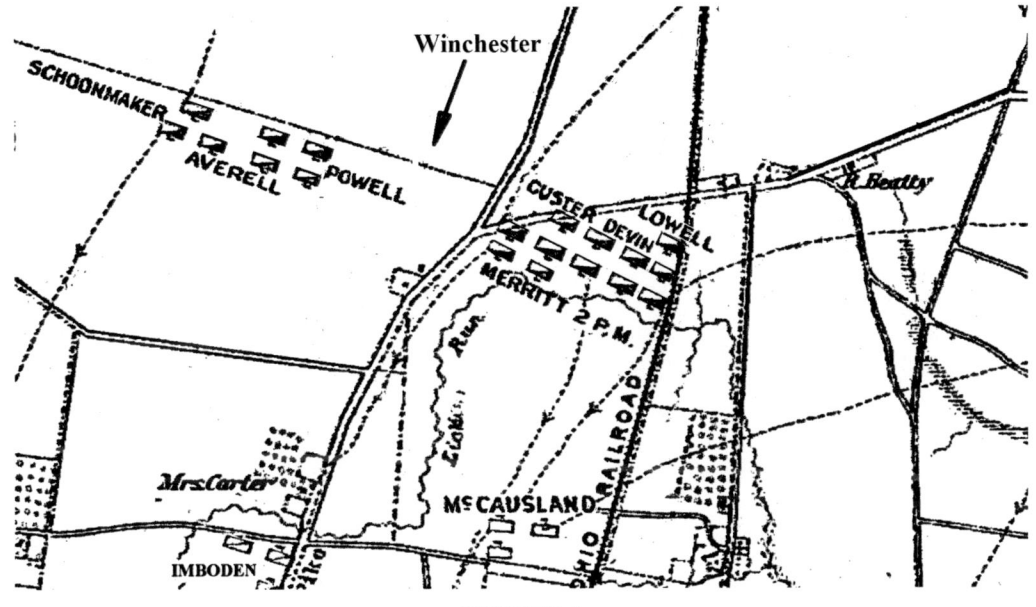

FIGURE 9

Two horses with their riders occupy slightly more space than a modern automobile. Visualize a modern mall or stadium lot with 4,500 cars, and you can appreciate the scale of the overwhelming cavalry force that was descending upon Winchester. Arrayed as they were, each brigade in line of squadron columns, three lines deep, their front extended more than half a mile.[1050] They advanced at a trot, covered by one continuous and heavy fire of skirmishers, using only carbines. Guidons fluttered, and in the sunlight there was presented one mass of glittering drawn sabers, while the bands played the national airs. James H. Kidd, then a colonel in Custer's 6th Michigan, remembers:[1051] "Officers vied with their men in gallantry and zeal. Even the horses seemed to catch the inspiration of the scene and emulated the martial ardor of their riders." They soon came upon McCausland's cavalry brigade, and Imboden's 18th, 23rd, and 62nd Virginia cavalry, which had taken a position near Mrs. Carter's house. Charles T. O'Ferrall, of the 23rd Virginia

---

1048. O.R. Vol. 43/I, p. 456; Figure 9, Atlas, plate 99, map 1, portion, altered.
1049. O.R. Vol. 43/II, pp. 65, 248. Cavalry returns for September, Torbert, 6,343, Averell, 4,758 present for duty. Wilson's division was not here, as it was posted below Abraham's (Abrams) Creek. Removing his strength, 2,977, results in the quoted total; O'Ferrall, p. 115; Munford, p. 451, claims 11,000.
1050. Kidd, p. 390; O.R. Vol. 43/I, p. 498; O'Ferrall, p. 115.
1051. Kidd, p. 391; O.R. Vol. 43/I, p. 456.

remembers: "But in the briefest time the Federal cavalry, in a compact mass and powerful in numbers, rushed upon us, and drove us rapidly and in disorder back upon the left flank of Early's infantry line." This Federal horde had pushed the Confederate cavalry to within three miles of Winchester by 3:00 o'clock, the very time that Crook, with his two divisions, the 1st under Thoburn, and the 2nd under Duval, had launched his drive. From the position he had taken on the right of the 19th Corps,[1052] Crook advanced along the north side of the Red Bud, and rapidly drove back Patton's infantry and Payne's brigade of Fitzhugh Lee's cavalry, which had been supporting Patton. Additional elements of Breckinridge's corps, Wharton's were moved by Early, as he writes "in double quick time" to his left and rear, but it would be to no avail.

At about 4:30, Thoburn's left had linked up with Duval, and then had moved on to the Confederate fortifications on the north side of Winchester, where Merritt, on the east side of the Pike, confronted Fort Collier, and Averell, on Merritt's right, was confronting Fort Jackson,[1053] in and near which were posted Munford's brigade of Wickham's cavalry, Wickham having succeeded Fitzhugh Lee, see "Federal Line" which the Confederate mapmaker Jedediah Hotchkiss shows in figure 10.

The "fragments of infantry" to the left in figure 10 represent the second line of Breckinridge, Gordon, and Rodes, which was established after their first, opposing Crook's Corps, had to withdraw. Soon, their whole front would give way, retiring south and west through Winchester.

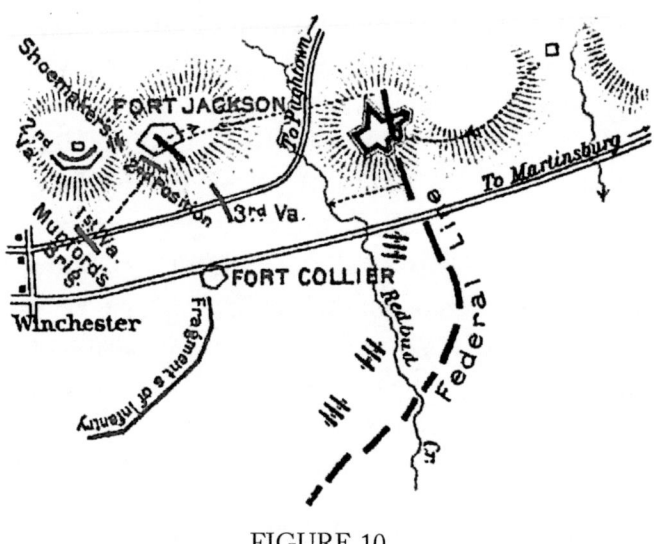

FIGURE 10

1052. Sheridan, Vol. 2, pp. 25–27; O.R. Vol. 43/I, p. 362; Early, *Sketch*, p. 425; Kidd, p. 391.
1053. Figure 10, Atlas plate 85, map 16, presented as it appears in the Atlas, if properly oriented, it would have to be rotated approximately 90° counterclockwise.

Capt. J. M. Garnett[1054] looked at his watch when he saw the Federals enter Winchester. It was 5:07.

Fortunately for Early, Ramseur's line had managed to maintain its organization as it moved south. It checked any attempt by Wilson's cavalry and elements of the 6th and 19th Corps, and even some parts of Merritt's cavalry, to close off Early's escape route. After sunset, which by today's calculations would have been 6:15,[1055] the Valley Pike remained open, and Early withdrew to the area of Newtown, about five miles south, which was not reached until after midnight. Crook reported that he had broken off the pursuit about two miles south.

Sheridan felt it imperative to report the results of the day to Grant, and General Crook conducted him to the home of Miss Rebecca Wright, where he met, for the first time, the lady, who as Sheridan says, "had contributed so much to our success…" Here, at a desk in her school room he wrote his dispatch. It found its way to the telegraph at Harper's Ferry, and was sent out at 11:40 a.m. on the 20th.[1056] It said:

> GENERAL: We fought Early from daylight to between 6 and 7 p. m. We drove him from Opequan Creek through Winchester and beyond the town. We captured 2,500 to 3,000 prisoners, 5 pieces of artillery, 9 battle-flags, all the rebel wounded and dead. Their wounded in Winchester amount to some 3,000. We lost in killed General David Russell, commanding division, Sixth Army Corps; wounded, Generals Chapman, McIntosh, and Upton. The rebels lost in killed the following general officers: General Rodes, General Wharton, General Gordon, and General Ramseur. We just sent them whirling through Winchester, and we are after them to-morrow.

Passing through Washington, the telegram became known to Stanton and Lincoln, and, it seems, everyone. The last phrase, "whirling through Winchester, and we are after them tomorrow" struck a chord, and it became a household word in a few hours.[1057]

Sheridan fails to mention his casualties, an estimate of which he must have had by that time. They were substantial:[1058]

6th Corps : 1,699
19th Corps: 2,074
8th Corps: 794
Cavalry: 451

The total at 5,018, must have been sobering, the huge number being nothing

---

1054. Garnett, p. 67; Early, *Sketch*, p. 426, Sheridan, Vol. 2, pp. 26–27.
1055. *aa.usno.navy.mil*, Winchester, VA; Garnett, p. 66; Nichols, p. 189; Newtown is now Stephens City; O.R. Vol. 43/I, p. 362.
1056. O.R. Vol. 43/II, p. 124; Sheridan, Vol. 2, pp. 29–30.
1057. Nicolay and Hay, Vol. 9, p. 305.
1058. O.R. Vol. 43/I, pp. 112–118.

to celebrate. He merely presents them as "about 4,500" in his memoirs.

Battery K–L had 3 men wounded, and since Franck Taylor wrote no official report of the battle, details of where or when are not known:[1059]

James Campbell	Pvt. 15 Nov.'62 Pensacola
John Kelly	Pvt. 26 Dec.'62 Pensacola
John C. Wood	Pvt. 16 Dec.'62 Pensacola

The praise of Sheridan from Washington was effusive, and he was promoted to the permanent rank of brigadier-general of the regular army, and to the permanent command of the Middle Military Department. Grant fired a one-hundred-gun salute from each of his armies at Petersburg and urged Sheridan to "push his success." He had fought what appeared to be a decisive battle and won. This was something that Lincoln had hoped for, but had been afraid of as well, for if Sheridan had gone down to defeat, as had every other Union general in the Valley before him, the news might have caused the administration to be defeated in the November elections.[1060]

None of the Democratic opposition to Lincoln could have quickly had the finer details of the battle, and would not have a chance to analyze them for some time after, even if they then cared. Early's bitter analysis of it has a ring of truth:

> A skillful and energetic commander…would have crushed Ramseur before any assistance could have reached him, and thus ensured the destruction of *my whole force*; and later in the day…with the immense superiority in cavalry which Sheridan had…would have destroyed my whole force and captured everything I had.

The losses which Early reported were 3,611, exclusive of those of his cavalry, which were 348 for the period of September 1st to October 1st.[1061] He hints that there were more, saying: "But many were captured, though a good many are missing as stragglers…" Though General Gordon was reported by Sheridan as killed, he was not.

Many of Early's desperately wounded and dying were left in Winchester when he retreated. The Union Hotel, which had been turned into a hospital, was filled. Other wounded were scattered almost everywhere: the courthouse, in churches, and in private homes; the Confederate surgeons remaining behind.

Now, with the Presidential campaign in full swing, the news of Sheridan's great victory at Winchester came along, and the process of the erosion of support for McClellan had begun. Of course, the Confederacy was interested in the outcome

---

1059. Haskin, p. 577, lists five, but the Battery L Monthly Return, from their station at Mt. Crawford, VA, for September lists only 3. The Regimental Return for September, where Battery K is listed, shows no casualties.
1060. Grant, Vol. 2, p. 332; Early, *Sketch*, p. 427.
1061. O.R. Vol. 43/I, p. 555; Park, R. E., pp. 93–95; Haines, A., p. 263.

of the election because in McClellan, there was the chance for a negotiated peace, or perhaps some softer treatment. Capt. William W. Chamberlain, of Company G, 6th Virginia Infantry, wrote in his memoirs:

> While I was on sick leave [he returned to duty on September 1st] the news from the North led us to believe that the Northern people were anxious to make peace, but a month or two later an entire change of sentiment seemed to have taken place. The Presidential Campaign was then in progress. Lincoln had been nominated by the Republicans, and McClellan by the Democratic Party.[1062]

He also recorded some gossip:

> General A. P. Hill and Mrs. Hill dined one day with General and Mrs. Walker, and in the course of the conversation I heard General Hill say that he hoped General McClellan would be elected, because if it were necessary to surrender, he would prefer to do so to McClellan.

Not only had the sentiment in the north changed, but here, a *corps commander* in Lee's Army of Northern Virginia was openly discussing *surrender*.

1062. Chamberlaine, W. W., p. 109.

Chapter 16

Fisher's Hill; The March Up The Valley; Terminated; Tom's Brook/Strasburg/Woodstock Races

Fisher's Hill

At daylight on the morning of September 20th, Early continued his retreat, leaving the sides of the Valley Pike strewn with muskets, knapsacks, canteens, and clothing. In the middle of the road were broken-down wagons, their teamsters having cut the harnesses and escaped on the horses. Early was permitted to fall back across Cedar Creek, and he briefly tried to fortify Hupp's Hill, but in the afternoon, he fell back to Fisher's Hill, below Strasburg. Here, he took position on his old defensive line, the one from which he had departed to follow Sheridan north on August 17th. This time, however, his line did not extend as far as it had in August, and it was more thinly manned.[1063]

Also at daylight, Merritt's Division, the 1st New York Dragoons in the advance, pushed briskly up the Pike, through Kernstown and Newtown, meeting the enemy cavalry at Middletown, which did not oppose them. Here Devin's brigade and Battery K–L were left to hold the town, and Custer and Lowell continued, arriving at Hupp's Hill, overlooking Strasburg, to discover Early's entrenchments on the south side of Tumbling Run.[1064] No further advance was attempted, pending the arrival of the rest of the army, and of Sheridan making a reconnaissance of the area. At about three o'clock the infantry came up, Emory and Wright marching in the open country on either side of the Pike, with Crook in the rear. Torbert then moved two of Merritt's brigades, Custer's and Lowell's, to the right to join Averell, who had pushed along the Back Road, to the west of, and parallel to the Pike. He was now in position near Cedar Creek, on Early's left. The position on Hupp's Hill, which Merritt's brigades vacated, was filled by Wright and Emory. Crook halted to the rear, north of Cedar Creek.

Wilson's division, when it had reached Middletown, turned toward Front

---

1063. Denison, p. 390; Nichols, p. 189; Early, *Sketch*, p. 429; O'Ferrall, p. 118; Long, SHS, Vol. 3, pp. 118–119; Bowen, p. 239.
1064. Bowen, J. R., p. 239; Sheridan, Vol. 2, pp. 33–34; Merritt, p. 510; Haines, p. 264; O.R. Vol. 43/I, pp. 223, 428, 441, 475.

Royal, chasing Wickham's cavalry, which Early had directed into the Luray Valley, through a narrow pass at Millwood, to try to prevent a flank attack upon his new position at Fisher's Hill.[1065]

By now, Sheridan had become convinced that any attempt at a frontal assault would be bloody and of questionable success. He resolved upon a flank attack, essentially what he had finally been forced into at Winchester, his frontal attack having taken almost all of the casualties that day. Again, it would be on the right, and it would involve Crook.[1066] Still cherishing the thought of cutting Early off, Sheridan decided to detach Torbert with Merritt's 1st and Reserve brigades to reinforce Wilson's move up the Luray, to ensure that Wickham could be driven out of Luray Pass, and then by crossing the Massanutten Mountain range near New Market, gain Early's rear.

Merritt's 2nd Brigade, Devin and Battery K–L, remained south of Middletown, guarding the rear.

The 21st was spent by Sheridan in reconnoitering the enemy's lines and repositioning his own, by seizing the high bank on the north side of Tumbling Run, which was accomplished by a "brisk fight," and afterwards the work of putting these heights in a defensible condition began. Trees were cleared to facilitate the fire of the artillery, and earthworks were thrown up. These positions, closely opposite Early's, are seen in figure 1,[1067] and are labeled as "SEIZED ON THE 21ST."

Early's defensive line is arrayed all along the high banks on the south (left) side of Tumbling Run. Wharton, now commanding Breckinridge's division, Breckinridge having been ordered to the command of the Confederate Department of Southwestern Virginia,[1068] was on his right, and extending to his left was Gordon, then Pegram commanding Ramseur's old division, then Ramseur, commanding Rodes' division, and at the end of the line, at Little North Mountain, was Lomax's cavalry, dismounted, a mere 300 men.[1069]

Early concluded not to retire, after observing all of Sheridan's activity to establish such a strong position opposite him. Early writes:[1070] "I began to think he was satisfied with the advantage he had gained and would not probably press it further…" Even now, Early seems to have regarded Sheridan as timid. So confident was he that Sheridan would not attack, he had taken all of the ammunition from his caissons and placed them nearby in the breastworks. This also seems to be proof that Early never saw Sheridan's most significant move, that of Crook, until it was

---

1065. Early, *Sketch*, p. 429; O.R. Vol. 43/I, pp. 518–519. Wickham succeeded Fitzhugh Lee, wounded at Winchester.
1066. Sheridan, Vol. 2, p. 35; O.R. Vol. 43/I, pp. 428, 441.
1067. Sheridan, Vol. 2, p. 39, altered; O.R. Vol. 43/I, p. 152.
1068. O.R. Vol. 43/II, p. 873.
1069. O'Ferrall, p. 118; Sheridan, Vol. 2, p. 34.
1070. Early, *Sketch*, p. 430; Sheridan, Vol. 2, p. 34; Bowen, J. R., p. 239.

too late.

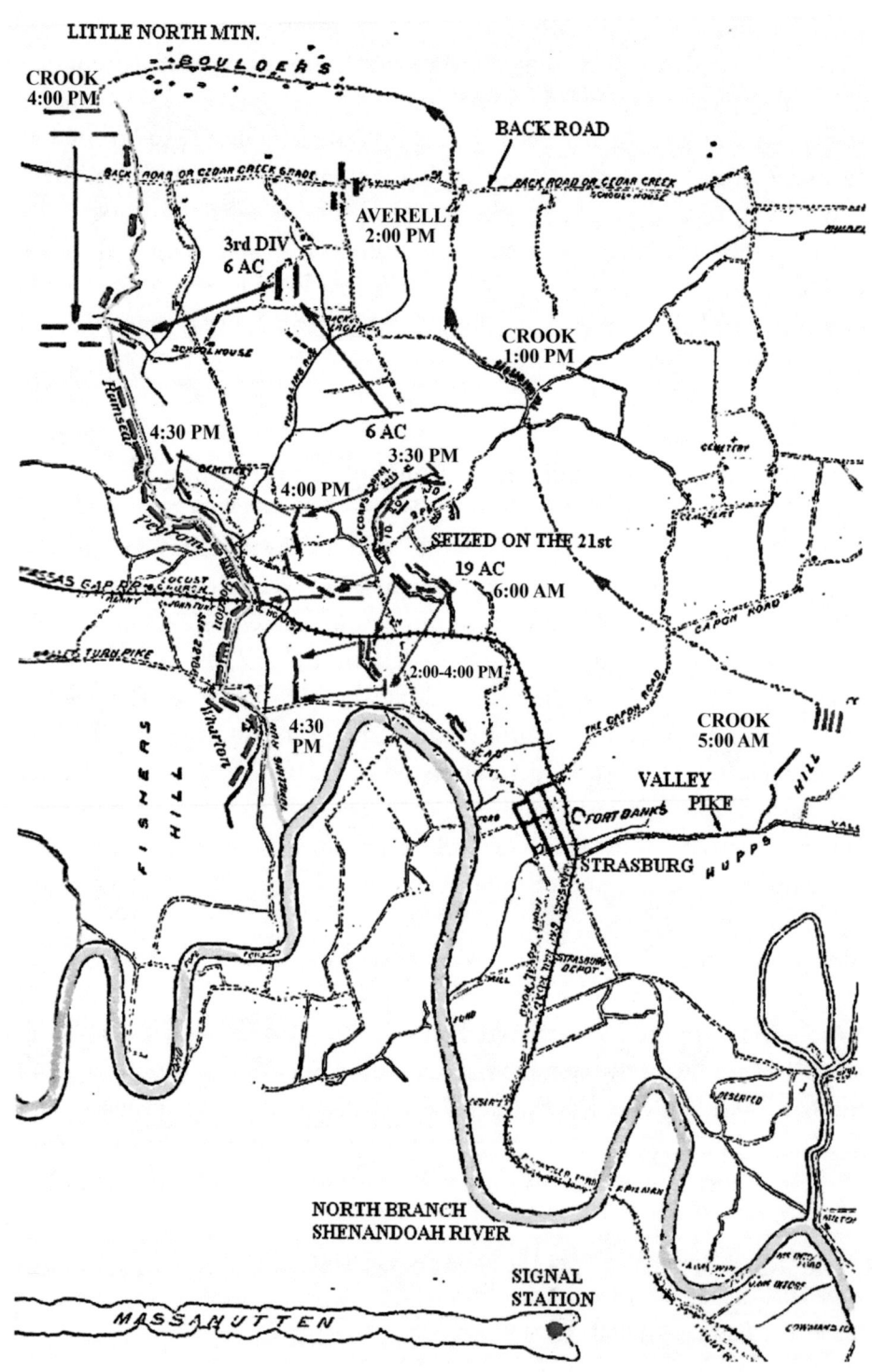

FIGURE 1

Knowing that Early could see every detail of his moves from the observation station on the top of Massanutten Mountain, and learning from a reliable guide that there were forest paths over which an army could move, presumably unseen, to Little North Mountain, Sheridan resolved to put Crook in motion. On the night of September 20th he moved Crook into some heavy timber on the far side of Hupp's Hill, near the Valley Pike, where he was ordered to hide all the day of the 21st.[1071] At daylight on the 22nd, he marched around the rear of the 6th Corps, under the cover of the intervening woods and ravines (the dotted trail in figure 1) to a concealed position near Back Road, which he reached by 1:00 p.m.

While this was taking place, Sheridan improved the positions of the 6th and 19th Corps, moving up closer to the Confederate works, which can be seen by the arrows and the times noted in figure 1. In the afternoon, Ricketts' 3rd Division of the 6th Army Corps was pushed out to the far right, near the end of the line of Early's infantry, in the vicinity of Averell's cavalry. While this very visible move occupied Early's attention, Crook completed his move to the base of Little North Mountain.[1072]

Finally, the 19th Corps was moved to a position nearer the Manassas Gap extension of the Virginia Central Railroad, its right resting on the railroad, and the 6th Army Corps was advanced closer still, to within about 700 yards of Early's line. It was now 4:00 p.m., and everyone was in position as planned.

Observing Rickett's move, Early says: "I discovered that another attack was contemplated…" He reacted by giving orders for his line to retire that night, after dark, but it was too late, Crook had begun his move[1073] along the rear of Early's line. As Crook's cheering men crossed the stretch of broken country between their hidden position and the dismounted cavalry of Lomax, Averell's 1st Brigade, dismounted, also charged forward, and though a piece of Confederate artillery poured grape and canister into them, they were not slowed, and the Confederate resistance began to crumble. Joined by Ramseur, a stand was made on a ridge about a mile from the base of the mountain. Wharton was ordered to the left as well as Pegram. Wharton never arrived, and when Pegram's division joined Ramseur, their lines became disordered. Crook, now united with Averell and Ricketts, then moved along in the Confederate rear with little resistance. The 6th Corps and the 19th then crossed Tumbling Run, and were soon scrambling up the heights, swinging successively into line, and Early was routed, losing 12 guns, 240 killed and wounded, 995 missing, and whatever property that was in his works.[1074] The high number of missing hints at increasing demoralization and disorganization, the situation offering the opportunity to walk away.

1071. Sheridan, Vol. 2, p. 35.
1072. Sheridan, Vol. 2, pp. 36–37; Haines, p. 265; O.R. Vol. 43/I, pp. 152–153.
1073. O.R. Vol. 43/I, pp. 223, 363–364; O'Ferrall, p. 119; Early, *Sketch*, p. 430; Farrar, pp. 385, 387.
1074. O.R. Vol. 43/I, pp. 64, 80, 153, 223–224, 283, 364, 555; Sheridan, Vol. 2, pp. 38–40, Early, *Sketch*, p. 430.

Crook followed the retreating Confederates for about two miles, where, in darkness, he halted, but Ricketts, the rest of the 6th Corps, and the 19th Corps, pursued Early all night, to Woodstock. Averell did not. At midnight, after his 2nd Brigade had guarded Crook's stragglers and captured equipment, he went into camp, and did not move again until daylight.[1075]

At about 5:30 p.m., Sheridan ordered forward the sole remaining brigade of Torbert's cavalry that had not been sent toward Front Royal, Devin's.[1076] To respond to Sheridan's order to chase Early, Devin and Battery K–L, would have to cover the seven miles south to Fisher's Hill before even beginning. Once arriving in the area of Strasburg, his pursuit was slowed by the infantry's presence in the road. Eventually, he says, "with great difficulty" he reached the head of the column, the 19th Corps, with Grover's Division leading. About five miles south of Fisher's Hill, they ran upon a creek, the opposite bank of which had been fortified. It was high, and was covered by woods, in which Early had placed a rear guard of artillery and infantry. The Confederate position was taken and the pursuit was resumed, after, according to Devin: "One section of my battery was placed in position and opened on the enemy's rear." In the absence of any report by Taylor describing who was involved, it could have been either the section often commanded by Taylor, or the second section, commanded by either of two junior officers from Battery K, W. C. Cuyler or John McGilvray. One section was equipped with Napoleons, and the other with 3-inch rifles.

The pursuit reached Woodstock by 3:00 a.m., and there the 19th Corps went into bivouac, though Devin's cavalry was ordered to continue, and pursue the enemy through to Mount Jackson.[1077] Without rest or breakfast, the pursuit moved on to Edinburg, on the way picking up a number of Confederate stragglers, and finding several burning wagons and an abandoned artillery piece.

Continuing on toward Mt. Jackson, about three miles south of Edinburg, they met Early's cavalry, and drove them through Hawkinsville to within about two miles of Mt. Jackson. Arriving there at about noon,[1078] they discovered a "large force of infantry" bivouacked around the town, and in line of battle. It was Gordon and Ramseur, see figure 2. Early had halted to allow his exhausted troops to rest, and to allow the sick and wounded, and the hospital supplies, all on the slower moving wagons, to be sent on to Staunton.

Battery K–L was ordered into position on a crest on the left of the Pike, and opened on the enemy line. Meanwhile, the 9th New York advanced as skirmishers, with the 6th New York in support, and a warm engagement ensued. At

---

1075. O.R. Vol. 43/I, p. 499.
1076. O.R. Vol. 43/I, pp. 92, 283, 475; Haskin, p. 213.
1077. Hall, Besley, and Wood, p. 225; O.R. Vol.43/I, pp. 476, 500.
1078. Cheney, p. 223; Hall, Besley, and Wood, p. 226; Nichols, p. 190. Figure 2 is map no. 22, Atlas, plate 85.

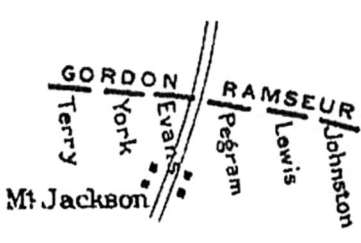

FIGURE 2

about 2:00 p.m., Averell, who had been following the column, finally arrived and ordered his two brigades into action, which continued until dark. At that point, Devin, having run out of ammunition, was ordered to retire.

It was here that Sheridan received the news that Torbert had fallen back to Front Royal, and as well, learned that Averell had gone into camp on the night of the 22nd and had left the pursuit to Devin.[1079] Furthermore, Averell had not pressed the action here at Mt. Jackson to Sheridan's satisfaction, Averell preferring to believe the report of a signal officer that a "brigade or division" was confronting him (which we see from figure 2 was true), and as a result, retired. Sheridan then sent him a note which, in part, said: "I do not advise rashness, but I do desire resolution and actual fighting, with necessary casualties, *before you retire.*"

Upon hearing that Averell had literally disobeyed his written order, Sheridan sacked him. If anyone had been privy to several events leading up to this, it could have been predicted. On September 1st, Grant had sent Sheridan a memo,[1080] which is here quoted in total:

> The frequent reports of Averell's falling back without much fighting or even skirmishing, and afterward being able to take his old position without opposition, presents a very bad appearance at this distance. You can judge better of his merits than I can, but it looks to me as if it was time to try some other officer in his place. If you think as I do in this matter, relieve him at once and name his successor.

To anyone who has been in military service, it is a given that when a superior even *hints* at what he desires, it had better be done. Grant's letter is hardly a hint, and now was the time. Perhaps the unjust part here was that Torbert was not also sacked.

Sheridan's situation at Fisher's Hill is reminiscent of Banks' at Bisland. Banks had sent Grover up the Teche, in order to cut off Taylor at Franklin. We know that the plan failed because Grover failed to block the cross road, but at least, Banks had waited. He had held off from making a strong frontal attack on Taylor, to hold him at Bisland, until finally the Clifton had returned with the word that Grover had landed above Irish Bend. Here, in contrast, the impatient Sheridan did not wait to make his attack until some word of Torbert's status was received. Not only should he be angry at Torbert for utterly failing, but at himself for not keeping aware of the situation. This was hardly the sound generalship expected of one of the icons of the war, and Early was allowed to escape once again.

1079. Sheridan, Vol. 2, pp. 42–43 (italics added); Early, *Sketch*, p. 432.
1080. O.R. Vol. 43/II, p. 3.

The March Up The Valley

Now began another episode of "the fox and the hounds," or, as Early terms it: "The March Up The Valley." After Averell broke off the engagement at Mount Jackson, Early fell back to Rude's Hill, figure 3,[1081] about two miles south of Mount Jackson, where he camped that night.[1082] On the morning of September 24th, Devin was ordered by Sheridan to push a regiment of his cavalry across the north branch of the Shenandoah and try to flank Early's right, and at the same time, Averell's division, now under the command of Col. William H. Powell, was sent to flank Early's left.

FIGURE 3   FIGURE 4

Devin's 1st New York Dragoons were sent across the bridge, it being intact, figure 4.[1083] Seeing that Early had begun to retreat, Devin deployed the rest of his division, and pressed forward on the trot, coming up on Early's line of battle at New Market, figure 5.[1084] A map of his entire route is shown in figure 6.

Battery K–L was ordered up to the front and placed on a ridge to the right of the road, in the fashion described by G. W. Nichols of the 61st Georgia:[1085]

> They would run cavalry batteries up on top of the hills and shell us severely.

From Devin:

> I opened with shell and spherical case shot, at the same time advancing the First New York as skirmishers.

1081. Atlas, plate 85, map no. 23.
1082. Early, *Sketch*, p. 432; Nichols, p.190.
1083. Duffey, p. 9. On October 2nd, when the Partisan Ranger McNeill ordered it burned, the local residents prevented it, fearing that Sheridan would retaliate by burning their homes. Note here the idyllic rural character, the macadamized road surface, and the telegraph poles. O.R. Vol. 43/I, p. 476; Sheridan, Vol. 2, p. 46.
1084. Figure 5, Atlas, plate 85, map no. 24.
1085. Nichols, p. 191.

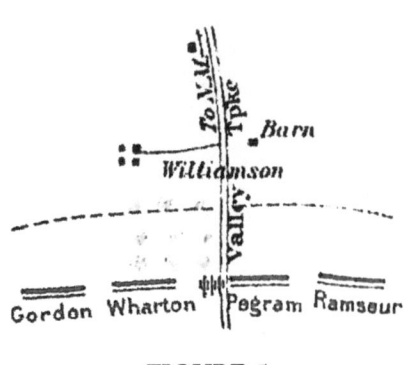

FIGURE 5

The enemy at once replied with a battery from a hill in my front. I had pressed up to within 500 yards, when the enemy retired precipitately through the town. I charged halfway through the main street, and on the left, but a hot fire from the inclosures [*sic*] and gardens forced me back. I now dismounted two squadrons of the First New York, cleared the town, charged through with the rest of my command, and found another line formed 300 yards beyond and retiring in excellent order. I again advanced my skirmishers and battery, and again the enemy retired. The chase continued in this manner to a point seven miles south of New Market, the enemy retiring from one position to another, while I pressed them so sharply with my skirmishers and Taylor's battery (I had nothing more) that I was frequently within 500 yards, and the enemy was compelled to retire in line. At dark, I was relieved by the infantry and went into camp…Nothing could surpass the gallantry with which my little force (less than 400 men) continued to press the enemy's line, though at times two miles from support. Lieutenant Taylor handled his guns most efficiently.[1086]

It was in this action that Pvt. Andrew Stoll was wounded.

Devin's men and Battery K–L were on a roll, as Devin rather proudly relates. Early's side of the story agrees, but reminds us of one very significant factor: "As the country was entirely open, and Rude's Hill an elevated position, I could see the whole movement of the enemy, and as soon as it was fully developed, I commenced retiring in line of battle, and in that manner retired through New Market to a point at which the road to Port Republic [the Keezeltown Road] leaves the Valley Pike, nine miles from Rude's Hill [figure 6.]"[1087]

On September 25th, Devin was ordered to advance to Harrisonburg, which he did, with the 9th New York in the advance. Though Lomax's cavalry had gone in that direction, he did not find it. It was later learned that Early had turned off on the Keezeltown Road toward Port Republic, and, though no one then knew it, it was because of Early's intention to join Kershaw, who had been ordered back to the Valley after Lee had heard about the Battle of Winchester.[1088] Devin then turned toward the little village of Port Republic.

On the same day, Torbert, with Merritt's two brigades and Wilson's division, finally arrived at New Market. As we have related, he had been ordered to cross

---

1086. O.R. Vol. 43/I, p. 476; Early, *Sketch*, p. 432; Sheridan, Vol. 2, pp. 46–47.
1087. Atlas, plate 85, Portion of map no. 1, altered; Battery L "September Monthly Report."
1088. Sheridan, Vol. 2, pp. 47–48; O.R. Vol. 43/I, pp. 476–477; Bowen, p. 242; Cheney, p. 223; Haskin, p. 213

the mountains into Front Royal, and with Wilson leading, they had run into Wickham's two brigades at Milford Creek on the 22nd, in a position which Torbert says, "was impossible to turn…" He reported: "Not knowing that the army had made an attack at Fisher's Hill, and thinking that the sacrifice would be too great to attack without that knowledge, I concluded to withdraw…"[1089] He returned to the vicinity of Front Royal, and while there, on the 23rd, he received Sheridan's peremptory order to turn around and advance up to Luray and cross into New Market.

FIGURE 6

Col. J. H. Kidd, of the 1st Michigan cavalry, remembers:

> Torbert made a fiasco of it. He allowed Wickham…with, at most two small brigades, to hold him at bay and withdrew without making any fight to speak of. I remember very well how the Michigan brigade lay in a safe position in rear of the line listening to the firing, and was not ordered in at all. If Custer or Merritt had been in command it would have been different.[1090]

J. R. Bowen, in the *History of the 1st New York Dragoons*,[1091] also notes:

> Speaking of this fizzle on the part of Torbert, Sheridan says: "I was astonished and chagrined. My disappointment was extreme. To this day I have been unable to account for Torbert's failure."

As it was, Custer and Lowell led the return trip, and on the morning of September 24th attacked Wickham, scattering his troopers, and the way ahead was opened. Kidd adds: "Even then the march was leisurely, and the two big divisions arrived in New Market too late."

1089. O.R. Vol. 43/I, p. 428.
1090. Kidd, p. 396.
1091. Bowen, J. R., p. 240.

Interestingly, Merritt made no official report, and in an article published in *Battles and Leaders*, Vol. 4, p. 510,[1092] he gives little detail, does not explain the expedition's purpose, and abruptly ends the matter, saying: "This design was not accomplished." Wilson's official report was also silent on all of the details about the embarrassing withdrawal. Custer made no report or is not found. Lowell's was factual, but brief. One statement of his, however, told it all: "September 22,... no attack on the enemy's position was ordered."

We have to conclude that Merritt's and Wilson's reports were an example of doubletalk designed to cover up an event that was embarrassing. Merritt and Wilson were West Pointers, as was Torbert. Lowell, a volunteer from the 2nd Massachusetts Cavalry, was not a part of the "old boy network," and was not influenced by it.

On September 26th, not yet knowing of Kershaw's marching back from Culpeper to the aid of Early (Lee had issued the order on the 23rd), and interpreting Early's turn toward Keezeltown as probably caused by Powell's push ahead to gain the Valley Pike at Lacey Springs, which would have flanked him, Sheridan ordered Merritt to catch up with Devin, with orders to advance as far as Port Republic, "to occupy the enemy's attention…"[1093] while Torbert, with Wilson's division and Lowell's Reserve Brigade, were ordered toward Staunton, from where he was to proceed to Waynesboro and blow up the railroad bridge and the tunnel at Rockfish Gap. On his return, he was to drive all of the cattle he could find, and "destroy all forage and breadstuffs, and burn the mills."

Terminated

The key word, "return" was in Sheridan's order to Torbert. Sheridan had begun to think about what to do after Early had escaped him at New Market. Driving Early further, likely to Brown's Gap, would take him across the Blue Ridge Mountains into eastern Virginia, and if he followed, he was certain that he would be urged to pursue further, through Charlottesville and beyond, on a line towards Richmond. It would pose such a burden on his supply train that a campaign there could not be supported without the opening of the Orange and Alexandria Railroad.

To repair and secure the railroad against guerilla attacks by the likes of Mosby would occupy many men. In addition, he would have to keep secure the gains already made, that of the Baltimore and Ohio Railroad, and the Chesapeake and Ohio canal. There was even the possibility that Grant could not sufficiently hold Lee besieged in Petersburg, who would then detach enough troops to overwhelm Sheridan's then much scattered force.[1094]

1092. O.R. Vol. 43/I, pp. 490, 519–520.
1093. Sheridan, Vol. 2, p. 49; O.R. Vol. 43/I, p. 49; O.R. Vol. 43/II, p. 878.
1094. Sheridan, Vol. 2, pp. 53–55; O.R. Vol. 43/I, p. 477; Hall, Besley, and Wood, p. 226.

It was not until the evening of September 29th that Sheridan was able to wire Grant what had happened after New Market, and Grant did not receive the telegram until October 2nd. In the meantime, on the morning of the 29th, Lincoln, with remarkable prescience, had telegraphed to Grant[1095] of his concern that Early would be reinforced, and that Sheridan was in danger. That afternoon, Grant responded that: "I am taking steps to prevent Lee sending re-enforcements to Early by attacking him here." Remarkably, Lee stood firm, not the least moved by Grant's attack, and not only did he not recall Kershaw, but in addition, had ordered Rosser out to aid Early. In fact, Lee urged Braxton Bragg[1096] now an advisor to Jefferson Davis, to see to it that everything be done to strengthen Early.

After introducing Grant to his objections to crossing over into east central Virginia in his September 29th telegram, Sheridan grew even more firm, and ended an October 1st report with: "I think that the best policy will be to let the burning of the crops of the Valley be the end of this campaign, and let some of this army go somewhere else."[1097] Later in the day, Sheridan received a dispatch from Halleck which asked about his "push forward to Staunton or Charlottesville."

This caused Sheridan some alarm, already Washington was assuming what he feared, and he immediately sent a lengthy dispatch to Grant, in which he repeated all of his previously announced doubts about a campaign into eastern Virginia. With a supply line of 135 to 145 miles, with his present means, he could not accumulate enough supplies to carry him over to the Orange and Alexandria Railroad. He also sent a response to Halleck, in which he proposed to terminate the campaign, and send the 6th and 19th Corps back to Grant, leaving only Crook to hold the Valley.

Here, Stanton stepped in, on October 3rd, not with a disapproval of Sheridan's plan, but, remarkably, with a question as to how to implement it.

Sheridan then resolved to move "at least as far as Strasburg," which was duly begun on October 6th. The infantry, passing down the Valley Pike, in rather a reverse of procedure, preceded the cavalry, which, ranging all the way from the Blue Ridge Mountains to the eastern slopes of the Alleghenies, was left to carrying out its orders of destruction.[1098] Henry P. Moyer, of the 17th Pennsylvania Cavalry, after

---

1095. O.R. Vol. 43/II, pp. 209, 879.

1096. Remember that Bragg, following his defeat at Chattanooga and his removal from command (O.R. Vol. 31/I, p. 3), became a military adviser to Jefferson Davis on Feb. 24th, 1864. (O.R. Vol. 32/III p. 3.)

1097. O.R. Vol. 43/II, pp. 196, 249–250.

1098. O.R. Vol. 43/I, pp. 430, 508; Vol. 43/II, pp. 218, 254; Norton, p. 95; Farrar, p. 395; Kidd, p. 396. Custer had briefly replaced Powell in command of the 2nd Division, but with the order to transfer Wilson to Sherman, Custer was moved to the command of the 3rd Division, and Powell again put in command of the 2nd Division; all within the space of a few days. Col. J. H. Kidd, of the 6th New York, then assumed the command of the 1st Brigade, 1st Division, Custer's old "Michigan Brigade."

placing a battalion of troops to protect a signal station on a small eminence near Staunton, took a minute to note the scene:

> The view was indeed a grand one, and in anticipation of what was soon to take place left impressions never to be forgotten.
>
> Looking southward…the eye falls on a broad valley…traversed by highways in all directions; towns, villages and churches forming local centers among farms, the improvements upon which were the best in Virginia and possibly in the South. From all points…small bodies of cavalry could be seen, by the aid of field glasses, on every public road, gradually spreading out…giving ample evidence of the thoroughness of their…execution of the order.[1099]

With the arrival of Kershaw's infantry, Cutshaw's artillery, and finally Rosser, who either proclaimed himself, or was proclaimed as the "savior of the Valley" with his Laurel Brigade,[1100] Early's strength had been brought back to roughly equal to what he had before the battle of Winchester.[1101] Determined to attack Sheridan, he sent out a reconnaissance on October 5th, only to discover, on the morning of the 6th, that Sheridan had retired. Thus would begin a "stern chase"; Rosser tearing at Custer's heels; Lomax rather tentatively nipping at Merritt's.

Custer, now in command of the 3rd Division, Wilson having been promoted and assigned to General W. T. Sherman in Georgia, took the Back Road; Merritt the Middle Road, and Powell down Page Valley to Luray.

Rosser's men had been eager to be ordered to the Valley; they were in high spirits, optimistic that they could avenge Sheridan's outrages. This, and tales of Rosser's recent deeds of daring apparently having preceded him, he hoped to trade upon them, and garner more recruits as he passed into the Valley. A poster found on the door of a grist mill near Port Republic bore this message from the young and dashing leader:

> PATRIOTS OF THE VALLEY: Once more to the rescue of your houses and firesides. Dream not of submission as long as the feet of the Northern vandals desecrate your own native soil. Temporary reverses have befallen our arms in this department; despair not. The government of your choice has declared its speedy redemption paramount to its present and final triumph, and confidently appeals to the patriotic impulses of the masses. Rally. Organize, and report mounted to:
>
> Rosser, Major-general.[1102]

1099. Moyer, p. 215.
1100. Fitzhugh Lee, having been wounded at Winchester, was still not able to take the field. Munford, SHS, Vol. 13, p. 133; Bowen, J. R., p. 244.
1101. Early, *Sketch*, p. 435.
1102. Moyer, pp. 214–215. At this time Rosser was 28 years old. O'Ferrall, p. 147. It is noted that Confederate cavalrymen were not issued horses, as "report mounted" here implies. They had to supply their own, for which they were given an allowance of 40 cents per day. O.R. Series

Rosser had just completed participating in Wade Hampton's Great Cattle Raid, which was an enormous embarrassment to the Army of the Potomac. In a three-day expedition, Hampton's men had raided a cattle corral some seven miles below Grant's Headquarters at City Point—obviously well within Union lines. They then successfully drove 2,486 cattle and 304 prisoners back behind Lee's lines. A cattle raid? Though this was cause for Confederate celebration, the desperate need for such a thing was telling. Lee's army[1103] had been on short rations for some time, and unless extraordinary measures such as this were taken, the result would be eventual starvation and surrender. Grant's siege was working.

At dawn on October 7th, Rosser's whole force was in the saddle, and he began his vigorous pursuit of Custer, and as mentioned, left Lomax to follow Devin. Rosser's aggressiveness increased as the miles passed, and as they watched the Valley go up in smoke. Many of his men were from the Valley, and Sheridan's acts filled them with rage.[1104] Custer then pushed on to Fisher's Hill, to within sight of the 19th Corps, before sundown.

Tom's Brook/Strasburg/Woodstock Races – *An all cavalry affair*

The pursuit ended that night with Rosser camped on the high ground on the south bank of Tom's Brook, with Merritt on the north side and Custer beyond, near Mt. Olive.

The occasion gave some of Rosser's older and more experienced officers the opportunity to express some sobering thoughts. First, they were some twenty-five miles ahead of Early's infantry, camped at New Market. There was no hope of support or relief. Then there was the size of Merritt's Division, indicated by the numerous campfires visible, and Custer was near. They tried to persuade him to withdraw during the night, but he was determined to stay, saying that, if pressed, he could withdraw quickly from "an enemy whom he had driven pell-mell for two days."[1105]

An accurate estimate of Rosser's force is important to what took place at Tom's Brook, and it is elusive. A calculation is derived from an incomplete Confederate Strength Report, for October, figure 7. It has been altered.[1106] The result is

---

4, Vol. 1, p. 340; Series 4, Vol. 3, p. 749.
1103. Foote, p. 241; McDonald, p. 292.
1104. McDonald, pp. 289–293, 299–301; Boudrye, pp. 177–178; Pickerill, pp. 166–167; Munford, SHS, Vol. 13, p. 136.
1105. McDonald, pp. 304, 305; Early, *Sketch*, p. 433; Clark, p. 237. O.R. Vol. 43/II, p. 248. Torbert's total strength, as of September, including his artillery, is listed as 6,885 officers and men.
1106. Figure 7, O.R. Vol. 43/II, pp. 556, 559, 612, 903. Lomax cites 800 as the total strength of Jackson's and Johnson's brigades, thus 400 is rationalized as added to the formerly blank "Present for Duty" column for Johnson. Likewise (Early, *Sketch*, p. 435), the remaining blank columns, Wickam's brigade (now commanded by Munford), and Lomax's brigade (now commanded

a total that could not have exceeded 2,740 officers and men present for duty. Other writers of the time, though none of them quote any authority, quote fewer than what has been derived here. McDonald in *A History of the Laurel Brigade* quotes, "less than 2,000."[1107] Thus, if Rosser had taken the advice of his older and more experienced officers, such as Lomax, West Point class of 1856,[1108] he would have skedaddled – his force was less than half that of the Union cavalry.

Command.	Present for duty.		Effective total present.	Aggregate present.	Aggregate present and absent.	Prisoners of war.	
	Officers.	Men.				Officers.	Men.
Lomax's division:							
McCausland's brigade†				670	2,796	40	353
Johnson's brigade		400*		652	2,873		
Jackson's brigade‡	55	386		528	2,559	25	627
Imboden's brigade				356	1,626		
Lee's (Rosser's) division:							
Rosser's (Funsten's, Dulany) brigade	59	754	725	954	2,651		
Wickham's brigade (Munford)		600*		1,505	3,557		
Lomax's (Payne's) brigade¶		600*		662	2,267	36	438
Total	114	2,740*	725	5,327	18,329	101	1,418

FIGURE 7

It is important to note that nearing Fisher's Hill, the Back Road and the Valley Pike are a little less than three miles apart. It would be a perfect place to combine Custer's and Merritt's divisions, and mount an attack.[1109] Annoyed with Torbert since his misadventure in the Luray, Sheridan brusquely dictated that he expected Torbert to "give Rosser a drubbing next morning or get whipped himself" and that the infantry would remain halted until the affair was over. Sheridan also informed him that he would keep an eye on the exclusively cavalry performance from nearby Round Top Mountain.

On the frosty morning of October 9th (there had been snow flurries the day before), Custer's division moved out from its position about six miles north of Tom's Brook at dawn. The advance guard, a battalion of the 5th New York cavalry, met Rosser's pickets near Mt. Olive, and after considerable skirmishing, both mounted and dismounted, drove them back to Rosser's established defenses, a line of low ridges running along the south bank of Tom's Brook.[1110] Near the base of

---

by Payne) has had 600 ascribed to them.
1107. McDonald, p. 305, gives Payne's brigade 300; Lomax reports less than 800 men effective, O.R. Vol. 43/I, p. 612.
1108. Lomax, Cullum, Vol. II, no. 1731. Rosser had attended West Point, and resigned from the class of 1861 at the outbreak of the war. O.R. Vol. 29/II, p. 772. He was a classmate of Custer; Whittaker, p. 258.
1109. Sheridan, Vol. 2, p. 56.
1110. O.R. Vol. 43/I, pp. 431, 520; Emerson, p. 58.

these, Rosser had placed a strong line of dismounted cavalry behind stone fences and barricades of rails and logs. On the crest of the ridge there was placed a battery of two guns, strongly supported.

The six miles of skirmishing had taken almost two hours, and it was not until after 8:00 that Peirce's battery was brought up to shell Rosser. The range from this position to the enemy was apparently too great,[1111] and the relatively ineffective shelling did nothing but cause Rosser to move a group of the led horses of his dismounted skirmishers out of sight. Rosser's guns did not respond at this time.

Merritt moved forward at 7:00; Kidd's 1st brigade, with the 6th Michigan on the right, moved along the north side of the brook to connect with Custer. Devin moved to the center and Lowell to the left, with his left overlapping the Valley Pike.[1112] When formed along the brook, their line of battle extended for more than the nearly three miles between the Back Road and the Pike. As Custer came up, the right of his 1st brigade, Pennington's, overlapped the Back Road.[1113]

Facing Custer, his left on the Back Road, was Munford, and behind him was Payne, in support of the artillery. To the right of Munford was Dulany, in command of the Laurel Brigade. Further along, were Lomax's two brigades, Johnson, and then Jackson, overlapping the Pike, opposite Lowell. The Confederate line was relatively lightly manned in the center, and Rosser's heaviest forces opposed Custer.[1114]

Now, there occurred one of the colorful actions which gave rise to many subsequent fireside stories. Posted to the front, facing Rosser, was Custer, "The Boy General," figure 8, slouch hat in hand, golden braids on his arms, with his golden curls hanging down to his scarlet necktie, and behind him his staff, all identifiable by their scarlet neckties, at the head of the 5th New York.[1115]

This could not be missed by Rosser, who looked down from his perch on the ridge. He turned to his staff, and said: "You see that officer down there…that's General Custer, the Yanks are so proud of, and I intend to give him the best whipping today that he will ever get, See if I don't."

1111. O.R. Vol. 43/I, pp. 520–521, 549. Custer attributes the ineffective fire to defective ammunition; Peirce to high wind. The light twelves in the battery were evidently near their one-mile effective range.
1112. Bowen, J. R., p. 244; O.R. Vol. 43/I, p.477.
1113. O.R. Vol. 43/I, pp. 520–522; McDonald, p. 305; O.R. Vol. 43/I, pp. 447, 460, 515; Munford, p. 136.
1114. Kidd, p. 402; O.R. Vol. 43/I, p. 612.
1115. Whittaker, p. 254, Boudrye, p. 178; Custer, Elizabeth, title page; Munford, p. 136.

FIGURE 8

Custer bowed in a knightly salute to his foe.[1116] He then donned his hat and returned to the line.

Custer's official report mentions none of this, stressing only that he was troubled by the ineffective fire of Peirce's Battery. He then ordered the 5th New York, 2nd Ohio, and 3rd New Jersey to advance as mounted skirmishers, allowing Pierce to be repositioned to within eight hundred yards of the enemy, the position shown in figure 9, where it shelled Rosser's (Thompson's) battery with "telling" effect, compelling Rosser to withdraw the two guns, after one was disabled.[1117] One of Peirce's light 12 pounders was also disabled, and all of the men of the 7th Michigan who were temporarily supporting the battery were wounded.

Seeing the stubborn resistance met by Pennington's skirmishers, Custer ordered the 18th Pennsylvania, supported by the 8th New York and the 22nd New York, to make a flanking move behind the hill upon which Rosser was positioned. To support the advanced line, he then ordered Wells' 2nd Brigade, which had been in reserve, forward along the Back Road.

In the meantime, Merritt had gone into position, directing Kidd's 1st Brigade, with Martin's 6th New York battery attached, to connect with Custer's line, and to make a flank attack on Dulany.[1118] At the same time, Devin was directed to cross Tom's Brook, and advance in the center, along what is now interstate 81, in between the two concentrations of the Confederate forces, along a ridge midway between the Back Road and the Pike, and into Dulany. Lowell was ordered to advance along the Pike, toward Lomax's two brigades, those of Johnson and Jackson.

1116. Whittaker, facing p. 258; O.R. Vol. 43/I, pp. 520, 549.
1117. McDonald, p. 305; Figure 9 drawn by the author.
1118. O.R. Vol. 43/I, pp. 431, 447, 460, 483.

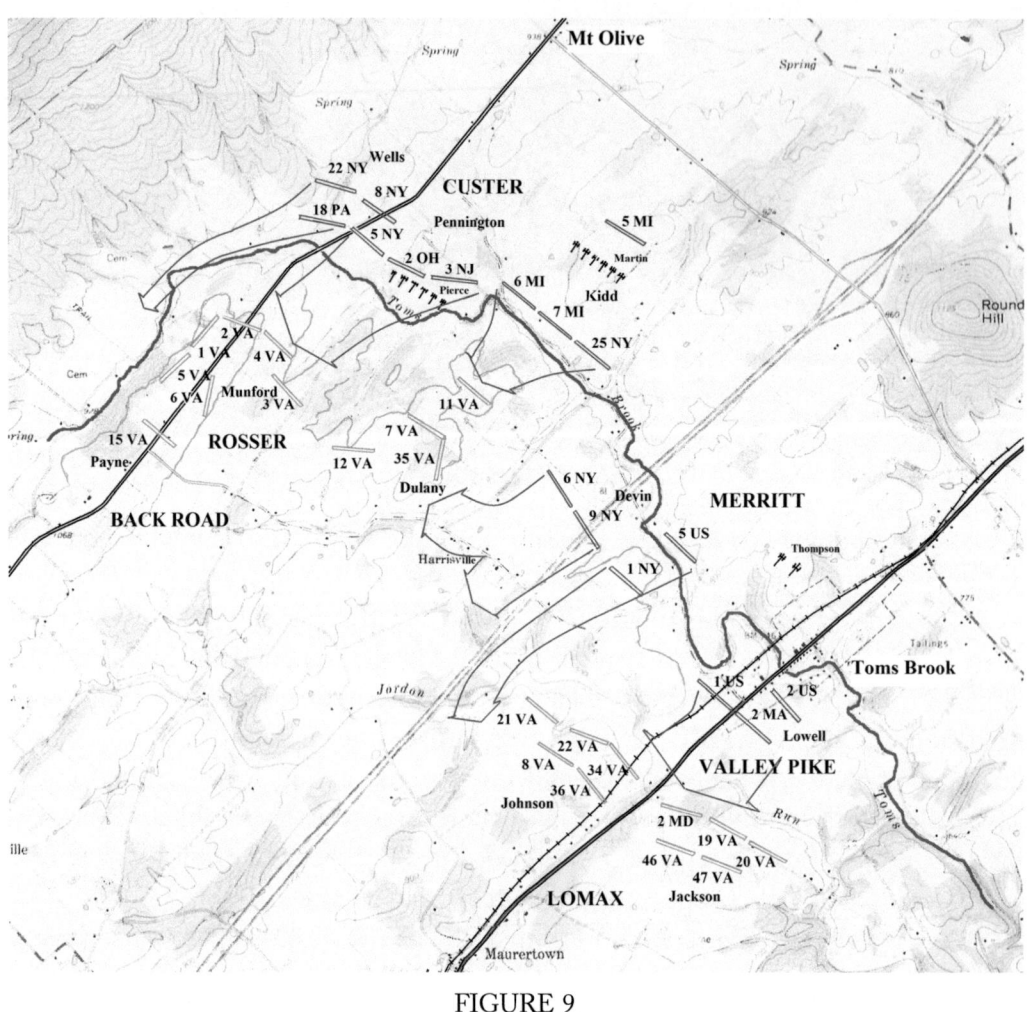

FIGURE 9

Lowell was met by an initial repulse. Seeing it, Merritt ordered Devin to detach two squadrons of the 1st New York Dragoons, while the 5th U.S. Cavalry was hurried from division headquarters to join them, and "curl around" Lomax's left.[1119] It is surmised that Battery K–L's bugler, Louis Rupprecht was wounded at this time.[1120] In the absence of any report by any member of Battery K–L, the only evidence of heavy fighting by any of Devin's commands was that of the 6th New York going after Dulany, and that of the 1st New York Dragoons in support of Lowell.

By now, it was approaching 9:00 a.m. The mounted charge of the 6th, and 7th Michiganders and the 25th New York, of Merritt's 1st Brigade (Kidd's), aided by the enfilading fire of Martin's battery, had pressured Rosser's right. On his left, the flank move had passed unobserved behind the hill to Rosser's left, and pushed rapidly to his rear, near where his hospital and camp of the night before were

---

1119. Bowen, J. R., p. 247.
1120. The monthly report only says: "Wounded in action near Strasburg."

located.[1121] The bugles then signaled the advance of Custer's line, personally led by Custer, the color bearer of the 5th New York by his side, straight into Rosser's position. Rosser could now hear the yells of the Yankee column in his rear, and on his right, the receding sounds of Lomax's guns. Payne had fallen back, and White's Battalion, on Dulany's right, fell back. There was no possible chance for Rosser than to move out now, and at a run.

He fell back about a half mile along the Back Road, to a wooded area, where Munford reformed the Laurel Brigade.[1122] Rosser planned to counterattack, but when a Yankee regiment, likely the 6th Michigan, which Col. Kidd reported as "in advance of the other two brigades of the First Division," drew up in full view with drawn sabers, Rosser quailed. To Munford, he confided: "We can't do it." He then fell back under fire, and after endeavoring to rally several times, reached a point, near Columbia Furnace, where the Yankee pursuit ended.[1123]

As to Devin,[1124] after detaching the 1st New York in support of Lowell, and after Rosser's lines were broken, he gathered up his dismounted skirmishers, and moved out at a trot. Leading the 9th New York, "Uncle Tommy" followed the line of a road west of the Pike.[1125] Reaching "Woodstock with but slight opposition," he then turned on to the Pike, where Torbert ordered him to take the advance and pursue Lomax to Edinburg.

Lowell's advance recovered quickly after being driven back only a "short distance." Lomax had decided to withdraw when he heard

> . . . the firing on General Rosser's front retiring rapidly, and stragglers coming from his command with the statement that his force was broken, I withdrew my force slowly, the enemy pressing.

It is noted that Munford and Rosser had decided to withdraw because they had heard the receding sounds of Lomax's guns. No military man likes to admit to deciding to retreat unless forced to, so the question remains as to who exactly withdrew first, Rosser or Lomax. Regardless, Lowell's 1st U.S., 5th U.S. and 2nd Massachusetts, with Devin's 1st New York, followed Lomax's deliberate retreat through the broken and wooded few miles leading up to Woodstock, where Lomax's left was threatened by the appearance of Devin. The country was now open, and Lowell ordered a charge, which fell upon B. T. Johnson's brigade, which, Lomax admits,

---

1121. Munford, p. 137; Boudrye, p. 179.
1122. O.R. Vol. 43/I, pp. 447, 521.
1123. Munford, p. 138; O.R. Vol. 43/I. pp. 431, 521. Custer's account and Munford's differ. Here, Merritt's is combined with Munford's. Merritt says beyond Columbia Furnace, Munford infers a point before Columbia Furnace.
1124. O.R. Vol. 43/I, pp. 483, 492, 612; Cheney, pp. 227–228; Bowen, p. 247.
1125. This road is mentioned in Cheney, p. 232. It could have been what is now Country Brook Road, or others nearby Saumsville, to Woodstock.

"was completely broken. I was unable to rally this command." Jackson's brigade was turned to meet Devin, and according to Lomax, "retired in good order." Outside of Edinburg, Torbert having ordered him to take up the pursuit, Devin arrived to find that Lomax had already passed through the town. Taking the 9th New York at the gallop, he pursued Lomax up the Pike, and at Hawkinsburg fell upon him, compelling him to leave one of his two remaining guns. Lomax then managed to throw off Devin at an intersection, sending the 9th down a side road. Learning of his mistake, Devin ordered the 6th New York to take the Pike. The 6th charged clear through Mount Jackson, to the river, where it was learned that a part of Early's infantry was ahead.[1126]

Devin held Mt. Jackson for an hour, while he took stock of the situation. The pursuit had gone on for more than 20 miles, 8 miles of that at a gallop, and the horses were beginning to break down. He decided to quit the pursuit and retire, thus ending the "Woodstock Races."[1127]

On returning, Devin learned that the 1st New York had discovered a park of 31 enemy wagons loaded with ordnance and stores. All were burned, according to Devin. Others, including six guns, were captured from Rosser by Custer. The guns turned out to be those that were abandoned by Battery K, at Ream's Station, during Wilsons Raid. The wagons included Rosser's headquarters wagons, which were guarded and sent back intact.[1128] This somewhat settled an old account. Custer got back many of his personal effects which had been captured by Wickham's Brigade at the Battle of Trevilian Station, earlier in June, and now Custer had custody of his old schoolmate's personal effects. Never failing to be the showman, Custer later appeared at headquarters wearing Rosser's best uniform.

Since Devin was never challenged by any substantial rear-guard action on the part of Lomax, Battery K–L was never called forward to unlimber and shell out any resistance. After Rupprecht was wounded back at Tom's Brook, they had merely gone along for the ride, and witnessed one of the great all-cavalry battles of the war. Battery L could, nevertheless, justifiably have "Strasburg" sewn into its battle flag.

The Union cavalry suffered only 57 casualties,[1129] 48 wounded and 9 killed. Confederate losses were never officially reported, but Sheridan's report of Confederate prisoners said " about 330." Rosser did not call for a report from his subordinates, and it was clear that Early was never fully informed of, as Col. T. T. Munford writes, "the extent of this disaster."

1126. Early, *Sketch*, p. 436; O.R. Vol. 43/I, p. 484.
1127. Sheridan, Vol. 2, pp. 59, 431.
1128. Rodenbough, Potter & Seal, p. 112; Crowninshield, *A History...* p. 27; Pickerill, p. 167.
1129. O.R. Vol. 43/I, p. 31; Munford, pp. 134, 139.

Chapter 17

Decisions, Decisions; Sheridan's Ride; Cedar Creek; Mosby; "Record" 12/64

Decisions, Decisions

The Battle of Tom's Brook did nothing to change Sheridan's mind about leaving the Valley, and the next day the whole army marched away from Fisher's Hill to the north side of Cedar Creek, just south of Middletown. On October 12th, the 6th Corps, save one brigade stationed at Winchester, was ordered to march to Alexandria.[1130] Sheridan had done this on his own initiative, heeding Grant's instructions to use his own judgment.

Grant's thoughts about future objectives now became confused by Halleck's habit of rewriting orders that should have been forwarded unaltered. The confusion had reached the point that Stanton stepped in and asked Sheridan to come to Washington for a conference. "I propose to visit General Grant, and would like to see you first."

Early had already heard that Sheridan was preparing to leave the Valley and had moved down, reaching Fisher's Hill on the morning of the 13th, to probe Sheridan's defenses. Rosser was sent out to test Custer's camp, and was quickly repulsed, and another party was sent forward to Hupp's Hill. A battery was brought forward which began shelling the camp of the 19th Army Corps. Crook's 1st and 3rd brigades, under Thoburn, responded, and a sharp fight[1131] developed between Thoburn and Conner's brigade of Kershaw's division. The Battle of Stickney Farm as it was called, resulted in the adversaries withdrawing – Thoburn to the north side of Cedar Creek, and Kershaw to Fisher's Hill.

The significant result of these encounters was that Sheridan concluded that Early was likely planning to resume the offensive, and writes: "To anticipate such a contingency I ordered the 6th Corps to return from its march to Ashby's Gap."[1132]

It must be noted that if Early had not announced his intention by these useless maneuvers, the 6th Army Corps would not have been present in any coming battle.

The 6th Corps duly arrived back at noon on the 14th, and went into camp west

---

1130. Sheridan, Vol. 2, pp. 59–61; Early, *Sketch*, p. 437; O.R. Vol. 43/II, p. 346.
1131. Wildes, pp. 197–199; Sheridan, Vol. 2, p. 61, Irwin, p. 406; Early, *Sketch*, p. 437; O.R. Vol. 43/II, p. 365.
1132. Sheridan, Vol. 2, p. 61.

of the Pike and north of the 19th Corps, which under Emory's guidance, had put up a defensive line.

Evidence of the urgency that Stanton had placed on seeing Sheridan (though Stanton was always impatient) is found in a telegram to General Auger who had come west to the terminus of the Manassas branch of the Virginia Central Railroad at Rectortown.[1133] "Has General Sheridan reached you yet?"

On the 15th, Sheridan, bound to obey Stanton's summons, rather reluctantly left Middletown, leaving General Wright of the 6th Corps in command. Sheridan was concerned whether Early would mount another attack, and had planned to beat him to the punch, as soon as the 6th Corps had returned. However, since Early had withdrawn to Fisher's Hill from the positions taken in the skirmishes of recent days, Sheridan concluded that he "could do us no serious hurt from there" and deferred his attack, hoping to get to Washington and "come to some definite understanding about my future operations."

As a part of the planned trip, Sheridan would include a raid suggested by Grant. Merritt's and Custer's cavalry were ordered to Front Royal, to join Powell, and push to Charlottesville[1134] to burn the Virginia Central Railroad Bridge over the Rivanna River, while he would go on to Rectortown. Sheridan knew that one railroad bridge in enemy held territory was not of much strategic value unless held, and at least miles of track were torn up as well, for a single bridge could be repaired in a matter of days. To prevent its rebuild would require the area to be held by a substantial force, which Sheridan opposed.

On the night of the 16th, upon his arrival at Front Royal, Sheridan received a dispatch and enclosure from Wright.[1135] The enclosure read:

> Be ready to move as soon as my forces join you, and we will crush Sheridan.
> Longstreet, Lieutenant-General

The message had been taken down as it was being flagged *from* the Confederate signal station on the top of Three Top Mountain. In cipher, it was translated by Sheridan's signal officers, who knew the Confederate code. What could this mean? Sheridan at first took it as a ruse, but to be on the safe side, he abandoned the cavalry raid toward Charlottesville.

He replied to Wright as follows:

> GENERAL, The Cavalry is all ordered back to you; make your position strong. If Longstreet's dispatch is true, he is under the impression that we have largely detached. I will go over to Augur, and may get additional news. Close in Colonel Powell, who will be at this point. If the enemy should make an advance

---

1133. Forsyth, p. 130; Sheridan, Vol. 2, pp. 62–63.
1134. O.R. Vol. 43/II, pp. 363, 508.
1135. Forsyth, pp. 132–133; O.R. Vol. 43/II, p. 51. The signal was from Early to Longstreet, a corps commander in Lee's Army.

I know you will defeat him. Look well to your ground and be well prepared. I will bring up all I can, and will be up on Tuesday, if not sooner.

If Early had scared Sheridan enough on the 13th to have resulted in his recalling the 6th Corps, now he was attempting to scare him into thinking that Longstreet was coming with reinforcements? The result was that Merritt's and Custer's Divisions were back at their old campsites, and Moore's brigade of Powell's division was closed in to Buckton Ford.

Sheridan then proceeded to Rectortown, now only accompanied by the four members of his staff, and his escort, the 2nd Ohio cavalry, of Custer's division.[1136] Arriving at about noon, Sheridan telegraphed Halleck, informing him of the intercepted signal dispatch, and asked whether it was known that any force had been detached from Lee. The answer came back, two hours later, that Grant knew of no troops having left Richmond, and reiterated that if Sheridan could leave his command "with safety, come to Washington, as I wish to give you the views of the authorities here." Having concluded that it was likely that the signal message was of no consequence, since, even if Longstreet had been detached from the Army of Northern Virginia, he could not have reached Early before Sheridan returned. Sheridan then went on, arriving at Washington on the morning of the 17th.[1137]

Early mentions nothing about the ruse in his *Autobiographical Sketch*, but says:

> I remained at Fisher's Hill until the 16th, observing the enemy, with the hope that he would move back from his very strong position on the north of Cedar Creek, and that we would be able to get him in a different position, but he did not give any indications of an intention to move, nor did he evince any purpose of attacking us, though the two positions were in sight of each other.

Early finally explained the matter in a private letter to Richard Irwin. Dated November 6th, 1890, Early admits that the signal was instituted by him, and was entirely fictitious; its object was to; "induce Sheridan to move back his troops from the position that they then occupied…"

One can only understand the logic of this foolery by recalling Early's previous commentary regarding Sheridan. Early evidently held such a low opinion of Sheridan's willpower, describing him with such words as "timid," "incapacity," and lacking in "skill or energy,"[1138] that he must have sincerely believed that Sheridan would react in the manner desired.

In fact, if Early had done nothing, Sheridan's strength would have been reduced by some 12,000 infantry and 7,000 cavalry.[1139] Dare it be mentioned how

---

1136. Nettleton, p. 657.
1137. Sheridan, Vol. 2, p. 66–72; Forsyth, pp. 134–135; Early, *Sketch*, pp. 437–438; O.R. Vol. 43/II, pp. 385–386.
1138. Early, *Memoir*, pp. 75–76; Early, *Sketch* (all subsequent "Early" references are to this), p. 427.
1139. O.R. Vol. 43/II p. 501.

different the outcome of the Battle of Cedar Creek would have been?

Meeting with Stanton and Halleck, Sheridan's objections to operating in eastern Virginia were agreed to, and two engineer officers were assigned to him to scout out a defensive line in the Valley that could be held while the bulk of the army could be returned to Petersburg. The meeting ended at noon. Concerned that he could return as quickly as possible, a special train was provided for his return trip, via the Baltimore and Ohio, and a cavalry escort was directed to meet him at Martinsburg. His party arrived there at about dark. Spending the night there, they started up the Valley Pike early the next morning, arriving at Winchester, where the two engineers conducted their survey.

On the 18th, Sheridan had sent a courier forward to Cedar Creek, with instructions to return with a report of affairs there. The word came back that everything was all right, the enemy was quiet, and that Grover was to make a reconnaissance the next morning. Sheridan writes: "I went to bed greatly relieved, and expecting to rejoin my headquarters at my leisure the next day."

Toward 6 o'clock the following morning, the officer on picket duty at Winchester came up to Sheridan's room at the Lloyd Logan home, figure 1, and reported that artillery firing could be heard from the direction of Cedar Creek. Asked if the firing was sustained or "desultory," the officer replied, "irregular and fitful." Sheridan assumed that it was Grover's reconnaissance that was heard, and sent the picket officer away.

Not able to return to sleep, Sheridan dressed. The officer returned to report that firing could still be heard. Again questioned, the officer replied that it did not sound like a battle. Regardless, Sheridan requested that breakfast be hurried up, and that the horses be readied. At about 8:45, Sheridan and his four aides were in the saddle. Riding "at a walk" through Winchester, they arrived at Mill Creek, a mile south, where they met their escort. By now, the sound of the artillery firing was an "unceasing roar."[1140] Moving on, to the crest of a hill, they fell upon the sight of a wagon train, halted, and in disarray; some wagons facing this way and others that. It had been warned to halt by news from the front that the army had been defeated. Moving further along, Sheridan met "hundreds of slightly wounded men" and hundreds of others, unhurt but demoralized, and when accosted, told of the army being broken up, and in full retreat.

FIGURE 1

1140. Sheridan, Vol. 2, pp. 73–77; Forsyth, p. 136; Figure 1, courtesy of the Handley Regional Library, Winchester, VA.

Sheridan's first reaction was an order to Colonel Oliver Edwards, commander of the 3rd Brigade of the 6th Army Corps, which occupied Winchester, to string his troops across the Valley, and stop all of the fugitives from passing, though the wagons could be passed through Winchester and parked on the north side.

FIGURE 2

What to do next? Wait and stop the rest of the whole army at Winchester? Waiting was not something ingrained in Sheridan, and he formulated the idea of riding to the front, to use whatever power his personal presence might accomplish to stop the retreat.[1141]

Sheridan's Ride

The distance from Winchester to Middletown is a little less than 12 miles, and to where the army had been encamped, south of Middletown, about 15.[1142] Presently, off dashed Sheridan. He was mounted on the 16-hand Morgan that Capt. Campbell of the 2nd Michigan had given him in Rienzi, Mississippi, after the Battle of Corinth, back in August of 1862, figure 2. He had the distinct advantage of riding a horse with great speed and endurance.[1143] His two aides, and the 20 men from the 17th Pennsylvania Cavalry, picked from his escort to accompany him, struggled to follow. The road had become so blocked with wagons, the wounded, and many others unhurt, who simply had fled danger, that they were forced to take to the adjoining fields. Waving his hat, Sheridan shouted: "Turn back, men! Turn back! Face the other way!

George A. Forsyth, one of the aides that accompanied him, testifies to the fact that this had but one result: "A wild cheer of recognition, an answering wave of the cap. In no case, as I glanced back, did I fail to see the men shoulder their arms and follow us."

Finally, he ran into the rear of the 1st Division of the 6th Corps, Getty commanding, and on the east side of the Pike, Lowell's brigade of Merritt's cavalry, a mile north of Middletown. The army had been pushed back about three miles from the line at Cedar Creek, though it was apparent that the enemy had been checked.

1141. Sheridan, Vol. 2, pp. 77–80.
1142. Forsyth, p. 138.
1143. Sheridan, Vol. 1, pp. 177–178; Vol. 2, pp. 81–82; Figure 2, Thustrup.

Cedar Creek

On October 16th, the indefatigable Rosser reported to Early that his scouts had discovered a campsite of Custer's on the Back Road, near Old Forge Farm, some three miles distant from the main Union line, which allegedly included Custer's headquarters.[1144] It would be an opportunity to "bag Custer." Once again, Early gave Rosser permission to attack, notwithstanding the fact that Rosser had already attacked and been repulsed on the 13th. A force of 500 picked men, consisting of Rosser's own cavalry and a mounted brigade of Grimes' North Carolina infantry, set out that night. Unfortunately, Custer's campsite had been moved and only a picket remained, of whom three officers and 33 men were captured.[1145] It was another of Rosser's failures, and worse, it put the cavalry on alert for a second time.

On the morning of the 17th, Early moved all of his troops to the front of his lines to cover Rosser's return. He also sent General Gordon, with a brigade, to determine if Hupp's Hill was fortified. It was, and Early now was faced with the decision to attack or retreat, "for want of provisions and forage." He decided to attack, but not being strong enough for a frontal attack, he would have to find an approach around "one of the enemy's flanks, and attack him by surprise if I could."[1146]

After Gordon's return from Hupp's Hill, Gordon, his chief of staff, Major Robert Hunter, Capt. Jedediah Hotchkiss, Early's topographical engineer, and brigade commander Gen. Clement Evans, were sent up to the signal station on Massanutten Mountain to examine Sheridan's dispositions. Also, General Pegram was ordered to go as near as he could to Cedar Creek, to scout out whether it was practical to make an attack from there.

Carefully studying the Union dispositions with a telescope, Gordon writes:[1147]

> It was unmistakably evident that General Sheridan concurred in the universally accepted opinion that it was impracticable for the Confederates to pass or march along the rugged and almost perpendicular face of Massanutten Mountain and assail his left. For he had left that end of his line with no protection save the natural barrier, and a very small detachment of cavalry [Moore] on the left bank on the river. His entire force of superb cavalry was massed on his right...

Thus, Gordon was convinced of the potential for success by attacking Sheridan's left, and his plan was adopted. Gordon's 2nd Corps, (Evans', Pegram's and Ramseur's divisions)[1148] was to cross the Shenandoah at Fisher's Hill, go around the end of Massanutten Mountain, and again cross the Shenandoah at McInturff's

---

1144. McDonald, pp. 308–310; Early, *Sketch*, p. 438; O.R. Vol. 43/I, pp. 580, 605. Old Forge Farm.
1145. O.R. Vol. 43/I, p. 100.
1146. Early, *Sketch*, p. 438.
1147. Gordon, pp. 333–334.
1148. O.R. Vol. 43/I, p. 580; Gordon, p. 335.

and Bowman's Fords, the route shown by the dotted line to the right in figure 3.[1149] It was so rough that it would have to be prepared by the pioneers of Ramseur's division prior to the attack. Once across, they would advance past the left flank of Crook's 8th Corps near the Cooley house, and press on to the Valley Pike in Crook's rear. Payne's Virginia cavalry was to accompany Gordon, advance among the disorganized Yankee infantry directly to the Belle Grove House, and capture the commander-in-chief and bring him back as a cavalry "trophy." (Remember Dwight's silly plan to have Grierson capture Gardner in his headquarters at Port Hudson?)

Referring to figure 3, Kershaw was to move along the Pike through Strasburg, shown by the dotted line to the north of Gordon's track, and at the old railroad crossing, turn off toward Bowman's Mill, crossing Cedar Creek at Roberts' Ford, and then advance forward to Thoburn's breastworks on the hill opposite, *only when he heard that Gordon had become engaged*. Wharton, followed by the artillery, was to move through Strasburg along the Pike toward Hupp's Hill, hide under the cover of the trees on its edge, and move on the enemy on his front and left, also as soon as Gordon had become engaged.

Rosser was to move on the Back Road, cross Cedar Creek, and engage the Yankee cavalry simultaneously with Gordon's attack. The remainder of Early's cavalry, Lomax's, was to move from the vicinity of Front Royal and support the attack once it had gained the Valley Pike. All watches were set. The attack by Gordon was to begin a little before daylight, at 5:00 a.m.[1150]

Gordon's 2nd Corps and Payne's cavalry began their march to be in position for the attack at 8:00 on the evening of the 18th. No officer was mounted, and none of the men carried canteens or anything that could make noise. Every man had been impressed with the gravity of the enterprise, and everyone spoke only in whispers. Leaving Strasburg,[1151] their guides brought them to the rugged wagon road, and the railroad bed. Finally, they reached the dim pathway the pioneers had prepared along the mountainside, where their long gray line was forced to move single file. Arriving at the two fords on the Shenandoah where their crossing was planned, they waited for more than an hour,[1152] resting on the bank. They could see the Union vedettes sitting on their horses, wholly unconscious of their presence, the low murmur of the river sufficient to mask any whisper.

1149. Figure 3, frontispiece map from: Crowninshield, *The Battle...* portion, reoriented north-south, and re-labeled; O.R. Vol. 43/I, pp. 580–581, 613; Early, *Sketch*, pp. 440–441; McDonald, p. 310; Gordon, p. 338.
1150. Early, *Sketch*, p. 440.
1151. O.R. Vol. 43/I, p. 580; Gordon, pp. 336–337. Walker, p. 134.
1152. Nichols, p. 194. They had all massed on the south bank by 3:00 a.m.

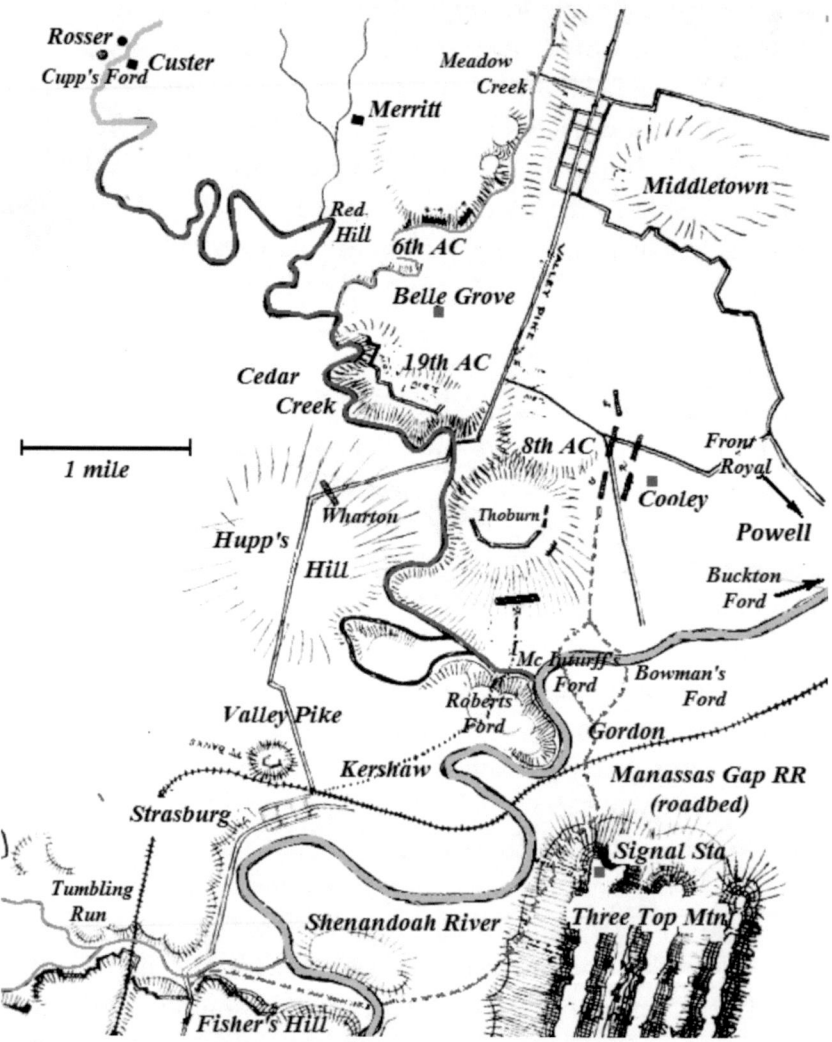

THE PLAN OF ATTACK – FIGURE 3

Meanwhile, after midnight, Kershaw's division, accompanied by Early,[1153] began its march along the Pike to Roberts' Ford, a little more than a foot deep, opposite the camp of Crook's 1st Division, commanded by Col. Joseph Thoburn. At 3:30, Kershaw's men were in position within sight of the Union campfires, and were directed by Early to halt, and await the order to cross and advance.

Wharton moved along the Pike to Hupp's Hill, with orders to remain hiding until the attack began. Carter's artillery was purposely delayed, to avoid the noise of its movement on the macadamized Pike.

Rosser, with the Laurel Brigade, now commanded by Col. O. R. Funsten, of the 11th Virginia, and Wickham's brigade, now commanded by Col. T. H. Owen, of the 3rd Virginia, had moved along the Back Road, to be ready to cross Cedar

---

1153. Early, *Sketch*, pp. 441–442: O.R. Vol. 43/I, p. 591, McDonald, p. 310; O.R. Vol. 43/II, p. 929.

Creek at Cupp's Ford, opposite Custer's camp.[1154]

Then followed a slight change; Pegram having observed that a new Federal work was being constructed since the earlier examination of the route Gordon was to take. He had seen the unfinished breastworks being built east of Thoburn by a regiment of Hayes' (the future President) 2nd Division. He recommended that Kershaw's force not wait for Gordon to be in position, which was to be near the rear of Hayes' line at the Cooley house; Gordon likely to have more difficulty in achieving the position. To this, Early agreed, and Pegram was told to inform Gordon. All of the attacks were now to be simultaneous.[1155]

Waiting on the precipitous bank of the Shenandoah, Gordon writes: "The minute-hand of the watch admonished us that it was time to move in order to reach the 8th Corps flank at the hour agreed upon…" The nearly full moon was still high in the sky, and would provide them with enough light to find their place on Crook's flank, at 5:00 a.m., which was an hour before twilight.[1156] Fortune was with Gordon. A private in the 12th Virginia Cavalry had captured a Yankee picket, who had given him the countersign. With it, the rest of the pickets were approached and captured. There was dead silence, there was no one left to give the alarm. Gordon now gave Payne's cavalry the order to plunge into the river and advance. That done, with Evans' Division leading, Gordon's men rushed into the cold breast-deep current, crossed, and pressed on at the double-quick. No alarm was given by Powell's 1st Brigade, under Moore, stationed a little more than two miles downriver, at Buckton Ford, though Moore's pickets were supposed to have been connected with those of Crook.

Not waiting for the sound of the firing expected from Payne's engagement with Crook's pickets at the river crossing, at 4:30 Early ordered Kershaw to advance.

In the interval, the firing of Rosser, and then Payne, was heard. Early then rode to Hupp's Hill to where Wharton and the artillery had been ordered. He found Wharton's skirmishers being shelled by a battery of Crook's command, L of the 1st Ohio, which was placed in a position overlooking the bridge at Cedar Creek. Kershaw's artillery was then brought up, and returned fire. Threatened in flank by Gordon's men, and of certain capture by Kershaw, the Ohio battery was forced to retreat down the Pike after only firing a few rounds.[1157]

Rosser, the remaining arm of the attack, had begun his move at about 4:00. Dismounted, Rosser's men met the pickets of the 7th Michigan Cavalry, of Merritt's 1st Division, who had been posted on the south side of Cedar Creek, east of Cupp's

---

1154. McDonald, p. 310; O.R. Vol. 43/I, p. 448.
1155. Early, *Sketch*, p. 442.
1156. *http://aa.usno.navy.mil/cgi-bin/aa*; Gordon, p. 338; Early, *Sketch*, pp. 442–443; O.R. Vol. 43/I, pp. 372–373, 589–591; Nichols, p. 194.
1157. Early, *Sketch*, p. 443; DuPont, p. 158.

Ford.[1158] Driving the Michigan pickets back across the creek, Rosser's men initially created a panic in the 7th Regiment. The confusion was aided by a fog which had settled in as the crisp, clear night approached dawn, but by now the whole brigade had been roused by the sounds of the scattering shots, and was soon mounted, with Martin's artillery limbered up. No bugle call had been necessary. As A. J. Kidd, the brigade commander, explains it, Rosser's raids of the 13th and 17th had kept them alerted, and "The Federal cavalry had recovered from their earlier habit of being 'away from home' when Rosser called. They were…'in' and ready to give him a warm reception."

The firing grew heavier, and from the hill where Custer's 3rd Division had camped, three-quarters of a mile away in the direction of Old Forge Farm, the bugle was sounded, telling that Custer, their old commander, had taken the alarm. The rest of Merritt's 1st brigade was promptly sent to the aid of the 7th. Seeing its deployment, Lowell, with the Reserve Brigade, arrived on the scene. The night before, they had been ordered by Wright to make a reconnaissance on the Back Road, and were already in the saddle when the attack began.[1159] Rosser did not press the attack. Quoting Kidd: "But contented himself with throwing a few shells from the opposite bank…" Rosser remained stalled for the rest of the day, and was never a factor in the battle.

The element of surprise, which Early would achieve elsewhere, would tend to cast a shadow on Wright, but his precautions had been thorough. On the 18th, Crook was ordered to send out a brigade on a reconnaissance, and it had found no evidence of the enemy in his "old camps."[1160] Crook thus concluded, in an interview with Wright that evening, that Early had retreated up the Valley. Nevertheless, Wright ordered Emory to start a division of the 19th Corps to move out the next morning on the Pike, up the Valley, and a brigade of cavalry, Lowell's, of Merritt's 1st Division, out on the Back Road, to move parallel to the infantry.[1161]

Though only one division of the 19th Corps, Grover's, was to be involved in the reconnaissance, the ever apprehensive Emory had given a standing order to the entire 19th Corps to be up, and under arms, that morning.

Returning to Kershaw, soon after 5:00, the Georgians of Bryan's brigade, commanded by Col. James P. Simms, without firing a shot, as ordered, had stealthily reached Thoburn's defensive works. Finally discovered, the attackers were fired upon by the Union pickets. The warning sounds did not give enough time for the 3rd Brigade to properly man the trenches, but it didn't matter. The 1st Brigade, having promptly formed up, and having repelled the initial attack, soon witnessed "the enemy inside the breastwork of the Fifty-fourth Pennsylvania Volunteers [of

---

1158. Isham, pp. 73, 74; Lee, pp. 100–101; Kidd, pp. 409–411; O.R. 43/I, pp. 100, 103.
1159. O.R. Vol. 43/I, p. 449.
1160. O.R. Vol. 43/I, pp. 158, 1042.
1161. O.R. Vol. 43/I, p. 284.

the 3rd Brigade, on their right], and also over the breastworks of the Fifth New York Heavy Artillery, vacated by the [that] regiment's being on picket duty."

The attackers soon discovered the vacancy—at the very center of the line—that had been assigned to the four companies of the 2nd Battalion of the 5th New York Heavy Artillery, and charging through, were able to lay enfilading fire in either direction. The entire Union line then collapsed, and the attackers were seen "sweeping," quoting Simms, through Thoburn's camp.

Four officers and 305 men of the 5th artillery were captured, and only about 40 men escaped.[1162] Only two of the 1st Brigade's regiments were able to fall back in good order, and were met at the Pike by Emory, where they joined a new line, at Kitching's camp site, that Wright and Emory were attempting to form.

The rest of Thoburn's division was driven on in disorder, figure 4. Simms, pausing only briefly to await the rest of Kershaw's division to close up—Connors, Humphreys, and Wofford's brigades—the four then advanced, as Simms reports: "Driving them like chaff before the wind…"

Thoburn was killed while trying to rally his troops.

When Kershaw had first fallen upon Crook, Emory immediately had headed toward the sounds of the firing.[1163] Though the fog prevented seeing the sweep of Gordon's or Kershaw's advance, he ordered Col. Stephen Thomas' 2nd Brigade of McMillan's 1st Division,[1164] to advance to the left of their breastworks, which faced Cedar Creek. Thomas was to cross the Pike and advance in a direction, figure 4, to support Thoburn. Across the Pike they plunged into a deep ravine, and up a wooded thicket beyond, where they formed up to meet the retreating men of the 8th Corps, and the advancing enemy. The intent was to give the fugitives a place to form up, but they failed to respond.[1165] In the haste of the slaughter, Thomas' men were unable to reload, and the combat became bayonet and rifle butt against the oncoming horde. The brigade, consisting of the 8th Vermont, 12th Connecticut, 160th New York, and 47th Pennsylvania, was overpowered and was swept back; leaving 85 killed, 246 wounded, and 167 missing. The 8th Vermont suffered 17 killed, 66 wounded, and 23 missing – 103 in all – out of 157 men present, or 67.9 percent casualties. This was the seventh highest percentage of any unit in the entire war.[1166] The brigade suffered the highest losses of any other that day. It had all happened in less than half an hour.

Thoburn's fleeing refugees and the remnants of Thomas' command now

---

1162. Early, *Sketch*, p. 443; O.R. 43/I, pp. 134–135, 379–383, 391–392, 591, 592; Figure 4, Merritt, *Battles and Leaders*, Vol. 4, p. 517.
1163. Irwin, p. 418.
1164. O.R. Vol. 43/I, pp. 134, 135, 284, 308. The casualty numbers for Thoburn's 1st Brigade total 110 killed and wounded. Compare this with the 498 casualties of Thomas.
1165. O.R. Vol. 43/I, pp. 133, 308, 309; Walker, A.F., p. 137.
1166. Fox, p. 36.

approached the campsite of Hayes' 2nd Division, which had been placed near the Pike as a reserve, about a mile north of Thoburn's line, figure 5.[1167] It had no breastworks though some were under construction about three-quarters of a mile to the southeast, the ones that Pegram had noticed.[1168] There were only 1,445 men in camp; one of Hayes' regiments was on picket, and another was guarding cattle south of Middletown. Still another had been assigned to build the breastworks.

Under the direction of Crook and the supervision of Wright, an attempt was made to form a defensive line along the Pike in the rear of, and just north of, Hayes' camp. This was the site of another defenseless camp (no entrenchments) – Kitching's. Unfortunately, the troops present were only detachments from his total, some 1,200, freshly arrived from the 22nd Corps, the defenses of Washington.[1169] The addition of those fleeing from the site of Thoburn's defeat, such as the 1st Ohio Battery, a part of DuPont's Battalion, and others like Wildes' 1st Brigade, gave hope that this line might hold.

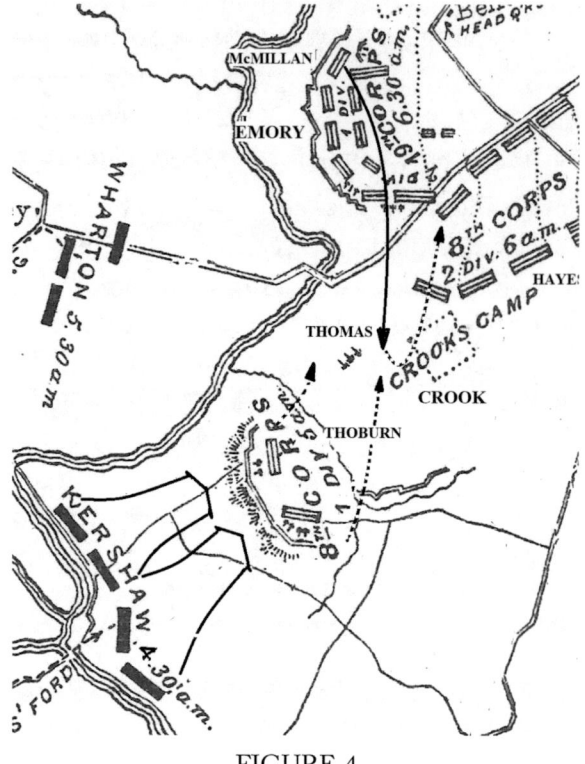

FIGURE 4

It was "early daylight" according to Hayes, something after 6:00 a.m. They were ordered to lie down. The heavy fog concealed anything at a distance, but firing in their front and on both flanks told them of the advancing foe.

Feeling that the improvised line would hold, Wright had ordered Ricketts, commanding the 6th Corps in Wright's absence, to send forward his 1st and 3rd Divisions,[1170] figuring that it would take them about 20 minutes to arrive from their bivouac on Red Hill about a mile to the northwest.[1171] It was not to be. About the time that Wright had given an order to close up the line on the 19th Corps, about a hundred yards to the right, the left of the line, Kitching's Brigade was seen to be falling back, flanked, as they were, by Gordon.

1167. Figures 5–7 and 9–11 are portions of a map, p. 517, from Merritt, *Battles & Leaders*....
1168. O.R. 43/I, pp. 403, 417.
1169. O.R. 43/I, p. 813; McNeily, SHS Vol. 32, p. 225.
1170. O.R. 43/I, p. 158.
1171. Stevens, H., p. 107; Irwin, p. 415.

The enemy now fell upon the 19th Corps' defenses, first striking Grover's Division. The 176th New York, 156th New York, and the 8th Indiana regiments were placed to cover the line facing the southeast, supporting Battery D of the 1st Rhode Island Artillery.[1172] In minutes, they received a random fire from an unseen enemy, which came from the heights formerly occupied by Crook's command. Soon, about 150 yards distant, the enemy line became dimly visible, extending from Cedar Creek to the left as far as the eye could see. Wharton[1173] had

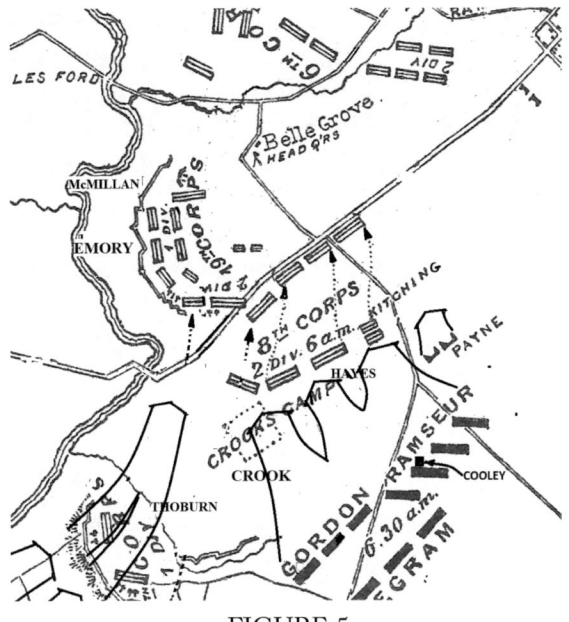

FIGURE 5

now come up, and joined the combined forces of Gordon and Kershaw, figure 6. The enemy swept into the regiments of Grover's 4th Brigade, and one gun of the 1st Rhode Island Battery was lost, with all of its horses killed, before it could answer Emory's order to withdraw. Next, the 1st Division, McMillan's, was hit, having gone into a position some 400 yards to the rear of the struggle on the Pike.[1174] A quote from the history of the 116th New York is instructive:

> Bullets began to reach us, but not from the front, and we saw plainly that if we were going to use *this* line of breastworks, which had cost us so much hard work, we must get upon the wrong side of them.

During this period, Ricketts was wounded, and Getty assumed command of the 6th Corps.[1175] The job of stabilizing the Union line would now fall to him.

Soon after the commencement of the attack, the 6th Corps had been ordered under arms, having no entrenchments, as it was considered a reserve,[1176] and it had moved forward and formed on the west side of Meadow Brook. As soon as formed up, the 2nd Division was to have advanced by its left to gain the Pike, and the 1st and 3rd Divisions had been ordered to come forward and join the 19th Corps near the line Wright was forming on the Pike. In Wright's words:

1172. O.R. 43/I, pp. 342, 346.
1173. Sumner, p. 145; O.R. 43/I, pp. 322–323.
1174. Stevens, H., p. 105; Clark, p. 240; O.R. 43/I, pp. 308–309; Irwin, pp. 425–428.
1175. O.R. 43/I, p. 159.
1176. O.R. 43/I, pp. 159, 193, 403–404.

As the two divisions of the Sixth Corps, ordered from the right of the line to the left, could reach that point within twenty minutes of the time that the line referred to was formed, and as the position taken up was a satisfactory one, I felt every confidence that the enemy would be promptly repulsed. In this anticipation, however, I was sadly disappointed. Seeing that no part of the original line could be held, as the enemy was already on the left flank of the Nineteenth Corps, I at once sent orders to the Sixth Corps to fall back to some tenable position in rear; and to General Emory ... that he should fall back and take position on the right of the Sixth.

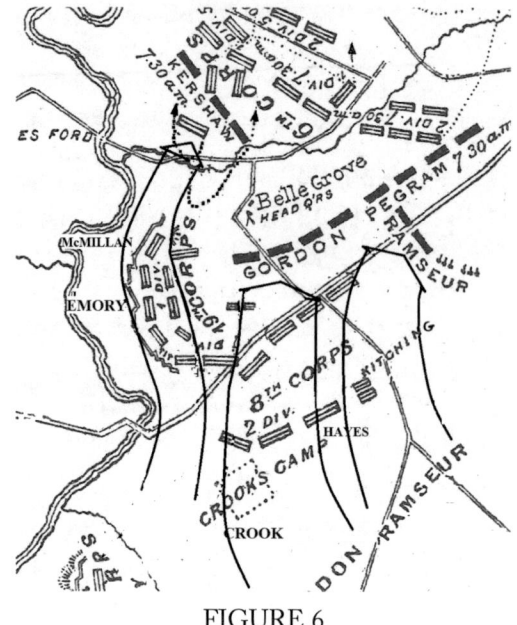

FIGURE 6

As can be seen in figures 6 and 7, the Confederate line, including Gordon's corps and Wharton's division, had advanced unopposed up the Pike, almost to the outskirts of Middletown by 7:30. It had pushed back the mix of Wright's defenders several hundred yards across the open country west of the Pike, as far as Belle Grove, which had been the planned objective of Payne's Cavalry, though the rapid advance of Gordon had completely obviated Payne's assignment.[1177] Kershaw had advanced right through the 19th Corps camp and up to a position which faced the old campground of the 6th Corps.

At this point, the losses to the infantry artillery had been substantial. The 1st Maine Battery had to abandon three of its guns, having had 49 horses killed. The 5th New York Battery lost three guns at the crossing of Meadow Brook. All told, the 19th Corps had lost 11 guns, Crook (DuPont's Battalion) seven, and the 6th Corps, which was to now feel the brunt of the attack, would soon lose six.[1178]

There was no fighting force yet unscathed except the 6th Corps and the cavalry, but Wright could not have known what little impact Rosser's attack had had on them. Wright had decided the next step, and it was withdrawal. With Gordon in control of the Valley Pike, and unopposed on the approach toward Middletown, Wright would soon be flanked on his left and rear. He gave the order, the objective point for the withdrawal was the hill upon which the Mt. Carmel cemetery rested, the extension of a ridge of high ground which runs west of Middletown. As it happened, it was adjacent to the camp of Merritt's 1st Cavalry Division. This deliberate

---

1177. Carpenter, pp. 208–209.
1178. Irwin, p. 421.

withdrawal, done in steps, was completed by the 6th Corps between 8:00 and 9:00, as is indicated in figure 7. The 19th Corps new position is shown encircled.

Fortunately for the visitor, both Belle Grove, figure 8, and the cemetery remain today, and are significant points to use for identifying the battle area. Note the open country.

Returning to the beginnings of the battle, not long after Rosser had attacked Merritt's camp, he withdrew to a position opposite Custer, at Cupp's Ford. An excerpt of Early's Official Report of the battle relates:[1179] "Rosser sent word that when he attacked the cavalry he encountered a part of the 6th Corps supporting it; that a very heavy force of cavalry had massed in his front, and that it was too strong for him, and that he would have to fall back. I sent word to him to get some position that he could hold…"

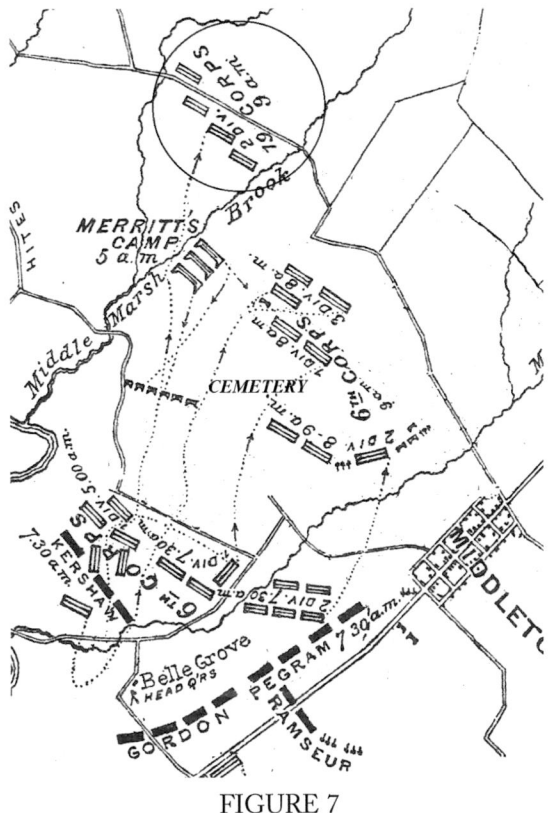

FIGURE 7

FIGURE 8

The "Savior of the Valley" henceforth remained hunkered down with only three of Custer's regiments opposing him for the rest of the day. Now, the fully awakened Union cavalrymen found themselves with nothing to do but strain to figure out what was happening to the rest of the army. They had not received any orders. Quoting from Col. J. H. Kidd, commanding the Michiganders, i.e.,

---

1179. O.R. Vol. 43/I, p. 562.

Merritt's 1st Brigade:[1180]

Colonel Lowell [Charles R. Lowell, Jr., commanding Merritt's Reserve Brigade] informed me that his orders were to support the Michigan men if they needed support. No help was needed at that time. I told him so. The enemy had been easily checked and, at the moment, had become so quiet as to give rise to the suspicion that he had withdrawn from our front, as indeed he had. A great battle was raging to the left, and in response to the suggestion that the army seemed to be retreating, he replied:

"I think so," and after a few moments reflection, said: "I shall return" and immediately began the countermarch. I said to him: "Colonel, what would you do if you were in my place?"

"I think you ought to go, too." he replied, and presently, turning in the saddle, continued: "Yes, I will take the responsibility to give you an order," whereat, the two brigades took up the march toward the point where the battle, judging from the sound, seemed to be in progress.

A startling sight presented itself as the long cavalry column came out into the open country overlooking the battleground. Guided by the sound, a direction had been taken that would bring us to the pike as directly as possible and at the same time would approach the union lines from the rear. This brought us out on a commanding ridge north of Middletown [the encircled position shown in figure 9]. This ridge…runs to and across the pike. The ground descends to the south a half mile, or more, then gradually rises to another ridge about on a line with Middletown. The Confederate forces were on the last named ridge, along which their batteries were planted, and their lines of infantry [Gordon, Kershaw, Ramseur, and Pegram] could be seen distinctly.

…The full scope of the calamity which had befallen our arms burst suddenly into view. The whole battlefield was in sight. The valley and intervening slopes, the fields and woods, were alive with infantry, moving singly and in squads. Some entire regiments were hurrying to the rear, while the Confederate artillery was raining shot and shell and spherical case among them…Some of the enemy's batteries were the very ones captured from us…but all these thousands, hurrying from the field, were not the entire army…There, between ourselves and the enemy—between the fugitives and the enemy—was a long line of blue, facing to the front, bravely battling to stem the tide of defeat. It was the old Sixth corps—the 'ironsides' of the Potomac army. Slowly, in perfect order, the veterans…were falling back contesting every inch of the way. One position was surrendered only to take another."

The successive attacks referred to were by Pegram, and when he was repulsed,

---

1180. Kidd, pp. 412–414. The 1st Brigade was called the Michigan Brigade, as it consisted of the 1st, 5th, 6th, and 7th Michigan regiments, though it also included the 25th New York. Kidd here disobeyed Merritt's order to use the 1st Michigan to picket the line against Rosser. O.R. 43/I, p. 449.

another by Wharton, who was sent to fill "a vacancy" in the Confederate line.[1181] Quoting Early:

> In a very short time, and while I was endeavoring to discover the enemy's line through the obscurity, Wharton's division came back in some confusion, and General Wharton informed me that, in advancing to the position pointed out to him by Generals Ramseur and Pegram, his division had been driven back by the 6th Corps...The fog soon rose for us to see the enemy's position on a ridge to the west of Middletown, and it was discovered to be a strong one...orders were given for concentrating all our guns on him.

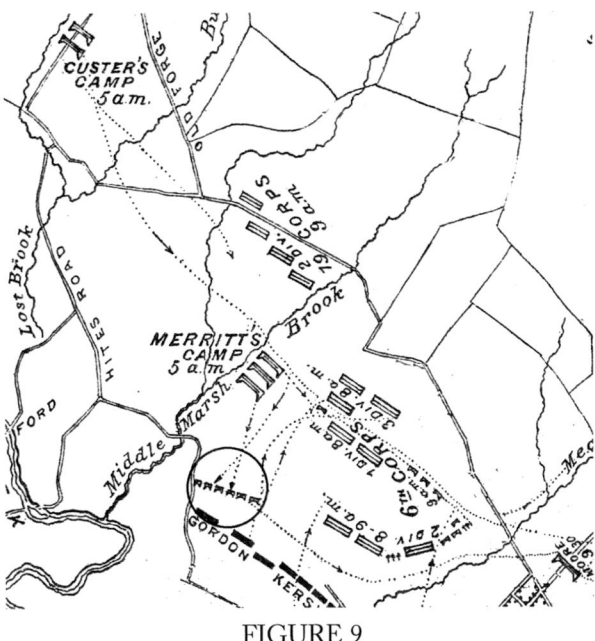

FIGURE 9

Early goes on to disclose the fact that he could see Federal cavalry where it was not expected, on his right, the east side of the Pike: "In the meantime a force of cavalry was advancing along the Pike, and through the fields to the right of Middletown..." This was Moore's brigade of Powell's division, the one that had been stationed at Buckton Ford. On his own initiative, Col. Moore had left his position[1182] some time prior to 8:00, and headed north to the sounds of the battle. His circuitous route of some 5½ miles is seen running along the right of figure 10. He went into position on the Pike at 9:30.

Soon after the time that a surprised Early discovered Moore's brigade, the cavalry present on the east side of the Pike would have included two regiments Torbert had placed there on his own initiative. Estimated to be as early as at 7:00, he had ordered his escort, the 1st Rhode Island Cavalry, and Merritt's escort, the 5th U.S. Cavalry, to move to the Pike for the purpose of trying to stem the northward flow of refugees from the 8th Corps. "About this time,"[1183] Devin's 2nd Brigade was ordered to the Pike by Merritt, long before Wright finally ordered "the whole cavalry force" to the left of the army, which was not until after 9:00. The history of the 1st New York Dragoons confirms that Devin "immediately saddled up and

---

1181. Early, *Sketch*, pp. 444–445; O.R. Vol. 43/I, pp. 226, 581.
1182. O.R. Vol. 43/I, p. 509.
1183. O.R. Vol. 43/I, pp. 433, 449.

moved out without breakfast."[1184] It is interesting to note that by that time, half of the cavalry had already begun their move to the left on their own initiative.

On departure, Devin ordered his only artillery battery, "Taylor's Battery" to report to Merritt's headquarters, and it was assigned to support the "infantry" which, of course, was the 6th Corps. The infantry had lost some 24 guns, and the remainder, according to Merritt, "...had gone unaccountably to the rear..." That remainder was a significant part of DuPont's Battalion, 11 guns, which had escaped from Kershaw's initial attack on Thoburn. They were in action on both sides of the Pike when Devin arrived. In fact, Devin assigned a squadron to support the 1st Ohio Battery.[1185]

As to Battery K–L it took positions in the battle front on the cemetery ridge, and moved successively as the 6th Corps moved – assumed to be those encircled positions (1–4) shown in figure 11.

Merritt is quoted in full as to the action:

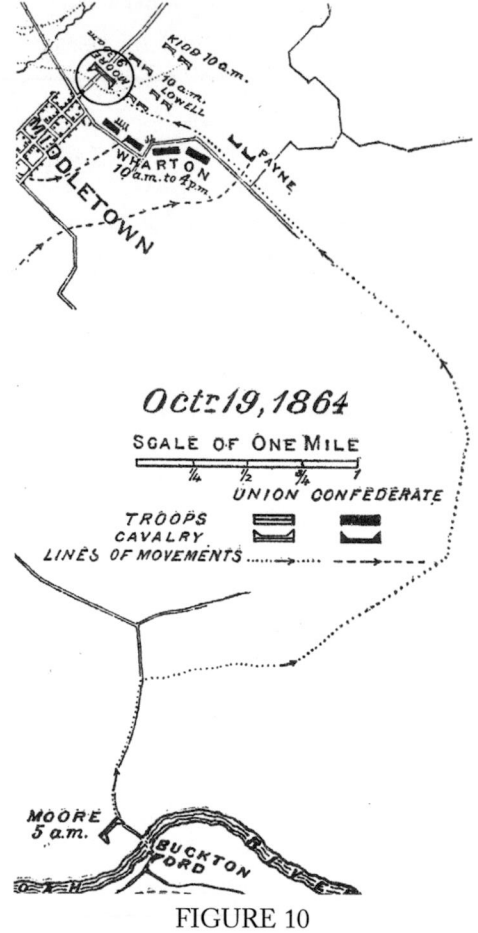

FIGURE 10

> On moving to the left General Devin ordered his battery to report to division headquarters, where Lieutenant Taylor, commanding, received orders to advance to an eligible position on the infantry line of battle, and use his pieces on the enemy till such time as it was unsafe to remain there. Great credit is due Lieutenant Taylor for the prompt and efficient manner in which he carried out this order. He was well advanced to the front of battle, without supports from his own command, and none save the thin and wavering line of infantry near his position. The artillery of the infantry had gone unaccountably to the rear, or had been captured by the enemy, and Taylor's was the *only battery* [italics added] for some time on that part of the field. It is thought that his rapid and destructive fire did much toward preventing a farther advance of the enemy on that flank in the early part of the day.

Torbert confirms Merritt's report and adds:

1184. Bowen, J. R., p. 252.
1185. O.R. Vol. 43/I, pp. 15, 417–421.

Was the last artillery to leave the front. Too much praise cannot be given to the officers and men of this battery for their coolness and gallantry on this occasion. When the infantry was forced back and was obliged to retire it joined its brigade on the right of the Pike, where it immediately went into action.[1186]

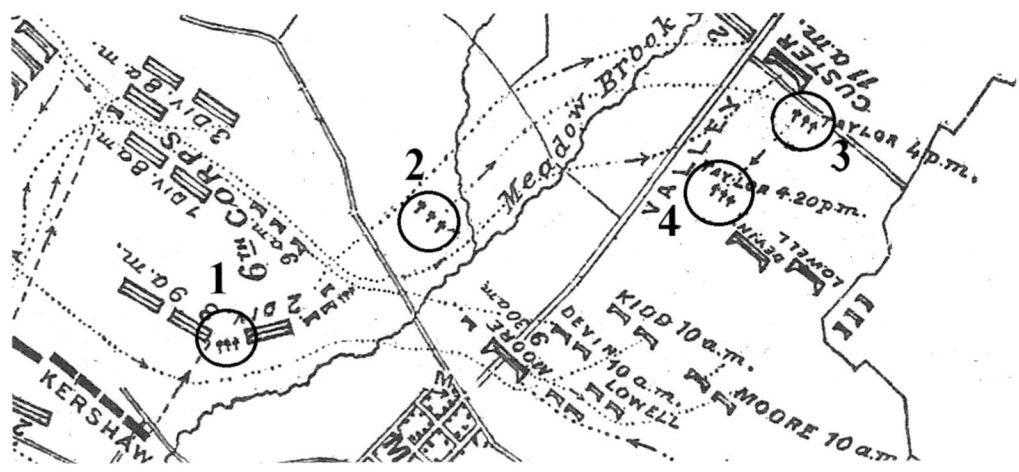

FIGURE 11

Before 11:00, the infantry had been ordered to again withdraw, and form on and intermediate position along Old Forge Road, on the west side of the Pike.[1187] Taylor was then forced to withdraw, and took a position behind some stone walls that Lowell had seized, which could be position 2 in figure 11, then taking a converging fire from "several" enemy batteries, disabling one three-inch gun, and killing "several men and a number of horses." Later, Devin ordered Taylor to move to a more sheltered position on the east side of the Pike, at the end of the final infantry line, number 3. At the time of Sheridan's order to advance, it is shown at number 4, at 4:20 p.m..

By the time that Wright had ordered Torbert to move his "whole force" to the left, it was only Custer's 1st Brigade that moved! As noted, Torbert had ordered Custer's 2nd Brigade (Wells),[1188] to remain and picket the area to hold Rosser in check, and Merritt's two brigades had long since moved out on their own. Moore, of Powell's Division, not ever hearing of Wright's plea, had also moved on his own, so the only element of the cavalry that was not utilized in the battle was a portion of Powell's 2nd Brigade.

Torbert's decision managed to stabilize the right, in that, Wells held the position "at great odds for five hours," collecting and reforming infantry refugees, who were rallied and began throwing up defensive works.[1189]

1186. O.R. Vol. 43/I, p. 433
1187. O.R. Vol. 43/I, pp. 478, 479.
1188. Jackson, p. 160.
1189. O.R. Vol. 43/I, pp. 521–522.

Early was concerned. Wharton's division was ordered to take a position, as Early says, to hold the enemy's cavalry in check. It is the first time that day that he utters a sentence with a tinge of the defensive in it. Discovering that the 6th Corps position on cemetery ridge could not be assaulted on the left, as its approach was "across an open flat and a boggy stream with deep banks," i.e., Middle Marsh Brook, he then ordered Gordon and Kershaw to attack on the right.[1190] Before the attack took place, the guns of Carter's artillery, 18 or 20 of them, which had been massed on the Pike since 8:00 a.m.,[1191] were put in position on the high ground on the Pike (interpreted to be near where the Cedar Creek Foundation visitor's center is today), which provided, once the fog lifted, a clear view of the 6th Corps position.

Early's bombardment having commenced, it was time to evacuate, and Wright so ordered both Getty and Emory.[1192] They were to fall back to the country lane known as Old Forge Road, a place affording no defensive position, but simply a feature upon which to unite the army, and gain time to find a better position further to the rear. The move experienced some confusion, the brigades losing sight of one another, the 1st and 3rd brigades of the 6th Corps moving 1,000 yards further north from where the 2nd Brigade struck the Pike. The 2nd remained at the Old Forge Road position no more than 20 minutes, and then "coolly marched in line of battle a mile further to the rear, when we found a position that General Getty considered suitable to form upon." This was the army's third, and final, position, and can be seen in figure 12.[1193] It was during the course of this move that Sheridan rode past. He notes:[1194] "Just south of Newtown I saw about three-fourths of a mile west of the Pike a body of troops, which proved to be Rickett's [3rd] and Wheaton's [1st] divisions of the Sixth Corps..." Moving on, he found Getty and the cavalry. The retreat was at an end, and so was Sheridan's ride. It was on a knoll about nine-tenths of a mile from the Middletown line, near what is now called Rienzi Knoll Lane.

Seeing the Union line retreating, Early ordered Gordon, Kershaw, and Ramseur forward as far as (quoting from Irwin) "...the cross-road beyond the cemetery [Old Forge Road]." Continuing, from Irwin:[1195]

> Early had now two courses of action open to him: one was to extricate his army away from its position, with its enemy directly in front and Cedar Creek in rear, before the Union commander could take the initiative; the other was to attack vigorously with all his force before the Union infantry should be able to complete the new line of battle now plainly in the act of formation.

1190. Early, *Sketch*, p. 445.
1191. O.R. Vol. 43/I, p. 599.
1192. O.R. Vol. 43/I, pp. 194–195, 226; Irwin, pp. 425–427; Stevens, H., p. 122.
1193. Walker, pp. 144–145; Figure 12, Atlas, plate 82, map 9, portion altered.
1194. Sheridan, Vol. II, p. 82.
1195. Irwin, p. 427.

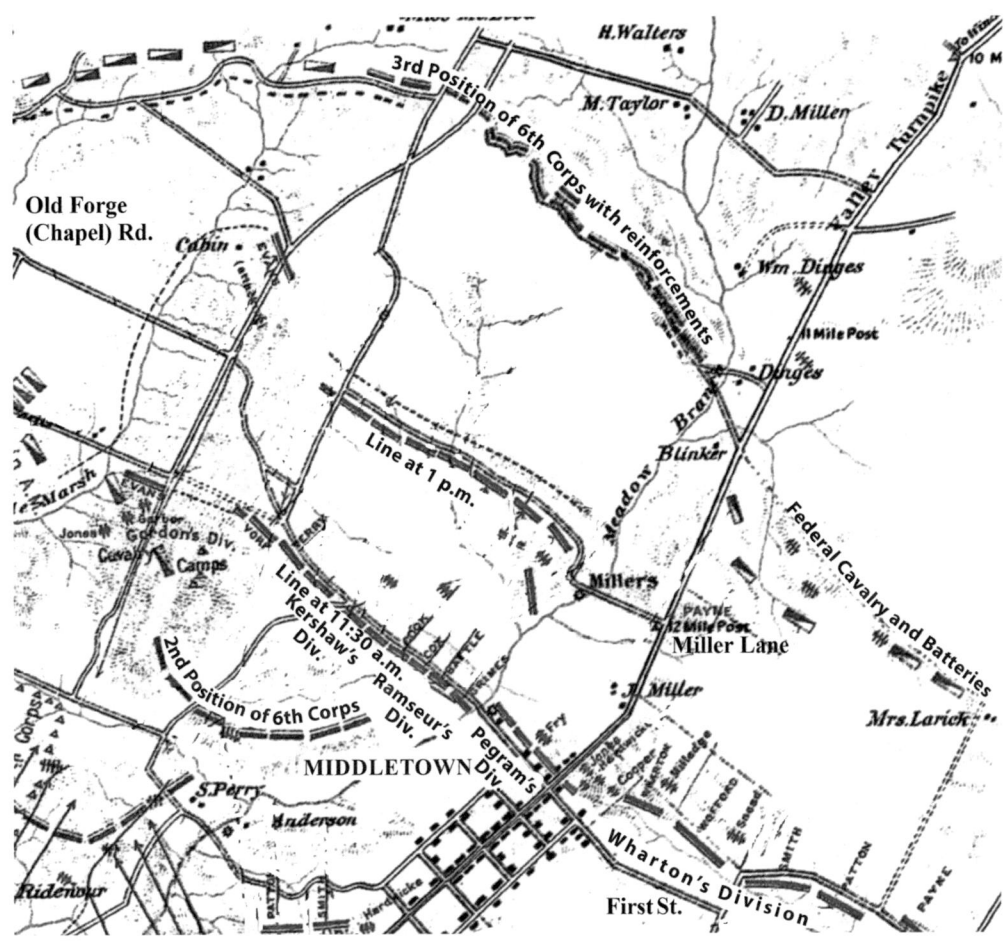

FIGURE 12

Early did neither. He ordered an attack, but with caveats. Referring to Gordon:

> I ordered him to take position on Kershaw's left, and advance for the purpose of driving the enemy from his new position—Kershaw and Ramseur being ordered to advance at the same time. As the enemy's cavalry on our left was very strong, and had the benefit of an open country to the rear of that flank, a repulse at this time would have been disastrous, and I therefore directed General Gordon, if he found the enemy's line too strong to attack with success, not to make the assault. The advance was made for some distance, when Gordon's skirmishers came back reporting a line of battle in front behind breastworks, and Gordon did not make the attack.

The advance referred to was pushed up as far as Miller's Lane; today called Cougill Road. The Miller house, which still stands; can be seen clearly in figure 12 ("Miller's"), a few hundred yards west of the Valley Pike. The times when the positions were reached are also shown; the Confederate line reached Old Forge Road by 11:30, and then had advanced to Miller's Lane by 1:00. The Confederates were now stalled.

Early is in error in stating that Gordon did not attack, and Gordon is faulty in skipping over the fact that he halted.

The failure to advance further caused recrimination in Southern literature after the war; Gordon accusing Early of "The Fatal halt at Cedar Creek" with Early and others defending. It is true that the halt allowed time for the Union line to be formed up, allowing the 19th Corps to join it, as well as many of the scattered refugees from the 8th Corps. It would soon be ready to make a stunning and victorious counterattack. However, it is not certain that had the Confederate attack been driven along without the pause that it would have succeeded.

Regarding the numbers of the Union troops still capable of opposing Early at that moment, there were about 10,000 cavalry, including their artillery,[1196] and the 18,500 infantry of the 6th and 19th corps, with a portion of their artillery. In comparison, according to Gordon, Early had an estimated 10,000 infantry. The numbers speak well in defense of Early.

The question about plundering touched a hot button. For example, a colonel in one of the North Carolina regiments in Grimes' Brigade insisted that none of the hats, blankets, boots, tents, etc., that were scattered about the 8th Corps camp were taken by anyone other than one man who had charges preferred against him.[1197] Grimes himself reports "little plundering..." Certainly, no one would pick up a blanket, to have to carry it throughout the day, but the assertion that no one would search out a canteen, since Gordon's men had taken none, tests credibility.

Here, eschewing any Union reports, or quotes from Union regimental histories, is a quote from Private G. W. Nichols, of the 61st Georgia, in Evans' Brigade of Gordon's Division:[1198]

> We took many prisoners and captured nearly all of their wagons, artillery, ambulances, horses, mules and a great deal of clothing, shoes blankets, tents, etc...We ran in pursuit of them till we had gotten about two miles from their camp, and then everything was halted. A great many of us *went back* to their camp after blankets, shoes, clothing, etc...
>
> It seemed that there were no Yankees in our front. Everything was quiet as death. Some of our boys went to sleep while the others were plundering the camps. I got two nice new tent flies, two fine blankets, a fine rubber cloth, two new overshirts, and two pair of new shoes. In fact, we could get anything we wanted except Yankee money.

1196. O.R. Vol. 43/II, pp. 248, 501. Powell's 2nd Brigade, not present. 8th Corps not included. Estimated 1,000 casualties deducted; Gordon, p. 343. Though Early had estimated he had 8,500 at Winchester, Kershaw's reinforcements were as many as (O.R. Vol. 43/II, p. 423) 4,000. Lomax's cavalry, estimated by Gordon at 1,700, has been deducted, as they never appeared at the battle; McNeily, J. S., SHS, Vol. 32, 1904, p. 226, gives Early's total infantry and artillery as 12,780; Irwin, p. 437, mentions others who claim 15,000 and up to 22,000.
1197. O.R. Vol. 43/I, pp. 600, 608.
1198. Nichols, p. 195; Author's italics.

While Private Nichols raises a question as to what came first, the halt or the plundering, he leaves no doubt that it happened on a massive scale, and he seems to exempt only those who fell asleep.

The Union line had begun to stabilize at its third and final position at about 11:00,[1199] before Sheridan had arrived; Wright had saved the situation, and members of his staff have claimed that he issued orders for a planned counterattack. That glory, however, would be remembered as Sheridan's. Upon his arrival, he immediately bought into the moves that Wright had made, though he apparently was unaware, at the moment, that it was Wright that had made them. He established his headquarters nearby.

He then sent his aide, Major Forsythe, to check with Colonel Lowell to see "whether he could hold on there," on the east side of the Pike.[1200] Lowell replied that he could. Sheridan only mentions Lowell, though it must have been obvious to him that nearly his entire cavalry force was there. Satisfied with Lowell's answer, he sent Custer back to the right. Wright was returned to the command of the 6th Corps, and all the others that moved up temporarily were returned to their old slots.

Seeing many of the retreated infantry returning, he directed them to take places in Getty's line. The 19th Corps was directed to take the right, between the 6th Corps and Middle Marsh Brook.

Torbert[1201] has this happening at 2:00, and that Custer got into position "just in the nick of time . . ." as the enemy had just succeeded in crossing infantry and cavalry over Cedar Creek. Custer charged and "drove them about a mile . . . behind their infantry support, from which they did not dare show . . . for the rest of the day."

This is assumed to be the point-in-time at which Early[1202] makes his fatal decision:

> It was now apparent that it would not do to press my troops further. They had been up all night and were much jaded. In passing over rough ground to attack the enemy in the early morning, their own ranks had been much disordered, and the men scattered, and it had required some time to re-form them. Their ranks, moreover, were much thinned by the advance of the men engaged in *plundering* [italics added, here again surfaces that controversial word regarding Confederate behavior] the enemy's camps. The delay which had unavoidably occurred had enabled the enemy to rally a portion of his routed troops, and his immense force of cavalry, which remained intact, was threatening both of our flanks in an open country, which in itself rendered an advance extremely hazardous.

That "immense force of cavalry" had broken him at Winchester in its classical

---

1199. Walker, p. 146; Haines, p. 278; O.R. Vol. 43/I, p. 201.
1200. Boudrye, p. 180; Sheridan, Vol. 2, pp. 84–85, 89–90; O.R. Vol. 43/I, p. 201.
1201. O.R. Vol. 43/I, p. 434.
1202. Early, *Sketch*, pp. 447–448.

role as attackers. Now, Early saw a line of dismounted horsemen with the most formidable armament the world had ever known. From J. S. McNeily of the 21st Mississippi Regiment, Kershaw's Division:[1203]

> ...Sheridan's mounted force of 7,000 men was to be reckoned with. He [Early] had not forgotten how his army had fled before this same powerful contingent at Winchester and Fisher's Hill. And it was menacing his right with a like overthrow.

Perhaps two hours had now passed, and at the suggestion of his aide, Sheridan showed himself by riding, hat in hand, along the entire front of the infantry line. The men now knew for sure that he was here, and who was in charge.

Returning to the east side of the Pike to see what the enemy "was doing," Sheridan writes that he concluded that Early was preparing for an attack, which came at about 1:00 p.m. This reference is taken to be the move forward from the Old Forge Road to Miller's Lane by Kershaw, and Evans' Brigade of Gordon's Division, further on the left, on the front of the 19th Corps, which was easily repulsed.[1204] As Sheridan relates: "This repulse...made me feel pretty safe from further offensive operations on their part, and I now decided to suspend the fighting till my thin ranks were further strengthened by the men who were continually coming up from the rear..."

Now, the ever thorough and cautious Sheridan took the time to once again probe the meaning of the mysterious signal from the top of Massanutten Mountain. Was Longstreet's Corps actually with Early? Merritt was ordered to attack an exposed Confederate battery and capture some prisoners. This done, the prisoners confirmed that the only portion of Longstreet's Corps that were present was Kershaw's. As Sheridan relates: "The receipt of this information entirely cleared the way for me to take the offensive."[1205]

Between 3:30 and 4:00 Sheridan was ready to give the order, which is quoted here in full:[1206]

> Cedar Creek, *October 19, 1864-3p.m.*
> The entire line will advance. The Nineteenth Corps will move in connection with the Sixth Corps. The right of the Nineteenth will swing toward the left, so as to drive the enemy upon the pike.
>
> P.H. SHERIDAN
> *Major-General*

The "swing toward the left" was meant to have the line pivot about its left and close on the Pike, much as a door on its hinge. The intent was to block the Pike,

---

1203. McNeily, S.H.S., Vol. XXXII, p. 227.
1204. O.R. Vol. 43/I, p. 285; Walker, p, 149; Sheridan, Vol. 2, pp. 86–87; Stevens, H., p. 130.
1205. Sheridan, Vol. 2, pp. 87–88.
1206. O.R. Vol. 43/II, pp. 33, 416.

preventing the retreat of the Confederates back to Strasburg. However, what actually happened was that enemy's right gave way after Lowell courageously charged; the charge in which he was killed.[1207] This tore loose the hinge, and the eagerness of the men, with the left insisting upon keeping pace with the right, the line instead advanced along the Pike. The old Union camps at Cedar Creek were regained, and there the infantry halted; but the cavalry were ordered to keep on.[1208]

The country being entirely open, Custer's impressive force of cavalry could be clearly seen by Early bearing down on his left flank, and quoting Custer: "Realizing the necessity of at once gaining the bridge, [across Cedar Creek] the disordered masses of the enemy, now completely panic-stricken, threw away their arms…and sought safety in ignominious flight."[1209] Though exhilarating to observe, the practical matter was that most of Early's army was able to escape, up to this point at least, as Custer admits, with a "small loss in prisoners." Sheridan's tactics had failed to trap Early here, just as it had failed at Winchester.

Some few prisoners were captured by Merritt's cavalry which had, only moments later, arrived at the bridge, and would soon unite with Custer in the chase.[1210] Custer had crossed Cedar Creek at a "difficult" ford, about a mile west of the bridge, and after meeting an enemy line of resistance behind a stone wall about a quarter of a mile beyond, his 1st Vermont and 5th New York broke through, and were able to reach the Pike, though darkness was fast approaching.[1211] Reaching a point about a half-mile beyond Strasburg, they found: "The road blockaded for miles with guns and wagons and ambulances filled with wounded. Whole batteries were captured, with guns, men and horses intact."

Devin, leading Merritt's Division, met the last resistance of the day, one volley, from a line of Confederate infantry on the south side of the Cedar Creek Bridge, before it broke for the woods as the 6th New York charged across the narrow 150-foot span.[1212] Shortly, they met a group from Custer's cavalry, and were confronted with all of what Custer had captured and left behind. Custer's 1st Vermont and 5th New York continued on through Strasburg, and approaching Fisher's Hill, at nearly midnight, the pursuit was ended.

The result of the battle was the recapture of all of the guns and camp equipage that had been lost earlier in the day, plus 24 guns, 1,200 prisoners, and an immense amount of wagons and equipment. The 1st Division and a brigade of the 2nd Division of the 19th Corps, after two hours rest, were sent out to Strasburg to aid in securing it.

1207. Sheridan, Vol. 2, p. 90.
1208. Walker, p. 153; Jackson, p. 160.
1209. O.R. Vol. 43/I, pp. 524, 562–563.
1210. Sheridan, Vol. 2, p. 90, O.R. Vol. 43/I, p. 525
1211. Jackson, p. 161; Walker, p. 153.
1212. O.R. Vol. 43/I, pp. 479, 526; Stevens, H., p. 140.

It was rather unabashedly declared a great victory by Sheridan, in his report to Grant, dated the 20th.[1213] Grant duly ordered a shotted (loaded with rounds) 100-gun salute from his armies, which was directed at Petersburg. Lincoln wrote Sheridan a personal note of gratitude, with "the thanks of the Nation." A few weeks later Sheridan was promoted to the rank of major-general in the regular army, having been elevated to brigadier only a month before.

	KILLED		WOUNDED		MISSING		TOTAL
	OFFICERS	MEN	OFFICERS	MEN	OFFICERS	MEN	
Sixth Army Corps	23	275	103	1,525	6	194	2,126
Nineteenth Army Corps	19	238	109	1,227	14	776	2,383
Army of West Virginia	7	41	17	253	10	530	858
Provisional Division	1	11	6	66	......	18	102
Cavalry	2	27	9	115	......	43	196
Grand total	52	592	244	3,186	30	1,561	5,665

FIGURE 13

The battle was not without cost. The Union losses totaled 5,665 in killed, wounded, and missing. A breakdown is given in figure 13.[1214] The Confederate losses were reported to Lee from Early, at New Market, on October 21st. He only reports killed and wounded, as "not more than 700 or 800." His later *Autobiographical Sketch*,[1215] which so often has been quoted herein, gave him the opportunity to reconsider, and he revises the total to 1,860, with "something over 1,000 prisoners."

Again, as at Winchester, Lieutenant Taylor wrote no report, or more precisely, none is found in the Official Records, and the same can be said of Lt. W. C. Cuyler, and Lt. John McGilvray. Though Devin reported three artillerymen killed, only two are in the Battery L "record": Pvt. Perry S. White of Battery K, and Pvt. Rowland Card of Battery L. Pvt. Michael Beckett of Battery K was wounded, and Pvt. Prosper Ferrari of Battery L is listed as wounded in the regimental record of casualties, though there is no record of his being wounded in the Battery L muster roll.[1216]

Sheridan offers no estimate of Early's losses. Gen. Hazard Stevens,[1217] quotes 2,250 for Early's killed and wounded, with 1,250 as prisoners, but offers no source, other than his authority as being present at the battle.

Though a great victory was won, as claimed by Sheridan, the cost was huge, and markedly disproportionate to Early's—even if General Stevens' figure is correct. In fact, 5,665 is more than twice the number of casualties taken by the U.S. 1st and

---

1213. O.R. Vol. 43/II, pp. 410, 423–424, 436. O.R. Vol. 43/I, pp. 32–34. A preliminary report was sent out on the night of the 19th, and supplementary ones on the 21st and 25th. Sheridan, Vol. 2, pp. 91–92.
1214. O.R. Vol. 43/I, pp. 137, 557, 564.
1215. Early, *Sketch*, p. 450.
1216. Little faith is held in this particular record, as Battery L's musician, Ludwig Rupprecht, is also listed as wounded. Actually, it was in the Battle of Tom's Brook, on October 9th.
1217. Stevens, H., p. 141.

29th Divisions in the landings at Omaha Beach on D-Day, the 6th of June, 1944.[1218] Only the Union could go on taking these numbers of casualties, because they had twice the number of cannon-fodder. However, even the Union could not take this forever. As we have seen, recruitment was getting nigh impossible and was now down to a trickle.

The losses taken by the 6th and 19th Corps indicates the heroism and determination of their defense, and what can happen in successive assaults on units if taken in detail.

Cedar Creek was the last battle in the Shenandoah for Battery K–L, and eventually would prove to be their last of the war. On the 27th of October, the battery was ordered into Reserve Camp at Pleasant Valley, Maryland. They arrived there on the 29th.

On Tuesday, November 8th, the people went to the polls, and Lincoln won re-election. What specific incident had influenced the election in Lincoln's favor was anybody's guess, maybe the people had always favored him, and all of the speculation was just wasted emotion. Perhaps it was the soldier vote from Ohio, cast in October, the results of which had been revealed. It was 48,000 for Lincoln to 7,000 for McClellan. Speaking to this, *Harper's Weekly* proudly said: "American soldiers are not fools."[1219] Nevertheless, the favorable way the war had gone had to have been a factor. Perhaps it was the capture of Atlanta on September 6th, or the Battle of Winchester on the 19th. Then, Cedar Creek was the most recent, and had provided concrete evidence that the South was losing.

On the 9th of November, Sheridan took up a defensive position south of Winchester, at Kernstown, meaning to go into winter quarters. The indefatigable Early, having recovered his bearings, followed as far as Middletown. Torbert was sent out on the 12th and fell upon Rosser, routing him, and though preparations for an infantry attack had been made, on the 13th Early fled back to New Market.[1220]

In mid-November, Early sent back Kershaw to Lee's army.[1221] Nothing could make a better statement about Grant's policy in the Valley than the fact that Lee had removed Early's reinforcements. The Valley was now burned out and of no value—just as he had planned it. The two contending armies had no reason to fight here. Not to be deterred, Rosser found a chink in the armor, attacking the Federal fort at New Creek, West Virginia, on November 28th, and though Sheridan had to send a division of Crook's 8th Corps there, he sent the rest of the 8th to Grant. By mid-December Lee had called back Early's 2nd Corps, leaving only Wharton and Rosser, and Sheridan had sent the 6th Corps to Grant. Only the 19th Corps remained at Kernstown, at what was called Camp Russell; but it would be

---

1218. *www.ddaymuseum.co.uk/d-d*
1219. *Harper's Weekly*, Nov. 12th, 1864, p. 722.
1220. Irwin, p. 440.
1221. Early, *Sketch*, p. 454; Sheridan, Vol. 2, pp. 98, 99.

brief—they left for Stephenson's Depot, the terminus of the railroad from Harper's Ferry, on December 30th. The practical reason for this move was that supplies could reach there by rail without the use of wagon transportation over winter's muddy roads.

Mosby

On October 3rd, Lt. John R. Meigs, a member of Sheridan's staff, and the son of Gen. Montgomery Meigs, the quartermaster of the army, was killed near Harrisonburg.[1222] Sheridan took retaliation by ordering all of the houses in the vicinity burned, but little else was done, as on October 6th Grant had ordered Sheridan north. On October 11th, a guerrilla band mortally wounded Sheridan's chief quartermaster and his medical inspector while they were being escorted to Winchester.

An angry Sheridan had not forgotten, and in the lull in operations that followed the Battle of Cedar Creek, he ordered Merritt into the Loudon Valley, "the hotbed of lawless bands,"[1223] where his orders were to burn all barns and mills and their contents, and drive off all of the livestock

Mosby comment: "his efforts only stimulated the fury of my men." Wounded and captured in December, he escaped, recovered, and by the end of February, 1865 he was back.

After Lee's surrender, he was offered the same terms, dodged the question, and instead disbanded his command and disappeared – into the sympathetic surroundings of the Valley. Finally, in February, 1866, he was arrested, and only then did he accept Lee's surrender terms, and was paroled by Grant.[1224]

"Record" 12/64
31 OCTOBER–31 DECEMBER, 1864 PLEASANT VALLEY, MARYLAND

Battery quartered in Reserve Camp, Pleasant Valley, Md. during the past two months.

Henry W. Closson	Capt. Abs. on det. Svc. In the field, Va. as Chief of Arty & Ord. of the Cav. of the Mid. Mil. Div. S.O. no. 45 Hdqtrs. Mid Mil Div Oct. 25, 1864.
Franck E. Taylor	1st Lt. Abs. on 20 days leave since Dec.18,'64.
John McGilvray	1st Lt. Battery K, in command.
Edward L. Appleton	1st Lt. Absent without leave.
Detached:	
John Meyer	Pvt. On det. Svc. With Battery G, 5th U.S. Artillery, New Orleans, La. since July 24,'64.

1222. Sheridan, Vol. 2, pp. 51,52, 99, 100.
1223. Humphreys, pp. 190-191; O.R. Vol. 43/II, p. 730; Mosby, p. 333.
1224. Mosby, pp. 359-364.

Amelius Straub	Pvt. Abs. on det. Svc. As Hospital Attendant since Nov. 22, '64.

Absent in confinement:

Patrick Gibbons	Pvt. At Ship Island, serving sentence of G.C.M. S.O. no. 18, Hdqtrs. 1st Div. 19th Army Corps Dec. 31, 1863.
John Lewery	Pvt. Abs. confined at New Orleans, La. Apprehended as a deserter.

Deserted:

Lyman Woodruff	Pvt. Deserted from Pleasant Valley, Md. Nov. 29, 1864. Apprehended in Frederick, Md. Dec. 1, 1864, and sent under guard to Company. Due U.S. $31. for expenses of arrest & for camp and garrison equipment, 52¢

Discharged:

William Parks	Pvt. At Pleasant Valley, Md. By virtue of S.O. no. 464, War Dept. A.G.O. Washington. Dec. 23, 1864.

Joined:

Charles Cooke	Pvt. Joined Co. from absent on Det. Svc. Dec. 30, 1864
Patrick Craffy	Pvt. Joined from absent sick, Dec. 23, 1864. Due U.S. for transportation $1.70.
Michael Kenny	Pvt. do.
Warren Shaw	Pvt. Joined Co. from abs. sick, on Dec. 30, 1864.
Lyman Woodruff	Pvt. Joined Co. from desertion, Dec. 15, 1864.

Strength: 75   Sick: 11

Sick Present: 3

Benjamin O. Hall	Pvt.
John McKenny	Pvt.
Ludwig Rupprecht	Bugler. Wounded in action near Strasburg, Oct. 9. On 30 days leave. Due U.S. for transportation $3.

Sick Absent: 8

James Campbell	Pvt. Sept. 19, '64 Wounded in action at Winchester, Va.
George Chase	Pvt. April 22, '63 At Brashear City, La.
William Crowley	Pvt. July 13, '63 At Baton Rouge, La.
Charles Jackel	Pvt. April 20, '64 At New Orleans, La.
John Kelly	Pvt. Sept. 19, '64 Wounded in action at Winchester, Va.
Churchill Moore	Pvt. April 20, '64 At New Orleans, La.
Andrew Stoll	Pvt. Sept. 25, '64 Wounded in action near New Market, Va.
John C. Wood	Pvt. Sept. 19, '64 Wounded in action at Winchester, Va.

The cooks are listed, and after each of their names is the remark: "Due cook $12. Error on last payroll." The errors on the cook's payrolls keep going on and on. Eventually, the paymaster will learn that they have been granted equal pay to that of a private.

Note that Henry Closson, though he is now a brevet lieutenant colonel, is still regarded as a captain in his permanent rank as commander of Battery L.

No doubt the continued absence of Appleton has been the subject of some discussion, and with all of Battery L's officers elsewhere, 1st Lt. John McGilvray, of Battery K, has moved into temporary command.

Note that there are really only two men recently listed as sick, out of the total of eleven. Five are listed from wounds, and four are from lingering diseases contracted while Battery L was in Louisiana. The better health of the whole of Sheridan's army had been noted in a report made by the surgeon-general, for the period from August 27th to December 31st.[1225] He notes: "This low rate of sickness, at a time when the troops were harassed and over-fatigued by repeated battles and skirmishes, can be attributed to the healthfulness of the climate and the quantity of good vegetable food procured solely from the country." At last, an intelligent utterance regarding diet.

His reference to "harassed," no doubt refers to continual skirmishes and major battles, but also to Mosby's guerillas.

Lyman Woodruff was the lone deserter during this period. He was recruited on February 27th, 1863, at Baton Rouge, from Company E of the 13th Connecticut Volunteers, to serve a term of three years. He had been mustered into the 13th Connecticut on February 5th, 1862. Having served without incident up to November 29th, 1864, when he deserted, his rationale for taking the risk of deserting is a bit of a mystery. Perhaps he was angry that he had not yet been paid the first installment of his $100 enlistment bonus, but many others were also still unpaid. His desertion was one of the most unsuccessful ones on record—he remained free only three days before being caught, while, as we have seen, dozens of others had never been caught, or had returned voluntarily and not been charged. It was true that the end of the war was being talked about, and perhaps he felt his job was done. However, he could not have picked a worse time. The war had moved away from this locale, and every petty official, both civilian and military, now had little else to distract them, so when Woodruff was seen at Fredericksburg he was quickly pounced upon. Recall that local police officials were entitled to collect a fee for apprehending a deserter. The fee for the constable or constables who picked up William F. Brown, Joseph H. Parslow, Patrick Craffy, and Owen A. Wren, at the Tarleton Plantation on Bayou Teche, back on October 1st, 1863, was $10. Here, poor Woodruff encountered the full weight of the bureaucracy – he was charged $31. At a general court-martial held on February 24th, he was sentenced to forfeit $10 of his monthly pay for four months. He had been held in confinement awaiting trial ever since he had been returned on December 15th. Two months in confinement satisfied the court, because he was not sentenced to additional jail time.

The pressure of the war having relaxed, at least for Battery K–L, a number of furloughs were granted, as recorded in the December monthly report, "by command of Major-General Sheridan." On December 8th, 1st Sgt. William E. Scott, Cpl. Miles McDonough, and privates James Ahern, John Burke, William Creed,

---

1225. O.R. Vol. 43/I, p. 145.

Prosper Ferrari, Henry H. Ward, Henry Williams, and Terence McGauley,[1226] each got 35 days. On the 14th, Sgt. Owen A. Wren, Cpl. George Howard, and privates Patrick Cummings, George Hadley, Reuben Townsend, and William V. Thompson each got 35 days. Cpl. Lewis Keller, and Bugler Louis Rupprecht got 30 days. Promptly on January 12th, 1865 (35 days from December 8th), Wren, McDonagh, Burke, Ferrari, Williams, Cummings, and Thompson were all declared deserters, despite the fact that those who had been given their furloughs on the 14th, Wren, Cummings, and Thompson, still had six days to go! Reuben Townsend was the only one in the latter group who was not declared a deserter until the 23rd of January.

Most of those furloughed left for Boston or New York, and a glance at their original enlistment record confirms that in most cases, this was "home." The record-keeping error was apparently recognized in the cases of Keller, and Rupprecht, who are listed without comment in the muster roll ending in February. In fact, for all of the others, the mistake hadn't mattered. Wren, McDonagh, Burke, Ferrari, Williams, and Cummings never returned and were never apprehended. Only Thompson and Townsend were unlucky; they were caught at Ellicott's Mills, Maryland, on February 7th, and returned to stand trial. They were found guilty and fined $10 out of each month's pay for three months, in addition to the $31 for the expenses of their arrest. To understand the harshness of this judgment, the inflation factor, as we have calculated earlier, was about 130. Thus, the two fugitives were each fined a total of $8,000, in today's dollars.

The creaking wheels of justice finally turned on May 17th, 1887, when the judgment for Wren, Cummings, and Thompson was removed by the Adjutant-General's Office. The mistake in the original record is duly noted. Unfortunately, the muster roll for 31 December–28 February is in such poor condition that the full notation is unreadable, yet it is clear that the records of other individuals were also reviewed. Wren's record was reviewed again on February 2nd, 1894, and his enlistment was changed to read: "Discharged at Washington, D.C., February 2/94, to date from January 12/65, by order of the Secretary of War and by reason of desertion." This is still rather confusing, as it does not make clear whether his discharge was honorable or dishonorable; if honorable, it would have enabled him to apply for a pension.

1226. McGauley has a variety of spellings in the various records. To date, it has been left as spelled on the particular record quoted. Here, we defer to the Massachusetts adjutant-general's list of its *Soldiers Sailors and Marines in the Civil War*, Massachusetts having been credited with his enlistment as part of its quota, apparently because his original enlistment was at Lynn, Massachusetts. Actually, he was from Jersey City, New Jersey.

Chapter 18

Rosser; "Record" 2/65; Sheridan Leaves the Valley; Richmond; Surrender; "Record" 4/65; The Faithful Few; The Three Cooks

Rosser

Rosser attacked the Federal fort at Beverly, West Virginia, on January 11th,[1227] the attraction being the large amount of food and stores kept there. The surprise of the sleeping garrison was complete and 572 surrendered. The stores secured, Rosser marched the prisoners the 100 miles back to Early's camp near Swope's Depot, west of Staunton, Virginia, through the snow and near zero weather "…many without overcoats and only partially clad. The frozen feet and hands of quite a number necessitated amputation." This raid and its senseless cruelty had no bearing on the greater course of Sheridan's plan to dismantle the Middle Military Department, or the course of the war, but is proof of the hatred engendered in the Valley by the Union policy of destruction.

On the 6th of January, Grover's Division of the 19th Corps was sent on its way to occupy Savannah,[1228] captured by Sherman on December 22nd.

"Record" 2/65
31 DECEMBER 1864–28 FEBRUARY 1865 PLEASANT VALLEY, MARYLAND

Battery Quartered at Pleasant Valley Reserve Artillery Camp.

Henry W. Closson	Capt. Commanding Battery. Relieved from det. Svc. in the field, Va. as Chief of Arty & Ord. of the Cav. Mid. Mil. Div. S.O. no. 141, Hdqrts. Cav. Corps. Mid. Mil. Div. Feb'y. 8th 1865.
Franck E. Taylor	1st Lt. Abs. on 20 days leave of absence since February 26th 1865.
Edward L. Appleton	1st Lt. Relieved from Gen. Recruiting Service and ordered to join his Battery. S.O. no. 315 War Dept. A.G.O. Washington Sept. 25th 1864. Absent without leave.
Detached:	
Franck E. Taylor	1st Lt. On detached svc. with Battery K 1st U.S. Artillery.

1227. Sheridan, Vol. 2, p. 100; Pellet, pp. 280–281.
1228. Irwin, p. 442, O.R. Vol. 44, p. 6; O.R. Vol. 43/I, pp. 88–89, 667–668; Clark, pp. 254–255.

John Meyer	Pvt. On detached service with Battery G, 5th U.S. Artillery, New Orleans, La. since July 24, '64.

Absent in Confinement:

Patrick Gibbons	Pvt. At Ship Island, serving sentence of G.C.M. S.O. no. 18. Hdqrts. 1st Div. 19th Army Corps Dec. 31, 1863.
John Lewery	Pvt. Abs. confined at New Orleans, La. Apprehended as a deserter.

Deserted:

Owen A. Wren	Sgt. Deserted January 12th 1865, while on furlough.
Miles McDonagh	Cpl. John Burke do.
Patrick Cummings	Pvt. Prosper Ferrari do.
William V. Thompson	Pvt. Deserted January 12th 1865, while on furlough. App. at Ellicott's Mills Md. and sent under guard to camp Feb. 7th 1865.
Reuben Townsend	Pvt. Deserted Jan. 25th 1865 from Pleasant Valley, Md. App. At Ellicottt's Mills Md. And sent under guard to camp on Feb. 7th 1865.
Henry Williams	Pvt. Deserted January 12th 1865, while on furlough.

Discharged:

Joseph Wilkinson	Pvt. Discharged by reason of expiration of service, and final statement given.
John G. Nitschke	Pvt. do.
Michael Kenny	Pvt. do.
Amelius Straub	Pvt. do.

Joined:

William V. Thompson	Pvt. Join Company from desertion Feb, 7th 1865.
Reuben Townsend	Pvt. do. do.

Strength: 65   Sick: 7

Sick Present: none

Sick Absent: 7

James Campbell	Pvt. Sept. 19, '64 Wounded in action at Winchester, Va.
George Chase	Pvt. April 22, '63 At Brashear City, La.
William Crowley	Pvt. July 13, '63 At Baton Rouge, La.
Charles Jackel	Pvt. April 20, '64 At New Orleans, La.
Churchill Moore	Pvt. April 20, 64 At New Orleans, La.
Andrew Stoll	Pvt. Sept. 24, 64 Wounded in action near New Market, Va.
John C. Wood	Pvt. Sept. 19, '64 Wounded in action at Winchester, Va.

The three cooks are listed, and as usual, are not counted in the "Strength."

Given the fact that Battery L had always been stationed in unhealthful southern climates since it was reorganized in 1854, it seemed that someone was always sick. For the first time in memory, and a search of early regimental records indicates that it may be for the first time *ever* since it was reorganized, that no one present is sick.

The company roster, (see Appendix) for the commencement of each year, has been prepared throughout this text from the 31 December–28 February muster roll, thus the 1865 roster status is as of the end of February. Since Wilkinson,

Nitschke, Michael Kenny, and Straub had all been discharged during February, they do not appear. Note the relatively few men present in the battery at this time, though we must be reminded that Battery L and Battery K are still officially consolidated, and would remain so until October.

It is hard to understand what those that deserted in January were thinking. All of them had reenlisted on July 18th, 1864, and were entitled to, but had not received, their $400 bounty. It would be gone forever if they did not voluntarily return, and soon.

Sheridan Leaves the Valley

Recall that on October 15th Grant had asked Sheridan to make a raid on the Virginia Central Railroad at Charlottesville at the time he was planning to travel to Washington, and that it was canceled by the circumstances of Early's mysterious signal from Massanutten Mountain.

Again, on the 19th of December, responding to Grant's urging, a similar raid was sent out, under Torbert.[1229]

The weather was terrible, the roads poor, and every movement was closely observed by Early. Custer was surprised by his old nemesis, Rosser, while bivouacked at Lacey Springs, and had to retreat down the Valley. Lomax checked the remainder of Torbert's force, and the frostbitten raiders returned on December 27th.

On the first of February, Sherman had begun his move north from Savannah, and had captured Columbia, South Carolina on February 17th. Grant now had the opportunity to put into motion his grand plan, which included another raid.

Grant's telegram to Sheridan, dated February 20th:

> GENERAL: As soon as it is possible to travel I think you will have no difficulty about reaching Lynchburg with a cavalry force alone. From there you could destroy the railroad in every direction… Sufficient cavalry should be left behind to look after Mosby's gang. From Lynchburg…you could strike south …and join Sherman. This additional raid, with [others] now starting from east Tennessee… Vicksburg…Eastport, Miss. … Canby from Mobile Bay… all pushing for Tuscaloosa, Selma, and Montgomery, and Sherman eating out the vitals of South Carolina, is all that will be wanted to leave nothing for the rebellion to stand upon.[1230]

General Winfield S. Hancock,[1231] who had been made available to command a "veteran volunteer corps" in November, was now assigned to head the Department

---

1229. Sheridan, Vol. 2, pp. 102-104; Humphreys, pp. 194-202.
1230. Grant, Vol. 2, p. 409.
1231. O.R. Vol. 42/I, p. 3; Vol.46/I, p. 2.

of West Virginia and "temporarily all of the troops of the Middle Military Division not under the immediate command of Major-General Sheridan". The next day, February 27th, Sheridan left Winchester and headed up the Valley Pike. He took with him only Custer's and Devin's divisions, one section of battery C-E of the 4th U.S Artillery, and one section of Battery M, 2nd U.S. Artillery,[1232] a total of 9,987 men. Battery M was one of the artillery companies that escaped from Texas with Battery L in 1861.

The weather was still rainy, and the spring thaw had raised the streams to almost higher than could be forded. At Woodstock, they found that Rosser had attempted to burn the bridge across the North River. The river was running lower the next morning and it was forded, hitting Rosser in flank, and dispersing him. They pressed on to Staunton, finding that the delay had allowed Early to escape.

Though Sheridan could have bypassed this remnant of Early's army, he decided to attack. Custer was ordered to take up the pursuit, and on March 2nd, found Early's small force placed in a strong position on a ridge outside of Waynesboro.[1233] A gap in his line toward the river was determined to be a vulnerable point, and Custer ordered three of Pennington's regiments, who were armed with Spencer carbines, to charge. Early's men were soon routed.

Quoting Early:

> I now saw that everything was lost...and had the mortification of seeing the greater part of my command being carried off as prisoners.

The battle netted 11 pieces of artillery, 200 wagons, and 1,600 prisoners. Only Rosser and his few men escaped.

At Lynchburg, on March 30th, Early received a telegram from Lee ordering him to turn over his command in Southwestern Virginia to Gen. John Echols, and that of the Valley to Lomax.[1234] His career was at an end.

Sheridan continued to Charlottesville, destroying the railroad bridges across the Rivanna River, and many miles of the Orange and Alexandria Railroad. He continued the scorched-earth policy to New Market and along the James River, but deciding that it was "impracticable" to attempt to join Sherman, he turned to join Grant.

He writes:

> ...feeling that the war was nearing its end, I desired to be in at the death.

He arrived at White House, in Grant's lines, on March 18th.[1235] The raid was an example of the urgency Grant had placed on ending the war. A less aggressive

---

1232. O.R. Vol. 46/I, pp. 475, 485, 488; Sheridan, Vol. 2, p. 113; McDonald, pp. 359,360.
1233. Early, *Sketch*, pp. 462-464; O.R. Vol. 46/I, pp. 485, 502-503, 516.
1234. Early, *Sketch*, pp. 465,466; O.R. Vol. 46/II, pp. 1041, 1184; Sheridan, Vol. 2, pp. 119-121.
1235. Longstreet, p. 591; O.R. Vol. 46/I, p. 480; Humphreys, pp. 214-216, 221,222.

commander might have waited for the weather to improve. It had rained for 16 out of the 20 days of the raid, and the roads were often partially frozen and always muddy. The ordeal so injured 4,000 horses that they had to be left behind, victims of hoof-rot, grease heel, and leg scratches.

Refitted, Sheridan went into camp at Hancock Station, in front of Petersburg on the 27th of March.

Richmond

Grant's final orders were issued on March 24th:

> On the 29th instant, the armies ... will be moved by our left, for the double purpose of turning the enemy out of his present position...and to ensure the success of the cavalry, under general Sheridan,...in its efforts to reach and destroy the South Side and Danville Railroad.

Meantime, Lee had planned a feint to distract Grant, allegedly to find dry roads for, in General Longstreet's words:[1236] "our march away, or for reinforcements to reach us." Lee still clung to the irrational hope that Johnston's army might join him, though he had just been defeated at the battle of Bentonville, and was now at Smithfield, North Carolina, a week's march away, and had only 15,000 men left.

The feint, planned by General Gordon, the author of the attack at Cedar Creek, was a breakout at Fort Stedman, on the eastern end of the Confederate defenses. It was bold, complex, and it failed miserably. It did not delay Grant's initiative for a minute, and preliminary moves for his attack began as scheduled.

Lee shifted forces to meet the threat, and the March 31st Battle of White Oak Road resulted. Sheridan was then ordered to move further west, to Five Forks. [1237]

He was repulsed, and had to retreat back to Dinwiddie. However, he used a tactic described by Grant:

> Here General Sheridan displayed great generalship. Instead of retreating ...He deployed his cavalry on foot, leaving only mounted men enough to take charge of the horses. This compelled the enemy to deploy over a vast extent of wooded and broken country, and made his progress slow.

The result was what was desired, the enemy was now out of his trenches, and with reinforcements, Sheridan attacked on the afternoon of April 1st. Five Forks was taken, and the Union forces were brought ever closer to the South Side Railroad. Lee again reacted as hoped, and in an attempt to retake Five Forks, he sent a part of Longstreet's command out from the lines at Richmond.

Regardless of darkness, Parke's 9th, and Wright's 6th Corps, were ordered

---

1236. Longstreet, p. 592.
1237. Grant, Vol. 2, pp. 439–442; Sheridan, Vol. 2, pp. 154, 155; O.R. Vol. 36/II, p.56.

to assemble near their assault positions, and after a massive artillery barrage, the assault began at 4:30 a.m. At nightfall, the outer works of Petersburg and the South Side railroad were in Union hands.

Lee was now compelled to abandon Richmond. It was 11:00 a.m. on April 2nd when Lee's dispatch reached Jefferson Davis at Sunday services. By 12:50 p.m., on April 3rd, troops were seen to be evacuating.[1238] The explosion of the magazine of Confederate Fort Drewry, and the general conflagration seen in the night sky confirmed the withdrawal, and Grant advised Weitzel, with his 25th Corps, to be prepared to assault in the morning.

What time had wrought – the engineer officer at Fort Pickens and the man who had once refused to command the black units enrolled by General Butler at New Orleans, was now a major-general in command of a corps of all black volunteers. Moving it was – that they were now the first of the victors to enter the capitol of an entity devoted to their slavery.

Surrender

Such place names as Amelia Springs, Jetersville, Rice's Station, Sailor's Creek, Cumberland Church, High Bridge, Farmville, and Appomattox Station adorn the week-long trek of Lee's disintegrating army. Some are where serious clashes took place, and others were only skirmishes. Sailors Creek, on April 6th, however, was "one of the severest of the war" according to Sheridan. "The enemy fought with desperation … and we, bent on his destruction, were no less eager and determined." [1239] The total number of the Confederate command captured that day is given by Gen. A. L. Long, of Lee's staff, as "about 10,000."

On April 7th, Lee was approached by members of his staff and advised to surrender. He also got a note from Grant,[1240] asking for surrender. Lee responded, asking for terms, and on the 8th Grant replied that Lee's army would be disqualified from taking up arms against the United States, until properly exchanged.

The negotiations would become more complicated than are remembered in popular history.

Later, in the afternoon of the 8th, when the Confederate advanced guard reached the neighborhood of Appomattox Court House, Lee learned that the supplies he had counted on at Appomattox Station had been captured – by Custer's cavalry.[1241] Only then did he agree to meet Grant on April 9th, at 10:00 a.m., "between the picket lines of the two armies," though at that time he had no

---

1238. Longstreet, p. 608; CWSAC VA091-097; Long, pp. 409-411; O.R. Vol. 46/I, p. 139; O.R. Vol. 46/III, pp. 496,509.
1239. Sheridan, Vol. 2, pp. 180, 185; Long, p. 414.
1240. Grant, Vol. 2, pp. 625-628.
1241. Sheridan, Vol. 2, pp. 189, 190; Longstreet, pp. 622-627; Long, pp. 420-422.

intention of doing so. The late hour of the 8th would allow a devious move. Lee ordered Fitzhugh Lee's cavalry transferred from the rear to the advanced guard, and then, with General Gordon commanding, a breakout was set for 1:00 a.m. on the 9th. It was fantasy, and it could not be put together; Gordon could do nothing without Longstreet's corps, which was then heavily engaged by Meade. Only then, did Lee send out a flag of truce, commenting to his staff:

> Then there is nothing left me but to go and see General Grant, and I would rather die a thousand deaths.

"Record" 4/65
28 FEBRUARY–30 APRIL 1865 WINCHESTER, VIRGINIA

The Company left Pleasant Valley, Maryland on the 24th day of April, 1865, and marched to Charlestown Virginia, distance twelve miles. Resumed the march the next morning, and marched to Winchester, Virginia. Joined the Provisional Brigade of Cavalry where the Company is now stationed.

Henry W. Closson	Capt. Commanding Battery.
Franck E. Taylor	1st Lt. Relieved from duty with Battery K 1st U.S. Artillery, and ordered on recruiting service. Orders no. 84, A.G.O. Washington, DC April 1st 1865.
Edward L. Appleton	1st Lt. Absent without leave since Sept. 22nd 1864.

Detached: none.

Absent in Confinement:

Patrick Gibbons	Pvt. At Ship Island, serving sentence of G.C.M. S.O. no. 18, Hdqrts. 1st Div. 19th Army Corps. Dec. 31, 1863.

Deserted:

John Lewery	Pvt. From absent in confinement at New Orleans, to deserted, date not known.

Discharged:

George Chase	Pvt. At Brashear City, for disability, date not known.
William Crowley	Pvt. At Baton Rouge, for disability, date not known.
Charles Jackel	Pvt. At New Orleans, for disability, Oct. 1, 1864.
John Meyer	Pvt. From Battery G 5th U.S. Artillery, New Orleans, by exp. of svc. Mar. 1, 1865.
Churchill Moore	Pvt. At New Orleans, for disability, date unknown.

Died:

James Campbell	Pvt. Oct. 6, 1864. From wounds received at the Battle of Winchester, Va.
John C. Wood	Pvt. Sept. 21, 1864 do.

Strength: 57 Sick: 2

Sick Present: none

Sick Absent: 2

Thomas Clinton	Pvt. Mar. 1, 1865 At Frederick City, Md.
Andrew Stoll	Pvt. Absent wounded since September 24th 1864, near New Market, Va.

With no transfers into it, the number of men in Battery K–L was insufficient to have been sent into action. Battery K at this time had only 45 men. Moreover, neither battery had any horses. According to the regimental return for April, 22 men have been requested from depot for Battery L, and 47 for Battery K. Roughly this same number had been posted on every regimental return since the first of the year. As has been mentioned before, recruitment had been becoming increasingly difficult. In April, there were only 10 recruits sent to the entire regiment, and they all were allotted to Battery C. This, no doubt, is why Lieutenant Taylor had been assigned to recruiting duty, the mysterious disappearance of Lt. Appleton having resulted in no new recruits.

There may have been great hopes of being assigned to Sheridan when he left the Valley, but this was unrealistic, with no horses and a total of only 102 men. Finally, in March, they were given 140 horses, but these were of no real use in any action, since no new recruits had come in, and as is shown above, the battery was losing men, not gaining them.

The Faithful Few

Of the 83 members of Battery L listed on the muster roll ending in February of 1861, 79 actually left Texas. On the march from Fort Duncan, Sgt. Charles Riley had to be left behind at Fort McIntosh, and privates Francis Hagan and John Bissell had deserted while the battery was at Fort Brown. Lt. James W. Robinson had resigned from the service; he had remained with the battery as a sutler; but could not now be counted as a member. The attrition continued, and brought the number of the "faithful" down to 11 at the close of the war, taken as July of 1865, a month after the last official surrender. Eight men deserted in July, and as usual, some returned later, but cannot be counted as members of the "faithful."

A listing of the "faithful" follows. Note that none of the officers of the battery are on it. Capt. Henry Closson, 1st Lt Franck Taylor, and 1st Lt. Edward Appleton were not assigned to Battery L until it was in Florida, and Appleton deserted in October of 1864.

1. James Ahern–18 October, 1860, Boston. This laborer from Boston, Massachusetts, joined Battery L as a member of the second group of recruits who arrived from Fort Columbus, New York, on December 5th, 1860. He was one of those who reenlisted under the war provisions of July 1864, and was mustered out at Fort Porter, Buffalo, New York on July 18th, 1867, as a corporal. His widow, Mary, applied for a pension on October 25th, 1889, but a corresponding file for application no. 407,090 has proven to be not available from the National Archives.

2. James Beglan–25 October, 1860, New York. A laborer, he joined Battery L as one of the second group of recruits. Like Ahern, he reenlisted in 1864, and was mustered out, as a private, then age 27, on July 18th, 1867, at Fort Porter.

He got into trouble in Baltimore while the company was stationed at Fort McHenry, and was held there by the "Civil Authorities" from October 1865 until March of 1866. Having returned in September, he was sentenced by a general court martial to forfeit one month's pay. Strangely, the whole time he had been AWOL he was not listed as a deserter, and his discharge was not delayed to make up the lost time.

He never married, and existed as a "clerk" in Brooklyn, until at age 60, on September 7th, 1899, he entered the Soldier's and Sailor's Home at Bath, New York. He applied for a disability pension, under the Act of June 27th, 1890. On the basis of a physical examination, he was granted a partial disability pension of $6 per month. At that time, the maximum he could have obtained was $10.

His Pension Record from 1899 onward reflects the changes wrought by new pension laws frequently enacted: May 9th, 1900; July 1st, 1902; February 6th, 1907; May 11th, 1912; and by his new applications, rejections, and appeals. He subsequently applied for a pension increase to the maximum of $10 and it was granted on June 22nd, 1904, after repeated correspondence between his pension lawyer, his examining physician, and the Department of the Interior, Bureau of Pensions. Of course, the Pension Application had to be verified by confirmation at the War Department—the Adjutant General's Office, Commissioner of Pensions. There, the records were kept that proved that the applicant had actually served, and where, and for how long, and whether he had been injured or sick. Each time a new application was made, the mountain of paperwork grew, as copies of previous correspondence were forwarded. Of course, all of this was to protect against fraud, and regardless of the good intent of the Congress, bureaucrats made the process slow and arbitrary. An industry had been created, and it lived on well into the 20th century.

On December 11th, 1906, Beglan was given a rate of $12, and on February 1st, 1910, a rate of $15. On May 20th, 1912, it was raised to $25, and on December 26th, 1914, it became $30. He died on July 19th, 1915.

3. *Owen Coyne*–10 December, 1860, Fort Duncan, Texas. Coyne was not from the groups of recruits enlisted in the east. He was 30 years of age, and this was his second enlistment; he had first served in Company G of the 1st Infantry, into which he had been recruited on July 4th, 1855, at Chicago. He had been discharged, by expiration of service at Fort Chadbourne, Texas, on July 13th, 1860. He then enlisted in Battery L. Because he did not take advantage of the huge reenlistment bonus of July 1864, it is concluded that he felt that at 40 years of age, 10 years of army life was enough, and when his enlistment expired, on December 11th, 1865, he was mustered out as a private, at Fort Schuyler, New York. No further record.

4. *Patrick Donnelly*–1 March, 1860, Boston. A farmer, he joined from the first group of recruits that arrived in April. At over five feet-ten, he stood taller than most. He must have been a man made of iron. For the entire war he was never

listed as sick, and was never disciplined, save being listed as "in confinement" while a fresh recruit. Finally, this steadfast soldier was promoted to corporal on April 13th, 1865. He had reenlisted at New Orleans in July of 1864, so his three-year term of service extended to 1867, while Battery L was stationed at Fort Porter, New York. There, he met Margaret Moran,[1242] whose family ran a grocery store ten blocks from the fort. His enlistment expired in July, and he was discharged as a sergeant. In August, he reenlisted in Battery L of the 4th Artillery, at Fort Delaware, Delaware City, Delaware. He was appointed a corporal on the same day, and promoted to sergeant the next month.

In December of 1867, Patrick and Margaret were married. His third term of service expired on August 13th, 1870, while he was serving at Fort Macon, Goldsboro, North Carolina. This ended his ten years in the army, and like Owen Coyne, he apparently felt that that was enough. The couple tried farming in Arkansas, but eventually returned to Buffalo, and, as described by his great-grandson, they opened a grocery store in the rough-and-tumble harbor area, near where Margaret's brother operated a saloon. The Donnelly's had six children.

Margaret died on October 4th, 1886. As Patrick's health declined with age, he first applied for a veteran's pension in 1890. Like so many of his comrades, he was at first denied by the bureaucracy, and like so many of his comrades, he would continue to re-apply for years, only to still be denied. In 1897, thoroughly disgusted with the system, he wrote a letter to the commissioner of pensions, which is here partially quoted. He closes a long review of his service, and the legal justifications for his claim, with the following:

> I have come to the conclusion that there is no pension law which covers my case or is of use or benefit to me…I shall go before no more Notary Publics nor Examining Doctors boards. It is evident that as the law stands it makes no difference whether a man served ten years or two months…I shall wait until the Government see the justice of my claim…

In 1899, he was granted a pension of $6 per month. He entered the Soldier's Home in Bath, New York, for two months in 1903, and for four months in 1904, and each time his health recovered and he was discharged. By 1910, his pension had increased to $15. He died on December 20th, 1910, and is buried at Holy Cross Cemetery, Lackawanna, New York.

His great-grandson, Mr. Paul Callsen, is the only known descendant of anyone in Battery L with whom the author has become acquainted.

5. *George F. Hadley*–1 March, 1860, Boston. A blacksmith, age 32, he joined

---

1242. Subsequent information supplied by Paul Callsen, Donnelly's great-grandson.

from the first group of recruits. By July of 1861, he was assigned on "Extra Duty" as blacksmith, and remained in that position for his entire term of service, though his official title was artificer. As the size of the battery grew, when it was converted, first to a mounted, and then to a light battery, he was joined in that capacity, in the war years of '63 and '64, by Henry Champion and Sirenus Kilburne. He reenlisted in July of 1864. Hadley has a special place in the history of Battery L, as he has been eulogized in Haskin's *History of the First Regiment of Artillery*, in a section which was written by Henry Closson, a portion of which is quoted here:[1243]

> And as I write of battery L there looms up the figure of "Hadley," a typical man of the sturdy rank and file, upon whose faithful shoulders so many generals were borne to honor and success. I seem again to see this honest old fellow, always cheerful, always ready, always at work. He was the company blacksmith, and considered himself responsible for the serviceable condition of the horses, and whenever a halt was made and the labors of other men ceased, Hadley still continued his. Whether sunshine or rain was beating down it made no difference; either found Hadley at it, fastening a nail here, loosening a shoe there, swearing at some careless driver, or petting some restive horse. On the march, he merrily plodded along through the mud or sand, and in camp the most conspicuous object was Hadley's brawny arms and bare head dodging about the horses' feet, and the rat-tat of his hammer was as regular as tattoo itself. And he died as he had lived—in harness—one of the crew drowned with their officer, in attempting, during a storm off Fort Niagara, to save a drifting boat.

The accident occurred on Lake Ontario, on May 4th, 1870. The "honest old fellow" was 42.

6. *Lewis Keller*–11 September, 1854, Baltimore, Maryland, butcher, age 21. In 1854–55 the battery was stationed at Key Biscayne, "in the field" in Florida, and he saw service in the Second Seminole War. The battery having been transferred to Fort Brown, Texas, he was stricken with yellow fever during the 1858 epidemic, but survived. He reenlisted on December 1st, 1859, at Newport, Kentucky. As has been noted, he was one of those appointed from Battery L as a 2nd Lieutenant in the newly created 2nd Louisiana Cavalry, one of the Corps d'Afrique units organized by General Banks. He served from November 29th, 1863, was wounded on May 6th, 1864, and promoted to 1st Lieutenant on May 7th, 1864. The 2nd Cavalry was reorganized in September of that year, and Keller was mustered out. He reenlisted in Battery L on October 11th, 1864.

He was married to Mary Ann Noonan, at Buffalo, New York, on June 2nd, 1867. On May 17th, 1867, he was appointed an ordnance sergeant, which was rather a special position given to veterans who had had at least eight years of

---

1243. Haskin, p. 372.

service, four of which was in the grade of a non-commissioned officer. This meant he could be assigned as a lone artillery storekeeper at any number of army installations. The assignment was essentially permanent, at the discretion only of the adjutant-general. Keller's assignment was Fort Douglas, Utah.

He continually reenlisted after he left Battery L in 1867, until his last enlistment expired in 1885. During this period, with his wife, he had served for 18 years at Fort Douglas with the 14th Infantry. Upon his retirement, they settled in Buffalo, New York, and the adopted home of several former members of Battery L. Ever faithful to the army, he was for a time employed as steward at the post exchange. He remained in Buffalo until he died at age 77, on July 10th, 1907.

7. *Philipp H. Schneider*–17 December, 1853, Lancaster, Pennsylvania, baker, age 33. Schneider claimed prior service in Company A, of the 1st Battalion of New Jersey Mexican War Volunteers, from August 1847 to February 1848. After his first enlistment in Battery L, he reenlisted at Fort Brown, Texas, on October 16th, 1858, and again at Baton Rouge in 1863. His record while serving with Battery L was spotless; all through the war, he was never sick, never deserted, and never disciplined. Though he was never injured in action, as luck would have it, on or about April 1st, 1865, while the battery was at Pleasant Valley, Maryland, he was thrown from his horse, causing injury to his head and back. He is not reported absent, sick, on the Battery L monthly report for April 1865, but is so reported for May. He is not reported sick on the April–June muster roll, a record-keeping deficiency which would, upon his later application for a disability pension, cause him difficulty, and cast doubt on the veracity of his claims. He was, in fact, in the hospital of the Provisional Cavalry Brigade, from May 5th, 1865, to June 10th, 1865.

On May 29th, 1866, while Battery L was stationed at Fort Schuyler, New York, he "went home" to Pennsylvania, and went to work as a baker. Due to his disabled condition, one employer admitted to paying him "two-thirds what I paid full hands because he was needy, but I do not consider him worth what I paid him." Unable to make a living, Schneider surrendered himself at Fort Hamilton, New York, on March 25th, 1868, more than 22 months after he had deserted. He was reassigned to Battery L, and by the intervention of Bvt. Lt. Col. Henry W. Closson, though he was no longer in command of Battery L, it was seen to it that Schneider was restored to duty without trial, for former services rendered, and good behavior. He was promptly discharged on disability, on June 11th, 1868, with an invalid pension of $2 per month.

He first made application for a pension increase on June 30th, 1880, and on a questionnaire from the Bureau of Pensions, he answered "No" to question no. 1, "Have you ever married?" His answer to question no. 5, "Have you any living children?" was "None that I know of." An increase to $4 was not granted him until March 30th, 1887. He, like James Beglan, spent the rest of his days in off-and-on correspondence with his pension lawyer, George Lemon, who charged fees that escalated to as much as $25 to process affidavits and declarations to support increases in his pension (for example, an affidavit was obtained from William V. Thompson, and another from George Kelly, still in the service in May 1885 at Fort Douglas, Utah). Upon his death at age 83, on January 8th, 1904, he was receiving $12.

8. William E. Scott–9 February, 1860, Boston. Clerk, age 24. He arrived at Fort Brown with the first group of recruits from depot. His record is without blemish, and though he was initially reported sick at Fort Brown and at Fort Duncan, and then again at Fort Pickens, after participating in the reconnaissance up Santa Rosa Island, he was never listed as sick again. Perhaps he had become "acclimatized." He was never wounded or injured. He was promoted to corporal on September 8th, 1863. He reenlisted on July 11th, 1864, and was advanced to 1st sergeant on October 25th, 1864. He remained in that capacity until his early discharge for disability (heart) at Fort Porter, New York, on March 6th, 1867.

On March 17th, 1867, he married Catharine Carson, whom he had met at New York while on furlough in May of 1865. Throughout his entire enlistment, he had used the alias Scott. His marriage license reveals that he was really William Edward Scott Simmonds.

The couple settled in Worcester, Massachusetts, where he was briefly listed as a cigar maker; the offshoot of a government sponsored program to teach disabled veterans a marketable trade. However, he quickly switched to insurance sales. A son, Robert, was born on February 9th, 1870. On May 25th, 1871, he was in a carriage accident, and was thrown to the street. Assured by a doctor that he would be recovered in a few days, he died on May 30th, age 34.

Having survived the entire war, he was killed in a carriage accident—a familiar ring to those students of history who are aware of the fate of Gen. George Patton.

9. Warren P. Shaw–26 October, 1860, Boston. Shoemaker, age 21. He was in the second group of recruits from depot who arrived at Fort Duncan. He is listed as sick, present, at Fort Pickens at the end of 1861 through February 1862, though was never listed as sick again. He did not desert with the party-goers in New Orleans in 1863, or at any other time. He was never disciplined. He declined to reenlist in 1864, and when his enlistment expired he was discharged on October 24th, 1865.

He returned home to Kingston, New Hampshire, to live with his parents and younger brother. He never married. He worked for the rest of his life in one of the many shoe factories in the New England area at that time and died on September 28th, 1893, aged 52.

10. William V. Thompson–13 September, 1860, Rochester, New York. Farmer, age 22. He was a member of the second group of recruits from depot. He was listed as sick, like so many others, at Fort Pickens, in 1861 and 1862. His pension file includes a detailed medical record, unlike many others. On July 31st, 1861, he was treated for "catarrhus" which is hay fever, despite the fact that it does not appear in the muster roll record. On August 9th, 1861, it was "constipation," and on August 18th it was "contusion," on September 6th, "diarrhea;" and on April 3rd, 1862, "conjunctivitis," or pink eye, an inflammation of the eyelid – hardly malaria, scurvy, or some other serious condition. He reenlisted on July 18th, 1864, and his spotless record continued throughout the Shenandoah Valley Campaign. At Reserve Camp, at Pleasant Valley, he was furloughed, and remained AWOL past its January 12th, 1865, expiration date. He was apprehended and returned on February 7th. As was noted in chapter 15, he had not been paid any installment of his $400 reenlistment bonus, and could never have collected if he had remained a deserter. On October 9th through 14th, 1865, he was treated for syphilis, and on February 4th to April 4th, 1866, for gonorrhea, two diseases delicately not mentioned anywhere in the muster roll records for the entire war period, but undoubtedly being cause while the battery was at Pensacola, New Orleans, and elsewhere. He was discharged promptly on July 18th, 1867 – evidence that he was not required to make up time lost while AWOL.

He returned to Pennsylvania and became a junk dealer. He married Mary A. Bedow, on March 13th, 1872. They had no surviving children. William applied for a pension under the Act of June 27th, 1890, but was rejected repeatedly for claims of disability due to malaria, and sundry other problems, including piles. He was finally allowed $6 per month on March 9th, 1903. This eventually rose as his situation deteriorated, and he was receiving $50 at the time of his death, on September 9, 1916, age 77.

11. Michael White–7 October, 1859, Boston. Laborer, age 21. Not from the fresh group of recruits, but on his first enlistment. He was only once listed as sick, at Pensacola, in the August–October 1862 roll. He was promoted to corporal on September 8th, 1863, and sergeant on March 2nd, 1864.

He is listed as wounded and missing on the February–April 1864 muster roll. Shot in the left breast, he was stunned and left for dead at the Battle of Pleasant Hill. Found alive by the Confederates, the bullet not having torn through him, but stopped by bone, he was marched off – as a prisoner – to Camp Ford, near Tyler, Texas. There, he got no treatment for his wound, and suffered from exposure for the next 13 months. "I was let live or die."

Exchanged prisoners from Camp Ford arrived at New Orleans on May 27th, 1865. He rejoined Battery L on August 15th, 1865, at Fort McHenry, Maryland, where his accounts were "settled up to that date." His first enlistment had expired on October 7th, 1864, and he was officially discharged. He then reenlisted for three years, serving until August 15th, 1868, when he was mustered out as a sergeant, at Fort Porter, New York. His record is spotless; he never indulged in the "vacations" that many took during the war at New Orleans and elsewhere and there is no record of his having been disciplined for any offence.

Freshly discharged, he left the Buffalo area for Ireland, and remained as a small farmer at Nuke, Arthurstown, County Wexford, for the rest of his life. He married Catharine Power on November 27th, 1872, and they had nine children. On July 16th, 1888, he was awarded a pension of $6 per month, which was raised to $8 on March 14th, 1891, and to $12 on May 10th, 1907. He died on September 16th, 1910, age 73.

In summary, all of the recruits of 1860 had left the Battery by August of 1870, and with the exception of Keller, none on the "faithful" list cared to make a lifelong career out of the army.

The Three Cooks

Henry Jefferson, Phillip Ewens, and Virgil Ayres, all served out their full enlistments with perfect records. All remained designated as cooks. None were advanced in the ranks. They were all discharged promptly on schedule, on October 31st, 1866, at Fort Porter, Buffalo, New York. They had come a long way.

Philip Ewens–After his discharge, he remained in Buffalo, married, and worked as a painter or whitewasher for the next 40 years. He never returned to the south. He died on August 5th, 1903, and is buried in Forest Lawn Cemetery.

Virgil Ayres[1244]–Like Ewens, after his discharge he remained in Buffalo, marrying and working there until his death in 1896. He is buried in Forest Lawn Cemetery.

Henry Jefferson–He vowed to never return to New Orleans, though he had relatives still living there after the war. He remained in Buffalo and raised a family. He was last recorded as living there in 1901, when he was 78. His place of burial is not known.

Battery L

The reader may have noted that many of the members were discharged at and remained in, or returned to, the Buffalo area. The story is compelling. The company was transferred from Winchester, Virginia to Fort McHenry, Maryland in August, 1865, and then to Fort Schuyler, New York, in October.

On June 1st, 1866, after participating in the funeral ceremonies of General Scott at West Point, they were notified of a possible special assignment for which they were to be ready to move at a moment's notice. Three days later they left Fort Schuyler for Buffalo. Three other batteries of the 1st Regiment, D, E, and H were also sent.

Their mission was to aid in the capture of a group of the Fenian Brotherhood, some 2,000 strong, who had crossed the Niagara River and occupied the Canadian village of Fort Erie. Their plan was to hold a portion of British Canada and use it as hostage to bargain for the independence of Ireland.

1244. Ayres pension file, National Archives, application no. 642824, certificate no. 433713.

Canadian volunteer forces rose to oppose them and at the Battle of Ridgeway the Canadians were defeated. An alarmed U.S. Government then intervened. The navy cut the raider's means of supply and the whole affair then began to break up. The artillery batteries were left with nothing more than to escort the prisoners captured by the USS *Michigan* from the landing place to the jail, and from the jail to the courthouse.

The only battery designated to remain in Buffalo was Battery L, where they occupied Fort Porter and later Fort Niagara, until October of 1872.

Appendix

1861 Roster
31 December 1860–29 February 1861 Muster Roll
FORT DUNCAN, TEXAS

1. Samuel K. Dawson   Capt. Leave of absence for 2 months, S.O. no. 4, Dept. of Texas, April 30, 1860. Extended for 6 mo. S.O. no. 124 A.G.O. June 20, 1860. Left Co. May 2, 1860.
2. William Silvey   1st Lt. Reg. Adjutant. O. no. 7 Hdqrs. 1st Arty, Ft. Dallas, Florida August 13, 1857
3. James W. Robinson   1st Lt. Det. Svc. on a train to the coast. Left Co. February 14, 1861.
4. Richard H. Jackson   2nd Lt. In Command since February 14, 1861.

1. Lewis Keller	1st Sgt.	1 Dec.'59 Newport, KY	1. Julius Becker	Cpl.	12 Oct.'59 New York
2. Thomas Conroy	Sgt.	1 Jan.'59 San Antonio, TX	2. Alexander J. Baby	Cpl.	10 Feb.'60 Boston
3. Thomas Newton	Sgt.	13 Dec.'58 Ft. Brown, TX	3. David J. Wicks	Cpl.	23 Oct.'59 New York
4. Charles Riley	Sgt.	7 Oct.'59 Boston	4. Andrew J. Beeler	Cpl.	9 Feb.'60 Boston

1. Ludwig Rupprecht   Musician   7 Feb.'60 New York   1. Isaac T. Cain   Artificer   4 Oct.'59 Boston

Privates

1. Ahern, James — 18 Oct.'60 Boston
2. Anglin, Edmond — 19 Oct.'59 New York
3. Beglan, James — 25 Oct.'60 New York
4. Bissel, John — 19 Oct.'59 New York
5. Brook, Thomas — 9 Feb.'60 Boston
6. Brown, William F. — 1 Nov.'59 Boston
7. Brunskill, William C. — 19 Oct.'59 New York
8. Buckley, John — 30 Sept.'58 New York
9. Burke, John — 7 Oct.'60 New York
10. Carroll, Patrick — 17 Dec.'60 Ft. Duncan, TX
11. Casey, John — 15 Oct.'59 New York
12. Connell, Jeremiah — 14 Sept.'59 Ft. Clark, TX
13. Cotterill, Edmond — 22 Sept.'60 Boston
14. Coyne, Owen — 10 Dec.'60 Ft. Duncan, TX
15. Craffy, Patrick — 27 Sept.'60 Boston
16. Creed, William — 27 Sept.'60 Boston
17. Cummings, Patrick — 25 Oct.'60 Boston
18. Curran, Robert — 6 Oct.'57 Newport, KY
19. Demarest, William — 19 Oct.'59 New York
20. Donnelly, Patrick — 1 Mar.'60 Boston
21. Farrell, Bernard — 11 Nov.'59 New York
22. Ferrari, Prosper — 22 Oct.'60 New York
23. Flint, Charles A. — 22 Sept.'60 Boston
24. Flynn, James — 12 Sept.'59 Ft. Clark, TX
36. Kenny, Michael — 8 Feb.'60 New York
37. Kutschor, Joseph — 19 Feb.'60 New York
38. Lighna, Louis — 18 Oct.'58 Ft. Brown, TX
39. McCarthy, James — 4 Oct.'60 Boston
40. McCoy, Daniel — 24 Oct.'60 New York
41. McDonagh, Miles — 17 Sept.'60 New York
42. McGaley, Terence — 18 Sept.'60 New York
43. McLaughlin, Edward — 30 Sept.'58 New York
44. McWaters, James — 16 July '57 Newport, KY
45. Meyer, John — 1 Mar.'60 New York
46. Murphy, John — 21 Feb.'60 Boston
47. Myers, Denis — 28 Sept.'58 Syracuse, NY
48. Nitschke, John G. — 6 Feb.'60 New York
49. O'Sullivan, Michael — 26 Sept.'60 Boston
50. Olvaney, Michael — 25 Oct.'60 New York
51. Parketton, William — 4 Oct.'60 New York
52. Poole, Thomas — 4 Nov.'57 Detroit, MI
53. Reedy, Michael — 1 Jan.'57 San Antonio, TX
54. Roper, John — 22 Oct.'60 New York
55. Schmidt, Heinrick — 26 Oct.'60 New York
56. Schneider, Philip H. — 16 Oct.'58 Ft. Brown, TX
57. Scott, William, E. — 9 Feb.'60 Boston
58. Shaw, Warren P. — 26 Oct.'60 Boston
59. Smith, Joseph — 11 Oct.'59 New York

25. Foley, Christopher	3 Nov.'59 Boston	60. Spangler, Charles	3 Sept.'58 New York
26. Friedman, George	7 Feb.'60 New York	61. Stoll, Andrew	24 Oct.'60 New York
27. Gilroyd, Thomas	26 Aug.'57 Newport, KY	62. Straub, Amelius	8 Feb.'60 New York
28. Golden, James	2 Feb.'60 Boston	63. Thompson, William V.	13 Sept.'60 Rochester, NY
29. Hadley, George	1 Mar.'60 Boston	64. Townsend, Reuben	27 Sept.'60 Boston
30. Hagan, Francis	21 Jan.'59 Ft. Brown, TX	65. White, Michael	7 Oct.'59 Boston
31. Harkins, James	10 Oct.'60 New York	66. Wilkinson, Joseph	4 Feb.'60 New York
32. Hey, Louis	3 Sept.'58 New York	67. William, Henry	28 Sept.'60 Rochester, NY
33. Holland, John	13 Oct.'57 Detroit, MI	68. Wright, Wallace D.	29 Sept.'59 Syracuse, NY
34. Howard, George	25 Oct.'60 New York	69. Wynne, William	9 Oct.'59 Detroit, MI
35. Jackel, Charles	26 Oct.'60 New York		

1862 ROSTER
From 31 December 1861–28 February 1862 muster roll
FORT PICKENS, FLORIDA

1. Henry W. Closson Capt.
2. Franck E. Taylor 1st Lt.
3. Edward L. Appleton 1st Lt.
4. T. K. Gibbs 2nd Lt.

1. Lewis Keller	1st Sgt	1 Dec.'59 Newport, KY	1. David J. Wicks	Cpl.	25 Oct.'59 New York
2. Thomas Newton	Sgt.	13 Dec.'58 Ft. Brown, TX	2. Andrew J. Beeler	Cpl.	9 Feb.'60 Boston
3. Julius Becker	Sgt.	12 Oct.'59 New York	3. Charles Spangler	Cpl.	3 Sept.'58 New York
4. Alexander J. Baby	Sgt.	10 Feb.'60 Boston	4. William Demarest	Cpl.	19 Oct.'59 New York

1. Louis Lighna Musician	13 Oct.'58 Ft. Brown, TX	1. Isaac T. Cain Artificer	4 Oct.'59 Boston

Privates

1. Ahern, James	18 Oct.'60 Boston	22. Flynn, James	12 Sept.'59 Ft. Clark, TX
2. Anglin, Edmond	19 Oct.'59 New York	23. Foley, Christopher	3 Nov.'59 Boston
3. Beglan, James	25 Oct.'59 New York	24. Friedman, George	7 Feb.'60 New York
4. Brook, Thomas	9 July '60 Boston	25. Galavan, Morris	14 Jan.'61 Boston
5. Brown, William F.	1 Nov.'59 Boston	26. Gilroyd, Thomas	26 Aug.'57 Newport, KY
6. Brunskill, William C.	19 Oct.'59 New York	27. Hadley, George F.	1 Mar.'60 Boston
7. Buckley, John	30 Sept.'58 New York	28. Hanney, James	8 Jan.'61 New York
8. Burke, John	17 Oct.'60 New York	29. Holland, John	13 Oct.'57 Detroit, MI
9. Carroll, Patrick	17 Dec.'60 Ft. Duncan, TX	30. Howard, George	25 Oct.'60 New York
10. Casey, Patrick	15 Oct.'59 New York	31. Jackel, Charles	26 Oct'60 New York
11. Connell, Jeremiah	14 Sept.'59 Ft. Clark, TX	32. Kenny, Michael	8 Feb.'60 New York
12. Cotterill, Edmond	22 Sept.'60 Boston	33. Kutschor, Joseph	13 Feb.'60 New York
13. Coyne, Owen	10 Dec.'60 Boston	34. Lanahan, John	9 Jan.'61 Rochester, NY
14. Craffy, Patrick	27 Sept.'60 Boston	35. Mansfield, Charles F.	16 Jan.'61 Boston
15. Creed, William	27 Sept.'60 Boston	36. McCarthy, James	4 Oct.'60 Boston
16. Cummings, Patrick	25 Oct.'60 New York	37. McCoy, Daniel	24 Oct.'60 Boston
17. Curran, Robert	6 Oct.'57 Newport, KY	38. Mc Donagh, Miles	17 Sept.'60 New York
18. Donnelly, Patrick	1 Mar.'60 Boston	39. McGaley, Terence	18 Sept.'60 New York
19. Farrell, Bernard	11 Nov.'59 New York	40. McLaughlin, Edward	30 Sept.'58 New York
20. Ferrari, Prosper	22 Oct.'60 New York	41. McSweny, David	11 Feb.'61 New York
21. Flint, Charles A.	22 Sept.'60 Boston	42. Mc Waters, James	16 July '57 Newport, KY

43. Meyer, John	1 Mar.'60 New York	59. Shapley, Morgan L.	18 Dec.'60 Buffalo, NY
44. Murphy, John	21 Feb.'60 Boston	60. Smith, Joseph	11 Oct.'59 New York
45. Myers, Denis	25 Sept.'58 Syracuse, NY	61. Stanners, Martin	19 Jan.'61 Boston
46. Nitschke, John G.	6 Feb.'60 New York	62. Stoll, Andrew	24 Oct.'60 New York
47. O'Sullivan, Michael	26 Sept.'60 Boston	63. Straub, Amelius	8 Feb.'60 New York
48. Olvany, Michael	25 Oct.'60 New York	64. Thompson, William V.	13 Sept.'60 Rochester, NY
49. Parketton, William	4 Oct.'60 New York	65. Tomson, John	9 Jan.'61 Boston
50. Richards, Franklin W.	6 Feb.'61 New York	66. Townsend, Reuben	27 Sept.'60 Boston
51. Riley, Charles	7 Oct.'59 Boston	67. Ward, Henry H.	8 Nov.'60 New York
52. Roper, John	22 Oct.'60 New York	68. White, Michael	7 Oct.'59 Boston
53. Rupprecht, Ludwig	7 Feb.'60 New York	69. Wilkinson, Joseph	4 Feb.'60 New York
54. Schaffer, William	18 Jan.'61 New York	70. Wilkson, Henry	24 Dec.'60 New York
55. Schmidt, Heinrick	26 Oct.'60 New York	71. William, Henry	28 Sept.'60 Rochester, NY
56. Schnieder, Philip H.	16 Oct.'58 Ft. Brown, TX	72. Wren, Owen A.	11 Dec.'60 Boston
57. Scott, William E.	9 Feb.'60 Boston	73. Wright, Wallace D.	28 Sept.'58 Rochester, NY
58. Shaw, Warren P.	26 Oct.'60 Boston	74. Wynne, William	9 Oct.'57 Detroit, MI

1863 ROSTER

From 31 December 1862–28 February 1863 muster roll
BATON ROUGE, LOUISIANA

1.	Henry W. Closson	Capt.
2.	Franck E. Taylor	1st Lt.
3.	Edward L. Appleton	1st Lt.
4.	James A. Sanderson	2nd Lt.

1. Lewis Keller	1st Sgt.	1 Dec.'59 Newport, KY	1. David J. Wicks	Cpl.	25 Oct.'59 New York
2. Thomas Newton	Sgt.	13 Dec.'58 Ft. Brown, TX	2. Charles Spangler	Cpl.	3 Sept.'58 New York
3. Julius Becker	Sgt.	12 Oct.'59 New York	3. William Demarest	Cpl.	19 Oct.'59 New York
4. Alex. J. Baby	Sgt.	10 Feb.'60 Boston	4. James Flynn	Cpl.	12 Sept.'59 Ft. Clark

1. Ludwig Rupprecht Musician 7 Feb.'60 New York 1. Isaac T. Cain Artificer 4 Oct.'59 Boston

Privates

1. Ahern, James	18 Oct.'60 Boston	69. Mansfield, Herbert E.	24 Feb.'63 Baton Rouge
2. Allen, James H.	17 Nov.'62 Pensacola	70. McCarthy, James	4 Oct.'60 Boston
3. Anglin, Edmond	19 Oct.'59 NewYork	71. McCoy, Daniel	24 Oct.'60 New York
4. Baker, John	15 Nov.'62 Pensacola	72. McDonagh, Miles	17 Sept.'60 New York
5. Beglan, James	25 Oct.'60 New York	73. McEnearny, Corneilus	16 Dec.'62 Pensacola
6. Bieber, Peter	24 Feb.'63 Baton Rouge	74. McGauley, Terence	18 Sept.'60 New York
7. Breen, Michael	24 Feb.'63 Baton Rouge	75. McGuiness, Angus	14 Nov.'62 Pensacola
8. Brooks, William	27 Feb.'63	76. McKinney, John	14 Nov.'62 Pensacola
9. Brown, William F.	1 Nov.'59 Boston	77. McLaughlin, Edward	30 Sept.'58 New York
10. Brunskill, William C.	19 Oct.'59 New York	78. McSweeny, Daniel	11 Feb.'61 New York
11. Buckley, John	30 Sept.'58 New York	79. Meese, Christian	12 Nov.'62 Pensacola
12. Burke, John	17 oct.'60 New York	80. Meyer, John	1Mar.'60 New York
13. Campbell, James	15 Nov.'62 Pensacola	81. Miller, John	12 Nov.'62 Pensacola

APPENDIX

14. Card, Rowland — 12 Nov.'62 Pensacola
15. Casey, John — 15 Oct.'59 New York
16. Champion, Henry — 12 Nov.'62 Pensacola
17. Chase, George — 11 Dec.'62 Pensacola
18. Clinton, Thomas — 24 Feb.'63 Baton Rouge
19. Comfort, James — 12 Nov.'62 Pensacola
20. Connell, Jeremiah — 14 Sept.'59 Ft. Clark, TX
21. Cook, Charles — 24 Feb.'63 Baton Rouge
22. Cotterill, Edmond — 22 Sept.'58 New York
23. Coyne, Owen — 10 Dec.'60 Ft. Duncan
24. Craffy, Patrick — 27 Sept.'60 Boston
25. Creed, William — 27 Sept.'60 Boston
26. Crowley, William — 17 Nov.'62 Pensacola
27. Cummings, Patrick — 25 Oct.'60 new York
28. Deal, Charles — 20 Dec.'62 Pensacola
29. Deering, John — 15 Nov.'62 Pensacola
30. Dickson, Clark — 23 Feb.'63 Baton Rouge
31. Donnelly, Patrick — 1 Mar.'60 Boston
32. Eisele, Joseph — 24 Feb.'63 Baton Rouge
33. Farrell, Bernard — 11 Nov.'59 New York
34. Ferrari, Prosper — 22 Oct.'60 New York
35. Flint, Charles A. — 22 Sept.'60 Boston
36. Flynn, Arthur — 12 Nov.'62 Pensacola
37. Foley, Christopher — 3 Nov.'59 Boston
38. Foote, Edward A. — 24 Feb.'63 Baton Rouge
39. Freidman, George — 7 Feb.'60 New York
40. Fudge, William — 14 Nov.'62 Pensacola
41. Galavan, Morris — 14 Jan.'61 Boston
42. Gibbons, Patrick — 16 Dec.'62 Pensacola
43. Hadley, George F. — 11 Mar.'60 Boston
44. Hall, Benjamin O. — 15 Oct.'62 Pensacola
45. Hanney, James — 8 Jan.'61 New York
46. Harrington, James R. — 27 Feb.'63 Baton Rouge
47. Harrison, George — 25 Dec.'62 Pensacola
48. Howard, Daniel — 24 Feb.'63 Baton Rouge
49. Howard, George — 25 Oct.'60 New York
50. Hubbard, Hiram — 19 Nov.'62 Pensacola
51. Hughes, Benjamin — 27 Feb.'63 Baton Rouge
52. Jackel, Charles — 26 Oct.'60 New York
53. Jessop, Francis — 12 Nov.'62 Pensacola
54. Kastenbader, John M. — 24 Feb.'63 Baton Rouge
55. Kelly, George — 16 Dec.'62 Pensacola
56. Kelley, John — 25 Dec.'62 Pensacola
57. Kenny, Michael — 8 Feb.'60 New York
58. Kenny, Theodore W. — 23 Feb.'63 Baton Rouge
59. Kilburne, Sirenus — 17 Nov.'62 Pensacola
60. Kutschor, Joseph — 13 Feb.'60 New York
61. Lanahan, John — 9 Jan.'61 Rochester, NY
82. Mint, William — 24 Feb.'63 Baton Rouge
83. Montgomery, Solomon J. — 14 Nov.'62 Pensacola
84. Moore, Churchill — 19 Nov.'62 Pensacola
85. Moore, Daniel — 24 Feb.'63 Baton Rouge
86. Moran, John H. — 28 Feb.'63 Baton Rouge
87. Morgan, Frank — 14 Nov.'62 Pensacola
88. Murphy, John — 21 Feb.'60 Boston
89. Myers, Denis — 25 Sept.'58 Syracuse
90. Nitschke, John G. — 6 Feb.'60 New York
91. O'Brien, Sholto — 24 Feb.'63 Baton Rouge
92. O'Sullivan, Michael — 26 Sept.'60 Boston
93. Olvany, Michael — 25 Oct.'60 New York
94. Orcutt, Ephraim — 19 Nov.'62 Pensacola
95. Parketton, William — 4 Oct.'60 New York
96. Parks, William — 17 Nov.'62 Pensacola
97. Pelky, Henry — 25 dec.'62 Pensacola
98. Parslow, Joseph H. — 15 Dec.'62 Pensacola
99. Pfiffer, George — 24 Feb.'63 Baton Rouge
100. Ranahan, Michael — 14 Nov.'62 Pensacola
101. Richards, Franklin W. — 6 Feb.'61 New York
102. Roper, John — 22 Oct.'60 New York
103. Schmidt, Heinrick — 26 Oct.'60 New York
104. Schneider, Phillip H. — 16 Oct.'58 Ft. Brown
105. Scott, William E. — 9 Feb.'60 Boston
106. Shapley, Morgan L. — 18 Dec.'60 Buffalo
107. Shaw, Warren P. — 26 Oct.'60 Boston
108. Smith, Hiram — 15 Dec.'62 Pensacola
109. Smith James H. — 16 Dec.'62 Pensacola
110. Smith, Joseph — 11 Oct.'59 New York
111. Smith, William H. — 16 Dec.'62 Pensacola
112. Stanners, Martin — 19 Jan.'61 Boston
113. Stewart, William — 12 Nov.'62 Pensacola
114.. Stoll, Andrew — 24 Oct.'60 New York
115. Straub, Amelius — 8 Feb.'60 New York
116. Thompson, William V. — 13 Sept.'60 Rochester
117. Tieghe, Michael — 25 Feb.'63 Baton Rouge
118. Tomson, John — 9 Jan.'61 Boston
119. Townsend, Reuben — 27 Sept.'60 Boston
120. Walton, Charles A. — 12 Nov.'62 Pensacola
121. Ward, Henry H. — 8 Nov.'60 New York
122. Welsch, Peter — 19 Nov.'62 Pensacola
123. White, Michael — 7 Oct.'60 Boston
124. Wilder, Joshua E. — 19 Nov.'62 Pensacola
125. Wilkinson, Joseph — 4 Feb.'60 New York
126. Wilkson, Henry — 24 Dec.'60 New York
127. Wilcox, Thomas M. — 15 Dec.'62 Pensacola
128. William, Henry — 28 Sept.'60 Rochester
129. Winn, Abram P. — 16 Dec.'62 Pensacola

62. Lashner, Joseph   16 Dec.'62 Pensacola   130. Winn, Joel T.   20 Dec.'62 Pensacola
63. Leonard, George F.   24 Feb.'63 Baton Rouge   131. Woodruff, Lyman   27 Feb.'63 Baton Rouge
64. Lewery, John   14 Nov.'62 Pensacola   132. Wood, John C.   16 Dec.'62 Pensacola
65. Lighna, Louis   13 Oct.'58 Ft. Brown   133. Wren, Owen A.   11 Dec.'60 Boston
66. Lowry, John   24 Feb.'63 Baton Rouge   134. Wright, Wallace D.   28 Sept.'58 Rochester
67. Mahoney, Thomas   14 Nov.'62 Pensacola   135. Wynne, William   9 Oct.'62 Pensacola
68. Mansfield, Charles F.   16 Jan.'61 Boston

1864 ROSTER
From 31 December 1863–29 February 1864 Muster Roll

1. Henry W. Closson   Capt.
2. Franck E. Taylor   1st Lt.
3. Edward L. Appleton   1st Lt.
4. James A. Sanderson   2nd Lt.

1. Julius Becker Sgt.   12 Oct.'59 New York   1. Michael White Cpl.   7 Oct.'59 Boston
2. David J. Wicks Sgt.   25 Oct.'59 New York   2. Edmond Cotterill Cpl.   12 July '62 New Orleans
3. William Demarest Sgt.   19 Oct.'59 New York   3. Charles E. Walton Cpl.   12 Nov.'62 Pensacola, FL
                                                   4. William E. Scott Cpl.   9 Feb.'60 Boston

1. Ludwig Rupprecht Musician 7 Feb.'60 New York   1. Isaac T. Cain Artificer 4 Oct.'59 Boston
2. Frank Morgan Artificer 14 Nov.'62 Pensacola, FL

Privates

1. Ahern, James   18 Oct.'60 Boston   45. Lowry, John   24 feb.'63 Baton Rouge
2. Anglin, Edmond   19 Oct.'59 New York   46. Mahoney, Thomas   14 Nov.'62 Pensacola
3. Beglan, James   25 Oct.'60 New York   47. Mansfield, Charles F.   16 Jan.'61 Boston
4. Brunskill, William C.   19 Oct.'59 New York   48. McCarthy, James   4 Oct.'60 Boston
5. Brown, William F.   1 Nov.'59 Boston   49. McDonagh, Miles   17 Sept.'60 New York
6. Burke, John   17 Oct.'60 New York   50. McEnearny, Corneilus   16 Dec.'62 Pensacola
7. Bieber, Peter   24 Feb.'63 Baton Rouge   51. McGaley, Terence   18 Sept.'60 New York
8. Breen, Michael   24 Feb.'63 Baton Rouge   52. McGuiness, Angus   14 Nov.'62 Pensacola
9. Campbell, James   15 Nov.'62 Pensacola   53. McKinney, John   14 Nov.'62 Pensacola
10. Card, Rowland   12 Nov.'62 Pensacola   54. Meese, Christian   12 Nov.'62 Pensacola
11. Champion, Henry   12 Nov.'62 Pensacola   55. Meyer, John   1 Mar.'60 New York
12. Chase, George   11 Dec.'62 Pensacola   56. Miller, John   12 Nov.'62 Pensacola
13. Clinton, Thomas   24 Feb.'63 Baton Rouge   57. Montgomery, Solomon J.   14 Nov.'62 Pensacola
14. Comfort, James   12 Nov.'62 Pensacola   58. Moore, Churchill   19 Nov.'62 Pensacola
15. Connell, Jeremiah   14 Sept.'59 Ft. Clark, TX   59. Moore, Daniel   24 Feb.'63 Baton Rouge
16. Cooke, Charles   24 Feb.'63 Baton Rouge   60. Moran, John H   23 Feb.'63 Baton Rouge
17. Coyne, Owen   10 Dec.'60 Ft. Duncan   61. Nitschke, John G.   6 Feb.'60 New York
18. Craffy, Patrick   27 Sept.'60 Boston   62. O'Brien, Sholto   24 Feb.'63 Baton Rouge
19. Creed, William   27 Sept.'60 Boston   63. O'Sullivan, Michael   26 Sept.'60 Boston
20. Crowley, William   17 Nov.'62 Pensacola   64. Olvany, Michael   25 Oct.'60 New York
21. Cummings, Patrick   25 Oct.'60 New York   65. Orcutt, Ephraim   19 Nov.'62 Pensacola
22. Donnely, Patrick   1 Nov.'60 Boston   66. Parks, William   17 Nov.'62 Pensacola
23. Eisle, Joseph   24 Feb.'63 Baton Rouge   67. Parslow, Joseph H.   15 Dec.'62 Pensacola
24. Ferrari, Prosper   22 Oct.'60 New York   68. Pfiffer, George   24 Feb.'63 Baron Rouge
25. Friedman, George   7 Feb.'60 New York   69. Schneider, Philip H.   16 Oct.'60 New York
26. Flynn, Arthur   12 Nov.'62 Pensacola   70. Shaw, Warren P.   26 Oct.'60 Boston
27. Flynn, James   12 Sept.'59 Ft. Clark, TX   71. Smith, James H.   16 Dec.'62 Pensacola

28. Foote, Edward A.	24 Feb.'63 Baton Rouge	72. Smith, Joseph	11 Oct.'59 New York
29. Gibbons, Patrick	16 Dec.'62 Pensacola	73. Smith, William H.	16 Dec.'62 Pensacola
30. Hadley, George F.	1 Mar.'60 Boston	74. Stewart, William	12 Nov.'62 Pensacola
31. Hall, Benjamin O.	15 Oct.'62 Pensacola	75. Straub, Amelius	8 Feb.'60 New York
32. Harrison, George	25 Dec.'62 Pensacola	76. Thompson, William V.	13 Sept.'60 Rochester, NY
33. Howard, Daniel	24 feb.'63 Baton Rouge	77. Tieghe, Michael	25 Feb.'63, Baton Rouge
34. Howard, George	25 Oct.'60 New york	78. Townsend, Reuben	27 Sept.'60 Boston
35. Hughs, Benjamin	27 Feb.'63 Baton Rouge	79. Ward, Henry H.	8 Nov.'60 New York
36. Jackel, Charles	25 Oct.'60 New York	80. Welsch, Peter	19 Nov.'62 Pensacola
37. Jessop, Francis	12 Nov.'62 Pensacola	81. Wilder, Joshua E.	19 Nov.'62 Pensacola
38. Kelly, John	25 Dec.'62 Pensacola	82. Wilkson, Henry	24 Dec.'60 New York
39. Kenny, Michael	8 Feb.'60 New York	83. Willcox, Thomas M.	15 Dec.'62 Pensacola
40. Kenny, Theodore W.	23 Feb.'63 Baton Rouge	84. William, Henry	28 Sept.'60 Rochester, NY
41. Kilburne, Sirenus	17 Nov.'62 Pensacola	85. Woodruff, Lyman	27 Feb.'63 Barton Rouge
42. Kutschor, Joseph	13 feb.'60 New York	86. Wood, John C.	16 Dec.'62 Pensacola
43. Lashner, Joseph	16 Dec.'62 Pensacola	87. Wren, Owen A.	11 Dec.'60 Boston
44. Lewery, John	14 Nov.'62 Pensacola		

1865 ROSTER

From 31 December 1864-28 February 1865

PLEASANT VALLEY, MARYLAND

1. Henry W. Closson Capt.
2. Franck E. Taylor 1st 1st Lt.
3. Edward L. Appleton 1st Lt.

1. William E. Scott 1st Sgt.	11 Jul.'64 New Orleans	1. George Howard Cpl.	Jul. 18,'64 New Orleans
2. William Demarest Sgt.	19 Oct.'59 New York		
3. Charles E. Walton Sgt.	Nov. 12,'62 New Orleans		
4. Lewis Keller Sgt.	Oct. 11,'64 New York		
1. Ludwig Rupprecht Musician	18 Jul.'64 New Orleans	1. Frank Morgan Artificer	14 Nov.'62 Pensacola

Privates

1. Ahern, James	18 Jul.'64 New Orleans	29. Mansfield, Hobart E.	24 Feb.'63 Baton Rouge
2. Beglan, James	18 Jul.'64 New Orleans	30. McGauley, Terence	18 Jul.'64 New Orleans
3. Beiber, Peter	18 Jul.'64 New Orleans	31. Meyer, John	1 Mar.'60 New York
4. Breen, Michael	24 Feb.'63 Baton Rouge	32. Meese, Christian	12 Nov.'62 Pensacola
5. Coyne, Owen	10 Dec.'60 Ft Duncan, TX	33. McKenny, John	14 Nov.'62 Pensacola
6. Craffy, Patrick	18 Jul.'64 New Orleans	34. McGinnis, Angus	14 Nov.'62 Pensacola
7. Creed, William	18 Jul.'64 New Orleans	35. McEnearny, Corneilus	16 Dec.'62 Pensacola
8. Campbell, James	17 Nov.'62 Pensacola	36. Mahoney, Thomas	14 Nov.'62 Pensacola
9. Crowley, William	17 Nov.'62 Pensacola	37. Moore, Churchill	19 Nov.'62 Pensacola
10. Chase, George	16 Dec.'62 Pensacola	38. Miller, John	12 Nov.'62 Pensacola
11. Clinton, Thomas	24 Feb.'63 Baton Rouge	39. Montgomery, Solomon	14 Nov.'62 Pensacola
12. Cooke, Charles	24 Feb.'63 Baton Rouge	40. Orcutt, Ephraim	19 Nov.'62 Pensacola
13. Comfort, James	13 Nov.'62 Pensacola	41. Pfiffer, George	24 Feb.'63 Baton Rouge
14. Donnelly, Patrick	19 Jul.'64 New Orleans		
15. Eisele, Joseph	24 feb.'63 Baton Rouge	42. Schnieder, Philip M.	16 Aug.'63 Baton Rouge
16. Flynn, Arthur	12 Nov.'62 Pensacola	43. Shaw, Warren P.	26 Oct.'60 Boston
17. Foote, Edward A.	24 Feb.'63 Baton Rouge	44. Stewart, William	12 Nov.'62 Pensacola
18. Gibbons, Patrick	11 Dec.'62 Pensacola	45. Stoll, Andrew	18 Jul.'64 New Orleans
19. Hadley, George F.	19 Jul.'64 New Orleans	46. Smith, James H.	16 Dec.'62 Pensacola
20. Howard, Daniel	24 Feb.'63 Baton Rouge	47. Thompson, William V.	18 Jul.'64 New Orleans

21. Hall, Benjamin O. 18 Dec.'62 Pensacola
22. Jackel, Charles 26 Oct.'60 New York
23. Kenny, Theodore W. 23 Feb.'63 Baton Rouge
24. Kilburne, Sirenus 17 Nov.'62 Pensacola
25. Kelley, John 17 Nov.'62 Pensacola
26. Kelly, George 16 Dec.'62 Pensacola
27. Lashner, Joseph 16 Dec.'62 Pensacola
28. Lewery, John 14 Nov.'62 Pensacola

48. Townsend, Reuben 18 Jul.'64 New Orleans
40. Teighe, Michael 25 Feb.'63 Baton Rouge
50. Ward, Henry H. 19 Jul.'64 New Orleans
51. Welsch, Peter 19 Nov.'62 Pensacola
52. Wilder, Joshua 19 Nov.'62 Pensacola
53. Wilcox, Thomas 19 Nov.'62 Pensacola
54. Wood, John C. 15 Dec.'62 Pensacola
55. Woodruff, Lyman 27 Feb.'63 Baton Rouge
56. White, Michael 7 Oct.'59 Boston

Sources

A primary reference for this work is *The War of the Rebellion: A Compilation of the Official Records of the Union and Confederate Armies,* published by the U.S Government Printing Office over a period from 1880 to 1901. It consists of four series and an atlas. Series I, 53 volumes, consists of Union and Confederate reports, and correspondence relating to military operations. Series II, eight volumes, relates to prisoners; Series III, five volumes, contains special reports; and Series IV, four volumes, consists of Confederate papers. The atlas consists primarily of battle maps.

By far the most references taken from the "O.R.," are from Series I, and for brevity, "Ser. I" is omitted in the footnote. By exception, one of the handful of footnotes which includes the series number is from one of the remaining three series, for example: O.R. Ser. III, Vol. 2, pp. x–xx.

Another significant source is the *Official Records of the Union and Confederate Navies in the War of the Rebellion,* published by the U.S. Government Printing Office over a period from 1894 to 1922. There are two series: I, volumes 1–27, and II, volumes 1–3. A typical footnote would be abbreviated: ORN Ser. 1, Vol. 21, p. x.

The framework of the book is based on the service records of Battery L, all available at the National Archives. They fall under RG 391.2.1, Records of the Artillery, 1815–1950. The three most heavily relied on are:

- Regimental Returns, Microfilm Series 727 – Returns from Regular Army Artillery Regiments
- Company Monthly Returns, available by special order
- Company Muster Roll Records, available by special order

Bibliography & Key to Footnotes

Adjutant General of the United States Army. Official Army Register, 1891.

Adjutant General, Massachusetts. *Record of the Massachusetts Volunteers, 1861–1865.* Vols. 1–2. Boston: Wright & Potter, 1868.

Adjutant General's Report, Massachusetts, by year, 1861–1865.

Adjutant General's Report, Supplement. Lansing, Michigan, 1863.

Adjutant General's Report. *Massachusetts Soldiers, Sailors, and Marines in the Civil War.* Vols. 1–8. Norwood, Massachusetts: The Norwood Press, 1930–1932.

Adjutant General's Report. Vol. 1–2. Concord, New Hampshire, 1865–1866.

An Historical Sketch of the 162nd New York Infantry. Albany: Weed Parsons & Co., 1867.

Army of the United States, War Dept. *Revised Regulations.* Philadelphia: J. G. L. Brown, August 10, 1861.

Babcock, Willoughby M., Jr. *Selections from the Letters and Diaries of Bvt. Brig. Gen. Willoughby Babcock of the Seventy-Fifth New York Volunteers.* University of New York Press, 1922.

Bacon, Edward. *Among the Cotton Thieves*. Detroit: The Free Steam Book and Job Printing House, 1867.

Battles and Leaders of the Civil War. 4 volumes. Edited by Robert U. Johnson and Clarence Buell. Century Press, 1887. (Contains articles authored by numerous participants in the war—refer to individual authors.)

Beecher, H. H. *Record of the 114th Regiment, N.Y.S.U.* Norwich, NY, JF Hubbard, 1866.

Benedict, G.C. "History of the 7th Vermont Volunteers." Chapter 21 in *Vermont in the Civil War*. Vol. 2. Burlington, VT: Free Press Association, 1888.

Bennett, A. J. *The Story of the First Massachusetts Light Battery*. Boston, MA: Deland & Barta, 1886.

Benson, S. F. *The Battle of Pleasant Hill Louisiana*. The Annals of Iowa Vol. 7, No. 7. Des Moines, IA: State Historical Society of Iowa. October, 1906.

Benson, S. F. Narrative: *Ben Van Dyke's Escape from the Hospital at Pleasant Hill, Louisiana*. The Annals of Iowa Vol. 7, No. 7. Des Moines, IA: State Historical Society of Iowa. October, 1906.

Bentley, W. H. *History of the 77th Illinois Vol. Infantry*. Peoria: E. Hine, Printer, 1883.

Billings, John D. *Hardtack and Coffee*. Boston: Geo. M. Smith & Co., 1887.

Biographical Directory of the United States Congress. bioguide.congress.gov/

Bissell, George. "Brief History of the Twenty-fifth Regiment, Connecticut Volunteers." *The Twenty-Fifth Regiment, Connecticut Volunteers in the War of the Rebellion*. Includes sections by Ellis, McManus, and Goodell. Rockville, CT: Press of the Rockville Journal, 1913.

Blessington, Joseph P. *The Campaigns of Walker's Texas Division*. New York: Lange Little & Co., 1875.

Board of Artillery Officers. *Instruction for Field Artillery*. Philadelphia: J. B. Lippincott & Co., 1860.

Bosson, C.P. *History of the 42nd Regiment Massachusetts Volunteers*. Boston: Mills Knight & Co., 1886.

Boudrye, Louis N. *Historic Records of the Fifth New York Cavalry* 2nd ed. Albany, NY: S.R. Gray, 1865.

Bowen, J. J. *The Strategy of Robert E. Lee*. New York: Neale Publishing, 1914.

Bowen, James L. *Thirty-seventh Regiment of Massachusetts Volunteers*, Holyoke, MA and NYC: Bryan & Co., 1884.

Bowen, James P. *Massachusetts in the War*. Springfield: Clark W. Bryan & Co., 1889.

Bowen, James R. *Regimental History of the First New York Dragoons*. Lyons, MI: Published by the Author, 1900.

Bringhurst, T. H., and Swigert, F. *History of the Forty-Sixth Regiment, Indiana Volunteer Infantry*. Compiled by the Regimental Association. Logansport, IN: Wilson, Humphries & Co., 1888.

Brown, J. H., ed. *The 20th Century Biographical Dictionary of Notable Americans*. Vol. 3. Boston, 1904.

Buffum, F. M. *Memorial of the Great Rebellion Being a History of the Fourteenth Regiment of New Hampshire*. Boston: Franklin Press, 1882.

Butler, Benjamin F. *Butler's Book*. Boston: A. M. Thayer, 1892.

Byers, S. H. M., *Iowa in War Times*, Des Moines: Coudit & Co. 1888.

Bynum, Tom. *Louisiana Militia Law.* Baton Rouge: Tom Bynum State Printer, 1862.

Carpenter, Geo. N. *History of the Eighth Regiment, Vermont Volunteers.* Boston: Deland & Barta, 1886.

Chamberlaine, Capt. W. W. *Memoirs of the Civil War.* Washington, DC: Press of Byron S. Adams, 1912.

Chase, Salmon P. *Diary and Correspondence of Salmon P. Chase.* Part 4. Letters from George S. Denison to Salmon P. Chase, May 15, 1862, to March 21, 1865. Washington: American Historical Association Annual Report for 1902. (Published in Vol. 2, 1903.)

Cheney, Newell. *History of the Ninth Regiment of New York Volunteer Cavalry.* Poland Center, NY: Martin Merz & Son, 1901.

Childers, Henry H. "Reminiscences of the Battle of Pleasant Hill." *The Annals of Iowa* Vol. 7, No. 7. Des Moines, IA: State Historical Society of Iowa, October, 1906.

Childs, G. W. *National Almanac and Annual Record for the Year 1863.* Philadelphia: G. W. Childs, 1863.

Clark, Orton S. *History of the 116th New York Volunteers.* Buffalo: Mathews & Warren, 1868.

Coffin, Charles Carleton. "The May Campaign in Virginia" Boston: *The Atlantic Monthly* Vol. 14. July 1864.

Congressional Globe. 38th U.S. Congress. 1st Session.

Croffut, W. A. and Morris, John M. *The Military and Civil History of Connecticut, The War of 1861–65.* New York: Ledyard Bill, 1869.

Crowninshield, B. W. "The Battle of Cedar Creek." A Paper read before the Massachusetts Military Historical Society, December 8, 1879. Cambridge, MA: Riverside Press, 1879.

Crowninshield, B. W. *A History of the First Regiment of Massachusetts Cavalry Volunteers.* Boston, MA: Houghton Mifflin, 1891.

Cullum, George. *Biographical Register of the Officers and Graduates of the U.S. Military Academy, at West Point, N.Y. from its Establishment in 1802 to 1890.* Vols. I–IV. Boston and New York: Houghton-Mifflin, Co., 1891–1910.

Custer, Elizabeth B. *The Boy General.* Edited by M.E. Hurt. New York: Scribner's, 1901.

CWSAC (Civil War Sites Advisory Commission). Battle Summaries. The American Battlefield Protection Program (ABPP). National Park Service.

Dahlgren, Rear Admiral John A., *Autobiography*, Peter E. Leubke, Ed., Washington, Naval Heritage and History command (no date).

Dana, Charles, A. *Recollections of the Civil War.* New York: Appleton & Co., 1902.

De Forest, J. W. "Sheridan's Battle of Winchester" *Harper's New Monthly Magazine* Vol. 30. New York, 1865.

DeBray, X. B., *A Sketch of DeBray's Twenty-Sixth Regiment of Texas Cavalry.* Southern Historical Society Papers (SHS), Vol. 13. Richmond, VA: re. J. William Jones, 1885.

Defenders of Port Hudson Association. "Port Hudson – Sketch of its Fortification, Siege and Surrender"– Compiled by the Defenders Association. Southern Historical

Society Papers (SHS), Vol. 14. Richmond, VA: Southern Historical Society, Rev. J. W. Jones, Secretary, 1886.

Denison, Frederic. *Sabers and Spurs*. Central Falls, RI: The First Rhode Island Veteran Cavalry Association, 1911.

Duffey, J. W. *McNeill's Last Charge*. Winchester, VA: G. F. Norton Pub. Co., 1912.

Duganne, A. J. H. *Twenty Months in the Department of the Gulf*. New York: J. P. Robens, 1865.

DuPont, H. H. *The Campaign of 1864 in the Valley of Virginia and the Expedition to Lynchburg*. New York: National Americana Society, 1925.

Duryee Zouaves. *Second Battalion, One Hundred and Sixty-fifth Regt. New York Volunteer Infantry*. Historical Committee, 1905. (See also *Album of the Second Battalion Duryee Zouaves*.)

Dyer, Frederick H. *A Compendium of the War of the Rebellion*. Des Moines, Iowa: The Dyer Publishing Co., 1908.

Early, Jubal A. "Winchester, Fisher's Hill and Cedar Creek." *Battles and Leaders of the Civil War*. Vol. 4. New York: The Century Co., 1884, 1888.

Early, Jubal A. *A Memoir of the Last Year of the War for the Independence of the Confederate States of America*. Toronto: Lowell & Gibson, 1866.

Early, Jubal Anderson. *Autobiographical Sketch and Narrative of the War Between the States*. Philadelphia and London: J. B. Lippincott Company, 1912.

Elliott, J. B., *Scott's Great Snake*. Entered according to Act of Congress in the year 1861. Map. Library of Congress.

Emerson, E. W. *Life and Letters of Charles Russell Lowell*. Boston, MA: Houghton-Mifflin, 1907.

Encyclopedia Britannica. 29 volumes. 1911.

Ewer, J. K. *History of the Third Massachusetts Cavalry*. Historical Committee of the Regimental Association, 1903.

Farragut, Loyall. *The Life of David Glasgow Farragut*. New York: D. Appleton & Co., 1882.

Farrar, S. C. The Twenty Second Pennsylvania Cavalry Pittsburgh, PA: Ringgold Cavalry Association, 1911.

Fitts, James F. "A June Day at Port Hudson" *The Galaxy*. Vol. 2, September 15th, 1866.

Flinn, Frank H. *Campaigning with Banks in Louisiana and with Sheridan in the Shenandoah Valley*. Boston, MA: W. B. Clark & Co., 1889.

Foote, F. H. "Recollections of Army Life with General Lee." *Southern Historical Society Papers (SHS) Vol. 31*. Richmond, VA: Southern Historical Society, Rev. J. W. Jones, Secretary, 1903.

Forstall, R. L. *Population of States and Counties . . . 1790-1990*, Dept. of Commerce, 1996.

Forsyth, G. A. *Thrilling Days in Army Life*. New York & London: Harper & Brothers, 1900.

Fox, William. *Regimental Losses in the American Civil War*. Albany, NY, 1889.

Garnett, Capt. J. M. "Battle of Winchester" *Southern Historical Society Papers* (SHS) Vol. 31. Richmond, VA: Published by the Society, 1903.

Gibbon, John. *The Artillerist's Manual.* 2nd Edition. New York: D. Van Nostrand, 1863.

Gilmor, Harry A. *Four Years in the Saddle* New York: Harper Brothers, 1866.

Gordon, Gen. John B. *Reminiscences of the Civil War.* New York: Charles Scribner's Sons, 1904.

Gould, J. M. *History of the First – Tenth – Twenty-Ninth Maine Regiment.* Portland, ME: Stephen Berry, 1871.

Gracey, S. L. *Annals of the Sixth Pennsylvania Cavalry.* E. H. Butler, Co., 1868.

Grant, U. S. *The Personal Memoirs of U.S. Grant.* Vols. 1–2. New York: J. J. Little & Co., 1885.

Grant, U. S. *Letters to a Friend.* New York & Boston, T.Y. Crowell & Co., 1897.

Haines, Alanson, A. *History of the Fifteenth Regiment of New Jersey Volunteers.* New York: Jenkins & Thomas, 1883.

Hall, Besley & Wood. *History of the Sixth New York Cavalry.* Worcester, MA: E. H. Blanchard Press, 1908.

Hall, Henry, and James. *Cayuga in the Field: A Record of the 19th N.Y. Volunteers, the 3rd New York Artillery, and the 75th New York Volunteers.* Auburn, NY: 1873.

Hanaburgh, D. H. *History of the 128th New York Regiment.* Poughkeepsie, NY: 1894.

Harding, Geo. C. *The Miscellaneous Writings of George C. Harding.* Indianapolis: Carlon & Hollenbeck, 1882.

Harper's New Monthly Magazine. Vol. 0022, Issue 129. February, 1861.

Harper's Weekly, 1861–1865.

Haskin, Wm., L. *The History of the First Regiment of Artillery.* Fort Preble, Portland, ME: B. Thurston & Co., 1879.

Heath, Wm. H. "Battle of Pleasant Hill Louisiana." *The Annals of Iowa,* Vol. 7, No. 7. Des Moines, IA: State Historical Society of Iowa. October, 1906.

Heitman, F. B. *Historical Register and Dictionary of the United States Army from its Organization Sept 29, 1798, to March 2, 1903.* Washington, D.C.: The National Tribune, 1890; reprinted U. S. Government Printing Office, 1903.

Henry, Guy V. *Military Record of Army and Civilian Appointments in the United States Army.* Vols. 1–2. New York: D. Van Nostrand, 1873.

Historical Sketch of the 162nd New York Volunteer Infantry. Albany: Weed Parsons & Co., 1867.

Hoffman, *Camp, Court, and Siege.* Washington, Harper & Brothers, 1837.

Hosmer, James K. *The Color – Guard.* Boston: Walker, Wise & Co., 1864.

Howe, H. W. *Passages from the Life of Henry Warren Howe.* Lowell, MA: Courier-Citizen Co., 1899.

Humphreys, Charles A. *Field, Camp, Hospital and Prison.* Boston, MA: G. H. Ellis Co., 1918.

Hunter, R. M. T. "The Peace Commission of 1865." *Southern Historical Society Papers.* Vol. 3. Richmond, VA: Rev. J. William Jones, D. D., 1877.

Irwin, Richard B. "Military Operations in Louisiana." *Battles and Leaders of the Civil War.* Vol. 3. New York: The Century Company, 1888.

Irwin, Richard, B. *History of the Nineteenth Army Corps.* New York: G. Putnam's Sons, 1893.

Isham, Asa. *An Historical Sketch of the Seventh Regiment, Michigan Volunteer Cavalry.* New York: Town Topics Publishing, 1892.

Jackson, H. N. *Dedication of the Statue to Brevet Major-General William Wells.* Burlington, VT: Privately printed, 1914.

Joel. Bible. (King James Version.)

Johns, Henry T. *Life with the 49th Massachusetts Volunteers.* Washington, D.C.: Ramsey & Bisbee, 1890.

Johnson, R., and Brown, J. H., eds. *The Twentieth Century Biographical Dictionary of Notable Americans.* Vol. 3. Boston: The Biographical Society, 1904.

Johnston, Joseph E. *"Jefferson Davis and the Mississippi Campaign."* Battles and Leaders of the Civil War. Vol. 3. New York: The Century Company, 1888.

Johnston, Joseph E. *"Jefferson Davis and the Mississippi Campaign."* The North American Review. December, 1866.

Kidd, J. H. *Personal Recollections of a Cavalryman.* Ionia, MI: Sentinel Publishing, 1908.

Lee, Wm. O. *Personal and Historical Sketches… of the Seventh Michigan Volunteer Cavalry 1862-1865.* Detroit, MI: 7th Michigan Cavalry Association, 1902.

Leslie, Mrs. Frank. *Frank Leslie's Famous Leaders and Battle Scenes of the Civil War.* New York: Mrs. Frank Leslie Publisher, 1896.

Library of Congress. Abraham Lincoln Papers.

Library of Congress. Name Authority File, http://id.loc.gov/

Little, H. F. *History of the 7th Regiment of New Hampshire Volunteers.* Concord, NH: Evans, 1896.

Long, A. L. *"General Early's Valley Campaign."* Southern Historical Society Papers (SHS) Vol. 3. Richmond, VA: Rev. J. Wm. Jones, 1877.

Longfellow, Henry Wadsworth. *Longfellow's Poems.* New York and Boston: Houghton, Mifflin, 1882.

Longstreet, James. *From Manassas to Appomattox.* Philadelphia: J.B. Lippincott Co. 1896.

Lufkin, E. B. *History of the Thirteenth Maine.* Bridgeton: H. A. Shorey & Son, 1898.

Maddocks, Eldon B. *History of the 26th Maine Regiment.* Bangor, ME: Chas. H. Glass & Co.,1899.

Mahan, D. H. *A Treatise of Field Fortification.* Richmond, VA: West & Johnston, 1862.

Marshall, T. B. *History of the Eighty-Third Ohio Infantry.* Cincinnati, OH: Eighty-Third Association, 1912.

McDonald, Wm. N. *A History of the Laurel Brigade.* Edited by B. C. Washington. Baltimore: Mrs. Kate McDonald, 1907.

McGregor, Chas. *History of the 15th New Hampshire Volunteers.* 15th Regiment Assn., 1900.

McKinney, E. P. *Life in Tent and Field.* Vol. 1. Boston: R. D. Badger, 1922.

McManus, Thomas P. *Battle Fields of Louisiana Revisited a Second Time.* Hartford, CT: Fowler & Miller. Co., 1897. (See also Bissell.)

McMorries, E.Y. *History of the First Regiment, Alabama Volunteer Infantry, C.S.A.* Montgomery, AL: The Brown Printing Co., 1904.

McNeily, Capt. J. S. *Southern Historical Society Papers, Vol. XXXII.* Richmond, VA: The Southern Historical Society, 1904.

"Memoirs." *The Galaxy Magazine.* Review of the *Memoirs of General W.T. Sherman.* September 1885.

Merritt, Wesley. "Sheridan in the Shenandoah Valley" *Battles and Leaders of the Civil War.* Vol. 4. New York: The Century Co., 1884, 1888.

Meyers, Augustus. *Ten Years in the Ranks.* New York: The Sterling Press, 1914.

"Military Training of the Regular Army" *Journal of the Military Service Institute of the U.S.* November 1889.

Miller, F. T., ed., *The Photographic History of the Civil War.* Volumes 1–10. New York: The Review of Reviews, Co., 1911.

Moors, J. F. *History of the Fifty-second Regiment, Massachusetts Volunteers.* Boston: Geo. H. Ellis, 1893.

Morris, Gouverneur. *History of a Volunteer Regiment.* New York: Veteran Volunteer Pub. Co., 1891.

Moyer, Henry P. *History of the Seventeenth Regiment of Pennsylvania Cavalry.* Lebanon, PA: Sowers Printing, 1911.

Munford, T. T. "Reminiscences of Cavalry Operations. - Operations Under Rosser." *Southern Historical Society Papers,* Vol. 13. Richmond, VA: Rev. J. Wm. Jones, 1885.

National Archives, Cantonment Duncan, RG 77, Fortification File, drawer 148, sheet 34;

National Archives, LC Railroad Maps.

Nettleton, Gen. A. B. "The Famous Fight at Cedar Creek." *The Annals of the War.* Philadelphia: The Times Publishing Co., 1879.

New York Adjutant General. *Registers of the 91st NY Infantry.*

New York Infantry. *An Historical Sketch of the 162nd Regiment N.Y. Vol. Infantry.* Albany: Weed Parsons & Co., 1867.

New York Times.

Newlin, W. H. *A History of the Seventy-Third Regiment of Illinois Infantry Volunteers.* Springfield, IL: Reunion Association, 1890.

Nichols, G. W. *A Soldier's Story of His Regiment (61st Georgia).* Jesup, GA: 1898.

Nicholson, A. O. P. *Message of the President to the Houses of Congress, 33rd Congress.* Part 2. Washington, 1854.

Nicolay, John G., and John Hay. *Abraham Lincoln, A History.* 10 volumes. New York: The Century Co.; 1886, 1890, 1904.

NOAA (National Oceanographic and Atmospheric Administration). Historical Map Collection.

Noel, Theophilus. *A Campaign from Santa Fe to the Mississippi of the Old Sibley Brigade.* Shreveport: Shreveport News Printing, 1865.

Noel, Theophilus. *Autobiography and Reminiscences of Theophilus Noel.* Chicago: Theophilus Noel Co., 1904.

Northcott, R. S., *The Annals of War – Union View of Exchange of Prisoners*, Philadelphia: The Times Publishing Co., 1879.

Norton, Henry. *Deeds of Daring, or History of the Eighth N. Y. Volunteer Cavalry.* Norwich, NY: Chenango Telegraph Printing House, 1889.

O'Ferrall, Chas. T. *Forty Years of Active Service*. New York and Washington: Neal Publishing, 1904.

Ould, Judge Robert. "The Exchange of Prisoners." *The Annals of the War*. Philadelphia: The Times Publishing Co., 1879.

Paine, Alanson. *The Fifteenth Regiment of New Jersey Volunteers*. New York: Jenkins and Thomas, 1883.

Park, Robert E. "Sketch of the Twelfth Alabama Infantry." *Southern Historical Society Papers* Vol. 33. Richmond, VA: Published by the Society, 1905.

Peck, Lewis. *A Brief Sketch of the 173rd Regiment, N.Y.V.* New York: Samuel P. Dill, 1868.

Pellet, E. P. *History of the 114th Regiment New York State Volunteers,* Norwich, NY: Telegraph & Chronicle Power Press Print, 1866.

Persec, A., B. M. Norman, and J. H. Colton. *Chart of the Lower Mississippi River.* Library of Congress.

Phisterer, Frederick. *New York in the War of the Rebellion.* Third edition. Albany: J.B. Lyon Co., 1912.

Pickerill, W. N. *History of the Third Indiana Cavalry.* Indianapolis, IN: Aetna Printing Co., 1906.

Pirtle, J. B. "The Battle of Baton Rouge," *Southern Historical Society Papers* (SHS). Vol. 8. Richmond, VA: Published by the Society, 1903.

Plummer, Albert. *History of the Forty-Eighth Regiment, M. V. M.* Boston: Press of the New England Druggist Co., 1907.

Pollard, E. A. *Life of Jefferson Davis,…* Philadelphia, Chicago, St. Louis, Atlanta: National Publishing Company, 1869.

Powell, W. H. *Army List.* New York: L.R. Hamersly & Co., 1896.

Powell, W. H. *Records of Living Officers in the U.S. Army.* Philadelphia: L.R. Hamersly & Co., 1890.

Rodenbough, T. F. *The Army of the United States.* Edited by T. F. Rodenbough and W. L. Haskin. New York: Maynard, Merrill & Co., 1896.

Rodenbough, T. F., Potter, Henry C., Seal, Wm. P. *History of the Eighteenth Regiment of Cavalry Pennsylvania Volunteers.* New York: Wynkoop, Hallenbeck Crawford Co., 1909.

Rodenbough, T.F. *The Army of the United States.* Edited by T.F. Rodenbough and W.L. Haskin. New York: Maynard, Merrill & Co., 1896.

Russell, William H. *My Dairy North and South.* Boston: T. O. H. P. Burnham, 1863.

Sanger, G. P., ed. *The Statutes at Large, Treaties, and Proclamations of the United States of America.* 18 volumes. Boston: Little, Brown and Company, 1789–1873.

Schaff, Morris. *The Spirit of Old West Point.* Boston and New York: Houghton-Mifflin, 1907.

Scharf, Thomas. *History of the Confederate States Navy from its Organization to the Surrender of its Last Vessel.* New York: Rogers and Sherwoods, 1887.

Schouler, W. B. *A History of Massachusetts in the Civil War.* Boston: Dutton & Co., 1868.

Scott, John. *Story of the Thirty Second Iowa Infantry Volunteers.* Nevada, IA: John Scott, 1896.

Scott, R. B. *The History of the 67th Regiment Indiana Infantry Volunteers*. Bedford, IN: Herald Book and Job Printing, 1867.

Scott, Winfield. *Memoirs of Lieut.-General Scott*. New York: Sheldon, 1864.

Selby, P. *The Lincoln-Conkling Correspondence*. Springfield, IL: Illinois Historical Society, 1908.

Sheridan, P. H. *Personal Memoirs of P.H. Sheridan*. Vols. 1–2. New York: C.L. Webster & Co., 1888.

Sherman, William T. "The Grand Strategy of the Last Year of the War." *Battles and Leaders of the Civil War*, Vol. 4. New York: The Century Co., 1884, 1888.

Sherman, William T. *Memoirs of General W. T. Sherman*. Second edition. New York: D. Appleton & Co., 1886.

Shorey, Henry A. *The Story of the Maine Fifteenth*. Bridgeton, ME: Bridgeton News Press, 1890.

Simpson, W. A. "The Second Regiment of Artillery." *The Army of the United States*, Edited by T. F. Rodenbough and W. L. Haskin. New York: Maynard, Merrill & Co. 1896.

Sliger, J. E. "How General Taylor Fought the Battle of Mansfield." *Confederate Veteran*, Vol. 31. Nashville: E. A. Cunningham, October, 1923.

Smith, Daniel P. *Company K First Alabama Regiment*. Prattville, AL: Burke & McFetridge, 1885. Reprint, Jackson County Genealogical Society, 1990.

Smith, W. G. *Life and Letters of Thomas Kilby Smith*. New York and London: G. P. Putnam's Sons, 1898.

Society of Survivors. *History of the Ram Fleet and the Mississippi Marine Brigade*. St. Louis, 1907.

Sprague, Homer B. *History of the 13th Infantry Regiment of Connecticut Volunteers*. Hartford, CT: Case, Lockwood & Co., 1867.

Stanyan, John M. *History of the Eighth New Hampshire Volunteers*. Concord, NH: I.C. Evans, 1892.

Stephens, Alexander. *A Constitutional View of the Late War Between the States*. 2 vols. Philadelphia, PA: National Publishing Co., 1868.

Stephenson, N. W. *An Autobiography of Abraham Lincoln*. Indianapolis: Bobbs-Merrill Co., 1926.

Stevens, General H. "*The Battle of Cedar Creek*." A Paper read before the Military Historical Society of Massachusetts, *December 8, 1879*. Cambridge, MA: Riverside Press, 1879.

Stevens, General H. *A Brief Sketch of the life of General Hazard Stevens*. Boston, MA: G. H. Ellis Co., 1908.

Stevens, Wm. B. *History of the Fiftieth Regiment of Infantry, Massachusetts Volunteer Militia*. Boston: Griffith-Stilling Press, 1907.

Sumner, G. C. *Battery D First Rhode Island Light Artillery in the Civil War*. Providence, RI: Rhode Island Printing Co., 1897.

Sutton, J. J. *The History of the Second Regiment West Virginia Cavalry Volunteers*. Portsmouth, OH: 1892.

Suydam, G. H. *Louisiana and Lower Mississippi Valley Collections*. Baton Rouge: LSU.

Taylor, Richard. *Destruction and Reconstruction.* New York: D. Appleton & Co., 1879.

Tenny, F. A. *War Diary of Luman Harris Tenny.* Cincinnati, OH: Evangelical Publishing House, 1914.

Thustrup, H.L., print. L. Prang & Co., 1886. Library of Congress.

Tiemann, Wm. F. *The 159th Regiment Infantry New York State Volunteers in the War of the Rebellion.* Brooklyn, NY: Wm. F. Tieman, 1891.

Townsend, L. R. *History of the 16th Regiment, New Hampshire Volunteer.* Washington, DC: N.T. Elliot, 1897.

U.S. Army Register, 1864.

U.S. Census Bureau, 1998. *Largest 100 Urban Places.*

U.S. Census, Ft. Duncan, Eagle Pass Post Office. August 4, 1860.

U.S. Congress. *Annals of the 50th U.S. Congress.* 1890.

U.S. Naval History and Heritage Command.

U.S. Navy. *Ordnance Instructions.* Washington: U.S. Government Printing Office, 1866.

U.S. War Department. *Revised Regulations for the United States Army.* 1861.

Walker, A. F. *The Vermont Brigade in the Shenandoah Valley.* Burlington, VT: The Free Press Association, 1869.

Watkins, Sam R. *Co. "AYTCH."* Chattanooga: Times Printing Co., 1900.

Welles, Gideon. "Admiral Farragut and New Orleans." *The Galaxy Magazine.* November 1871.

Welles, Gideon. *Diary of Gideon Welles.* Vols.1–2. Boston: Houghton-Mifflin, 1911.

Wheeler, Joseph. "Bragg's Invasion of Kentucky." *Battles and Leaders of the Civil War.* Vol. 3. New York: The Century Company, 1887.

Whitcomb, Caroline E. *History of the Second Massachusetts Battery of Light Artillery 1861-1865.* Concord, NH: The Rumford Press, 1912.

Whittaker, Frederick. *A Complete Life of General George A. Custer.* New York: Sheldon & Co., 1876.

Wilds, T. F. *Record of the One hundred and Sixteenth Regiment Ohio Volunteers.* Sandusky, OH: L. F. Mack & Bro., Printers, 1884.

Willis, Henry A. *The Fifty-Third Regiment Massachusetts Volunteers.* Fitchburg, MA: Blanchard & Brown, 1889.

Wilson, James H., *Under the Old Flag*, Vol. 1, New York and London: D. D. Appleton & Co., 1912.

Wilson, James Grant, and John Fiske, eds. *Appletons' Cyclopedia of American Biography.* 6 volumes. New York: D. Appleton & Co., 1887–1889.

Wilson, J. T. *Black Phalanx.* Hartford, CT: The American Publishing Co., 1890.

Wood, J. H. *The War.* Cumberland, MO: The Eddy Press, 1916.

Woods, J. T. *Services of the Ninety-Sixth Ohio Volunteers.* Toledo: Blade Printing, 1874.

Woods, W. B. *Cases Argued and Determined in the Circuit Court for the Fifth Judicial Circuit.* Vol. 1. Chicago: Callaghan & Co., 1875.

Woodward, Joseph T. *Historic Record and Complete Biographic Roster Twenty-First Maine Volunteers.* Augusta, Maine: Charles Nash & Sons, 1907.

Index

Page numbers followed by *fig*. indicate illustrations.

A

Abbay, George F., 167
Agnus, Felix, 199
Ahern, James, 5, 20, 36, 44, 69, 88, 235, 315, 414, 423, 432–434, 436–437
Ahern, Mary, 423
Alabama Infantry Regiment, 1st, 170, 182, 195–196, 199, 230; Company K, 188, 195; 7th, 62; 20th, 23 rd, 66; 49 th, 208
Albatross, USS, 123–124, 158, 244
Alden, James, 155
Allen, H. A., 57*fig.*
Allen, Henry W., 288
Allen, James A., 127
Allen, James H., 102, 128, 434
Allendorf, Christian, 88, 102
Anaconda Plan, 18, 39, 39*fig.*, 40, 62, 72, 238
Anderson, J. Patton, 50–51
Anderson, James, 3, 5
Anderson, Richard H., 50, 52, 52n157, 59n176, 61, 64–65, 335–338, 340, 345–346
Anderson, Robert, 14, 33–34
Andrew, John A., 41, 98, 150, 189
Andrews, George L., 232
Anglin, Edmund, 3, 5–6, 44, 69, 314, 343, 432–434, 436
Appleton, Edward L., 53–55, 63–64, 68–70, 83, 86, 88, 101, 103, 108, 126, 135, 138, 162, 233, 241, 244–245, 251, 256–257, 286, 290–291, 295, 308, 312–314, 319, 343–344, 412–413, 416, 422–423, 433–434, 436–437
Appomattox Court House, 42; Station, 421
Aransas Pass, 250
Arizona, USS, 131, 151–152, 244
Arizona Infantry Regiment, 2nd (Confederate), 282;
3 rd, 282
Arkansas, CSS, 91, 95
Arkansas Infantry Regiment, 1st (Confederate), 282; 10th, 182, 196; 15th, 170, 174, 182, 184, 196, 208, 229; 23rd, 185
Arkansas Light Battery, 6th (Confederate), 282
Armstrong, James, 29–30
Army Corps, 5th, 336; 6th, 321–322, 328, 337–338, 342, 349, 352, 355–359, 363, 369–370, 376, 385–387, 389, 396–399, 402, 404, 407, 411, 420; 7th, 299; 8th, 351, 357, 363, 401, 406, 411; 9th, 420; 10th, 336;
13th, 131, 245–248, 253–255, 262, 266, 268–269, 271, 279–281, 301; 16th, 265–266, 280, 283, 287, 291; 17th, 263; 19th, 54, 205n632, 236, 240, 247, 253–255, 257, 262, 264, 266, 271–272, 277, 283, 311n912, 312–313, 319, 323, 328–329, 335, 349, 352, 355–359, 363, 369–370, 376, 378, 385–386, 394, 396–399, 406–408, 411; 22nd, 396; 25th, 42
Army of Northern Virginia, 237, 327, 336, 387; of the James, 319; of the Ohio, 74, 82; of the Potomac, 46–47, 87, 93, 97, 139, 317, 319, 325, 328–329, 334, 336, 378; of the Shenandoah, 328–331; of the Tennessee, 74, 82, 177, 261, 262n781, 283; of the Valley, 352; of Virginia, 97; of West Virginia, 323, 328, 355
Arnold, Lewis G., 15, 25–26, 36, 52, 60, 67–70, 84, 87, 89, 96, 181, 219, 232
Atchafalaya River, 110, 112, 117, 129, 152–153, 160, 170, 263, 311–312
Atlantic, USS, 35
Attakapas Region, 112, 129
Augur, Christopher C.
  defense of Baton Rouge, 118–119, 130
  demonstrations towards Port Hudson, 166
  Godfrey's cavalry and, 157; leave of absence, 238
  Port Hudson attack, 165–168, 170, 172, 174, 176–179, 181–182, 187, 189, 192–193, 196, 206, 211, 216; reserve battery for Sheridan, 329
Averell, William, 323–324, 326, 328–329, 333, 335, 348, 354–356, 360, 361n1049, 362, 369–372
Ayres, Virgil, 243, 245, 252, 309, 316, 430

B

Babcock, Willoughby M., Jr., 184–185, 213, 358
Baby, Alexander J., 3, 55, 57*fig.*, 69, 242, 244, 432–434
Bache, A. D., 40
Bacon, Edward, 191n601, 199, 217–219
Bagby, Arthur P., 282, 288, 300, 303, 309
Bailey, Joseph, 205, 205n632, 302, 306–307, 311
Bainbridge, Edmund C., 140, 180, 185
Baker, Edward D., 53
Baker, Eugene M., 350
Baker, John, 102, 127, 234, 252, 434
Balls' Bluff, Va., 53
Baltic (steamer), 62

Baltimore, Md., 321
Baltimore and Ohio Railroad, 329, 348, 375, 388
Banks, Nathaniel P.
  abandonment of Teche country, 160–161
  arrival in New Orleans, 98–99, 99*fig.*, 151
  command of combined forces, 159–160
  communication with Steele, 300–301
  Confederate forts on Mississippi, 112
  confiscation of horses, 150–151
  Department of the Gulf and, 98, 100, 107, 246, 312
  Grant and, 114, 125, 130, 150, 177, 301–302
  Louisiana constitutional convention and, 260–261
  Mississippi River water passage, 98, 109, 153, 159
  nine-month regiments of, 203n627; prevention of Maryland secession, 41, 97
  refusal to bury Native Guards, 195–196
  request for troops from Grant, 236–237
  response to Fort Butler attack, 226
  retreat to Grand Ecore, 301–305, 307, 310
  Rio Grande expedition, 247–250, 253, 255, 264
  Sabine Crossroads, 270, 270n800, 272, 273n810
  Shenandoah Valley commands, 97
  Sherman's troops and, 300
  on troop conditions, 109
  truce at Port Hudson, 195
  units of, 100–101
Barnard, J. G., 40
Barnes, James, 181
Barrancas Barracks, Pensacola Bay, Fl., 2, 15, 29–30, 59
Barre's Landing, Bayou Cortableau, La., 152, 152*fig.*, 153–154, 161–162, 253
Barrett, Richard, 112, 135, 138
Barrett, William M., 199
Barry, William F., 42
Bartlett, William F., 194, 199
Bates, Edward, 34
Batson, William, 258
Battery Cameron, 47, 56, 60
Battery L
  1861 Roster, 432–433; 1862 Roster, 433–434; 1863 Roster, 434–436; 1864 Roster, 436–437; 1865 Roster, 437–438
  command by Closson, 54, 64, 69–70
  consolidation with Battery K, 317–319, 329, 337, 418
  desertions , 6, 23, 423; in New Orleans, 236, 243–244, 309; in New York, 316
  enlistment of escaped slaves, 245
  furloughs for, 414–415
  Jackson expedition, 216
  letters to families of dead, 127–128, 128*fig.*
  losses of, 257, 312–313, 319, 423
  as mounted battery, 103
  ordered to New York, 312–317
  Ordnance Rifles and, 332
Battle at Southwest Pass, 56
Battle of Baton Rouge, 169; of Bentonville, 420; of Cedar Creek, 388–412; of Champion's Hill, 156; of Corinth, 389; of Fisher's Hill, 367, 368*fig.*, 369–371, 371*fig.*; of Fort Morgan, 339–340, 344–345; of Fredericksburg, 240; of Globe Tavern, 336; of Grand Coteau, 253–254, 254*fig.*, 255, 273; of Irish Bend, 142–143, 143*fig.*, 144, 144*fig.*, 145–148, 156, 264; of Kock's Plantation, 233; of Mansfield (Pleasant Grove), 253, 276–280, 356; of Mansura, 310–311
of New Market, 374–375; of Pittsburg Landing (Shiloh), 74, 82–83; of Plains Store, 167, 170
of Pleasant Hill, 280–283, 283*fig.*, 284, 285*fig.*, 286–291, 292*fig.*, 293–296, 297*fig.*, 298, 429; of Port Royal Sound, 62; of Puebla, 239; of Ridgeway, 431; of Sailor's Creek, 421; of Smithfield Crossing, 340–341, 341*fig.*, 342; of Stickney Farm, 385; of Tom's Brook, 378–382, 382*fig.*, 383–385; of Trevilian Station, 384; of White Oak Road, 420–421; of Williamsburg, 131; of Wilson's Farm, 268–269; of Winchester, 343, 373, 411;
Bayou's Boeuf, 131, 223–225; Bourbeau, 254; Courtableau, 152, 162; Goula, 224; Pierre, 266, 269–270; Plaquemine, 117; Rapides, 264; Sara, 160, 166, 168–169, 175; St. Patrick, 268–269, 272; Teche, 110, 112–113, 113*fig.*, 114, 126, 134, 135*fig.*, 136, 138*fig.*–139*fig.*, 140–141, 152–153, 414
Beall, L. J., 164, 196, 221
Beauregard, P. G. T., 32, 37, 67, 74, 82, 97, 319
Becker, Julius, 3, 55, 60, 63, 69, 83, 127, 162–163, 216, 222–223, 226, 232, 235, 343, 432–434, 436
Bee, Hamilton, 280, 282, 288, 303
Beeler, Andrew J., 3, 53–54, 59–60, 63, 69, 432–433
Beglan, James, 5, 69, 163, 235, 315, 423–424, 427, 432–434, 436–437
Bell, Henry H., 90, 248
Belle Grove Plantation, 391, 398–399, 399*fig.*
Benedict, Lewis, 217, 219, 277, 283, 283*fig.*, 286–287, 290–291, 293–294
Benjamin, George, 3, 5
Benjamin, Judah P., 62, 66–67, 74
Benjamin, William H., 351
Benton, USS, 92, 92*fig.*
Berryville, Va., 323, 337, 345–346, 348–349
Berwick Bay, La., 110, 112, 117, 130–133, 161, 223–225, 247–248
Berwick City, La., 150, 161, 199, 201, 224
Bieber, Peter, 108, 434, 436–437
Biloxi, Miss., 62–63
Birge, Henry, 134, 142, 145, 147, 178–179, 215, 222, 228, 303, 358

Bisland Plantation, 133, 139
Bissell, John, 3, 5–6, 21, 23, 423, 432
Black, Jeremiah S., 24
Black soldiers
    Banks failure to bury, 195–196, 231
    Confederate refusal to exchange former slaves, 230
    funeral processions for, 231, 232*fig.*
    imprisonment of, 116
    Native Guards, 86, 151, 197
    Northern press on, 196–197
    Port Hudson attack, 188, 194, 196, 222–223
    valor of, 197
    white officers and, 151
Blair, Austin, 332
Blair, Henry W., 191, 203
Blair, Montgomery, 33
Blue Ridge Mountains, 344, 375–376
Blunt, Mathew M., 26–27
Boardman, Frederick A., 199
Boonsboro, Md., 321
Boonville, Miss., 82
Boutte Station, La., 225
Bowen, J. R., 374
Bowling Green, Ky., 74
Bowman's Ford, 391
Bradbury, Albert W., 351
Bradley, Theodore, 142, 145–146
Bragg, Braxton
    Army of Mississippi and, 74, 83
    Battle at Southwest Pass and, 56, 61
    bombardment of Fort Pickens, 64–65
    Confederate troops from Vicksburg and, 238
    defeat at Chattanooga, 262, 376n1096
    Department of Alabama and West Florida command, 53, 63
    Department of the Trans-Mississippi command, 62
    end of truce at Fort Pickens, 47
    as Jefferson Davis advisor, 376
    navy raid on the *Judah*, 48
    Pensacola area command, 32, 37–38
    raid on Santa Rosa Island, 50, 52, 55
    request for troops and ammunition, 38, 45–46
    secession of Virginia and, 36
    Van Dorn's cavalry and, 157
    weapons at Pensacola, 331–332
    withdrawals from Pensacola, 65–67
Brannan, John M., 15, 24, 26, 87
Brashear City, La., 130–132, 153–154, 161, 167, 223–225, 247
Braxton, Carter M., 348, 352, 358
Brazos Santiago, Tex., 2, 20, 248
Breckinridge, John C., 11, 74, 95, 95n325, 320–321, 333, 338, 348–349, 352, 352n1012, 354, 357, 360, 362, 367

Breen, Michael, 108, 234, 252, 434, 436–437
Brent, J. L., 282, 287
Brook, Thomas, 3, 20, 36, 44–45, 48, 54, 63, 69, 83, 86, 88, 432–433
Brooklyn, USS, 14, 31, 37, 80, 339
Brooks, William, 108, 235, 242, 251, 434
Brown, Harvey, 33, 35–37, 42–43, 47–50, 52–53, 55, 61, 64, 67, 331
Brown, William F., 3, 20, 36, 44, 69, 88, 234, 242, 308, 315, 414, 432–434, 436
Bruce, William, 88, 10
Brunskill, William C., 3, 21, 44, 48, 54, 63, 69, 83, 86, 89, 102, 108, 127, 163, 235, 243, 252–253, 256–257, 259, 308–309, 316, 343–344, 432–434, 436
Bryan, Michael K., 219, 394
Buchanan, Franklin, 113–114
Buchanan, James, 11, 13–14, 24, 30–31, 33
Buchel, Augustus C., 282, 288, 295
Buckley, John, 54, 162–163, 174, 432–434
Buell, Don Carlos, 74, 82
Bullen, Joseph D., 226
Bunker Hill, W. V., 333, 333*fig.*, 337–338, 338*fig.*, 340–341, 348–349, 355
Burbridge, John Q., 253–25
Burke, John, 5, 69, 127, 315, 414, 432–434, 436
Burnside, Ambrose, 65, 240
Burt, Charles A., 186, 194, 215
Butler, Andrew, 97
Butler, Benjamin F.
    Army of the James, 319
    capture of Hatteras Inlet batteries, 41, 47
    Department of New England and, 73
    Department of the Gulf and, 87, 98
    expedition to New Orleans, 65, 68, 72–78, 100
    expedition to Vicksburg, 80, 90, 92
    Farragut and, 117
    fear of Baton Rouge attack, 87, 93, 95
    fear of New Orleans attack, 95
    as Federal Commissioner of Exchange, 230
    Lincoln's offer to return to New Orleans, 110
    New Orleans loyalty oath and, 151
    occupation of New Orleans, 79, 86, 96–97
    prevention of Maryland secession, 41
    recruitment of northeastern volunteers, 41
    relief from command, 98
    Weitzel and, 110–111
Butte La Rose, La., 117, 133, 151, 201

C

Cailloux, Andre, 231
Cain, Isaac T., 3, 44, 69, 103, 127, 343, 432–434, 436
Cairo, USS, 40, 40*fig.*
Calhoun, USS, 113, 131, 151–152
Callsen, Paul, 425
Cameron, Simon, 34, 272, 274–275, 277, 27

Camp's Barry, Washington, D.C., 314, 317–318;
  Brown, Santa Rosa Island, 50–51; Cooper, Tex.,
  105; Emory, 259; Ford, 275, 279, 279n823, 429;
  Groce, Tex., 225; Kearney, La., 233, 236; Parapet,
  La., 87; Parole, Md., 129; Pratt, 255; Russell,
  411; Stevens, La., 113; Verde, Tex., 16, 105;
  Walton, Fla., 70
Campbell, James, 102, 234, 344, 364, 389, 413, 417,
  422, 434, 436–437
Campbell, John A., 49
Canby, E. R. S., 312, 312n916, 345
Cane River, 303–304
Capehart, Henry, 350
Card, Rowland, 343, 410, 435–436
Carleton, William, 258
Carney, David, 258
Caro, Mary E., 84n288
Carondelet, USS, 305
Carpenter, George N., 182
Carr, Edward, 3, 5
Carr, Gouverneur, 192
Carrion Crow Bayou, 150, 248, 253, 253n753,
  254–255
Carroll, Patrick, 86, 432–433
Carroll's Mill, La., 268, 273
Carrollton, La., 168, 236
Carruth, William W., 140
Carson, Catharine, 428
Carter, J. W., 392, 404
Carter, T. H., 352
Casey, John, 3, 69, 162–163, 227, 235, 432, 435
Casey, Patrick, 433
Cass, Lewis, 11
Castle Pinckney, Charleston, S.C., 12n16
Cavalry
  Battery L support for, 329
  conversion of infantry to, 150–151
  grain and forage for, 109–110, 150, 222, 267, 331
  light batteries and, ix
Cedar Creek, Va., 366, 385, 388–395, 397, 409, 411
Chalfin, S. F., 57*fig.*
Chalmers, James R., 50
Chamberlain, William W., 365
Chambersburg, Penn., 326, 333
Champion, Henry, 102–103, 235, 243, 251, 253,
  426, 435–436
Champion No. 3 (pump boat), 304
Champion No. 5 (pump boat), 304
Chapin, Edward P., 179–180, 192–194, 199
Chapman, George H., 351, 363
Charles Town, W.V., 329, 337–338, 345, 347
Charleston Arsenal, 14
Charleston Harbor, 30
Charlottesville, Va., 386

Chase, George, 102, 127, 163, 225, 235, 243, 252,
  256–257, 259, 308–309, 316, 344, 413, 417, 422,
  435–437
Chase, Salmon P., 34, 97
Chase, William H., 30
Chattanooga, Tenn., 262
Che Kiang (steamer), 105, 236
Cheneyville, La., 311
Chesapeake and Ohio Canal, 375
Chicago Mercantile Battery, 272–273
Chickasaw, USS, 340
Chickering, Thomas E., 199
Childers mansion, 284, 287
Chillicothe, USS, 281, 305
Chilton, R. H., 46
Choctaw, USS, 223
Churchill, Thomas, 278, 280, 282, 284, 287, 291,
  293–294, 300
Citizen Defense Committee (New Orleans), 76
City Belle, USS, 309
City of Vicksburg, CSS, 115
Clack, Franklin, 138, 138n458, 142–143, 272n808
Clark, John H., 282
Clark, Orton S., 176–177, 238, 358
Clark, Thomas S., 191, 191n601, 199, 216–219
Clifton, USS, 131, 133, 133n443, 134, 141, 151–152
Clifton, Va., 345
Clinton, La., 165, 168
Clinton, Thomas, 108, 259, 296, 298, 422, 435–437
Clinton, USS, 249
Closson, Henry W.
  Battle of Fort Morgan, 339–340, 344; of Mansfield, 278; of Mansura, 311; of Pleasant Hill, 283
  as Brevet Major, 236
  Cane River engagement, 303–304
  chief of artillery, 19th Corp, 87, 135, 311n912,
    312–313, 343
  defenses of New Orleans and, 87, 106
  Fort Pickens and, 56, 58–60
  on George Hadley, 426
  letters to families of dead, 127, 128*fig.*
  Port Hudson attack, 222
  reconnaissance on Santa Rosa, 70
  Reserve Artillery and, 262
  Teche campaign and, 147
Cloutierville, La., 303
Cobb, Howell, 11
Cockefair, James F., 283
Cohen, Patrick, 220
Cold Harbor, Va., 177
Colorado, USS, 48, 55
Colorado Infantry Regiment, 10th, 138
Colored Infantry, 77th, 87
Columbia, S. C., 418
Columbia Furnace, Va., 383, 383n1123
Columbus, Ky., 74, 76

Comfort, James, 102, 163, 234–235, 252, 435–437
Conant, Charles E., 221
Confederacy
   capital moved to Virginia, 38
   conscription age, 335
   eastern theater and, 45
   firing on Fort Sumter, 33
   forts on Mississippi, 111*fig.*, 112
   impressment by, 85
   increase in troops, 32, 32*fig.*
   interception of mail from Washington, 29
   prisoner parole and exchange, 129
   raid on Washington, 319–323
   abandons Pensacola, 65–69, 84–85
   Sam Houston refusal of allegiance to, 18n38
   spies in the North, 14, 62
   Sumter and Pickens Truce, 30–32
   surrender of Pensacola naval base to, 29–30
   Union control of coastal points, 18, 39–40, 62
   weapons and, 331–332, 342
   Weed Convention and, 30
Confederate Army of Mississippi, 74, 82–83
Confederate Congress, 31
Confederate Conscription Act, 85
Confederate Department of Southern Mississippi and East Louisiana, 95
Confederate Department of Southwestern Virginia and East Tennessee, 352n1012, 367
Confederate Partisan Rangers, 175, 230, 326, 340
Confederate War Department, 62, 67, 76
Connecticut Infantry Regiment, 9th, 101; 12th, 101, 139, 179, 183, 205, 210, 213–214, 395; 13th, 100, 134–137, 142–143, 145–147, 179, 186*fig.*, 187, 215, 222, 229; 23rd, 101; 24th, 100, 132, 179, 213; 25th, 100, 132, 136, 138, 142–147, 179, 186, 186*fig.*, 215;
26th, 100, 180, 190–191, 217, 219; 28th, 101, 105, 201, 210, 213–214
Connell, Jeremiah, 69, 251, 256, 258, 343, 432–433, 435–436
Conner, James, 385
Connor, James, 258–259
Connors, Patrick, 258, 395
Conroy, Elizabeth, 19
Conroy, Thomas, 3, 19, 54, 60–61, 432
Consolidated Crescent Regiment, 272, 272n808, 288
Cooke, Augustus P., 151
Cooke, Charles, 108, 343, 413, 435–437
Cooper, Pvt., 59
Cordon, John, 201
Corinth, Miss., 67, 74–75, 81–83
Cornay, Florian O., 132–134, 144–146, 304
Corps d'Afrique Regiments, 151, 232, 248, 262, 266, 280. *See also* Louisiana Native Guards
Corpus Christi Pass, 249

Corse, J. M., 300
Cortina, "Marauder" Juan N., 1, 249
Cotterill, Edmond, 102, 108, 234, 244–245, 251, 253, 256, 258, 308, 314, 343, 432–433, 435–436
Cotton, CSS, 112–113
Counselman, Jacob, 318
Covington, USS, 309
Cowles, David S., 191, 199
Cox, Chambers, 208
Coyne, Owen, 69, 424–425, 432–433, 435–437
Craffy, Patrick, 5, 21, 234, 243, 252, 308, 315, 344, 413–414, 432–433, 435–437
Craven, Thomas, 80
Creed, William, 5, 69, 256, 315, 414, 432–433, 435–437
Creole (steamer), 87, 89
Crescent, USS, 323
Cricket, USS, 281, 304
Crook, George, 323, 325, 328–329, 334, 345–347, 349–351, 355–357, 360, 362–363, 366–367, 369–370, 376, 385, 391–394, 396–398, 411
Crowley, William, 102, 127, 163, 235, 243, 252, 256, 259, 308–309, 316, 344, 413, 417, 422, 435–437
Crowninshield, Casper, 350
Crump's Hill, 266
Crusader, USS, 37
Cullum, George, 81
Culpeper, Va., 326, 335
Cummings, Patrick, 5, 20, 36, 44–45, 48, 315, 415, 417, 432–433, 435–436
Cunningham, Edward, 310
Cunningham, Peter, 5
Cupp's Ford, 393–394, 399
Curran, Robert, 54, 69, 83, 88–89, 432–433
Currie, L. D. H., 212
Curtin, Patrick, 258
Custer, George A.
   3rd Brigade, 337, 376n1098, 377
   Battle of Cedar Creek, 390, 393–394, 399, 403, 407, 409; of Tom's Brook, 378–381, 381*fig.*, 383, 383n1123, 384
   destruction of railroads, 418–419
   to Front Royal, 386–387
   march up Shenandoah Valley, 374–375, 377
   near Cedar Creek, 366
   pursuit of Early, 419
     pursuit of Rosser, 378, 380, 385, 399
   Smithfield Crossing, 340–341
   Winchester, 350, 352–353, 353n1018, 354–356, 360–361
Cutshaw, Wilfred E., 347, 377
Cuyler, W. C., 370, 410

D

Dahlgren, John, 332
Dailey, Darby, 258

Dana, Charles, 198–199, 323
Dana, N. J. T., 248
Daniel Webster (steamer), 22, 22*fig.*
Danville Railroad, 420
Dauphin Island, 339
David's Ferry, 309
Davis, Charles H., 40, 90–93, 114
Davis, George W., 258
Davis, Jefferson
    Battle of Pleasant Hill and, 299
    Bragg as military advisor to, 376, 376n1096
    on Butler occupation of New Orleans, 96
    Confederate Government and, 30, 38
    defenses of New Orleans and, 75, 77
    intervention of Confederate forces for Port Hudson, 164, 164n532, 165
    Lee's abandonment of Richmond and, 421
    reinforcements at Mobile, 261
    removal of Bragg from command, 262
    on the secretary of war, 4
    on trials for Federal officers commanding black soldiers, 230
Davis, Theodore, 37
Dawson, Samuel K., 1–2, 7, 11, 18, 20, 26, 44, 48, 54, 432
Day, Nicholas, 135, 215, 331–332
Deal, Charles/Enos, 102–103, 242, 244, 339, 435
DeBray, X. B., 282, 286n838, 288, 295, 303
Decrow's Point, Tex., 250
Deep Bottom, 324, 327, 336
Deering, John, 102–103, 235, 242, 435
Demarest, William, 3, 69, 83, 163, 235, 244, 309, 314, 319, 343, 432–434, 436–437
Denison, George S., 97
Department of Alabama and West Florida, 53, 63, 67
Department of Arkansas, 312
Department of Florida, 67
Department of the Missouri, 73
Department of New England, 73
Department of the Ohio, 73
Department of the Susquehanna, 324–325
Department of Texas, 8, 12, 16
Department of the Gulf, 73, 87–88, 96, 98, 100, 107, 124, 151, 178, 246, 312–313
Department of the Potomac, 73
Department of the South, 67–68, 87
Department of the Trans-Mississippi, 62–63, 67, 148, 164–165
Department of the West, 40, 45–46
Department of Virginia, 41
Department of Washington, 324–325
Department of West Virginia, 320, 324–325, 329, 418–419

Devin, Thomas, 337, 340–341, 350, 352–353, 355, 366–367, 370–373, 375, 378, 380–384, 401–403, 409–410
Di Cesnola, Louis P., 341–342
Diana, USS, 113, 130, 130n429, 140–142, 145, 147
Dickey, William H., 262, 266, 280
Dickson, Clark, 108, 163, 235, 435
Dinwiddie, Va., 420
Dinwiddie Court House, 318
Dix-Hill Cartel, 129
Dodge, Charles E., 3, 5
Donaldsonville, La., 224–226, 232–233, 236
Donnelly, Patrick, 3, 21, 69, 315, 424–425, 425*fig.*, 432–433, 435–437
Douglas, Stephen A., 11
Dow, Neal, 89, 180, 190–191, 199
Draft riots, 169
Du Pont, Henry A., 351
Duane, James C., 35
Dubois, Cesar, 258
Dudley, N. A. M., 180, 192–194, 208, 272
Duffy, Thomas, 3, 5, 60
Duganne, A. J. H., 279n823
Dulany, R. H., 380–381, 383
Duncan, Ashbell F., 350
DuPont, Samuel F., 40, 62, 396, 402
Durland, Coe, 350
Duryea, R. C., 148, 180, 185, 202n626, 210
Duval, Isaac, 351, 362
Dwight, William
    in Alexandria, 160, 168n548
    background of, 131
    Battle of Irish Bend and, 143, 144*fig.*, 146–147; of Mansfield and, 277–278; of Pleasant Hill and, 280, 283, 283*fig.*, 286–287, 289–291, 294; of Williamsburg and, 131
    Port Hudson attack, 124, 156, 168n548, 175–176, 178–179, 182–183, 187–188, 196, 199, 203, 206, 208, 216–220
    pursuit of Taylor, 149
    Teche campaign and, 133–134
    Winchester, 351, 358–360

E

Eagle Pass, Texas, 8, 12
Early, Jubal A.
    advance on Washington, 313, 320–328
    attack on B&O Railroad, 348
    Battle of Cedar Creek, 390–394, 399, 401, 404–406, 406n1196, 407–411; of Fisher's Hill, 369–370; of Smithfield Crossing, 340–342; of Stickney Farm, 385
    Bunker Hill retreat and, 333–334, 337–338, 348
    Custer's campsite and, 390
    escape at New Market, 373–375

Lee requests for divisions, 344–345, 347
Lee's order to turn over command, 419
reconnaissance of Sheridan, 377
reconnaissance of Torbert, 418
reinforcements at Fisher's Hill, 335–337
retreat to Fisher's Hill, 366–367, 369, 386–387
retreat to Mount Jackson, 372–373
retreat to New Market, 411
return of divisions to, 348–349
return of Kershaw to Lee, 411
ruse of Longstreet Corps, 386–387, 408
Shenandoah Valley Campaign, 345–346
Shenandoah Valley resources and, 326–327
Sheridan attack on, 419
Signal Knob observation station, 335
skirmish with Wilson's 3rd Division, 337–338
  Winchester, 352–356, 358, 360, 362–364
Woodstock Races, 383–384
Eastport, USS, 302–304
Ebel, Christopher, 44, 48–49
Echols, John, 419
Edwards, James M., 258
Edwards, Oliver, 389
Egan, John, 318
Eisle, Joseph, 109, 127, 308, 435–437
Elkin's Ford, 299
Ellet, Alfred, 114–115, 158, 263, 265
Ellet, Charles, 40, 91
Elliott, W. L., 82
Emerson, Frank, 268–270, 272, 275
Emma, USS, 309
Emory, William H., 119–120, 130–131, 139–142, 150, 168n548, 225–227, 245, 259, 262, 266, 268, 271, 275–280, 281n830, 283–284, 287, 291, 294, 311–313, 323, 329, 337, 351, 355–356, 359, 366, 386, 394–395, 397, 404
Empire Parish (steamer), 168
Essex, USS, 166, 202
Estrella, USS, 113–114, 131, 134, 151–152
Evans, Clement A., 358, 390, 393, 406, 408
Evens, Phillip, 243, 252, 309, 316
Everett, Charles, 199
Everett's Battery, 91
Ewens, Philip, 245, 430
Ewing (steamer), 51

F

Farragut, David G.
  attack on Mobile, 244
  background of, 77
  expedition to New Orleans, 68, 73, 75, 77–79
  expedition to open Mississippi, 79, 90–93
  Grant on Lake Providence, 154, 154n507
  loss of *Indianola* and, 117–118
  Mobile Bay expedition, 339–340, 345
  occupation of Baton Rouge, 99
  occupation of New Orleans, 79
  plan to redeploy gunboats from New Orleans, 110, 114
  plans to take Vicksburg, 80–81
  Port Hudson attack, 117–119, 121–123, 123*fig.*, 124–125, 166, 175, 182, 206
  raising flag on U.S. Mint, 96
  Red River Campaign and, 158–159
  request for additional troops, 92, 99
  request for coal and rams, 158
  West Gulf Blockading Squadron, 68, 77
Farrell, Bernard, 3, 21, 88, 234, 432–433, 435
Fearing, Hawkes, 179, 183, 210
Feret, James, 228
Ferguson, John, 258–259, 296, 298
Ferrari, Prosper, 5, 69, 256, 315, 415, 432–433, 435–436
Ferris, Samuel P., 210
First Battle of Manassas (Bull Run), 45, 47, 53
First Confiscation Act, 45, 47
Fisher's Hill, Va., 334, 334*fig.*, 335, 337, 343, 366–367, 370, 374, 378–379, 385, 390, 409
Fiske, Stuart W., 133
Fitts, James F., 214
Fitzpatrick, Benjamin, 31
Five Forks, Va., 420
Fleming, Daniel H., 258
Flint, Charles A., 5, 44, 69, 102, 108, 127, 128*fig.*, 432–433, 435
Florida Regiment of Volunteers, 1st, 50
Flowing Spring, W.V., 337
Floyd, John B., 8, 11–14, 331
Flynn, Arthur, 102, 234, 242, 244, 314, 435–437
Flynn, James, 3, 19, 69, 235, 252, 343, 432–434, 436
Foley, Christopher, 3, 21, 36, 69, 83, 234, 433, 435
Foote, A. H., 39, 66, 77, 79, 81, 90, 309
Foote, Edward A., 109, 127, 435, 437
Ford Davidson, Miss., 67n211
Forsberg, Augustus, 352, 353n1018, 354
Forsyth, George A., 389, 407
Fort Barrancas, Fla., 28–30, 32, 36, 36*fig.*, 84, 84*fig.*, 107
Fort Beauregard, Bay Point, S.C., 62
Fort Beauregard, La., 112
Fort Bisland, La., 105, 110, 112, 131n434, 133, 140, 142
Fort Brown, Brownsville, Tex., 1–3, 6, 16, 20, 23, 423, 426–428
Fort Buchanan, 224
Fort Burton, La., 112, 151–152
Fort Butler, La., 112n380, 224–226, 254
Fort Chadbourne, Tex., 424
Fort Clark, Hatteras Inlet, 41
Fort Clark, Tex., 2
Fort Collier, Va., 362
Fort Columbus, N.Y., 423

Fort Crawford, Wi., 138
Fort Delaware, Del., 425
Fort DeRussy, La., 112, 115, 115n390, 152, 159, 263
Fort Donelson, Tenn., 66, 74, 90
Fort Douglas, Utah, 427–428
Fort Drewry, 421
Fort Duncan, Eagle Pass, Tex., 5, 7–9, 9*fig.*, 10, 10*fig.*, 11, 16–20, 23, 244, 423–424, 428
Fort Esperanza, 250
Fort Fisher, Wilmington, N. C., 177
Fort Gaines, Dauphin Island, Ala., 53, 339–340
Fort Garland, Co., 138
Fort Grigsby, 241
Fort Hamilton, N.Y., 22, 35, 39, 253, 427
Fort Hatteras, Hatteras Inlet, 41
Fort Heiman, Ky., 74
Fort Henry, Tenn., 66, 74
Fort Hindman, USS, 281, 304–306
Fort Independence, Boston, 15, 25
Fort Jackson, La., 76–78, 106, 362
Fort Jefferson, Tortugas, Fla., 15, 20, 22, 24–25, 25*fig.*, 26–27, 36, 43, 45
Fort Jesup, 266
Fort Leavenworth, Kan., 2, 17
Fort Macon, Goldsboro, N. C., 425
Fort Marcy, N.M., 138
Fort McHenry, Baltimore, Md., 35, 424, 429
Fort McIntosh, Laredo, Tex., 7, 20, 104–105, 423
Fort McRee, Perdido Key, Fla., 24n48, 28–30, 42, 43*fig.*, 59–61, 67, 84
Fort Monroe, Va., 2, 7, 31, 40–41, 65, 93, 98
Fort Morgan, Mobile Point, Ala., 29, 53, 339–340, 344–345
Fort Moultrie, Charleston, S.C., 12, 12n16, 13–14
Fort Niagara, N. Y., 426, 431
Fort Pickens, Pensacola, Fla., 27*fig.*, 28*fig.*, 43*fig.*
   armament of, 41–43
   batteries of, 56–57, 57*fig.*, 58, 58n171, 58*fig.*, 59
   Battery L at, 43–45, 48, 53–56, 60
   bombardment of Confederate installations, 55–61, 61n184
   change in command at, 67
   Confederate bombardment of, 64–65
   Confederate demands for surrender of, 30
   Confederate raid on, 50–53
   Confederate withdrawals and, 68–69
   garrison of, 2, 13, 15, 24, 26, 29–30
   Judah raid and, 48
   plan of, 27–28
   secret expeditions to, 33, 35–38
   truce at, 31–32
Fort Pillow, Tenn., 77
Fort Porter, Buffalo, N. Y., 423, 425, 429–431
Fort Saratoga, 323
Fort Schuyler, N.Y., 314, 424, 427, 430
Fort Semmes, 250

Fort Smith, Ark., 299
Fort Snelling, Minn., 138
Fort St. Philip, La., 77–78, 106
Fort Stedman, Va., 420
Fort Stevens, Washington, D.C., 321–322, 322*fig.*, 323
Fort Sumter, Charleston, S.C., 2, 13–14, 23, 30–31, 33–34, 37, 46
Fort Taylor, Key West, Fla., 2, 15, 22, 24, 26, 36, 85
Fort Union, N.M., 138
Fort Wagner, Morris Island, S.C., 177
Fort Walker, Hilton Head, S.C., 62
Fortress Monroe, 323–324
Fox, Gustavus, 33–35, 72
Franklin, La., 130, 138, 141–142, 145–148, 256, 259
Franklin, William B., 240–241, 245, 247–248, 250, 253, 255, 257, 262, 264, 266–269, 271, 274–275, 295, 309, 312, 324
Frederick, Md., 321, 327
Fremont, John C., 45, 47
French, William H., 2, 8, 19, 21, 35
Friedman, George, 3, 69, 83, 86–88, 102, 108, 127, 163, 233, 242, 251, 256, 258, 308, 315–316, 433, 435–436
Front Royal, Va., 335, 337, 345, 367, 370–371, 374, 386, 391
Fuchs, Henry, 258
Fudge, William, 102, 234–235, 242, 435
Fuller, Edward W., 133
Fulton (steamer), 84n285
Funsten, O. R., 392

G

Gaines' Mill, Va., 94
Galavan, Morris, 44, 69, 256, 433, 435
Galloway, A. Power, 212
Galveston, Tex., 99–100, 124
Gantt Plantation, 153–154
Gardner, Franklin
   agreement to surrender, 228–229
   Banks request for medical supplies, 220
   command of Port Hudson, 109, 121–122, 164–166
   defense of Port Huron, 208
   Dwight's assault and, 217
   escape plan at Port Hudson, 170, 173
   message on surrender of Vicksburg, 228
   need for men at Port Hudson, 166–168
   refusal to surrender, 206
   ruse to allow escape of Confederate officers, 228–229
   truce at Port Hudson, 175, 195
   on wounded Union soldiers, 220–221
Garnett, James, 360, 363
Gaubaudan, Edward C., 154
General Price, USS, 228

General Quitman, CSS, 116
General Rusk (steamer), 20, 22
Genesee, USS, 124, 166
Georgia Infantry Regiment, 61st, 372, 406
Georgia Regiment of Volunteers, 5th, 50
Gerfie, Henri Louis, 245
Gertrude, USS, 244
Gettsburg, Pa., 237
Getty, George W., 338, 351, 358, 389, 397, 404, 407
Gibbons, Patrick, 102, 234, 256, 258, 308, 314, 343, 413, 417, 422, 435, 437
Gibbons Plantation, 172, 180
Gibbs, Alfred, 340, 350
Gibbs, T. K., 54, 63, 68–69, 83, 433
Gilbert, Benjamin F., 351
Gilmor, Harry A., 326, 340
Gilroyd, Thomas, 83, 86, 89, 433
Glorieta, N.M., 74
Godfrey, Edward S., 157
Golden, James, 3, 26, 44, 433
Gooding, Oliver P., 140–141, 179, 183, 266, 280, 283, 303
Gordon, John B., 334, 341–342, 348–349, 352, 354–358, 362–363, 370, 390–391, 393, 396–398, 404–406, 406n1196, 408, 420, 422
Gorgas, Josiah, 38, 332
Graham, William, 318
Grand Duke, CSS, 133
Grand Ecore, La., 160, 281, 298, 301–302
Grand Gulf, 156, 158–159
Grand Lake, 117, 132–133, 133n441, 141, 224
Granger, Gordon, 339–340, 345
Grant, John V., 351
Grant, Ulysses S.
    appointment of Sheridan, 327–329
    Army of the Tennessee and, 74, 82, 263
    arrival at Bruinsburg, 164
    assault of Richmond, 421
    on Averell's falling back, 371
    Banks and, 114, 125, 130, 150, 301–302, 302n887
    on Banks' plan to capture Shreveport, 304–305
    Banks request for troops from, 236–237
    Battle of Cedar Creek, 410
    Battle of Pittsburg Landing (Shiloh), 74, 81–82
    besieging of Lee in Petersburg, 375–376, 378
    capture of Fort Davidson, 67n211
    capture of Vicksburg, 228
    cooperation with Banks, 154–156, 159–160, 238
    Corinth and, 83
    destruction of railroads, 418–420
    diversionary raids, 324, 327, 336
    on Early, 323, 335
    as general-in-chief, 328
    Grierson's Raid and, 157–158
    Halleck's rewriting of orders, 385
    joint operation with Foote, 66
    on Lake Providence, 154, 154n507
    as lieutenant-general, 261–262
    on merger of army departments, 324–325
    Military Division of the Mississippi, 246
    offer of services to Union, 46
    parole of Lee, 412
    Red River Campaign and, 262
    report of Battle of Winchester, 363–364
    request for Lee's surrender, 421
    review of the 13th Corps, 247
    Rio Grande expedition and, 250
    Siege of Petersburg and, 320
    on sinking of *Indianola*, 118
    surrender of Fort Donelson, 68
    Taylor attacks on supply points, 223
    Teche campaign and, 130
    Texas expedition and, 245, 247, 262n779
    Vicksburg attacks, 154, 156, 159, 177, 198
    Wilson's Raid and, 317
Gray, Henry, 144*fig.*, 145–147, 272n808, 282
Great Cattle Raid, 378
Great Western Railroad, 130
Green, Thomas, 141, 149, 224–226, 254–255, 264, 266, 270, 280, 282, 288, 294, 300
Greenleaf, Halbert S., 153–154
Green's Plantation, 120
Gregg, David M., 336
Gregg, John, 164–165
Gregg's Brigade, 109
Grey Jacket (schooner), 244
Grierson, Benjamin
    Arrival in Baton Rouge, 158*fig*
    Battle of Plains Store, 167, 170
    capture of Confederate steamers, 170
    foraging cavalry, 222–223
    Grant request for, 155–156, 198, 238
    Port Hudson attack, 180, 201
    prisoners taken during raid, 166
    raid through Mississippi and Louisiana, 157–158
Griffiths Plantation, 172, 208
Grimes, Bryan, 390, 406
Grover, Cuvier
    background of, 138–139
    Battery K-L and, 335
    Battle of Cedar Creek, 394, 397
    Battle of Fisher's Hill, 370
    divisions of, 100
    Grand Ecore and, 301
    occupation of Baton Rouge, 99
    occupation of Savannah, 416
    Port Hudson attack, 119–120, 120*fig.*, 126, 168–170, 173–175, 178, 178n576, 179–181, 185–187, 188n592, 189, 205–206, 208, 215
    Port Hudson divisions, 200

pursuit of Taylor, 149–150
reconnaissance of Cedar Creek, 388
Red River Campaign and, 262, 264–266
Rio Grande expedition and, 253
Shenandoah Valley and, 337
Sheridan and, 329
Teche campaign and, 130–143, 145, 147–148
Battle of Winchester, 351, 356, 358–360
Gulf Blockading Squadron, 48, 56, 62

H

Hadley, George, 3, 44, 69, 103, 315, 415, 425–426, 433, 435, 437
Hagan, Francis, 3, 5–6, 21, 23, 244, 423, 433
Hahn, Michael, 260
Hall, Benjamin O., 102, 108, 127, 133n443, 163, 235, 314, 344, 413, 435, 437–438
Hall, William, 144*fig.*
Halleck, Henry W.
   Banks and opening of Mississippi, 98–99, 107, 109, 113, 117
   Banks on need for Nims' Battery, 150
   Banks on Port Hudson siege, 227, 236
   Butler and Pensacola, 87
   as commander of combined departments, 325
   on cooperation between Banks and Grant, 159
   on defeat of Early, 323
   denial of Grant's leave, 246
   failure at Corinth, 74–75, 81–83, 94, 237
   Farragut's request for troops at Vicksburg, 91–93
   on Franklin, 324
   as general-in-chief, 95
   Grant on Banks, 302, 302n887
   Grant on sinking of *Indianola*, 118
   on parole of Confederate officers, 129
   Red River Campaign and, 260–262
   refusal of troops to Lincoln, 94
   rewriting of orders, 385
   Sheridan and, 327–328, 376, 387–388
   Steele and Shreveport, 299
   successes in the west, 94
   Texas expedition and, 238, 240, 247–248, 311
   Vicksburg expedition and, 79
Hallett, Joseph L., 121, 123
Hallonquist, James, 51
Halltown, Va., 338
Hamilton, Andrew J., 98–99
Hamlin, Charles, 313
Hampton, Wade, 318, 336, 378
Hampton Roads, Va., 65, 76–77, 93
Hancock, Md., 326, 335
Hancock, Winfield S., 336, 418
Hanney, James, 44, 60, 252, 433, 435
Hanney, Michael, 54
Hardee, William J., 74, 229
Harkins, James, 5, 44, 48–49, 433

Harper's Ferry, Va., 321, 325, 328–329, 338, 347, 363, 412
Harriet Lane, USS, 34, 79, 100, 110, 116
Harrington, James R., 109, 127, 435
Harrington, Michael, 102
Harrison, George, 102, 108, 234, 315, 435, 437
Harrison, John, 3, 5
Harrison, Thomas, 300
Harrisonburg, Va., 373, 412
Hartford, USS, 81, 92, 92*fig.*, 121–125, 155, 158–159, 175, 339
Hartwell, James A., 258
Haskin, W. L., 310, 426
Hastings, Smith H., 350
Hatch, Edward, 157
Hatteras Inlet, N.C., 40–41, 47
Hawkinsburg, Va., 384
Hay, John, 94
Hayes, Rutherford B., 351, 393, 396
Haynes, James, 5
Haynes Bluff, Mississippi, 157
Hebard, George, 181
Hehn, Henry, 19
Helen (steamer), 84
Henderson's Hill, 264
Henry, William, 69
Herron, Francis J., 236–237
Hesseltine, Charles, 258- 259, 296, 298
Hesson, Michael, 258
Heuberer, Charles E., 60
Hey, Louis, 54, 61, 4
Hildt, John, 52, 57*fig.*, 59n175, 60
Hill, A. P., 365
Hill, Bennett H., 2, 21
Hill, Sylvester, 283
Hiscock, George, 258
History of the 1st New York Dragoons (Bowen), 374
History of the 13th Infantry Regiment of Connecticut Volunteers (Sprague), 215
History of the 15th New Hampshire Volunteers (McGregor), 201, 202n625
History of the 114th Regiment New York State Volunteers (Pellet), 142
History of the 116th New York Volunteers (Clark), 238
History of the Eighth New Hampshire Regiment (Stanyan), 221
History of the Eighth Regiment, Vermont Volunteers (Carpenter), 182
History of the First Regiment, Alabama Volunteer Infantry (McMorries), 195, 229
History of the First Regiment of Artillery, The (Haskin), 426
History of the Laurel Brigade, A (McDonald), 379
History of the Nineteenth Army Corps (Irwin), 118, 124, 148

History of the Sixteenth New Hampshire Volunteers (Townsend), 152
Holcomb, Richard E., 133, 135, 181, 192–193, 215–216
Holland, John, 21, 69, 88–89, 433
Hollyhock, USS, 224
Holmes, Philip, 146, 164, 164n532
Holmes, Theophilus H., 233
Holt, Joseph, 14, 16, 24, 30
Hood, Charles, 258
Hosmer, James K., 133n443, 137n453
Hotchkiss, Jedediah, 362, 390
Houston, Sam, 17–18
Howard, Daniel, 109, 163, 235, 316, 435, 437
Howard, George, 5, 69, 244, 315, 415, 433, 435, 437
Hubbard, Hiram, 102, 215, 242, 252, 435
Hubbard, Lucius F., 283
Hughes, Benjamin, 109, 163, 235, 315–316, 435, 437
Hughes, John T., 222
Hull, Walter C., 351
Humphreys, Charles A., 354
Humphreys, J. M., 395
Hunt, Henry J., 42
Hunter, David, 68, 302, 304, 320–321, 323, 325–327, 327n955, 328–329, 333
Hunter, Robert, 390
Hunter, Sherod, 224
Hupp's Hill, Va., 334, 366, 369, 385, 390, 392–393

I
Iberville (steamer), 166
Illinois, USS, 35
Illinois Cavalry, 2nd, 272; 6th, 157; 7th, 157, 175
Illinois Infantry Regiment, 47th, 283; 49th, 283; 58th, 283, 287, 293; 77th, 273; 94th, 249; 117th, 283; 119th, 283, 287
Illinois State Militia, 46
Imboden, J. D., 230, 355, 361
Imperial (steamboat), 233
Indiana Battery, 1st, 192, 272–273; 9th, 286, 293; 16th 283
Indiana Cavalry, 16th, 272
Indiana Heavy Artillery, 1st, 121, 140, 179n577, 181, 185, 199, 211, 213, 240
Indiana Infantry Regiment, 8th, 249, 397; 21st, 101; 46th, 275, 279; 60th, 273; 67th, 273; 89th, 283, 293
Indiana Light Battery, 3rd, 283; 9th, 283
Indianola, Texas, 18, 20, 250
Indianola, USS, 115, 115*fig.*, 116–118
Iowa Cavalry, 2nd, 82, 157
Iowa Infantry Regiment, 14th, 283, 288–290, 294; 24th, 275; 26th, 249; 27th, 283, 288; 28th, 275; 29th, 223;

32nd, 283, 288–289, 291, 294; 35th, 283
Irish Bend, La., 133, 141–142, 143*fig.*
Iroquois, USS, 81
Irwin, Richard B., 124–125, 133n441, 139, 148, 166, 176, 181, 187–188, 204, 221–222, 225, 229, 231, 240, 264, 275, 279, 294, 301, 311, 387, 404
Island No. 10, 74
Itasca, USS, 81

J
J. A. Cotton (steamer), 110, 133
Jackel, Charles, 5, 60, 69, 256, 259, 308, 316, 344, 413, 417, 422, 433, 435, 437–438
Jackson, C. M., 229
Jackson, John K., 50–51
Jackson, Richard H., 19–20, 26, 44, 49, 54, 56, 57*fig.*, 70, 104, 432
Jackson, Stonewall, 93
Jackson, William L., 358, 378n1106, 380–381, 384
Jackson Crossroads, 222
Jackson Railroad, 80
Jaecke, Daniel, 4–5
James River, 319, 419
Jeanerette, La., 105, 149
Jefferson, Henry, 243, 245, 252, 309, 316, 430
Jerrard, Simon G., 215
Jessop, Francis, 103, 234, 252, 315–316, 435, 437
John Warner, USS, 309
Johns, Henry T., 178, 193–194
Johnson, Benjamin W., 170, 173, 184, 208
Johnson, Bradley T., 326, 333, 335, 340, 358, 378n1106, 380–381, 383
Johnson, Reverdy, 97
Johnston, Albert Sidney, 62, 66–67, 76
Johnston, Joseph E., 74, 156–157, 164, 164n532, 165, 177, 198, 223, 229, 261, 420
Jones, Charles E., 244
Jones, Samuel P., 65, 67, 71, 79
Jones, Thomas M., 71
Joseph Whitney (steamer), 25
Judah (schooner), 47–48, 50
Juliet, USS, 304

K
Kastenbader, John M., 108, 163, 235, 242, 435
Kautz, August, 317
Kearneysville, W. V., 338, 340
Keith, John A., 199
Keller, Lewis, 2, 44, 60, 69, 162, 226, 245, 251, 253, 319, 343, 415, 426–427, 432–434, 437
Kelly, George, 103, 242, 308–309, 428, 435, 438
Kelly, John, 103, 163, 234–235, 308, 344, 364, 413, 435, 437–438
Kennebec, USS, 244, 339
Kenny, Joseph, 5

Kenny, Michael, 4, 54, 69, 127, 314, 343, 413, 417–418, 432–433, 435, 437
Kenny, Theodore W., 108, 435, 437–438
Kentucky Infantry Regiment, 19th, 273
Kernstown, Va., 74, 325, 334, 366, 411
Kershaw, Joseph B., 337–338, 345–347, 373, 375–377, 385, 391–395, 397–398, 402, 404–405, 408, 411
Key West Barracks, 2, 25
Keyes, Erasmus, 34
Kidd, A. J., 394
Kidd, James H., 332, 350, 361, 374, 376n1098, 380–381, 383, 399
Kilburne, Sirenus T., 103, 426, 435, 437–438
Kimball, William K., 143, 178–179, 211
Kineo, USS, 124, 166
Kingman, John W., 203
Kingsley, Amos N., 258
Kingsley, Thomas G., 191
Kinney, Joseph, 4, 287
Kinsman, USS, 113
Kirby, Edmund, 48, 54
Kitching, J. Howard, 395–396
Knox, S. L., 229–230
Kutschor, Joseph, 4, 21, 44, 69, 86, 163, 225, 242–243, 251, 253, 256, 308, 315, 432–433, 435, 437

L

La Grange, Tennessee, 157
Lacey Springs, Va., 375
Lafourche River, 112
Lake Providence, 154
Lanahan, John, 44, 234, 433, 435
Lancaster, USS, 158
Landram, W. J., 269, 272–273, 275
Lane, Walter P., 282
Langdon, Loomis L., 2, 7, 60
Larrabee, Thomas, 258
Lashner, Joseph, 103, 436–438
Laughlin, Edward M., 242
Laurel Brigade, 377, 380, 383, 392
Laurel Hill (steamer), 126, 130–131, 131*fig.*, 133, 133n443, 134
Lawler, Michael K., 253, 311
Lay, George W., 13
Lee, Albert L., 257, 262, 264, 266–270, 272, 279
Lee, Fitzhugh, 318, 337–338, 345, 352, 358, 360n1047, 362, 377n1100, 422
Lee, Paul Lynch, 229
Lee, Robert E.
　abandonment of Richmond, 421
　Deep Bottom front, 324, 327, 375
　defeat at Gettysburg, 237–238, 333
　Department of Texas, 8
　Fort Stedman breakout plan, 420
　Great Cattle Raid and, 378
　movement towards Washington, 319–320, 322
　occupation of Sharpsburg, 333
　reinforcements for Early, 335–336, 375–376
　requests for troops from Early, 344–345, 347, 411
　Shenandoah Valley resources and, 327
　surrender of, 412, 421–422
Lee, Samuel P., 80
Leetown, W. V., 338, 340
Lemon, George, 427
Leonard, George F., 108, 234, 436
Lewery, John, 103, 234, 256, 258, 308, 315, 343, 413, 417, 422, 436–438
Lexington, USS, 223, 281, 305, 307
Liddell, John R., 300
Lighna, Louis, 3, 19, 234, 236, 242, 252–253, 253n752, 432–433, 436
Light House Point, 318
Lincoln, Abraham
　Battle of Cedar Creek, 410
　blockade of the South, 23
　call for volunteers, ix, 45–46
　caution to Banks on excessive *impedimenta*, 109–110, 150, 267, 331
　concern with Texas, 238–239, 260, 311
　concerns of Early reinforcements, 376
　discovery of quasi armistice, 31, 33
　Early's raid and, 322–323
　expeditions to Sumter and Pickens, 33–35
　inauguration of, 11, 21, 31
　lack of general-in-chief, 93
　Proclamation of Thanks, 75
　re-election campaign, 364–365, 411
　report of Third Battle of Winchester, 363
　request for troops from Halleck, 94
Lincoln, James M., 258
Little North Mountain, 367, 369
Livingston, Alexander, 3, 5
Livingston, Rhett, 351
Locke, M. B., 182–184
Loeb, Sigmund, 88, 102
Logan, John L., 165, 168
Logan, Lloyd, 388, 388*fig.*
Loggy Bayou, La., 281, 281*fig.*, 299–300, 303
Lomax, Lunsford L., 348, 352, 358, 360n1047, 367, 369, 373, 378, 378n1106, 379, 379*fig.*, 381, 383–384, 391, 406n1196, 418–419
Long, A. L., 421
Longstreet, James, 386–387, 408, 420, 422
Loudon Valley, Va., 412
Louisiana, CSS, 77–78
Louisiana Artillery, 100–101
Louisiana Battalion, 8th, 79
Louisiana Battery, 6th (Confederate), 282, 287
Louisiana Cavalry, 1st, 254
　Company A, 101, 179

Company B, 101
Company C, 100
Company E, 119
Louisiana Cavalry, 2nd, 132–133, 253, 426
Company B, 112
Louisiana Heavy Artillery, 12th, 175
Louisiana Infantry Regiment, 1st, 86, 101, 133–135, 138, 147, 179, 183, 185, 210, 215, 236; 2nd, 100, 180, 193, 199, 216; 12th (Confederate), 164; 28th (Confederate), 145–146, 272n808, 282, 288
Louisiana militia, 76
Louisiana Native Guards, 1st, 75–76, 101, 175–176, 179, 185, 187–188, 195–197, 222n679, 231, 248; 2nd, 101, 194, 196–197; 3rd, 101, 175–176, 179, 185, 187–188, 195, 197, 222n679, 231; 6th, 232; 7th, 232; 8th, 232; 9th, 232; 10th, 232;16th, 248
Louisiana Partisan Rangers, 9th, 170, 196
Louisiana Regiment of Heavy Artillery, 1st, 86
Louisiana State Penitentiary, 107*fig.*
Louisiana Volunteer Cavalry, 2nd, 319
Louisiana Volunteers, 27th, 79
Louisville, USS, 305
Lovell, Mansfield, 67, 76–77, 79
Lowell, Charles R., 334, 350, 352–355, 366, 374–375, 380–383, 389, 394, 400, 403, 407, 409
Lowry, John, 108, 234, 242, 252, 315, 436
Lucas, Thomas J., 268–269, 272, 281, 283, 295, 310–311
Lull, Oliver M., 199
Lundenberg, Henry, 19
Luray Valley, 335, 367
Lyle, O. P., 185
Lynch, Thomas, 258
Lynch, William F., 283, 287
Lynchburg, Va., 322, 418–419

M

Mack, Albert G., 175, 180
Madison, George T., 280
Magruder, John B., 2, 99–100
Mahone, William, 318
Mahoney, James, 234
Mahoney, Thomas, 102, 436–437
Maine Battery, 1st, 101, 140, 180, 185, 359, 398
Maine Infantry Regiment, 12th, 100, 132, 179, 187, 260; 13th, 101, 248–249, 278, 283; 14th, 100, 180, 190, 218; 15th, 87, 101, 103, 248–249, 273, 277, 283; 21st, 180, 193–194, 216; 22nd, 100, 135, 138, 144*fig.*, 161, 200, 205, 210, 215; 24th, 190; 26th, 99–100, 119, 142, 145–147, 161, 200, 215, 331; 28th, 217, 219, 29th, 277, 283, 286, 289–290; 30th, 283, 291
Major, James P., 223–224, 282, 288, 294, 303, 309
Mallory, Stephen R., 31, 77, 79, 89, 107n368
Malvern Hill, 94
Manassas, CSS, 56, 56*fig.*, 77–78

Manassas Gap, 335, 369
Mansfield, Charles F., 44, 69, 86, 89, 102, 108, 127, 163, 235, 243, 252, 256–257, 433, 436
Mansfield, Herbert B., 108, 434, 437
Mansfield, La., 266, 269–270, 276–279
Mansura, La., 311
Marcy, George O., 351
Maria Wood, USS, 70
Marksville, La., 310
Marland, William, 255
Marshall J. Smith (schooner), 244
Martin, Andrew, 258
Martin, Joseph W., 381–382, 394
Martinsburg, W. V., 321, 326, 338, 348–349, 352, 355, 388
Mary (steamer), 84
Mary T, CSS, 133, 152
Maryland Cavalry, 1st, 318; 2nd, 326
Massachusetts, USS, 42
Massachusetts Battery, 2nd (Nims'), 91, 92*fig.*, 100, 109, 119, 122, 132, 143, 146–147, 149, 161, 210, 236, 240, 255, 257, 272–273, 275, 383; 4th, 180, 199, 257; 6th, 101, 188, 257; 13th, 258, 296, 296n875, 312–313
Massachusetts Cavalry, 2nd, 375
Company A, 100
Company B, 101
Massachusetts Cavalry, 3rd, 151, 201, 270, 272; 31st, 272
Massachusetts Infantry Regiment, 2nd, 150; 4th, 101, 201, 210, 212, 212n648, 232; 26th, 101; 30th, 101, 167, 180, 210; 31st, 140, 179, 210, 212; 37th, 333, 359; 38th, 140, 175, 179, 185, 199, 210–212; 41st, 98, 151, 200; 42nd, 99–100, 185; 48th, 167, 180, 192–193, 208, 208n637, 217, 219–220; 49th, 106, 167, 180, 193–194, 199, 216
Company G, 166
Massachusetts Infantry Regiment, 50th, 180, 192, 208, 217, 219, 232; 52nd, 100, 132, 137n453, 153, 161, 200, 222; 53rd, 140, 179, 185, 210–212, 226; 54th, 177
Massachusetts Infantry Regiments, 73
Massachusetts Mounted Rifles, 151
Massachusetts Mounted Rifles, 41st, 161
Massachusetts Unattached Calvary, 2nd, Company B, 100
Massanutten Mountain, 367, 369, 390, 408, 418
Matagorda Island, Tex., 250
Matamoras, USS, 249
Matamoros, Mexico, 125, 259
Mathews, Johnny, 184
Maxey, Samuel B., 164
Maynadier, William, 318
McCarrick, John O., 258

McCarthy, James, 5, 21, 36, 69, 89, 234, 308, 432–434, 436
McCausland, John, 321, 326, 333, 335, 338, 355, 361
McClellan, George, 46–47, 72–73, 75, 94, 97, 345, 364–365, 411
McClellan, USS, 55, 248
McClernand, John, 154, 301, 309, 311
McCostello, Michael, 258
McCoy, Daniel, 5, 44, 69, 89, 108, 127, 432–434
McCoy, James, 21, 36
McCrea, Tulley, 318
McCulloch, Benjamin, 16
McDonagh, Miles, 69, 163, 235, 315, 414, 417, 432–434, 436
McDonald, John, 258
McDonald, William, 379
McDonough, Miles, 5, 44
McEnearny, Cornelius, 102, 163, 235, 252, 434, 436–437
McFarland, W., 57*fig.*
McGauley, Terence, 5, 69, 251, 415, 415n1226, 432–434, 436–437
McGilvray, John, 353, 370, 410, 412–413
McGinnis, Angus, 103, 108, 253, 253n753, 437
McGregor, Charles, 201
McGuiness, Angus, 434, 436
McIntosh, John B., 346, 351–352, 363
McInturff's Ford, 390
McKean, William, 55–56, 59, 78, 331
McKenny, John, 103, 316, 344, 413, 437
McKenzie, James, 5
McKerrall Plantation, 143, 147
McKinney, John, 234, 434, 436
McLaflin, Edward, 211, 213
McLaughlin, Edward, 21, 69, 127, 432–434
McLaughlin, John, 258
McMillan, James W., 277, 280, 283, 287, 291, 395, 397
McMorries, E.Y., 195, 229
McNamara, Edward F., 318
McNeill, John, 230, 372n1083
McNeily, J. S., 408
McPherson, James B., 70, 262, 262n781, 264–265
McSweeney, Daniel, 44, 256, 433–434
McWaters, James, 44, 69, 86, 89, 432–433
McWilliams Plantation, 133, 135
Meade, George Gordon, 237, 317–318, 324, 422
Meese, Christian, 102, 163, 234–235, 252, 434, 436–437
Meherrin Station, 318
Meigs, John R., 412
Meigs, Montgomery C., 25, 34–35, 412
Merle, Francis, 5
Merrill, Alfred K., 258
Merritt, Wesley, 328–329, 334, 337–338, 340–342, 349–350, 352–354, 356, 360, 362, 366–367, 373, 375, 377, 379–382, 383n1123, 386–387, 389, 393, 398–400, 402, 409, 412
Mervine, William, 48
Meyer, John, 21, 36, 44, 314, 343, 412, 417, 422, 432, 434, 436–437
Meyers, David, 36, 44
Michigan, USS, 431
Michigan Brigade, 1st, 399–400, 400n1180
Michigan Cavalry, 1st, 354, 374; 2nd, 82, 328, 389; 6th, 201, 361, 380, 382–383; 7th, 353–354, 381–382, 393–394
Michigan Infantry Regiment, 6th, 80, 95, 100, 180, 190, 199, 216–218
Middle Department, 324–325
Middle Military Division, 54, 328–330, 330*fig.*, 364, 416, 419
Middletown, Va., 366, 385, 389, 396*fig.*, 398, 400–401
Miles, W. R., 167, 174–175
Miles' Legion, 229
Milford Creek, 374
Military Division of the Mississippi, 246, 262n781, 311
Mill Creek, 388
Mill Springs, Ky., 74
Miller, George, 258
Miller, John, 102, 235, 434, 436–437
Miller's Lane, 405, 408
Miller's Point, 133n441
Milliken's Bend, 223
Minnesota, USS, 37
Minnesota Infantry Regiment, 5th, 283
Mint, William, 108, 234, 242, 435
Misener, James B., 258
Mississippi, CSS, 77, 79
Mississippi, USS, 77–78, 123, 123*fig.*, 124, 124n417
Mississippi Battery, 167
Mississippi Infantry Regiment, 1st, 182, 184; 6th, 121, 164; 9th, 50, 66; 15th, 164; 21st, 408; 39th, 187, 195–196
Mississippi Marine Brigade, 115, 158, 263, 265
Mississippi River, 39–40, 55, 73, 98, 111*fig.*, 112, 117, 153, 156, 158, 160, 233, 311
Missouri Battery, 6th, 273, 283
Missouri Cavalry, 6th, 272
Missouri Infantry Regiment, 24th, 283–284, 288, 288n845, 289–290; 33rd, 283
Missouri Light Artillery, 1st, Company M, 315–316
Mobile, Ala., 88, 93, 230, 261–262, 339
Mobile and Ohio Railroad, 82
Molineaux, E. L., 119, 122, 146, 236
Monett's Bluff, 303
Monocacy River, 321–322
Monocacy Station, 327–328

Monongahela, USS, 124, 166, 175, 206, 249, 339
Montgomery, Solomon V., 102, 163, 258, 308, 314, 435–437
Moore, Alpheus S., 350, 387, 393, 401, 403
Moore, Churchill, 102, 259, 309, 316, 344, 413, 417, 422, 435–437
Moore, Daniel, 108, 163, 235, 256, 315, 435–436
Moore, Dennis, 234
Moore, Risdon M., 283
Moore, Thomas Overton, 67, 75–77, 245
Moors, J. F., 132, 222
Moran, John H., 109, 163, 234–235, 242, 244, 252, 308, 315–316, 435–436
Moreno, Francesco, 107, 107n368
Morgan, Frank, 102–103, 234, 256, 435–437
Morgan, Joseph S., 161, 168–169, 199, 215
Morgan's Ferry, 223
Morganza, La., 160, 160*fig.*, 312
Morning Light, USS, 130
Morris, John, 4–5
Mortar Flotilla, 78, 88, 91, 93, 166
Morton, John E., 180, 185
Mosby, John S., 230, 375, 412, 414
Moseley, B., 287
Moss Plantation, 269
Mound City, USS, 305
Mount Airy, 335
Mount Jackson, 370–373, 384
Mount Pleasant, La., 226
Mouton, Alfred, 140–141, 147, 154, 222, 270, 272, 272n808, 278, 288
Mower, Joseph A., 263–264, 281, 283, 286–287, 291, 311
Moyer, Henry P., 376
Mt. Carmel Cemetery, 398–399
Mt. Olive, 378–379
Mt. Pleasant Road, 217–220
Mumford, William, 96, 230
Munford, Thomas, 362, 380, 383, 383n1123, 384
Murphy, Edward, 258
Murphy, John, 4–6, 69, 234, 432, 434–435
Murphy, Michael, 4–5, 19, 23, 23n47
Murtaugh, John, 3–4
Mustang Island, Tex., 249
Myers, Denis, 69, 83, 108, 242, 432, 434–435
Myers, John, 4

N

Nashville, Tennessee, 74
Nassau, USS, 89
Natchez (steamer), 107
Natchitoches, La., 161, 265, 265*fig.*, 266–267, 299–300
Native Guards, 86
Navy Bureau of Ordnance, 332
Neaffie (steamer), 51, 52n157, 59, 59n176

Nelms, CSS, 59, 59n176
Nelson, John A., 176, 178–179, 187–188, 356, 358
Nelson, William, 352
Neosho, USS, 281, 305–306
Nerson's Woods, 143, 147
Nettleton, A. Bayard, 351
New Bern, N.C., 74
New Creek, W. V., 411
New Falls City, CSS, 299
New Hampshire Cavalry, 2nd, 272
New Hampshire Infantry Regiment, 8th, 101, 142, 179, 183, 196–197, 199, 210, 212, 221
  Company G, 151
  Company G, mounted, 179
New Hampshire Infantry Regiment, 15th, 100, 106, 178, 180, 190–191, 201–202, 203n627, 217–218, 218*fig.*, 219, 227–228
  Company D, 190
New Hampshire Infantry Regiment, 16th, 100, 151–152, 156, 201
New Iberia, La., 150, 255, 257
New Jersey Cavalry, 3rd, 381
New Jersey Mexican War Volunteers, 1st, 427
New Madrid, Mo., 74
New Market, Va., 372–373, 373*fig.*, 374–376, 378, 411, 419
New Orleans, La.
  1st Louisiana Native Guards and, 75
  Anaconda Plan and, 40, 72
  Battery L in, 236
  Butler expedition to, 65, 68, 75–76, 79
  Butler occupation of, 79, 86, 96–97
  Confederate protection of, 66, 75, 77
  defenses of, 101
  Emory's request for reinforcements, 227
  Farragut and Butler expedition, 77–79, 81
  Lovell's evacuation of, 79
  occupation of, 79, 86
  requests for reinforcement, 86–87
  secret Navy plan for, 72–73
  Sherman and, 130
  Star of the West and, 14
  starvation in, 77, 96
  woman order in, 96
New Orleans–Opelousas Railroad, 223
New York Battery, 5th; 6th, 381; 14th, 283; 18th, 100, 167, 175, 180, 240; 21st, 181; 25th, 257, 287–288, 290, 296
New York Cavalry, 1st, 383–384; 2nd, 283; 4th, 35; 5th, 379–381, 383, 409; 6th, 384; 8th, 381; 9th, 384; 25th, 353–354, 382
New York Dragoons, 1st, 366, 372, 382, 401
New York Harbor, 67
New York Heavy Artillery, 5th, 395

New York Infantry Regiment, 6th, 42, 43*fig.*, 50–52, 59–60, 68–69, 84, 89, 100, 124, 130, 133, 135, 138, 147, 169, 201, 370, 376n1098, 382, 409
  Company A, 43
  Company B, 43
  Company D, 69
  Company E, 43
  Company G, 51
New York Infantry Regiment, 9th, 370, 373, 383; 22nd, 381;70th, 131; 75th, 49, 62, 65, 68, 84, 87, 101, 113, 139–140, 179, 183–185, 199, 204–205, 208, 210, 213–214, 358; 90th, 161, 200, 210, 215; 91st, 85, 87, 100, 103, 105–106, 179, 183–185, 210, 213–214, 244; 110th, 100, 161, 210, 217; 114th, 100, 139, 142, 150, 161, 179n578, 201, 210, 213–215, 277, 283, 286, 289, 312, 323, 359; 116th, 167, 176, 180, 193, 216, 247, 278, 283, 319, 358, 397
  Company B, 192
  Company G, 192
New York Infantry Regiment, 128th, 101, 173, 180, 191, 199, 203, 217, 219
  Company A, 190
  Company C, 190
New York Infantry Regiment, 131st, 100, 135, 138, 179, 183, 185, 205, 210, 215, 331, 359; 133rd, 179, 210, 212, 301–302; 153rd, 276, 283, 286, 289, 323; 156th, 100, 210, 359, 397; 159th, 100, 109, 119, 134–135, 142, 145–147, 150, 179, 186, 186*fig.*, 194, 215, 236; 160th, 139, 179, 183, 210, 213–214, 283, 395; 161st, 100, 167, 180, 208, 210, 276–278, 283, 289–290; 162nd, 100, 217, 219, 283, 291; 165th, 101, 140, 180, 190–191, 191n601, 192, 199, 217, 219, 279, 290, 290n855, 291; 173rd, 100, 179, 210, 212, 283, 291; 174th, 100, 167, 180, 208, 210; 175th, 140, 161, 217, 219; 176th, 397; 177th, 101, 180, 190, 217, 219; 178th, 283, 286
New York Light Artillery, 25th, 283–284
Newlan, Thomas, 287, 293
Newton, Thomas, 3, 19, 60, 69, 89, 242, 244–245, 251, 366, 432–434
Newtown, Va,, 334, 348–350, 363
Niagara, USS, 56, 60
Niblett's Bluff, 251
Nichols, Edward A., 258
Nichols, G. W., 372, 406–407
Nichols, George S., 350
Nichols, W. A., 20
Nickerson, Frank S., 180, 190–191
Nicolay, John, 94
Nims, Ormond F., 272
Nims's Battery. *See* Massachusetts Battery, 2nd (Nims')
Nitschke, John G., 4, 69, 108, 163, 235, 417–418, 432, 434–436

Noel, Theophilus, 148, 310
Norfolk Navy Yard, 41
North Carolina Expedition, 65
North Star, USS, 99*fig.*
Northern Pacific Railroad, 138
Northern Virginia Campaign, 139
Nottaway Station, 318

O

Oaklawn Manor, 136*fig.*, 137
O'Brien, James, 192–194
O'Brien, Sholto, 108, 163, 234–235, 242, 252, 259, 286, 296, 298, 435–436
Ocean Queen, USS, 87
O'Donnell, Edward, 19
O'Ferrall, Charles T., 355, 361
Ohio Battery, 1st, 393, 396, 402
Ohio Cavalry, 2nd, 381, 387
Ohio Infantry Regiment, 56th, 275, 309; 83rd, 254, 275, 309; 96th, 254–255, 277; 120th, 309
Ohio militia, 47
Old Forge Road, 403–405, 408
Olvany, Michael, 5, 48, 63, 83, 315–316, 432, 434–436
Oneida, USS, 80–81
O'Neil, Henry C., 258
Opelousas, La., 130, 150, 153–154, 159, 253
Opequan Creek, 340, 342, 346, 348–349, 355, 357
Orange and Alexandria Railroad, 375–376, 419
Orcutt, Ephraim, 102, 108, 163, 235, 435–437
Ord, E. O. C., 247
Osage, USS, 281, 305–306
O'Sullivan, Michael, 5, 21, 36, 69, 234, 242, 256, 258, 308, 315–316, 432, 434–436
Ould, Robert, 230, 243
Owasco, USS, 248
Owen, T. H., 392
Ozark, USS, 305

P

Paine, Charles J., 199
Paine, Halbert
  command of 3rd Division, 168n548, 169, 175
  divisions of, 179–180, 199, 201
  Port Hudson attack, 170, 172–175, 178, 182–185, 208–211
  shot during assault, 211–212, 220
  Teche campaign and, 141–142
Palmer, James S., 159, 175
Parketton, William, 5, 21, 36, 69, 234, 432, 434–435
Parks, William, 102, 259, 296, 298, 413, 435–436
Parslow, Joseph, 102, 127–128, 243–244, 252, 259, 308, 315, 342, 414, 435–436
Parsons, Mosby, 280, 282, 294, 300
Pascagoula Bay, 62
Pass Cavallo, Tex., 250, 301

Patton, George S., 352, 355, 355n1023, 360, 362
Pawnee, USS, 34
Payne, William, 362, 380, 383, 391, 393, 398
Pea Ridge, Arkansas, 74
Pegram, John, 369, 390, 393, 396, 400–401
Peirce, Charles H., 351, 380–381
Pelky, Henry, 102, 242, 435
Pellet, Elias P., 142
Pemberton, John C., 157, 164, 164n532, 165–166, 177, 223
Peninsular Campaign, 75, 93–94, 139
Pennington, Alexander C. M., 380–381, 419
Pennsylvania Cavalry, 17th, 376, 389; 18th, 381
Pennsylvania Infantry Regiment, 6th, 340, 342; 47th, 278, 283, 395
Pennsylvania Volunteers, 54th, 394
Pensacola, Fla., 15, 27, 29, 37, 61, 64–65, 84–89, 101
Pensacola Bay, 28–29, 29*fig.*
Perkins, Washington, 204
Personal Memoirs of U.S. Grant, The, 324–325
Petersburg, Va., 319–320, 324, 326–327, 388, 421
Pfiffer, George, 109, 435–437
Phillips, John W., 351
Piedras Negras, Mexico, 10
Pinola, USS, 244
Piper, Joseph, 258
Pittsburg Landing, Tenn., 74, 81–82
Pittsburgh, USS, 305
Plains Store, La., 167, 174, 192–193, 206, 208, 208n637, 219–220
Plaquemine, La., 117, 224
Pleasant Grove, La., 275–276, 276*fig.*, 277–279
Pleasant Hill, La., 266, 268, 278–281, 281*fig.*, 282
Pleasant Valley, Md., 422, 427
Plunkett, Charles T., 199
Pocahontas, USS, 34
Point Isabel, Tex., 250
Polignac, C. J., 272, 278, 280, 282, 288, 294, 300, 309–311
Polk, Leonidas, 74
Poole, Thomas, 21, 36, 54, 61, 432
Pope, John, 82, 97
Port Hudson, La.
    Augur's feint, 216
    Banks and Grant cooperation on, 154–156, 159–160
    Banks attack on, 164, 166, 169
    batteries of, 203*tab.*
    Battery L and, 40, 119, 122, 126, 162, 174, 202n626, 210, 216, 220, 226–227, 232
    battery positions, 173–174, 182, 184–185, 188, 192–193, 199, 201–202, 202n625, 220, 227*fig.*
    Battle of Plains Store, 167
    black units in battle, 195–197, 197*fig.*, 222
    capture of supply ships for, 115
    cease fire, 194–195
    Confederate defense of, 164–166, 168, 170, 171*fig.*, 172, 172*fig.*, 173–178, 181–183, 183*fig.*, 184–187, 187*fig.*, 188–194, 201–202, 204–209, 211–216, 218–220, 226–228
    Confederate losses at, 196, 229
    Confederate prisoners and, 229–230
    Confederate troops at, 109, 112
    control of the Mississippi and, 40, 160
    Dwight's assault, 216–218, 218*fig.*, 219–220
    Farragut passing of, 117, 121–123, 123*fig.*, 124–125, 158
    first and second campaigns, 120*fig.*
    navy assistance, 202, 202*fig.*
    organization for the assault, 179–181
    Paine's assault, 210–212, 214
    paroled prisoners and, 165–166, 229–230
    surrender of, 228–231, 231*fig.*, 236
    Teche campaign and, 130
    truce at, 195–196, 198, 221
    Union losses at, 196, 221, 229
    Union raid on, 109, 117, 119–120, 168, 168*fig.*, 169–171, 171*fig.*, 172–173, 173*fig.*, 174, 174*fig.*, 175–179, 179*fig.*, 180, 180*fig.*, 181–183, 183*fig.*, 184–186, 186*fig.*, 187, 187*fig.*, 188–190, 190*fig.*, 191–194, 198–207, 207*fig.*, 208–209, 209*fig.*, 210, 220–222, 226–228
    water passages around, 114, 153
    Weitzel's assault, 213–216
Port Republic, Va., 373
Port Royal Expedition, 177–178
Port Royal, S.C., 40
Porter, David Dixon
    bombardment of Fort DeRussy, 159
    bombardment of Fort Jackson, 78
    command of *Powhatan*, 35
    cooperation with Grant, 154
    fake gunboat and, 118
    on loss of *Indianola*, 116, 118
    Mississippi Marine Brigade and, 114–115, 158
    mortar flotilla and, 78, 88, 90–91
    orders to Vicksburg, 90–91
    passage to Alexandria, 304
    patrol of Red River, 159, 263, 265–266, 299
    prize money and, 305, 305n901
    rapids at Alexandria, 305, 305*fig.*, 306–307
    Red River transport, 261, 263
    secret Navy plan to attack New Orleans, 72
    shallows near Grand Ecore, 301–303, 305–307
    Taylor pursuit of, 300, 304
    unsupported fleet and, 281, 303
    Western Gunboat Flotilla and, 114, 166
Porter, Fitz John, 21–22
Porter, James, 137
Porter, Mary, 137
Porter Plantation, 135–137, 141–143

Powell, William, 372, 375, 376n1098, 377, 386–387, 393, 401, 403, 406n1196
Power, Catharine, 430
Powers, F. P., 167
Powhatan, USS, 34–35, 37
Prairie D'Ane, Ark., 299–300
Prentiss, Benjamin, 233
Prescott, T. C., 197
Price, Sterling, 266, 278, 294, 300
Priest Cap, La., 170, 206, 208, 213
Prince, Edward, 157, 170
Profit's Island, La., 119*fig.*
Provost, William Y., 109
Purington, George A., 351

Q

Queen of the West (ram), 115–116, 133
Quinn, Timothy, 350
Quinnebaug, USS, 131

R

Ramseur, Stephen, 333, 341, 349, 352, 355–356, 358, 363, 367, 369–370, 391, 401, 404–405
Ranahan, Michael, 102, 234, 435
Randal, Horace, 270, 282
Randolph, George W., 76, 164n532
Ransom, Dunbar R., 351
Ransom, T. E. G., 249–250, 250n748, 262, 266, 270, 270n800, 272–273, 275–276
Rawles, Jacob B., 181, 273
Rawlins, John, 324
Raynor, William H., 275
Reading, George, 296n875
Ream's Station, 318, 384
Recruits, 3–6
Rectortown, Va., 386–387
Red Bud Run, 356–358
Red Chief, CSS, 170, 302
Red Hill, 396
Red River, 95, 115–116, 118, 124–125, 152–154, 156, 159–160, 261, 301, 303, 305, 305*fig.*, 306, 306*fig.*, 307, 307*fig.*, 309
Red River Campaign, 265*fig.*
 Banks' troops for, 262–267
 Battle of Mansfield (Pleasant Grove), 253, 276–280
 Battle of Pleasant Hill, 281–283, 283*fig.*, 284, 285*fig.*, 286–291, 292*fig.*, 293–296, 297*fig.*, 298
 failure of, 311
 Halleck and, 261–262
 Sabine Crossroads attack and, 269–275
 Taylor and, 148
 Wilson's Farm Battle, 268–269
Redding, George W., 258–259, 298, 308
Redding, J., 296
Reedy, Michael, 54, 61, 432

Reilly, Terence, 351
Reily, James, 138, 141–143, 147
Renshaw, Francis B., 100
Reserve Camp, Pleasant Valley, Md., 411–412
Rhode Island Artillery, 1st, 397
Rhode Island Cavalry, 1st, 350, 401; 2nd, 222; 3rd, 283, 286
Richards, Franklin W., 44, 69, 102, 163, 434–435
Richardson, James P., 199
Richmond, USS, 55–56, 60, 118, 119*fig.*, 122, 124, 124n417, 155, 166, 202
Richmond, Va., 336, 421
Richmond and Danville Railroad, 318
Ricketts, James B., 321, 351, 355, 358–359, 369–370, 396, 404
Rienzi Knoll Lane, 404
Riley, Charles, 4, 19–21, 26, 44, 54–55, 63, 69, 83, 86, 89, 102, 104–105, 129, 149, 243, 423, 432, 434
Riley Plantation, 170
Ringgold Barracks, Tex., 2, 7, 20–21, 105, 250
Rio Grande City, Tex., 250
Rio Grande Expedition, 247–249, 249*fig.*, 250, 253
Ripley, James, 332
Rivanna River, 386, 419
Rivers, Harry, 258
Rivers, James H., 258
Roanoke Island, N.C., 65, 74
Roanoke Station, 318
Robb, Hamilton, 279
Roberts, Thomas, 258
Roberts' Ford, 392
Robertson, James, 52, 52n157, 59n175
Robeson, William P., Jr., 351
Robinson, James W., 2, 7, 11, 19–20, 26, 423, 432
Robinson, John K., 268
Robinson, William, 19
Rockfish Gap, 375
Rockville, Md., 321
Rodenbough, Theophilus M., 350
Rodes, Robert E., 333, 348–349, 352, 357–358, 360, 362–363, 367
Rodgers, J. I., 313
Rodgers, John, 39, 132, 135, 138, 142, 145–147, 180
Rodman, William M., 199
Rogers, John, 353n1018
Roper, John, 5, 83, 234, 432, 434–435
Rosecrans, William S., 246, 325
Ross Landing, La., 121
Rosser, Thomas L.
 attack on Beverly, W.V. fort, 416
 attack on New Creek, W.V. fort, 411
 Battle of Cedar Creek, 390–394, 398–399, 400n1180, 403
 Battle of Tom's Brook, 378, 380–384
 Great Cattle Raid, 378

joining Early, 376–377, 377n1102
Laurel Brigade and, 377, 383
pursuit of Custer, 378
reconnaissance of Custer, 385
at Tom's Brook, 379
West Point and, 379n1108
Rowanty Creek, 318
Rowland, Card, 102
Rude's Hill, Va., 372, 372*fig.*, 373
Ruggles, Daniel, 95, 109
Rupprecht, Ludwig, 4, 69, 315, 344, 382, 413, 415, 432, 434, 436–437
Russell, David A., 351, 355, 359, 363
Russell, John H., 48
Russell, William, 37–38, 331
Rust, Albert, 121, 164
Ryan, James, 4–5

S

Sabine, USS, 37
Sabine Crossroads, 269–271, 271*fig.*, 272–274, 274*fig.*, 275–276, 276*fig.*, 277–278
Sabine Pass Expedition, 240–241, 245, 247, 267, 267n795
Sabine River, 241, 269
Sachem, USS, 159
San Antonio, Tex., 12, 16, 20
Sanderson, James A., 88, 102, 108, 126, 135, 163, 233, 241, 251, 256, 258–259, 286, 293, 296, 298, 434, 436
Sandy Creek, La., 170, 175–176, 182, 184, 187–188
Santa Rosa Island, Fla., 27, 29*fig.*, 35, 37, 42, 47, 50–51, 51*fig.*, 52, 62, 69–71, 84
Savannah, Ga., 416, 418
Schaffer, William, 44, 54, 63, 69, 83, 88, 434
Schlatter, Robert E., 6
Schmidt, Heinrick, 5, 69, 234, 432, 434–435
Schneider, Philip H., 69, 163, 234–235, 427–428, 432, 434–437
Schoenfeld, Charles, 4–5
Schoonmaker, James M., 350
Schwartz, Edward, 350
Sciota, USS, 81, 244
Scott, Catharine Carson, 428
Scott, John, 288–289, 295
Scott, William E., 4, 21, 69, 89, 241, 244, 251, 315, 414, 428, 432, 434–437
Scott, Winfield, 12–16, 18, 25, 29, 31, 33–34, 38, 40–41, 47, 81, 95, 262, 430
Scurry, William R., 282
Secession, 11–13, 15, 17
Sedgely, Robert, 258
Seeley, Francis, 59n175, 59n176
Seiver's Ford, 352–353
Selma, CSS, 339
Semmes, Raphael, 141–142, 145, 147

Sewall, James, 258
Seward, William H., 33–34, 97, 110, 239, 260, 311
Seymour, Charles J., 169, 350
Shapley, Morgan S., 44, 234, 434–435
Sharpe, Jacob, 359
Shaw, Warren P., 5, 54, 63, 69, 344, 413, 428, 432, 434, 436–437
Shaw, William T., 280, 283, 283*fig.*, 284, 286–288, 288n844, 289–291, 294
Shelby, W. B., 187–188, 195
Shenandoah River, 337, 390–391, 393
Shenandoah Valley, 326, 328–331, 334, 334*fig.*, 335
Shenandoah Valley Campaign, 337–348
Shepherdstown, W. V., 338
Shepley, George, 260
Sheridan, Phillip H.
   attack on Early's army, 419
   Cedar Creek, 388–390, 403–404, 407–411
   Fisher's Hill, 367, 369–371
     New Market, 374–376
     Sailor's Creek, 421
     Smithfield Crossing, 340–342
     Stickney Farm, 385
     Tom's Brook, 379–382, 385
     White Oak Road, 420–421
   conference with Stanton, 385–388
   contact with Rebecca Wright, 346–347
   destruction of railroads, 419
   Hancock Station camp, 420
   health of army, 414
   Middle Military Division commander, 325, 327–330, 330*fig.*, 331, 364, 416, 419
   New Creek attack, 411
   Piedras Negras and, 10
   preference for cavalry, 328
   promotion to major-general, 410
   pursuit of Early, 372–376
   raid on Weldon and Southside Railroads, 317
   recall of 6th Corps, 385–387
   reconnaissance of Early's divisions, 345–348
   reconnaissance of Fisher's Hill, 366–367
   retaliation for death of Meigs, 412
   retreat to Halltown, 339–340
   return of Early's divisions and, 348–349
   ride to Middletown, 389, 389*fig.*
   Rosser pursuit of, 377–379
   Shenandoah Valley Campaign, 333–335, 337–341, 345–348
   termination of Shenandoah campaign, 375–376
     Winchester, 348–350, 352–364
   Tom's Brook attack, 379–380
   Woodstock Races, 383–384
Sherman, Thomas W.
   6th New York regiment and, 89
     Plains Store, 166, 168
     Port Royal Sound, 62

defenses of New Orleans and, 130, 150, 168n548
divisions of, 180–181
failure to assemble at Port Hudson, 187, 189
Port Hudson attack, 170, 172, 173*fig.*, 174, 177–178, 181–182, 189–192, 196, 199
Sherman, William T., 100, 114, 157, 260–261, 262n781, 300–302, 305, 340, 345, 377, 416
Ship Island, Mississippi, 63, 65, 73–74, 77, 88, 98
Shipley, Alexander N., 47, 57*fig.*, 60
Shorey, Henry A., 273, 273n810
Shorter, John G., 63
Shreveport, La., 150, 262, 266, 269, 281, 299–300, 304–305
Sias, Chauncy R., 258
Sibley, Caleb C., 104–105, 154
Sibley, Henry H., 139–140
Siege of Petersburg, 320
Sigel, Franz, 320–321
Signal, USS, 309
Signal Knob, 335
Silver Spring, Md., 321
Silvey, William, 1–2, 7, 11, 19–20, 26, 44, 48, 432
Simmesport, La., 156, 160, 162, 168–170, 263, 310–311
Simmonds, William Edward Scott, 428
Simms, James P., 394–395
Simonds, John, 258
Slack, Charles B., 233, 241, 251–252, 273
Slaughter Plantation, 172–173, 189–191, 201, 216–217
Slemmer, Adam J., 15, 29–32, 39
Slidell, John, 31
Smith, Abel, 192, 199
Smith, Andrew J.
  Battle of Pleasant Hill and, 280–281, 281n830, 283–284, 286–287, 289, 291, 293–295
  protest at Banks abandoning wounded, 295
  pursuit of Taylor, 299
  Red River Campaign and, 263–268
  retreat to Grand Ecore, 300–302, 309
  return to Vicksburg, 312
    Yellow Bayou attack, 311
Smith, Caleb B., 34
Smith, E. Kirby, 148, 156, 164, 223, 223n683, 227, 230, 266, 270, 294, 299, 310
Smith, Elisha B., 213–214
Smith, George S., 355
Smith, Hiram, 102, 234–235, 242, 435
Smith, James H., 102, 199, 435–437
Smith, Joseph, 4, 21, 83, 127, 314, 343, 432, 434–435, 437
Smith, Martin L., 79–80
Smith, Melanchton, 78
Smith, Robert S., 350
Smith, T. Kilby, 263, 265, 281, 281n830, 299–300, 303

Smith, Thomas, 352
Smith, William, 258
Smith, William H., 102, 259, 259n766, 286, 296, 298, 435, 437
Smithfield, N. C., 420
Smithfield, W. V., 340
Smithfield Crossing, 340–341, 341*fig.*, 342
Snow, Lt., 273
South Atlantic Squadron, 62
South Carolina Infantry Regiment, 8th, 346, 352
South Carolina, USS, 42
Southern Historical Society Papers, 192
Southside Railroad, 317, 317*fig.*, 318, 420–421
Spangler, Charles, 60, 69, 242, 244, 433–434
Sprague, Homer, 137, 147, 215
Springfield Landing, La., 120–121, 167–168, 170, 266, 281
St. Francisville, La., 169, 169*fig.*
St. Joseph's Island, Tex., 250
St. Louis, USS, 37
St. Martinsville, La., 150, 161
St. Mary's Cannoneers, 304
St. Mary's, USS, 131–132, 134, 136, 137n453
St. Maurice (steamer), 168
Stafford Plantation, 162
Stagg, Peter, 350
Stanners, Martin, 44, 127, 163, 225, 243, 256, 434–435
Stanton, Edwin M., 24, 72–73, 93, 115, 238, 263, 324–325, 328, 363, 376, 385–386, 388
Stanyan, John M., 221
Star of the South (steamer), 42
Star of the West (steamer), 14–15, 30–31
Starlight, CSS, 170, 302
Staunton, Va., 321, 370, 375, 377, 416
Staunton River, 318
Steedman, I. G., 170, 172–173, 175–176, 182, 183*fig.*, 185, 188–189, 196, 199, 205, 207
Steele, Frederick, 261, 294, 299–301, 312
Stephenson's Depot, Va., 325, 325*fig.*, 338, 340, 345, 349, 349*fig.*, 352, 358, 360–361, 361*fig.*, 412
Stevens, Hazard, 410
Stevens, John, 258
Stewart, William, 102, 435, 437
Stickney, Albert, 225
Stoll, Andrew, 5, 26, 44, 69, 163, 234–235, 308–309, 315, 344, 373, 413, 417, 422, 433–435, 437
Stone, Charles P., 226, 286–287, 293
Stone, R. C., 282
Stone, Richard, 4, 6
Stony Creek Depot, Va., 318
Story of the Maine Fifteenth, The (Shorey), 273
Strasburg, Va., 384, 391, 409
Straub, Amelius, 4, 19, 127, 163, 233, 242, 251, 256, 258, 308, 413, 417–418, 433–435, 437
Straub, Margaret, 19

Stringham, Silas H., 41
Strong, George C., 73n224
Summit Point, W. V., 337, 345, 349, 352
Sumner, Charles, 110
Sumner, S. B., 199
Switzerland, USS, 158
Sykes (steamer), 131n434

T

Taft, Elijah D., 351
Tappan, James C., 280, 282, 294, 300
Tarleton Plantation, 414
Taylor, Franck E., 56, 57*fig.*, 60, 63–64, 68–69, 83, 86, 88, 101, 108, 126–127, 130n429, 135, 162, 174, 226, 233, 241, 251, 256–257, 286, 293, 295–296, 308, 314, 340, 342–344, 351, 364, 370, 403, 410, 412, 416, 422–423, 433–434, 436–437
Taylor, Franck, Sr., 64
Taylor, George W., 199
Taylor, Richard
  attack on Berwick Bay, 223–226
  attack on Grant's supply points, 223
  Banks pursuit of, 148–149, 149*fig.*, 150
  Grand Coteau, 254–255
  Irish Bend, 144–148, 156
    Mansfield and, 277–279
  Pleasant Hill and, 279–281, 283, 283*fig.*, 284, 287–288, 291, 293–295, 299
  Wilson's Farm and, 268
  blockade of Red River, 309
  Cane River and, 303
  chased from La Fourche, 233
  command of the District of Louisiana, 111
  cutoff road and, 141–142, 148
  engagement at Bayou Teche, 112–114
    Bisland and, 110, 133
    Burton attack, 151
    Butte La Rose and, 117
    DeRussy and, 115
  forts on Mississippi, 112, 125
  pursuit of Porter, 300, 304
  pursuit of Steele, 299–300
  reconnaissance of Texas expedition, 248, 251, 253
  Red River Campaign and, 148, 264–266
  return to Alexandria, 161
  Sabine Crossroads attack and, 270, 272, 275
  Teche campaign and, 129–130, 130n429, 132–133, 138, 140–142, 148
  Vermilion River and, 149
  Yellow Bayou attack, 311
Teche campaign
  Banks and, 125, 130, 132, 139–141, 143, 148
  Battery L and, 130–132, 134–135, 143, 146–148
    Irish Bend, 142–143, 143*fig.*, 144, 144*fig.*, 145–148

Bayou Teche, 130*fig.*, 134*fig.*
  cutoff road and, 138, 138*fig.*, 141, 148
  failure of, 161
  Bisland advance, 131n434, 133–134, 139, 139*fig.*, 140–142
  Grover and, 130–143, 145, 147–148
  McWilliams Bridge and, 135
  plan to remove Taylor from Attakapas region, 129–130
  Porter Plantation and, 136–137
Tecumseh, USS, 339
Teighe, Michael, 163, 314, 344, 438
Tennessee, CSS, 339
Tennessee, USS, 248
Tennessee River, 66
Terre Noir Bayou, 299
Terrell, Alexander, 282
Terry, Edward, 202, 210–211
Texas Battery, 4th, 282; 7th, 282, 287; 9th, 282; 12th, 282
Texas Cavalry, 1st, 100–101, 282, 288; 2nd, 224; 3rd, 280; 4th, 133, 146, 148, 293; 13th, 282; 15th, 282; 17th, 282; 26th, 282, 288; 31st, 282; 34th, 282; 37th, 282
Texas expedition, 245–246, 246*fig.*
Texas Infantry Regiment, 4th, 282; 5th, 141, 282; 7th, 282; 12th, 282; 13th, 282; 18th , 282; 22nd, 282
Texas invasion, 238–240, 245–248, 311
Texas Mounted Riflemen, 1st Regiment, 16
Texas Partisan Rangers, 1st, 282; 2nd, 282
Texas Volunteers, 98
Thayer, Henry B., 4, 6
Thayer, Sylvanus, 299
Thibodeaux, La., 224
Battle of Winchester, 350–353, 353*fig.*, 354–356, 357*fig.*, 358–361, 361*fig.*, 362, 362*fig.*, 363–364
Thoburn, Joseph, 351, 362, 385, 391–395, 395n1164, 396
Thomas, Charles, 258
Thomas, Lorenzo, 17
Thomas, Stephen, 178–179, 183–185, 395
Thompson, Jacob, 11
Thompson, John L., 351
Thompson, Mary A. Bedow, 429
Thompson, William V., 5, 48, 54, 63, 69, 315, 415, 417, 428–429, 433–435, 437
Thompson's Creek, La., 169, 169*fig.*, 170, 175, 302
Three Top Mountain, 335, 386
Tieghe, Michael, 109, 235, 435, 437
Time (steamer), 50, 52, 59, 59n176
Timmins, John, 258
Tompkins, Charles H., 351
Tom's Brook, 378–384
Tomson, John, 44, 69, 86, 256, 434–435

Torbert, Alfred, 329, 334, 337–338, 345, 349–350, 353n1018, 355, 361n1049, 366–367, 370–371, 373–375, 378n1105, 379, 383, 401, 403, 407, 411, 418
Toucey, Isaac, 31
Tower, Zealous B., 15
Townsend, E. D., 243
Townsend, L.R., 152
Townsend, Reuben, 5, 44, 69, 89, 314–315, 344, 415, 417, 433–435, 437–438
Tumbling Run, 366–367, 369
Turner, William H., 350
Twiggs, David E., 12, 12n17, 13, 16, 20, 73, 76, 79
Twohig, John, 8
Tyler, Tex., 275
Tyler, USS, 39, 40*fig.*

U

Ulmann, Daniel, 232
Union
   Anaconda Plan, 18, 39, 39*fig*, 40, 62, 72, 238
   call for volunteers, 45–46
   control of the Mississippi and, 233, 238, 311
   defective weapons and, 135
   engagement at Bayou Teche, 112–113, 113*fig.*, 114
   forces in Texas, 238–239
   frontal assaults and, 177
   prisoner parole and exchange, 129
   recruiting needs, 319
   soldier's rations in, 163–164
   Southern secession and, 12–13
   weapons and, 331–333
United States Artillery, 1st, 148, 383
   Battery A, 2, 56, 101, 180, 185, 253, 312–313
   Battery B, 2
   Battery C, 2, 100
   Battery D, 2, 430
   Battery E, 2, 430
   Battery F, 22, 54, 101, 180, 185, 210, 310–313
   Battery G, 2
   Battery H, 2, 430
   Battery K, 22, 317–318, 329, 335, 337, 342–343, 353, 353n1017, 384, 402–403, 413, 423
   Battery L. *see* Battery L
   Battery M, 22
   Company A, 7, 31, 59, 257
   Company B, 15, 24
   Company F, 8, 16, 41, 104
   Company G, 15, 29–30, 39, 49
   Company I, 2
   Company K, 2, 8, 16
   Company L, 16
   Company M, 2, 21
United States Artillery, 2nd
   Battery C, 180, 312–313
   Battery H, 101
   Battery K, 101
   Battery M, 419
   Company A, 35, 42
   Company C, 15, 25
   Company D, 340
   Company H, 35
   Company K, 35
   Company M, 16, 21–22, 35, 42
United States Artillery, 4th, Battery C-E, 419
United States Artillery, 5th, 383
   Battery G, 101, 167, 181, 273
   Battery M, 356
United States Cavalry, 1st, 281, 310, 329, 398; 2nd, 340, 354; 3rd, 310, 333; 5th, 382, 401; 6th, 321
United States Colored Heavy Artillery, 10th, 87
   Company F, 87
United States Infantry, 1st, 16; 3rd, 61
   Battery C, 22, 35
   Battery E, 22, 35
   Company C, 47, 87
   Company E, 60, 87
United States Infantry, 8th, 23, 116; 14th, 427
Universe (steamer), 308
Upton, Emory, 351, 363
Urban, Gustavus, 350
U.S. Volunteers, 67
Utah Expedition, 138

V

Valley Pike, 334, 357, 361, 361*fig.*, 363, 366, 369, 375–376, 379–380, 388, 391–392, 398, 401–402, 404–405, 407–409
Valverde Battery, 287
Van Dorn, Earl, 63, 67, 83, 91, 95, 157
Van Petten, John, 214
Van Zandt, Jacob, 178–179, 183–185
Vance, Joseph W., 269, 272, 275
Vanderbilt, USS, 42
Varner, Harry, 258
Vermilion Bayou, 149–150, 247, 254
Vermilionville, La., 255
Vermont Battery, 1st, 100, 173, 181, 296, 311; 2nd, 167, 180–181, 192
Vermont Cavalry, 1st, 409
Vermont Infantry Regiment, 1st, 205; 7th, 89, 101, 103; 8th, 101, 113, 139, 179, 183, 210, 213–214, 395
Vicksburg, Miss.
   canal and, 90–91, 91*fig.*
   Confederate defense of, 165, 223
   Confederate fortification of, 80
   falling river in, 92–93
   Farragut and, 79–81, 90–92
   Grant attacks on, 154, 156, 159, 177, 198
   Grierson's Raid and, 157

insufficient troops for, 92–93
Lovell's force and, 79
Porter's mortar schooners and, 91, 93
Sherman's attack on, 114
surrender of, 40, 228, 236
Villepigue, John B., 60
Vincennes, USS, 55, 55n166
Vincent, William G., 132–134, 137–138, 140, 143, 147, 264
Virginia, USS, 244, 248
Virginia Cavalry, 12th, 393; 18th, 361; 23rd, 355, 361; 62nd, 355, 361
Virginia Central Railroad, 369, 386, 418
Virginia Infantry Regiment, 3rd, 392; 6th, 365; 11th, 392; 22nd, 355n1023; 37th, 237
Vodges, Israel, 31, 33, 35–37, 51–52, 60

W

Waite, Carlos A., 16–18, 20
Walker, John G., 161, 223, 227, 254, 275, 280, 282, 288, 291, 294, 300
Walker, L. P., 37, 75
Wallace, Lewis, 321
Waller, Edwin, 141
Walsh, Richard, 6
Walton, Charles E., 103, 244, 435–437
Ward, Henry A., 21, 44, 163, 235, 315, 415, 434–435, 437–438
Ward, Henry H., 54
Warren, Gouverneur, 336
Warrenton, Miss., 80, 116, 158
Warrington, Fla., 29–30, 60–61, 64, 88, 107
Washburn, C. C., 250, 253n753, 254–255
Washburn, H. D., 250
Washburne, Elihu B., 46, 262
Washington, D. C., 319–323, 336
Washington, George, 262
Washington Navy Yard, 332
Waul, Thomas, 282
Weapons, 331–333
Webb, CSS, 116, 152
Webb Plantation, 162
Weed Convention, 30
Weir, Julian V., 350
Weitzel, Godfrey
3rd Division command, 168n548
in Alexandria, 160
assault of Richmond, 421
Grand Coteau, 255
Irish Bend and, 143
as engineer at Fort Pickens, 110–111
as expert on lower Mississippi, 73n224, 148
Port Hudson attack, 170, 172–173, 175–178, 178n576, 180–186, 188n592, 189, 196, 201–202, 205, 208–209, 211, 213–216
pursuit of Taylor, 148, 150, 231

Rio Grande expedition and, 253
Teche campaign and, 112–113, 130–131, 131n434, 139–141, 148
Texas expedition and, 245
Welch's Spring, W. V., 337
Weldon Railroad, 317, 317*fig.*, 318, 336
Welles, Gideon, 33, 40, 55, 72, 88, 93, 114, 116–118, 159, 237, 263, 339
Wells, William, 351, 381, 403
Welsch, Peter, 103, 163, 234, 308, 435, 437–438
Welsh, Benjamin C., 258
West Gulf Blockading Squadron, 77, 117, 248
Western Flotilla, 114
Western Gulf Blockading Squadron, 68
Western Gunboat Flotilla, 77, 166
Westfield, USS, 100, 110
Wharton, Gabriel C., 352, 354–355, 363, 367, 369, 391–392, 397–398, 401, 404, 411
Wheaton, Frank, 404
Wheeler, Charles, 241, 251
White, Catharine Power, 430
White, Elijah V., 383
White, Michael, 4, 69, 89, 244, 259, 296, 298, 308, 315, 343–344, 429–430, 433–436, 438
White, Perry S., 410
White Post, Va., 345
Wickham, John, 362, 367, 374, 378n1106, 384
Wicks, David J., 4, 60, 69, 235, 244, 259, 343, 432–434, 436
Wilcox, Thomas M., 108, 163, 435, 437–438
Wilder, Joshua E., 103, 108, 435, 437–438
Wildes, Thomas F., 396
Wilkins, John C., 258
Wilkinson, Joseph, 4, 69, 163, 235, 417, 433–435
Wilkson, Henry, 21, 315–316, 434–435, 437
William, Henry, 5, 69, 433, 435, 437
Williams, Alpheus, 97
Williams, Henry, 234, 259, 309, 315, 415, 417, 434
Williams, James, 6
Williams, Thomas, 80–81, 90–91, 95
Williston, Edward, 340, 342, 351
Wilson, James H., 317–318, 328, 332–333, 337, 346, 349, 351–352, 355–358, 361n1049, 363, 367, 373–375, 377
Wilson, John (Benjamin Hughes), 315–316
Wilson, Thomas W., 103
Wilson, William "Billy," 42, 42*fig.*, 89, 130–131, 169
Wilson's Creek, Miss., 53
Wilson's Farm, 268–269
Wilson's Raid, 317, 317*fig.*, 318–319, 384
Winchester, Va., 320–321, 325, 334, 334*fig.*, 335, 337, 345–346, 348–350, 354–355, 357, 357*fig.*, 361–364, 367, 388–389, 412, 419
Wingfield, J. H., 170, 175, 187, 196
Winn, Abram F., 103, 127–129, 435
Winn, Joel T., 103, 127–129, 436

Winn, Mary, 129
Winona, USS, 81
Wisconsin Infantry Regiment, 4th, 80, 100, 141, 151, 179, 199, 201, 205n632, 210–212; 8th, 283; 23rd, 254, 272–273; 29th, 275
Wissahickon, USS, 81
Witcher, John S., 350
Withers, Jones M., 53, 63, 66
Wofford, William, 395
Wood, James, 237
Wood, John C., 344, 364, 413, 417, 422, 436–438
Woodruff, George A., 44, 48, 54
Woodruff, Lyman, 109, 413–414, 436–438
Woods, John C., 103
Woodstock Races, 383–384
Woodward, Joseph T., 194
Wool, John E., 40
Woolsey, Florida, 61, 107
Wren, Owen A., 21, 69, 234, 243, 252, 315, 414–415, 417, 434, 436–437
Wright, Horatio, 321–323, 325, 327, 329, 338, 345, 351, 355, 366, 386, 394–398, 401, 403–404, 407
Wright, Rebecca, 346–347, 363
Wright, Wallace D., 19–20, 26, 44, 69, 102, 108, 127, 163, 235, 242, 433–434, 436
Wyandotte, USS, 37
Wynne, William, 44, 69, 102, 163, 234–235, 244, 433–434, 436

Y

Yates, Richard, 46
Yazoo, USS, 314, 317
Yazoo City, Miss., 236
Yellow Bayou, 311
Yokely Bayou, 143
Yokely Bridge, 142, 147

Z

Zouaves, 192, 192*fig.*, 219n669, 290n855

About the Author

Mr. Simmonds received a BS degree in Mechanical Engineering from Worcester Polytechnic Institute in 1958. After service in the army (he notes that ROTC was "the thing expected of us"), he began employment in machinery research and development.

He received an MS in Mechanical Engineering from Northeastern University in 1967. His entire career remained in the design and development of special machinery for the plastics and automotive industries—the kind of equipment that most people never see—and which is frequently dismissed as "automation."

Surprisingly, rather than remaining hunched over a drawing board (later a computer) the job required considerable international travel. This was an unexpected benefit in what he describes as "an otherise stressful existence."

Retirement has meant a rather fateful step into genealogy, which led to the writing of this book—his great-grandfather served in Battery L before, during, and after the Civil War under the alias of William E. Scott. Unfortunately, no family lore has been passed down to explain why.